Freshwater Mussels of Florida

Freshwater Mussels of Florida

James D. Williams, Robert S. Butler, Gary L. Warren, and Nathan A. Johnson

With special thanks to
Sherry L. Bostick
Richard T. Bryant

Sponsored by
Florida Fish and Wildlife Conservation Commission

The University of Alabama Press / Tuscaloosa

∞
The paper on which this book is printed meets the minimum requirements of American National Standard for Information Sciences—Permanence of Paper for Printed Library Materials, ANSI Z39.48-1984.

Library of Congress Cataloging-in-Publication Data

Williams, James D. (James David), 1941–
  Freshwater mussels of Florida / James D. Williams, Robert S. Butler, Gary L. Warren, and Nathan A. Johnson.
       pages cm
  Includes bibliographical references and index.
  ISBN 978-0-8173-1847-5 (hardcover : alk. paper) — ISBN 978-0-8173-8779-2 (e book)
  1. Freshwater mussels—Florida. I. Title.
  QL430.6.W553 2014
  639'.4209759—dc23
                                            2013040134

Publication made possible in part by generous contributions from the Florida Fish and Wildlife Conservation Commission; the Florida Department of Environmental Protection; the Institute of Food and Agricultural Sciences, University of Florida, Gainesville; the Jacksonville Shell Club, Florida; the U.S. Fish and Wildlife Service, Region 4 Office, Atlanta, Georgia; and the U.S. Geological Survey, Southeast Ecological Science Center, Gainesville, Florida.

We take great pleasure in dedicating this book to

Thompson H. Van Hyning (1861–1948)

and

William H. Heard (1935–)

in recognition of their significant contribution to our understanding of the freshwater mussel diversity of Florida over the past century.

CONTENTS

Preface

Just a half century ago, there were fewer than a dozen individuals in the United States actively conducting research on freshwater mussels. With passage of the federal Endangered Species Act (ESA) in 1973, interest in the plight of mussels and other non-game aquatic organisms in the United States began to build. It was soon realized that as many as 30 mussel species were extinct and many more were declining at an alarming rate. This wave of extinction and decline ignited efforts by malacologists and university, state, and federal aquatic biologists to compile mussel lists for various drainages, identify habitat threats, and evaluate conservation status for North American taxa. Publication of *Unionacean Clams of North America* by Jack B. Burch in 1975 represented a monumental step forward in the effort to document the U.S. mussel fauna. This was the first widely available comprehensive work on North American freshwater mussels that included a key to species. Today there are books addressing the mussel faunas of many states in the eastern United States (e.g., Parmalee and Bogan 1998; Williams et al. 2008).

Investigation of the Florida mussel fauna began in the early 1800s by naturalists living in the northeastern United States. In the 1800s Isaac Lea, a conchologist from Philadelphia, Pennsylvania, described 18 nominal taxa from Florida based on shells that he obtained from other mollusk collectors. In the late 1800s and early 1900s, father and son naturalists Samuel H. Wright and Berlin H. Wright, who lived in New York, described 33 nominal species of Florida mussels. Charles T. Simpson of the U.S. National Museum of Natural History (USNM), Smithsonian Institution, also described new species from Florida but more importantly provided an early critique of previous work. Interest in Florida mussels waned from the early 1900s to the mid-1950s. Renewed interest came from a 1950s mollusk survey in the Gulf Coast rivers of Alabama, Florida, and Georgia by William J. Clench and Ruth D. Turner from the Museum of Comparative Zoology (MCZ) at Harvard University. During the 1960s and 1970s, Richard I. Johnson, also from the MCZ, produced several publications on mussels in Florida and adjacent states.

The *Identification Manual of the Freshwater Clams of Florida* was published by William H. Heard (1979a) of Florida State University. This work represented the first statewide treatment of a mussel fauna in the southeastern United States and heralded the beginning of renewed interest in Florida mussels. The Florida mussel fauna is considerably less diverse than those of Alabama and Georgia. However, Florida's relatively lower species diversity is offset by extraordinary levels of variation in shell morphology. This morphological diversity and difficulty in species delineation continues to be a challenge. The Florida land mass experienced dramatic changes during the Pleistocene Epoch, and the present-day fauna, especially in the peninsular region, is geologically young. The fact that Florida is adjacent to inland areas with higher molluscan diversity, has both Gulf and Atlantic Coast rivers, and a variety of aquatic habitats unparalleled in the southeastern United States provides an interesting setting for the study of mussel biology, ecology, and evolution.

Acknowledgments

Preparation of this document during the past five years has required the support and assistance from individuals, universities, museum institutions, and federal and state agencies. We would like to express our appreciation to those individuals who have contributed to the completion of this project. We sincerely apologize if we have inadvertently omitted someone from these acknowledgments.

First, we would like to acknowledge the superb assistance and support of Sherry Bostick. She played a key role in managing all aspects of the preparation of the manuscript, including entering data, editing text, and formatting layout. She also scanned, edited, and arranged all photographs and illustrations included herein. Some of these tasks were major challenges, but she always rose to the occasion and did so with patience and good humor. It is difficult to express in words our appreciation to Sherry for her efforts and support to see this book come to fruition.

Funding

Funding for the Florida mussel book project was provided by the Florida Fish and Wildlife Conservation Commission, State Wildlife Legacy Program, through a State Wildlife Grant. The grant was administered by Brian Branciforte, Laura Morris, and Stacey Whichel. Additional administrative support was provided by Lisa Clancy, Thomas Eason, Jim Estes, and Dick Krause.

Publication of the Florida mussel book was supported by funding from Florida Fish and Wildlife Conservation Commission; Florida Department of Environmental Protection; Institute of Food and Agricultural Sciences, University of Florida, Gainesville; Jacksonville Shell Club, Florida; U.S. Fish and Wildlife Service, Region 4 Office, Atlanta, Georgia; and U.S. Geological Survey, Southeast Ecological Science Center, Gainesville, Florida.

Text Review

We extend our utmost appreciation and gratitude to an outstanding group of outside reviewers. Drafts of chapters were provided to reviewers often with a request for a short turnaround time. We greatly appreciate reviews by the following individuals: Steve Ahlstedt, Art Bogan, Mike Gangloff, Jeff Garner, Wendell Haag, John Harris, Bill Heard, Jess Jones, Harry Lee, John Pfeiffer, Roger Portell, Sandy Pursifull, Matt Rowe, and Nancy White.

Museum Support

We are grateful to the museum curators, collection managers, and other staff for assistance during the collection of mussel distribution data. They were helpful and provided open access to mussel collections, catalogue data, lab equipment, and archival photographs. They were always patient and accommodating to us during our visits and responded to requests for loan of specimens. We express our sincere gratitude to the following institutions and individuals: Academy of Natural Sciences of Philadelphia—Paul Callomon, Judy Goldberg, and Amanda Lawless; Auburn University Museum—Mike Gangloff and Brian Helms; Carnegie Museum of Natural History—Tim Pearce and

John Rawlins; Cincinnati Museum of Natural History—Francisco Borrero; Field Museum of Natural History—Jochen Gerber; Florida Museum of Natural History, University of Florida—Kurt Auffenberg, Amanda Bemis, Gustav Paulay, Roger Portell, Griffin Sheehy, John Slapcinsky, and Fred Thompson; Illinois Natural History Survey—Kevin Cummings and Christine Mayer; Mississippi Museum of Natural Science—Bob Jones; Museum of Comparative Zoology, Harvard University—Adam Baldinger, Ken Boss, Samantha Edelheit, and Richard Johnson; North Carolina State Museum of Natural Sciences—Art Bogan and Jamie Smith; Ohio State University Museum of Biological Diversity—David Stansbery and Tom Watters; U.S. National Museum of Natural History, Smithsonian Institution—Paul Greenhall, Robert Hershler, and Tyjuana Nickens; University of Michigan Museum of Zoology—Liath Appleton, Jack Burch, Taehwan Lee, and Diarmaid O'Foighil; McClung Museum of Natural History and Culture, Paul W. Parmalee Malacology Collection, University of Tennessee—Gerry Dinkins.

Field Assistance

Since our arrival in Florida in the late 1980s, we began having discussions of someday writing a state mussel book. During the past 25 years, we had assistance in the field and received mussels and collection data from a number of individuals. The efforts of the following individuals contributed greatly to our understanding of distribution and conservation status of mussels in Florida: Steve Ahlstedt, Herb Athearn (deceased), Jamie Barichivich, Jennifer Bernatis, Holly Blalock-Herod, Marc Blouin, Sherry Bostick, Jayne Brim Box, Mary Brown, Michael Buntin, Matt Burgess, Noel Burkhead, Stuart Butler (deceased), Matt Callahan, Ron Cicerello, Andre Daniels, Gerry Dinkins, Todd Fobian, Ben Forest, Butch Forrer, Ann Foster, Dick Franz, Paul Freeman, Pam Fuller, Sam Fuller (deceased), Mike Gangloff, Jeff Garner, Andy Gerberich, Jim Godwin, Steve Golladay, Margaret Gunzburger, Wendell Haag, Hannah Hamilton, Byron Hamstead, Dennis Haney, Kate Harriger, Bill Heard, Jeff Herod, Karen Herrington, Amy Hester, Gary Hill, Mike Hill, Randy Hoeh, Ted Hoehn, Tim Hogan, Steve Howard, Mark Hughes, Colette Jacono, Howard Jelks, John Johansen, Dan Johnson, Paul Johnson, Bob Jones, Anne Keller, Nikki Kernaghan, Jeff Kline, John Knight, Sabrina Krisberg, Ricardo Lattimore, Harry Lee, Eddie Leonard, Bob Lewis, Bill Loftus, Glen Long, Rob Mattson, Henry McCullagh, Stuart McGregor, Drew Miller, J.B. Miller, Marc Minno, Sara Mirabilio, Paul Moler, Rachel Muir, Eric Nagid, Eric Nelson, Leo Nico, Dawson Nowling, Christine O'Brien, Noel Ocampo (deceased), Les Parker, Jeff Powell, Sandy Pursifull, Lance Riley, Rob Robins, Stacy Rowe, Shane Ruessler, Beth Schilling, Pam Schofield, Colin Shea, Doug Shelton, Tracey Smith, Bill Smith-Vaniz, Mike Spelman, Channing St. Aubin, Jennifer Staiger, Carson Stringfellow (deceased), Doug Strom, Will Strong, Bill Tate, Travis Tuten, Mark Vogel, Jackie Wagner, Steve Walsh, Doug Weaver, Dave Werneke, Jason Wisniewski, Jerry Ziewitz, and Greg Zimmerman.

Information

Assembling information for this book required contacting many individuals who had knowledge of mussels, particularly those in Florida and adjacent states. The information came in the form of publications, reports, unpublished data, correspondence, and verbal conversations over the past several decades. We appreciate the time, effort, and patience of the following individuals: Steve Ahlstedt, Herb Athearn (deceased), Jim Austin, Dave Berg, Dick Biggins, Jayne Brim Box, Michael Buntin, David Campbell, Gail Carmody, Ron Cicerello, Cliff Coney (deceased), Kevin Cummings, Steve Fraley, Andrea Fritts, Sam Fuller (deceased), Mike Gangloff, Jeff Garner, Mark Gordon, Wendell Haag, John Harris, Paul Hartfield, Andy Hartzog, Bill Heard, Karen Herrington, Randy Hoeh, Bob Howells, Paul Johnson, Jess Jones, Karen Kandl, Steve Karl, Harry Lee, Henry McCullagh, Stuart McGregor, Clayton Metcalf (deceased), Dick Neves, Christine O'Brien, Lisa Preister, Sandy Pursifull, Morgan Raley, Harvey Rudolph, Shane Ruessler, Doug Shelton, Dave Stansbery, Dave Strayer, Carson Stringfellow (deceased), Malcolm Vidrine, Tom Watters, Jason Wisniewski, and Greg Zimmerman.

Photographs/Artwork

We are grateful to individuals that contributed photographs and illustrations used in this book. Richard (Dick) T. Bryant provided most of the photographs of Florida mussels included herein. His patience and attention to detail is obvious when admiring his photographs. Also, Dick's ability to artfully depict very small specimens has contributed immensely to this book. We also thank Zach Randall for providing numerous additional photographs of Florida mussels. The National Science Foundation award, DEB 0845392, to David Reed provided the Visionary Digital (Palmyra, Virginia) camera system utilized by Zach Randall. Other photographs were provided by Jennifer Bernatis, Jayne Brim Box, Michael Buntin, Noel Burkhead, Paul Callomon, Paul Freeman, Robert Hershler, Paul Johnson, Christine O'Brien, Sandy Pursifull, Shane Ruessler, Thomas Tarpley, Ken Womble, Florida Fish and Wildlife Conservation Commission, Florida Geological Survey, Florida Museum of Natural History, Museum of Comparative Zoology at Harvard University, North Carolina State Museum of Natural History Archives, and St. Johns River Water Management District. The outstanding scanning electron micrographs of glochidia were taken by Sabrina Krisberg and Eric Nelson and edited and arranged by Sherry Bostick. We would also like to express our appreciation to Susan Trammell for creating the excellent illustrations of mussel soft anatomy and shell morphology.

Mapping Assistance

The task of creating the species dot distribution maps using GIS software was accomplished by Brian Beneke. Brian also prepared the Florida hydrologic units map. Production of the shaded range maps of each species was artfully done by Sherry Bostick. We thank John Johansen for the tedious and laborious task of checking or determining latitude and longitude data for several thousand museum records of Florida mussels. Wade Ross provided the base map for the major rivers of Florida.

Authors' Personal Recognitions

I (JDW) express my gratitude to Herbert Boschung, affectionately known as "Bo," my graduate school advisor, mentor, and friend. Bo nurtured and stimulated my interest in ichthyology, natural history, and conservation during my tenure as a student at the University of Alabama. I would like to thank Dave Stansbery and Sam Fuller (deceased) for introducing me to the world of mussels and tolerating my questions over the years. I would also like to thank my colleagues Art Bogan and Jeff Garner (my Alabama mussel book coauthors) for listening and responding to my many thoughts and questions during the past 20 years. They have been extremely patient considering the fact that many of our phone conversations took place after hours and often late in the evening. I also wish to thank my research assistant, Sherry Bostick, for providing much needed assistance in support of almost every aspect of my research during the past two decades.

I (RSB) thank my professors at Eastern Kentucky University for exposing me to the amazing world of aquatics. Donald Batch introduced me to mussels, Branley Branson fostered my interest in fishes, and Guenter Schuster taught me insects—in addition to being my primary mentor and long-time friend. John Williams (deceased), with Guenter, expanded my interest in mussels while brailing down the Ohio River in 1981. After learning of my plans to work for Florida Game and Fresh Water Fish Commission (currently Florida Fish and Wildlife Conservation Commission) in 1986, Batch gave me a copy of Heard's mussel identification manual (Heard 1979a), which was my introduction to Florida mussels. This singular publication provided the impetus to strive to become an expert on the Florida fauna. For that, I owe him a lot (at least a gratis copy of the book!). There were other tremendous influences in my life as a biologist. My late parents, Evelyn and Stuart Butler, cultivated an early love of nature. Ron Cicerello kept me focused on field biology while I floundered during indecisive years between degrees. Steve Ahlstedt really taught me how to dig shell. And I can't begin to express how much Jim Williams has contributed to my career development. My good friend Wendell Haag—who knows more about mussels than anyone—deserves special credit. While writing this book, I learned so much reviewing his book (Haag 2012), and Wendell freely allowed us to incorporate information herein. I am eternally grateful to them and many others. Lastly, I want to thank my dear wife, Bridget Downey, for her unending support and tolerance during long weeks on collecting trips and writing marathons in Gainesville.

I (GLW) thank Richard W. "Lari" Larimore, Stephen O. Swadener, Mark Wetzel, Warren Brigham (deceased), Michael Wiley, and Don Fox for providing the career inspiration and guidance that ultimately resulted in my contribution to the *Freshwater Mussels of Florida*. I thank Darrell Scovell of the Florida Fish and Wildlife Conservation Commission, whose leadership and support was instrumental in codifying the Florida Mussel Rule, thus prohibiting commercial harvest of freshwater mussels in Florida and raising awareness of the importance and fragility of mussel populations in the state. I express my gratitude to Jim Estes and Dick Krause of the Florida Fish and Wildlife Conservation Commission for providing the opportunity and freedom to contribute to this book. My efforts in the production of this book are dedicated to my family, especially my father, Wallace L. Warren (deceased), my mother, Nancy L. Davis, and my grandmother Helen E. Brady (deceased), and to those concerned with conservation of mussels everywhere.

I (NAJ) am grateful to all my colleagues, professors, friends, and family who have supported my academic career. I acknowledge and thank my undergraduate and master's advisor at Virginia Tech, Eric Hallerman, for helping me realize my calling to study aquatic sciences and for teaching me the power of genetics as a tool for conservation and recovery. I owe my roots in malacology to Dick Neves and Jess Jones, who began teaching me about the fascinating life histories of freshwater mussels and the plight of the habitats these organisms occupy. It was also at that point in my career when I became keenly interested in freshwater mussel systematics and conservation genetics. I thank Yniv Palti and Caird Rexroad for nurturing my background in molecular methods and for showing me how to execute large-scale molecular projects. I also thank my supervisors, Howard Jelks and Gary Mahon, along with our Center Director, Ken Rice, for supporting my time and travel during the last two years of this project and providing me the opportunity to start a mussel research program at the U.S. Geological Survey Southeast Ecological Science Center. Lastly, I am extremely grateful to my PhD committee chair, Jim Austin, and committee members, Mark Brenner, Tom Fraser, Gustav Paulay, and Jim Williams, who I have had the pleasure of working with over the past five years while studying the diversity and distributions of freshwater mussels in Florida and throughout the southeastern United States.

Abbreviations

The following abbreviations are used in the text for state and federal agencies, museum institutions, and nongovernmental organizations, as well as standard technical terms and measurements.

ACF	Apalachicola, Chattahoochee, and Flint
ALMNH	Alabama Museum of Natural History, University of Alabama, Tuscaloosa
AMNH	American Museum of Natural History, New York, New York
ANSP	Academy of Natural Sciences of Philadelphia, Pennsylvania
AUM	Auburn University Museum, Auburn, Alabama
BP	years before present
cfs	cubic feet per second
cm	centimeter
CMNH	Carnegie Museum of Natural History, Pittsburgh, Pennsylvania
CMNML	Canadian Museum of Nature, Mollusks, Ottawa, Ontario (formerly National Museums of Canada [NMC])
DO	dissolved oxygen
DNA	deoxyribonucleic acid
DUI	doubly uniparental inheritance
ESA	Endangered Species Act
FDEP	Florida Department of Environmental Protection
FLMNH	Florida Museum of Natural History, University of Florida, Gainesville (used when referring to the museum)
FMCS	Freshwater Mollusk Conservation Society
FMNH	Field Museum of Natural History, Chicago, Illinois
FWC	Florida Fish and Wildlife Conservation Commission
Ga	billion years ago
ha	hectare
ICZN	International Commission of Zoological Nomenclature, The Natural History Museum, London, England
km	kilometer
km^2	square kilometer
L	liter
m	meter
m^2	square meter
Ma	million years ago
MCZ	Museum of Comparative Zoology, Harvard University, Cambridge, Massachusetts
MIS	Marine Isotopic Stage
mg	milligram
mm	millimeter
MNHN	Muséum national d'Histoire naturelle, Paris, France
msl	mean sea level
mtDNA	mitochondrial DNA

NCSM	North Carolina State Museum of Natural Sciences, Raleigh
OSU	Ohio State University, Columbus
ppt	parts per thousand
SMF	Senckenberg Forschungsinstitut und Naturmuseum, Frankfurt am Main, Germany
UF	University of Florida, Florida Museum of Natural History, Gainesville (used only with catalogue numbers)
UMMZ	University of Michigan Museum of Zoology, Ann Arbor
USACE	U.S. Army Corps of Engineers
USFWS	U.S. Fish and Wildlife Service, Department of the Interior
USGS	U.S. Geological Survey, Department of the Interior
USNM	U.S. National Museum of Natural History, Smithsonian Institution, Washington, DC
WMD	water management district
μm	micrometer

List of Bivalve Mollusk Taxa

The following list includes generic names, binomials, authors, and dates of bivalve mollusks used herein with the exception of those in the synonymy sections of species accounts. Authors and dates are not included in the text except with the heading and type species of the generic accounts and the heading and synonymy section of each species account. Common names are included for all valid binomials and trinomials. For taxa not currently recognized as valid, the senior synonym is given in place of a common name.

Family/Scientific Name	Author and Date	Common Name/Synonym
Corbiculidae	Gray 1847	Cyrenidae
Cyrenidae	Gray 1840	
Corbicula	Megerle von Mühlfeld 1811	
Corbicula fluminea	(Müller 1774)	Asian Clam
Polymesoda	Rafinesque 1820	
Polymesoda caroliniana	(Bosc 1801)	Carolina Marshclam
Polymesoda maritima	d'Orbigny 1842	Southern Marshclam
Dreissenidae	Gray 1840	
Dreissena	van Beneden 1835	
Dreissena bugensis	Andrusov 1897	Quagga Mussel
Dreissena polymorpha	(Pallas 1771)	Zebra Mussel
Mytilopsis	Conrad 1857	
Mytilopsis leucophaeata	(Conrad 1831)	Dark Falsemussel
Mytilopsis sallei	(Récluz 1849)	Santo Domingo Falsemussel
Mactridae	Lamarck 1809	
Rangia cuneata	(Sowerby 1831)	Atlantic Rangia
Margaritanidae	Ortmann 1911	Margaritiferidae
Margaritiferidae	Henderson 1929	
Margaritifera	Schumacher 1816	
Margaritifera falcata	(Gould 1850)	Western Pearlshell
Margaritifera margaritifera	(Linnaeus 1758)	Eastern Pearlshell
Margaritifera marrianae	Johnson 1983	Alabama Pearlshell
Mycetopodidae	Gray 1840	
Anodontites trapesialis	Lamarck 1819	Table Latin Floater
Sphaeriidae	Deshayes 1855	
Eupera	Bourguignat 1854	
Eupera cubensis	(Prime 1865)	Mottled Fingernailclam
Musculium	Link 1807	
Musculium lacustre	(Müller 1774)	Lake Fingernailclam
Musculium partumeium	(Say 1822)	Swamp Fingernailclam
Musculium securis	(Prime 1852)	Pond Fingernailclam
Musculium transversum	(Say 1829)	Long Fingernailclam
Pisidium	Pfeiffer 1821	
Pisidium adamsi	Stimpson 1851	Adam Peaclam

Family/Scientific Name	Author and Date	Common Name/Synonym
Pisidium casertanum	(Poli 1791)	Ubiquitous Peaclam
Pisidium compressum	Prime 1852	Ridgebeak Peaclam
Pisidium dubium	(Say 1817)	Greater Eastern Peaclam
Pisidium punctiferum	(Guppy 1867)	Striate Peaclam
Pisidium nitidum	Jenyns 1832	Shiny Peaclam
Sphaerium	Scopoli 1777	
Sphaerium occidentale	Prime 1853	Herrington Fingernailclam
Sphaerium striatinum	(Lamarck 1818)	Striated Fingernailclam
Unionidae	Rafinesque 1820	
Alasmidonta	Say 1818	
Alasmidonta arcula	(Lea 1838)	Altamaha Arcmussel
Alasmidonta mccordi	Athearn 1964	Coosa Elktoe
Alasmidonta triangulata	(Lea 1858)	Southern Elktoe
Alasmidonta undulata	(Say 1817)	Triangle Floater
Alasmidonta wrightiana	(Walker 1901)	Ochlockonee Arcmussel
Amblema	Rafinesque 1820	
Amblema costata	Rafinesque 1820	*Amblema plicata*
Amblema elliottii	(Lea 1856)	Coosa Fiveridge
Amblema neislerii	(Lea 1858)	Fat Threeridge
Amblema perplicata	Conrad 1841	*Amblema plicata*
Amblema plicata	(Say 1817)	Threeridge
Anodon	Oken 1815	
Anodonta	Lamarck 1799	
Anodonta couperiana	Lea 1840	Barrel Floater
Anodonta cygnea	(Linnaeus 1758)	Swan Mussel
Anodonta ferussacianus	Lea 1834	*Anodontoides ferussacianus*
Anodonta globosa	Lea 1841	*Pyganodon grandis*
Anodonta hartfieldorum	Williams, Bogan, & Garner 2009	Cypress Floater
Anodonta heardi	Gordon & Hoeh 1995	Apalachicola Floater
Anodonta imbecillis	Say 1829	*Utterbackia imbecillis*
Anodonta suborbiculata	Say 1831	Flat Floater
Anodontoides	Simpson 1898	
Anodontoides denigrata	(Lea 1852)	Cumberland Papershell
Anodontoides elliottii	(Lea 1858)	*Anodontoides radiatus*
Anodontoides ferussacianus	(Lea 1834)	Cylindrical Papershell
Anodontoides radiatus	(Conrad 1834)	Rayed Creekshell
Carunculina	Simpson 1898	
Carunculina parva	(Barnes 1823)	*Toxolasma parvum*
Carunculina vesicularis	(Lea 1872)	*Villosa amygdalum*
Elliptio	Rafinesque 1819	
Elliptio ahenea	(Lea 1843)	Southern Lance
Elliptio arctata	(Conrad 1834)	Delicate Spike
Elliptio buckleyi	(Lea 1843)	*Elliptio jayensis*

Family/Scientific Name	Author and Date	Common Name/Synonym
Elliptio caloosaensis	(Dall 1898)	no common name
Elliptio chipolaensis	(Walker 1905)	Chipola Slabshell
Elliptio complanata	(Lightfoot 1786)	Eastern Elliptio
Elliptio crassidens	(Lamarck 1819)	Elephantear
Elliptio crassidens incrassatus	(Lea 1840)	*Elliptio crassidens*
Elliptio dariensis	(Lea 1842)	Georgia Elephantear
Elliptio dilatata	(Rafinesque 1820)	Spike
Elliptio downiei	(Lea 1858)	Satilla Elephantear
Elliptio fraterna	(Lea 1852)	Brother Spike
Elliptio fumata	(Lea 1857)	Gulf Slabshell
Elliptio hazlehurstianus	(Lea 1858)	Satilla Spike
Elliptio icterina	(Conrad 1834)	Variable Spike
Elliptio jayensis	(Lea 1838)	Florida Spike
Elliptio maywebbae	B.H. Wright 1934	*Elliptio jayensis*
Elliptio mcmichaeli	Clench & Turner 1956	Fluted Elephantear
Elliptio monroensis	(Lea 1843)	St. Johns Elephantear
Elliptio nigella	(Lea 1852)	Winged Spike
Elliptio occulta	(Lea 1843)	Hidden Spike
Elliptio pachyodon	Pilsbry 1953	no common name
Elliptio pullata	(Lea 1856)	Gulf Spike
Elliptio purpurella	(Lea 1857)	Inflated Spike
Elliptio strigosa	(Lea 1840)	*Elliptio arctata*
Elliptio waltoni	(B.H. Wright 1888)	*Elliptio ahenea*
Elliptoideus	Frierson 1927	
Elliptoideus sloatianus	(Lea 1840)	Purple Bankclimber
Fusconaia	Simpson 1900	
Fusconaia apalachicola	Williams & Fradkin 1999	*Reginaia apalachicola*
Fusconaia barnesiana	(Lea 1838)	*Pleuronaia barnesiana*
Fusconaia burkei	(Walker 1922)	Tapered Pigtoe
Fusconaia ebenus	(Lea 1831)	*Reginaia ebenus*
Fusconaia escambia	Clench & Turner 1956	Narrow Pigtoe
Fusconaia flava	(Rafinesque 1820)	Wabash Pigtoe
Fusconaia rotulata	(B.H. Wright 1899)	*Reginaia rotulata*
Fusconaia succissa	(Lea 1852)	*Quadrula succissa*
Glebula	Conrad 1853	
Glebula rotundata	(Lamarck 1819)	Round Pearlshell
Glochidium parasiticum	Rathke 1797	no common name
Hamiota	Roe & Hartfield 2005	
Hamiota altilis	(Conrad 1834)	Finelined Pocketbook
Hamiota australis	(Simpson 1900)	Southern Sandshell
Hamiota perovalis	(Conrad 1834)	Orangenacre Mucket
Hamiota subangulata	(Lea 1840)	Shinyrayed Pocketbook
Jugosus	Simpson 1914	
Lampsilis	Rafinesque 1820	

Family/Scientific Name	Author and Date	Common Name/Synonym
Lampsilis altilis	(Conrad 1834)	*Hamiota altilis*
Lampsilis australis	Simpson 1900	*Hamiota australis*
Lampsilis binominata	Simpson 1900	Lined Pocketbook
Lampsilis claibornensis	(Lea 1838)	*Lampsilis straminea*
Lampsilis excavatus	(Lea 1857)	*Lampsilis ornata*
Lampsilis floridensis	(Lea 1852)	Florida Sandshell
Lampsilis haddletoni	Athearn 1964	*Obovaria haddletoni*
Lampsilis jonesi	van der Schalie 1934	*Ptychobranchus jonesi*
Lampsilis ornata	(Conrad 1835)	Southern Pocketbook
Lampsilis ovata	(Say 1817)	Pocketbook
Lampsilis perovalis	(Conrad 1834)	*Hamiota perovalis*
Lampsilis siliquoidea	(Barnes 1823)	Fatmucket
Lampsilis straminea	(Conrad 1834)	Southern Fatmucket
Lampsilis straminea claibornensis	(Lea 1838)	*Lampsilis straminea*
Lampsilis straminea straminea	(Conrad 1834)	*Lampsilis straminea*
Lampsilis subangulata	(Lea 1840)	*Hamiota subangulata*
Lampsilis teres	(Rafinesque 1820)	Yellow Sandshell
Lampsilis wrightiana	Frierson 1927	*Villosa amygdalum*
Lasmigona	Rafinesque 1831	
Lasmigona subviridis	(Conrad 1835)	Green Floater
Leptodea fragilis	(Rafinesque 1820)	Fragile Papershell
Margaritana	Schumacher 1817	
Medionidus	Simpson 1900	
Medionidus acutissimus	(Lea 1831)	Alabama Moccasinshell
Medionidus conradicus	(Lea 1834)	Cumberland Moccasinshell
Medionidus mcglameriae	van der Schalie 1939	*Leptodea fragilis*
Medionidus penicillatus	(Lea 1857)	Gulf Moccasinshell
Medionidus simpsonianus	Walker 1905	Ochlockonee Moccasinshell
Medionidus walkeri	(B.H. Wright 1897)	Suwannee Moccasinshell
Megalonaias	Utterback 1915	
Megalonaias boykiniana	(Lea 1840)	*Megalonaias nervosa*
Megalonaias nervosa	(Rafinesque 1820)	Washboard
Megalonaias nickliniana	(Lea 1834)	Nicklins Pearlymussel
Micromya	Agassiz 1852	
Monodonta undulata	Say 1817	*Alasmidonta undulata*
Mytilus cygnea	Linnaeus 1758	*Anodonta cygnea*
Nephronaias gundlachi	(Dunker 1858)	no common name
Obliquaria (Quadrula) quadrula	Rafinesque 1820	*Quadrula quadrula*
Obovaria	Rafinesque 1819	
Obovaria arkansasensis	(Lea 1862)	Southern Hickorynut
Obovaria choctawensis	(Athearn 1964)	Choctaw Bean
Obovaria haddletoni	(Athearn 1964)	Haddleton Lampmussel
Obovaria retusa	(Lamarck 1819)	Ring Pink

Family/Scientific Name	Author and Date	Common Name/Synonym
Obovaria rotulata	(B.H. Wright 1899)	*Reginaia rotulata*
Obovaria unicolor	(Lea 1845)	Alabama Hickorynut
Plectomerus	Conrad 1853	
Plectomerus dombeyanus	(Valenciennes 1827)	Bankclimber
Pleurobema	Rafinesque 1819	
Pleurobema athearni	Gangloff, Williams, & Feminella 2006	Canoe Creek Clubshell
Pleurobema clava	(Lamarck 1819)	Clubshell
Pleurobema collina	(Conrad 1837)	James Spinymussel
Pleurobema mytiloides	Rafinesque 1820	*Pleurobema clava*
Pleurobema pyriforme	(Lea 1857)	Oval Pigtoe
Pleurobema reclusum	(B.H. Wright 1898)	*Pleurobema pyriforme*
Pleurobema simpsoni	Vanatta 1915	*Pleurobema pyriforme*
Pleurobema strodeanum	(B.H. Wright 1898)	Fuzzy Pigtoe
Pleuronaia	Frierson 1927	
Pleuronaia barnesiana	(Lea 1838)	Tennessee Pigtoe
Potamilus	Rafinesque 1818	
Ptychobranchus	Simpson 1900	
Ptychobranchus fasciolaris	(Rafinesque 1820)	Kidneyshell
Ptychobranchus foremanianus	(Lea 1842)	Rayed Kidneyshell
Ptychobranchus jonesi	(van der Schalie 1934)	Southern Kidneyshell
Pyganodon	Crosse & Fischer 1894	
Pyganodon cataracta	(Say 1817)	Eastern Floater
Pyganodon gibbosa	(Say 1824)	Inflated Floater
Pyganodon grandis	(Say 1829)	Giant Floater
Quadrula	Rafinesque 1820	
Quadrula asperata archeri	(Frierson 1905)	Alabama Orb
Quadrula infucata	(Conrad 1834)	Sculptured Pigtoe
Quadrula kieneriana	(Lea 1852)	Coosa Orb
Quadrula kleiniana	(Lea 1852)	Florida Mapleleaf
Quadrula nobilis	(Conrad 1853)	Gulf Mapleleaf
Quadrula quadrula	(Rafinesque 1820)	Mapleleaf
Quadrula succissa	(Lea 1852)	Purple Pigtoe
Quincuncina	Ortmann 1922	
Quincuncina burkei	Walker 1922	*Fusconaia burkei*
Quincuncina infucata	(Conrad 1834)	*Quadrula infucata*
Reginaia	Campbell & Lydeard 2012	
Reginaia apalachicola	(Williams & Fradkin 1999)	Apalachicola Ebonyshell
Reginaia ebenus	(Lea 1831)	Ebonyshell
Reginaia rotulata	(B.H. Wright 1899)	Round Ebonyshell
Strophitus	Rafinesque 1820	
Strophitus spillmanii	(Lea 1858)	*Strophitus subvexus*
Strophitus subvexus	(Conrad 1834)	Southern Creekmussel
Toxolasma	Rafinesque 1831	

Family/Scientific Name	Author and Date	Common Name/Synonym
Toxolasma corvunculus	(Lea 1868)	Southern Purple Lilliput
Toxolasma cylindrellus	(Lea 1868)	Pale Lilliput
Toxolasma lividum	(Rafinesque 1831)	Purple Lilliput
Toxolasma lividus	(Rafinesque 1831)	*Toxolasma lividum*
Toxolasma minor	(Lea 1843)	*Toxolasma paulum*
Toxolasma parvum	(Barnes 1823)	Lilliput
Toxolasma parvus	(Barnes 1823)	*Toxolasma parvum*
Toxolasma paulum	(Lea 1840)	Iridescent Lilliput
Toxolasma paulus	(Lea 1840)	*Toxolasma paulum*
Toxolasma pullus	(Conrad 1838)	Savannah Lilliput
Toxolasma sp. cf. *corvunculus*	undescribed	Gulf Lilliput
Toxolasma texasense	(Lea 1857)	Texas Lilliput
Toxolasma texasensis	(Lea 1857)	*Toxolasma texasense*
Tritogonia	Agassiz 1852	
Unio	Retzius 1788	
Unio (Elliptio) nigra	Rafinesque 1820	*Elliptio crassidens*
Unio blandingianus	Lea 1834	*Uniomerus carolinianus*
Unio caloosaensis	Dall 1898	*Elliptio caloosaensis*
Unio conradicus	Lea 1834	*Medionidus conradicus*
Unio dombeyana	Valenciennes 1827	*Plectomerus dombeyanus*
Unio ebenus	Lea 1831	*Reginaia ebenus*
Unio excultus	Conrad 1838	*Uniomerus tetralasmus*
Unio hebes	Lea 1852	*Uniomerus carolinianus*
Unio heros	Say 1829	*Megalonaias nervosa*
Unio incrassatus	Lea 1840	*Elliptio crassidens*
Unio jayensis	Lea 1838	*Elliptio jayensis*
Unio kirklandianus	S.H. Wright 1897	*Hamiota subangulata*
Unio lehmanii	S.H. Wright 1897	*Elliptio occulta*
Unio lividus	Rafinesque 1831	*Toxolasma lividum*
Unio ornatus	Conrad 1835	*Lampsilis ornata*
Unio ovatus	Say 1817	*Lampsilis ovata*
Unio phaseolus	Hildreth 1828	*Ptychobranchus fasciolaris*
Unio prasinatus	Conrad 1866	*Elliptio jayensis*
Unio pyriformis	Lea 1857	*Pleurobema pyriforme*
Unio retusa	Lamarck 1819	*Obovaria retusa*
Unio rivicolus	Conrad 1868	*Uniomerus carolinianus*
Unio rotundatus	Lamarck 1819	*Glebula rotundata*
Unio sloatianus	Lea 1840	*Elliptoideus sloatianus*
Unio striatus	Rafinesque 1820	*Plethobasus cooperianus*
Unio striatus	Lea 1840	*Pleurobema pyriforme*
Unio subangulatus	Lea 1840	*Hamiota subangulata*
Unio subluridus	Simpson 1892	*Elliptio jayensis*
Unio tenuisculus	Frierson 1911	*Elliptio jayensis*
Unio trapezoides	Lea 1831	*Plectomerus dombeyanus*

Family/Scientific Name	Author and Date	Common Name/Synonym
Unio trigonus	Lea 1831	*Fusconaia flava*
Unio vesicularis	Lea 1872	*Villosa amygdalum*
Unio villosus	B.H. Wright 1898	*Villosa villosa*
Uniomerus	Conrad 1853	
Uniomerus carolinianus	(Bosc 1801)	Eastern Pondhorn
Uniomerus columbensis	(Lea 1857)	Apalachicola Pondhorn
Uniomerus declivis	(Say 1831)	Tapered Pondhorn
Uniomerus obesus	(Lea 1831)	*Uniomerus carolinianus*
Uniomerus tetralasmus	(Say 1831)	Pondhorn
Utterbackia	Baker 1927	
Utterbackia imbecillis	(Say 1829)	Paper Pondshell
Utterbackia peggyae	(Johnson 1965)	Florida Floater
Utterbackia peninsularis	Bogan & Hoeh 1995	Peninsular Floater
Utterbackiana	Frierson 1927	
Villosa	Frierson 1927	
Villosa amygdalum	(Lea 1843)	Florida Rainbow
Villosa australis	(Simpson 1900)	*Hamiota australis*
Villosa choctawensis	Athearn 1964	*Obovaria choctawensis*
Villosa lienosa	(Conrad 1834)	Little Spectaclecase
Villosa nebulosa	(Conrad 1834)	Alabama Rainbow
Villosa subangulata	(Lea 1840)	*Hamiota subangulata*
Villosa umbrans	(Lea 1857)	Coosa Creekshell
Villosa vanuxemensis	(Lea 1838)	Mountain Creekshell
Villosa vanuxemensis umbrans	(Lea 1857)	*Villosa umbrans*
Vlllosa vanuxemensis vanuxemensis	(Lea 1838)	*Villosa vanuxemensis*
Villosa vibex	(Conrad 1834)	Southern Rainbow
Villosa villosa	(B.H. Wright 1898)	Downy Rainbow
Veneridae	Rafinesque 1815	
Macrocallista nimbosa	Lightfoot 1786	Sunray Venus

Chapter 1

Introduction

Freshwater mussels known to occur in Florida's inland waters belong to the family Unionidae (class Bivalvia), which is widespread in North America and also occurs in most of Africa, Asia, and Europe. The family is represented by about 365 species and 61 genera in North America (Williams et al. in review). The Florida mussel fauna, 60 species (58 native and 2 introduced) in 23 genera, is less diverse than that of adjacent Alabama and Georgia. There are several species whose range is predominantly in Florida waters, but only one, *Villosa amygdalum*, is confined entirely to the state. Florida mussels are known to occur in a wide variety of natural habitats including blackwater creeks and rivers, large silty rivers, lower ends of spring runs, and sand-bottom lakes, as well as artificial habitats such as canals and reservoirs. However, some streams in the state are naturally devoid of mussels. Absence of mussels in these watersheds is possibly due to acidic and nutrient-poor waters or intrusion of saline waters.

The abundance of mussels in most of Florida's inland waters made them an important natural resource for Native Americans. Mussels were harvested in large quantities and provided a relatively easy and dependable source of food. The large shell mounds along some north Florida rivers (e.g., Apalachicola) illustrate this point. Not all shells were discarded; some were used as tools or ornaments. Early American settlers also harvested shell but primarily for pearls. In the late 1800s and early 1900s, Americans harvested shell for the pearl button industry. During this period demand for shell threatened some mussel populations until plastic was developed and used to replace mother-of-pearl buttons. There was some interest in developing a shell harvest operation on the Chipola River, but there is no evidence that it was ever a successful venture. In the 1950s, harvest of shell in the Mississippi basin was renewed for the cultured pearl industry. Currently, the lack of demand for cultured pearls has resulted in the decline of shell exports.

In addition to the mussels (Unionidae), four additional bivalve families are known from inland waters of Florida. There are two species of marshclams, Cyrenidae (formerly Corbiculidae), in the state. One is a native estuarine bivalve, *Polymesoda caroliniana*, which also inhabits the tidal freshwater reaches of rivers. The other is an invasive species, *Corbicula fluminea*, which occurs in most freshwater habitats throughout the state. Dreissenidae, the falsemussels, are found primarily in estuaries, but at least one species, *Mytilopsis leucophaeata*, may be found in the lower reaches of rivers. This family also has two well-known freshwater, invasive species, *Dreissena bugensis* and *Dreissena polymorpha*, neither of which are known to occur in Florida. The Mactridae, surfclams, primarily inhabit marine and estuarine waters, but one species, *Rangia cuneata*, occurs in the lower reaches of most Florida rivers. The Sphaeriidae—fingernailclams, peaclams, and pillclams—are small (less than 20 mm in length) and most are widely distributed. There are 12 species in 4 genera found in Florida where they occur in a variety of natural and artificial freshwater habitats (Heard 1979a).

Taxonomy and evolutionary relationships of taxa in the Unionidae have historically been a contentious topic among malacologists and remain so today. Past classifications at the species, genus, and subfamily levels were based on shell morphology along with some anatomical characters. In addition to morphological characters, current researchers use molecular techniques to evaluate phylogenetic relationships. Research in the past decade has resulted in realignment of several species in the genera *Fusconaia*, *Quadrula*, and *Quincuncina* (Lydeard et al. 2000) and description of the new genus *Reginaia* (Campbell and Lydeard 2012a). Like most tools, these are not always perfect and work well for delineating species in some genera (e.g., *Lampsilis*, *Toxolasma*, *Villosa*) but provide little or no resolution for others (e.g., *Elliptio*).

A classification of North American unionids put forth more than a century ago, based on shell morphology (adult and larval) and soft anatomy, recognized four subfamilies—Anodontinae, Lampsilinae, Margaritaninae, and Unioninae (Ortmann 1910a, 1911). Ortmann (1912) revised his classification by elevating Margaritaninae to family level and continued to recognize Unionidae with three subfamilies. After 1912 Margaritanidae was variously recognized as a distinct family or subfamily of Unionidae. The family name Margaritanidae was corrected by Henderson (1929) to Margaritiferidae, but the concept of the family remained the same. There is a general consensus today that both Margaritiferidae and Unionidae do represent distinct families. Characteristics used to separate the derived Unionidae from the ancestral Margaritiferidae include: (1) excurrent and supra-anal apertures are separated by a bridge of tissue in unionids that is absent in margaritiferids; (2) unionid gills are more completely fused with the mantle and have water tubes formed by interlamellar connections as continuous septa oriented parallel to the gill filaments, while margaritiferid gills are not fused to the mantle posteriorly and lack water tubes but have interlamellar connections scattered or in irregular rows in an oblique orientation to the gill filaments; and (3) lateral mantle attachment muscles are absent in unionids but present in margaritiferids.

In the past decade, phylogenies of selected unionid genera, tribes, and subfamilies have been proposed based primarily on molecular data (Serb et al. 2003; Campbell et al. 2005; Graf and Cummings 2007; Campbell and Lydeard 2012a, 2012b). The most recent classifications of the family were those proposed by Bieler et al. (2010), Carter et al. (2011), and Whelan et al. (2011).

Acceptance of the subfamily alignments has been mixed and questions remain concerning the alignment of some genera at the tribe level, particularly within the Ambleminae. A recent genetics analysis aligned *Uniomerus* with the tribe Quadrulini instead of Pleurobemini where it has usually been placed (Campbell and Lydeard 2012a); however, herein it is retained as a member of Pleurobemini until additional supporting evidence is provided. While the new genus *Reginaia* described by Campbell and Lydeard (2012a) is distinct and recognizable, its alignment with an existing tribe of Ambleminae is unclear. Uncertainty of generic relationships and species boundaries will likely lead to further instability in nomenclature in the short term. Clearly, additional research will be necessary to refine the relationships within Florida Unionidae. The contrasting higher taxonomic arrangements for Florida Unionidae are presented in Figure 1.1.

Many Florida mussel species, like those of other central and eastern states, are under severe threat of extinction due primarily to habitat loss and alteration. Conservation efforts for mussels are complicated by their unusual life cycle, which typically (about 98 percent of species) includes a larval stage that is parasitic on the gills and/or fins of one or more fish species. Many mussel species exhibit a high degree of host specificity, adding another hurdle for successful reproduction. This means that restoration and conservation actions must focus not only on mussels but also their hosts and respective habitats. Conservation activities must consider ecological factors such as temperature, DO, stream flow, food resources, and other requirements for the mussel and fish faunas. These efforts are currently hampered by a lack of knowledge of host fishes for many mussels. This highlights the need for intense research to answer basic ecological and biological questions about mussels and their host fishes.

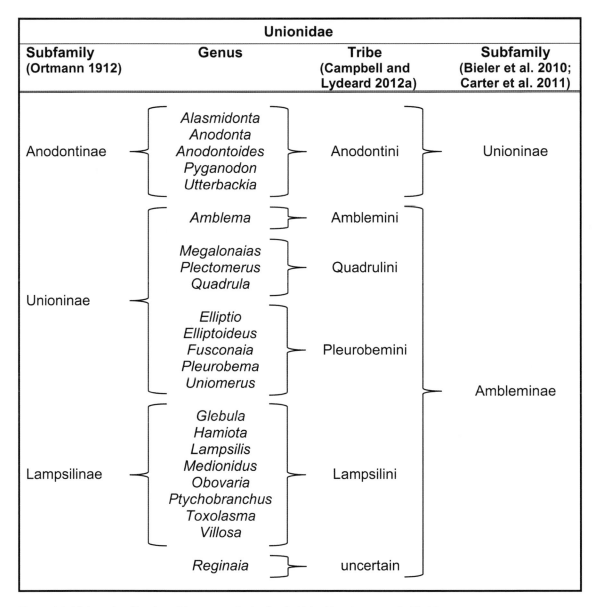

Figure 1.1. Higher classification of the genera in the family Unionidae that occur in Florida.

Florida freshwater ecosystems from the panhandle to the Everglades, as well as their associated mussel faunas, are in need of immediate conservation attention. A serious threat to Florida mussels and other freshwater organisms is loss of groundwater. Out of sight, but not out of reach, groundwater is the primary source of water to maintain streams during periods of low rainfall. The increasing quantities being withdrawn for agricultural and municipal uses will ultimately result in reduced flows for some stream channels and complete dewatering of others (Golliday et al. 2004; Torak et al. 2010; Peterson et al. 2011; Rugel et al. 2011). Captive propagation techniques are being developed that may result in production of large numbers of juvenile mussels for augmentation and reintroduction, but this alone will not succeed without habitat restoration and minimum flow requirements that protect and sustain mussel populations.

Increased interest by state, federal, and nongovernment conservation organizations in protection and management of freshwater habitats is a positive development. Without habitat restoration and increased conservation of water resources, populations will continue to decline and species will continue to go extinct. The purpose of this book is to furnish information to natural resource managers and the scientific community, as well as the lay public, on the plight of Florida's mussel fauna while providing a foundation on which future ecology, biology, phylogeny, and conservation research efforts can be based.

Chapter 2

History of Mussel Exploration

Research on Recent Mussels: 1834–2012

Interest in Florida's natural history began early in the 1700s, but more than a century passed before the mussel fauna attracted attention. The first publication involving a Florida mussel was the description of *Unio blandingianus* (= *Uniomerus carolinianus*) (Figure 2.1) based on a single shell reportedly given to William Blanding, a physician and naturalist from South Carolina, by a Native American in St. Augustine (Lea 1834a, 1834b). The shell was from the St. Johns River, but the exact locality was not known. Four years later, Lea (1838a) described *Unio jayensis* from "Florida." Not surprisingly, this species is one of the most common mussels in peninsular Florida.

Figure 2.1. *Unio blandingianus* (= *Uniomerus carolinianus*), the first freshwater mussel described from Florida waters. The description was by Isaac Lea and appeared in the *Transactions of the American Philosophical Society* in 1834. Photograph by James D. Williams.

Isaac Lea (1792–1886), who lived in Philadelphia, Pennsylvania, and was active in the ANSP, was the most active worker describing Florida mussels during the mid-1800s. Lea published 19 descriptions of Florida mussels between 1834 and 1872, including his last named species, *Unio vesicularis* (= *Toxolasma paulum*), from Lake Okeechobee. All of Lea's descriptions were based on material sent to him by others, and he apparently never personally made collections in Florida.

During his career in conchology, Lea amassed a large malacological collection, most of which was donated to the USNM. This included most of the specimens illustrated in his publications and those subsequently designated as types (Johnson 1956, 1971, 1974, 1980, 1998; Johnson and Baker 1973; Boyko and Sage 1996). Other museums that have specimens from the Lea collection include ANSP, MCZ, and some in Europe. Lea's original descriptions and subsequent elaborations were usually published in *Transactions* and *Proceedings of the American Philosophical Society* and *Journal* and *Proceedings of the Academy of Natural Sciences of Philadelphia*. Periodically, he would privately reproduce collections of species descriptions, which originally appeared in *Transactions* and *Journal*, in his *Observations on the Genus Unio* (Bogan and Bogan 2002). Between 1834 and 1874, Lea published a total of 13 volumes of *Observations*, plus 3 indices (Lea 1867a, 1869, 1874a). He also published four editions of his synopsis, which had lists of taxa and provided his own classification (Lea 1836, 1838b, 1852a, 1870). During his career Lea described more North American mussels than any other conchologist.

However, in Florida Lea was second to Berlin H. Wright, who named 28 new taxa from the state (Figure 2.2).

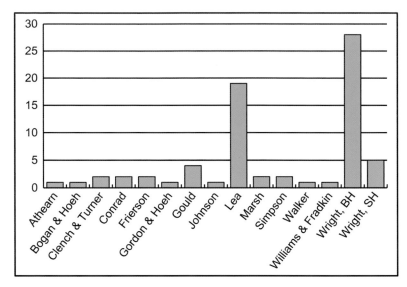

Figure 2.2. Number of mussel taxa described from Florida by various authors between 1834 and 2010.

The mid- to late 1800s was an active period for collecting and describing Florida mussels (Figure 2.3). Among the earliest records of unionids from southwest Florida were *Elliptio jayensis* and *Villosa amygdalum*, which were obtained by dredge in the canal connecting Lakes Hicpochee and Okeechobee (Heilprin 1887). Timothy Abbott Conrad, who worked in the ANSP, took extended trips to collect fossil and recent mollusks for himself, as well as other naturalists, many of whom contributed funds to support his travels. In the winter of 1842, he participated in a government expedition led by Major Powell to survey Tampa Bay, Florida (Conrad 1846). En route the steamer was forced to make repairs in the St. Johns River near Mayport, Florida, and Conrad "embraced the opportunity to collect shells at Hasard" (Wheeler 1935). Conrad only described two Florida species, *Unio prasinatus* (= *Elliptio jayensis*) and *Unio rivicolus* (= *Uniomerus carolinianus*).

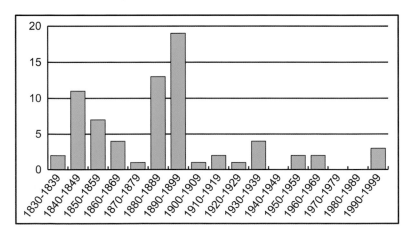

Figure 2.3. Number of mussel taxa described from Florida per decade between 1830 and 1999. There have been no species described from Florida since 1999.

Charles Torrey Simpson (1846–1932) was a biologist who worked at the USNM from 1889 to 1902. He retired to Florida where he continued to work as a naturalist and author and was awarded a ScD degree from the University of Miami in 1927 at the age of 80. Simpson (1892b) commented on the artificial nature of unionid classification, a point on which many of his colleagues agreed. The imminent naturalist is best known for his comprehensive *Synopsis of the Naiades, or Pearly Fresh-Water Mussels* (Simpson 1900a) and a three-part monograph entitled *A Descriptive Catalogue of the Naiades, or Pearly Fresh-water Mussels* (Simpson 1914). Simpson's synopsis attempted to dismantle the genus *Unio* in North America and recognized genera based in part on soft anatomy of gravid females. His catalogue was the first effort to compile descriptions of mussel taxa worldwide. These contributions provided valuable insight into evolutionary relationships of genera in the family Unionidae.

Simpson's efforts to define genera and species were not always appreciated by contemporary conchologists. In the opening line of an essay on "What is a species?" he mentioned the current practice of naming everything by the "so-called new school of conchologists," clearly recognizing the problem of excessive description of species (Simpson 1889, 1892a). Although not specifically mentioned, it is fairly clear that Simpson was referring to Berlin H. and Samuel H. Wright, who were naming shell after shell from Florida as new species of *Elliptio*. In a paper entitled "The *Unio* muddle," B.H. Wright (1893) defended naming new species based on what he interpreted as different shell morphology but readily admitted that some are difficult to distinguish depending on size and sex of the individual. In a series of short notes in *The Nautilus* during 1893, there were several exchanges between B.H. Wright, Simpson, and others regarding validity of *Elliptio* species. There was a suggestion to convene a meeting of distinguished conchologists to determine validity of various unionid taxa. The last paper on the subject (Wheeler 1893) referred to Simpson as "Mr. Ego" and suggested that he might arrive at the meeting with "carpet bag, microscope and manuscript," which could easily result in quarrelsome scientists throwing shells at each other. Based on the current interpretation of *Elliptio* (Williams et al. 2011), Simpson was clearly correct regarding the excessive description of Florida mussels.

In his review of the Unionidae of Florida and southeastern states, Simpson (1892a) stated that one goal was "to considerably reduce the number of so called species." His ensuing critique of previous work in Florida called attention to the large number of *Elliptio* species that were synonyms of previously described taxa. While recognizing both *Elliptio buckleyi* and *Elliptio jayensis* as distinct species (*E. buckleyi* was synonymized with *E. jayensis* by Williams et al. 2011), Simpson placed seven previously recognized species of Florida *Elliptio* in their synonymies. Simpson's publication provided descriptions, distributional notes, and synonyms for selected species and was the most detailed account of Florida unionids at that time.

While Simpson described only two taxa from Florida, he collected extensively and provided an overview of resident freshwater mollusks (Simpson 1892a, 1893a). His initial paper, read before the Davenport [Iowa] Academy of Science in 1886, consisted of an annotated list of Florida mollusks and was based on collections made during his four-year residence in Bradenton, Manatee County, Florida (Simpson 1893a). The list included 21 species in the genus *Unio* plus *Anodonta couperiana*. Also included was a reference to thousands of *Unio hebes* (= *Uniomerus carolinianus*) in Wares Creek in

Bradenton. Simpson (1896) was the first to note the difference in umbo sculpture between the conchologically similar *Villosa* (as *Unio amygdalum*) and *Toxolasma* (as *Unio parvus* group).

The Wrights, Samuel Hart Wright (1825–1905) and his son Berlin Hart Wright (1851–1940), were actively involved in southeastern natural history during the late 1800s and early 1900s. Between 1883 and 1934, they described 52 new mollusk taxa from the southeastern United States (Johnson 1967a), and 33 of these were mussels from Florida (Figure 2.2). Berlin H. Wright was far more active than his father in describing Florida mussels, naming 28 taxa between 1883 and 1934, with 18 of these—17 *Elliptio* and 1 *Villosa*—from the St. Johns River drainage. The Wrights (1887, 1888) also published a series of notes on Florida unionid distribution and shell morphology. Many of the Wrights' species descriptions were published without figures; however, Simpson (1900b) provided the first illustrations of some of these taxa and Johnson (1967a) photographed almost all of the Wrights' types.

In addition to species descriptions, S.H. Wright (1891) also published a paper on the similarity of the mussels in peninsular Florida to those of adjacent states. This represents one of the earliest reports on geographic distribution of the eastern Gulf Coast mussel fauna. Berlin H. Wright (1888) developed a checklist of North American Unionidae, based primarily on Lea's synopsis, with the addition of subsequently described species. His checklist was revised based primarily on Simpson's (1900a) synopsis and republished with Bryant Walker (Wright and Walker 1902).

An overview of Florida unionids was published by Hermann von Ihering (1850–1930), a Brazilian zoologist born in Germany. He was founder and first director of the São Paulo Museum and published on a wide variety of subjects in zoology. His only publication on Florida unionids concluded that the fauna was depauperate compared with other regions of eastern North America and correctly recognized that most Florida unionids were derived from the Mississippi basin fauna (Ihering 1895). This was an enlightened deduction given that he never visited the state and only had access to Florida shells that were given to him by U.S. conchologists.

Most of the Florida mussel descriptions in the 1800s were species confined to the peninsular region, with very little attention given to the panhandle in the northwestern portion of the state (Figure 2.4). This is evident when examining the distribution and number of taxa, 57 of 72 (79 percent), described from the state prior to 1900. The large number from the St. Johns River drainage is due in part to the ease of access in the late 1800s and the tremendous shell variation exhibited by species of the genus *Elliptio*.

Herbert Huntington Smith (1851–1919) was a naturalist and conchologist who moved to Wetumpka, Alabama, in 1903. He served as curator of the ALMNH at the University of Alabama, Tuscaloosa, from 1910 until his untimely death in 1919. Smith personally made only a few collections in Florida; however, he organized others to sample mollusks in the state, including brothers Joseph B. and Charles A. Burke, who collected in northwestern Florida and southeastern Alabama from 1915 to 1918. The large number of unionid collections made by the Burkes is critical to our understanding of the distribution and former abundance of panhandle mussels. Specimens collected by the Burke brothers were originally sent to Smith and housed in the ALMNH where some lots were subsequently sent to other mollusk collections. The Smith collection at the

ALMNH was transferred to the FLMNH at the University of Florida, Gainesville, in the 1980s.

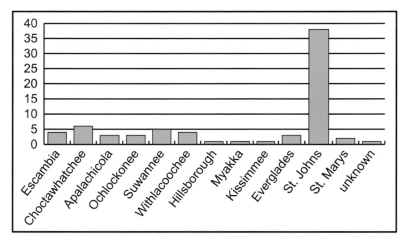

Figure 2.4. Number of mussel taxa described from major drainages in Florida.

Arnold Edward Ortmann (1863–1929) was curator of invertebrate zoology at the CMNH in Pittsburgh, Pennsylvania, from 1903–1927. Apparently Ortmann never collected in Florida, but he coauthored the description of the genus *Quincuncina*, which included two Florida taxa that are now recognized as species of *Fusconaia* and *Quadrula* (Lydeard et al. 2000). The first attempt to describe details of the soft anatomy of unionids found in eastern Gulf Coast rivers was presented in a series of papers by Ortmann (1923a, 1923b, 1924a). These descriptions demonstrated the importance of soft anatomy characters in delineating generic and subfamily relationships of unionids.

Lorraine Scriven Frierson (1861–1933), a cotton planter and conchologist who lived in Louisiana, worked on mussels throughout the southeastern United States between 1900 and 1927. He is best known for *A Classified and Annotated Checklist of North American Naiades*, which included a synonymy of North American species, generic realignments, and descriptions of several new taxa (Frierson 1927). He introduced the name *Elliptoideus* as a subgenus of *Elliptio* and noted its similarity to *Plectomerus* but added "close observation shows essential differences." Frierson described two species from Florida, *Unio tenuisculus* (= *Elliptio jayensis*) and *Lampsilis wrightiana* (= *Villosa amygdalum*).

Thompson H. Van Hyning (1861–1948) served as the first director of the Florida State Museum (currently the FLMNH) from 1914 to 1943 (Figure 2.5). While he was responsible for all collections in the museum, he was most enthusiastic about mollusks. He extended this enthusiasm to his family by naming his first two children, Clio and Arca, after genera of marine mollusks (Roberts and Moorhead 1914). Several marine and freshwater mollusks were named in honor of Van Hyning, as well as one freshwater fish.

In 1940 Van Hyning completed a manuscript entitled *A check-list of the Mollusca of Florida*. While it was never published, a copy is on file in the FLMNH and the Division of Mollusks library at the USNM. This 30-page manuscript included a list of 1,238 molluscan taxa from the state, including 90 unionid species plus numerous subspecies, along with a few species that clearly did not occur in the state. Perhaps most interesting was his commentary regarding shell collectors and traders. The best shells were bought and sold by collectors, making it difficult to acquire quality shells for the museum. He was clearly upset and stated that "the Florida shell game is becoming pretty much of a racket among speculative collectors." Van Hyning (1917) pointed out the usefulness of umbo sculpture to distinguish *Toxolasma minor* (= *Toxolasma paulum*) from the similar *Villosa villosa*.

Figure 2.5. Thompson H. Van Hyning, first director of the Florida State Museum, University of Florida, Gainesville. Photograph courtesy of the Florida Museum of Natural History.

In the early 1930s, Van Hyning became reacquainted with former University of Florida student Daniel B. Gillis, who lived near the Choctawhatchee River in Walton County. In the ensuing months, Van Hyning devised a plan to pay Daniel Gillis, his brother Harold W. Gillis, and friend Clayton Metcalf, to sample mussels in the Choctawhatchee River and ship the shells to the FLMNH. The Gillis brothers were later joined by Louie Madison Rushing and others who made valuable collections, some of which were never duplicated (e.g., *Amblema plicata* from the Choctawhatchee River) and others that have only recently been duplicated (e.g., *Ptychobranchus jonesi* in Florida). Daniel Gillis died in 1934, but Harold Gillis and others continued to sample streams in west Florida and south Alabama for several years. These collections represent the most important records of mussels from the Choctawhatchee River drainage in the early 1900s. This information was obtained in 1992–1993 during visits with C. Metcalf, L.M. Rushing, and H.W. Gillis in Defuniak Springs (R.S. Butler and J.D. Williams unpublished data).

Henry van der Schalie (1907–1986) was a malacologist and curator of mollusks at the UMMZ, who published papers on southeastern unionids during the 1930s and 1940s. His most significant contribution to Florida unionids was on Chipola River mussels in northwestern Florida (van der Schalie 1940). While the publication appeared in 1940, most of the referenced specimens were collected between 1915 and 1918 by the Burke brothers. In a prior publication, van der Schalie described *Lampsilis jonesi* (= *Ptychobranchus jonesi*) from the Pea River, a tributary to Choctawhatchee River in south Alabama (van der Schalie 1934). The terms "West Floridian" or "Appalachicolan" were first proposed by van der Schalie and van der Schalie (1950) for the mussel assemblage occurring in Gulf Coast basins from Escambia River east to Tampa Bay. Others have used these terms, but the geographic limits have varied (Burch 1975a; Butler 1990; Johnson 1970, 1972; Parmalee and Bogan 1998; Haag 2010).

William James Clench (1897–1984) was curator of mollusks at the MCZ from 1926 to 1966. During his tenure Clench was joined by various colleagues for periodic excursions to the southeastern United States to sample mollusks. His most notable foray into Florida was the summer of 1954 when he was joined by Ruth Dixon Turner (1914–2000) (Figure 2.6), also from the MCZ. Their survey was stimulated by interest at the University of Florida for a study of the area to be impounded by the USACE's Jim Woodruff Lock and Dam on Apalachicola River (Clench 1955). The Florida State Museum (currently FLMNH) and University of Florida Biology Department secured National Science Foundation and National Park Service funding to support a survey of the area to be destroyed by the impoundment. Clench and Turner expanded the survey to include all rivers from the Escambia east to the Suwannee, which led to the classic publication *Freshwater Mollusks of Alabama, Georgia, and Florida from the Escambia to the Suwannee River* (Clench and Turner 1956). Sampling between the Escambia and Chipola Rivers during this survey was the most intensive since those of the early 1900s. Clench and Turner discovered and described a new species, *Elliptio mcmichaeli*, named for Australian malacologist Donald F. McMichael, one of the survey participants.

Figure 2.6. William J. Clench and Ruth D. Turner, malacologists at the Museum of Comparative Zoology at Harvard University, Cambridge, Massachusetts. Photographs courtesy of Museum of Comparative Zoology, Harvard University, Cambridge, Massachusetts.

Herbert David Athearn (1923–2011) was born in Fall River, Massachusetts, and began collecting mollusks at the age of 16 (Figure 2.7). After serving in World War II, he lived in Boston where he volunteered at the MCZ. An ardent collector and student of freshwater mollusks, he moved to Cleveland, Tennessee, in 1955, having determined that this was the center of molluscan diversity in North America and relocating there would facilitate weekend and vacation collecting trips. Upon his arrival he began to collect aquatic mollusks throughout the southeastern states, including parts of Florida, continuing at a steady pace until about 2000 when his health no longer permitted long field excursions. Many of his collections were made prior to the invasion of the nonindigenous *Corbicula fluminea* and before construction of some of the large dams on southeastern rivers.

Figure 2.7. Herbert D. Athearn at age 17 in Boca Chica Key, Florida, holding a coral head (left). Photograph from H.D. Athearn archives at the North Carolina State Museum of Natural Sciences, Raleigh. Athearn at age 77 at his home in Cleveland, Tennessee, holding the William J. Clench Memorial Award from the Freshwater Mollusk Conservation Society, 2001 (right). Photograph by Paul D. Johnson.

Athearn's first mussel collection in Florida was in the Escambia River at the Florida Highway 4 crossing west of Jay on 23 April 1952. Between 1952 and 2000, Athearn visited more than 400 sites in Florida to sample mollusks, and most included unionids. He maintained a personal collection, the Museum of Fluviatile Mollusks, in his home until 2007, when it was donated to the NCSM (Smith et al. 2007). Among Athearn's contributions were the descriptions of two new species, *Lampsilis haddletoni* (= *Obovaria haddletoni*) and *Villosa choctawensis* (= *Obovaria choctawensis*) (Athearn 1964). Athearn was among the first to direct attention to reductions in freshwater mollusk populations in the southeastern United States and to comment on the threat to native mussels posed by *Corbicula fluminea* (Athearn 1967). In 1970 he participated in the American Malacological Union (now the American Malacological Society) symposium on conservation of rare and endangered mollusks of North America and continued to address the serious decline of southeastern aquatic mollusks (Athearn 1970). The focus of his last publication was distribution records of unionids of eastern Gulf Coast rivers (Athearn 1998). In 2001 the FMCS presented Athearn with its William J. Clench Memorial Award for his extraordinary contributions to the field of malacology, particularly his large personal collection of aquatic mollusks. Athearn's field experience and thousands of collections provide a tremendous foundation on southeastern aquatic mollusk distributions, former abundance, population trends, and ecology.

Richard Irwin Johnson (1925—), a malacologist in the MCZ, published numerous papers on distribution and taxonomy of mussels, several of which focused on Florida fauna (Figure 2.8). His first Florida publication was the description of *Utterbackia peggyae* (Johnson 1965), which was followed by three papers providing information on several species in eastern Gulf Coast rivers of west Florida and south Alabama overlooked by previous workers (Johnson 1967b, 1968, 1969a). Johnson included mussels from the St. Marys River in his monograph of unionids of the southern Atlantic slope region (Johnson 1970). Johnson (1972) published a comprehensive summary on peninsular Florida mussels. This included some museum records and the results of a survey that was conducted by Johnson, his family, and Harvard student Samuel L.H. Fuller during the summer of 1962. Their collections were primarily confined to lakes along the central Florida ridge.

Figure 2.8. Richard I. Johnson in the Division of Mollusks library, Museum of Comparative Zoology, Harvard University, Cambridge, Massachusetts, December 2009. Photograph by James D. Williams.

Samuel Liberty Harvey Fuller (1942–2001), born in Suffield, Connecticut, began collecting mussels along Windsor Locks Canal during his youth. In 1964 he graduated from Harvard University where he spent many hours studying mollusks in the MCZ under the guidance of the curator of mollusks, William J. Clench. In 1968 Fuller took a position in the ANSP Department of Limnology (Johnson 2002), which provided him with the opportunity to conduct mussel surveys in Gulf and Atlantic Coast rivers. It was on one of these surveys that he observed the folded gill of a gravid female *Lampsilis jonesi* and immediately realized it belonged to the genus *Ptychobranchus* (Fuller and

Bereza 1973). Fuller was lead author on a manuscript (26 pages, 7 figures) coauthored with D.J. Bereza, G.M. Davis, and M.F. Vidrine on the natural history of *Elliptoideus sloatianus*, which was completed while at the ANSP but unfortunately was never published. Fuller left the ANSP, moved to Florida, and worked with James D. Williams in the USFWS laboratory in Gainesville from 1990–1993, where he participated in west Florida mussel surveys. Fuller was one of the most knowledgeable individuals on mussel biology, ecology, and taxonomy in North America.

William Henry Heard (1935—) was born in Michigan and attended the University of Michigan where he completed his doctoral degree in 1963 (Figure 2.9). During his graduate career he made an extended trip to the southeastern United States where he collected freshwater mollusks in west Florida and south Alabama. Since 1962, Heard has held a position in the Department of Biological Sciences at Florida State University, Tallahassee. Heard was active in mussel conservation and was invited to participate in the American Malacological Union 1970 symposium on rare and endangered mollusks where he gave a presentation on status of the freshwater mollusk fauna in the eastern United States (Heard 1970). In 1974, under contract to the USFWS Office of Endangered Species, he conducted a status survey of mussels in eastern Gulf Coast rivers of the southeastern states (Heard 1975a), which was the first systematic survey of eastern Gulf Coast rivers since the mid-1950s. In the proceedings of a conference on the Apalachicola River, Heard published the earliest compilation of Apalachicola mussels (Heard 1977).

Figure 2.9. William H. Heard, professor emeritus in the Department of Biological Sciences at Florida State University, Tallahassee, 2008. Photograph by Ken Womble.

While in Florida, Heard's research has focused on reproductive and systematic biology of mollusks (Heard 1975b, 1979b). He supervised several graduate students who researched a variety of topics on the biology of freshwater mollusks. In 1979 he published an identification manual of the freshwater mussels of Florida for the Florida Department of Environmental Regulation (Heard 1979a) that was the first comprehensive guide for a state mussel fauna in the southeastern United States. In 2001 Heard was honored with the Freshwater Mollusk Conservation Society Lifetime Achievement

Award. The mussel *Anodonta heardi* was named in his honor (Gordon and Hoeh 1995), as well as two species of mayflies that he discovered living within the shells of freshwater mussels in Thailand. Heard remains active in his position as professor emeritus and continues to participate in research on freshwater mollusks.

Two Florida conchologists, William Henry McCullagh Jr. (1938—) and Harry G. Lee (1940—), have made a large number of mussel collections in the state. These collections represent a major contribution to our understanding of the diversity and distribution of Florida unionids. In his youth, McCullagh was interested in the outdoors, and starting at about age 10, visits to local beaches, Sanibel Island, and the Florida Keys were seminal events in a lifelong interest in shell collecting. He attended Emory University in Atlanta, Georgia, where he earned a medical degree. After a period of military service, he returned home in 1971 to set up his medical practice in Jacksonville, Florida.

McCullagh's interest in land and freshwater mollusks began in earnest in 1974 when Lee, whom he had met during medical training in Atlanta, joined him as a partner in medical practice. Almost immediately, McCullagh and Lee began field trips in the Southeast collecting freshwater mollusks. The first areas sampled were local streams, Black Creek in the St. Johns River drainage and the Santa Fe and New Rivers in the Suwannee River drainage. By the end of 1974, they had surveyed most of the drainages from St. Johns River west to Apalachicola and Chipola Rivers.

Since retiring, McCullagh continues to pursue regional mollusks, which has resulted in a computerized collection now tabulated at about 3,500 lots of freshwater mussels and about 1,500 lots of freshwater and land snails. Hundreds of duplicates have been deposited in the FLMNH and the OSU Museum of Biological Diversity in Columbus. In 2003 McCullagh was given the William J. Clench Memorial Award by FMCS for outstanding contributions to malacology, including his significant collections of aquatic mollusks.

In July 1987 James David Williams (1941—) moved from the USFWS Office of Endangered Species in Washington, DC, to the USFWS National Fisheries Research Center (currently USGS Southeast Ecological Science Center), in Gainesville, Florida. Williams and graduate students from the University of Florida, on whose committees he served as adjunct professor, began mussel surveys of Florida panhandle rivers. The first was the Apalachicola basin survey, which was undertaken to determine the conservation status of several endemic mussels that were candidates for federal listing as endangered or threatened species. The project was managed by doctoral student Jayne Brim Box and included examination of historical Apalachicola basin collections housed in various museums (Brim Box and Williams 2000). Additional research on reproductive biology of four rare Apalachicola basin mussels was conducted by master's student Christine A. O'Brien (O'Brien and Williams 2002).

Other survey projects at the Gainesville laboratory examined mussel diversity and distribution in the Suwannee, Escambia, Yellow, and Choctawhatchee Rivers. A project in the New River, a tributary of Suwannee River, focused on community ecology of mussels (Blalock-Herod 2000) and substrate utilization by *Corbicula fluminea* (Blalock and Herod 1999). The Choctawhatchee River survey involved Holly N. Blalock-Herod, Jeffrey J. Herod, Stuart W. McGregor, and Williams (Blalock-Herod et al. 2005) and gave rise to a project examining the conservation status, distribution, and reproductive characteristics of *Hamiota australis* (Blalock-Herod et al. 2002). During 2002–2004 Williams and Blalock-Herod initiated a survey of the Suwannee River drainage in Florida and Georgia to examine potential critical habitat of *Pleurobema pyriforme* and determine the conservation status of *Medionidus walkeri*. In 2001–2002 Stephen J. Walsh and Williams surveyed fishes and mussels in 16 springs in north-central Florida state parks (Walsh and Williams 2003). Almost all mussels collected by Williams and associates were deposited in the FLMNH.

Additional research projects on mussel biology, including the effects of contaminants, were performed at the USFWS/USGS Gainesville laboratory during the 1990s by Anne Keller and early 2000s by Timothy Gross. Host fishes for nine species of Florida mussels were determined by Keller and Ruessler (1997a), with the goal of developing techniques to produce large numbers of juvenile mussels for use in contaminant studies (Keller and Ruessler 1997b). Research by Gross and his staff focused on endocrine disruptor contaminants in aquatic systems and more specifically on fate and potential effects of paper mill effluents on mussels (Kernaghan et al. 2004).

Robert Stuart Butler (1954—) arrived in Kissimmee, Florida, in September 1986, to take a position with the Florida Game and Fresh Water Fish Commission (currently FWC), Lake Restoration Section. Although Butler's position involved management of fisheries resources in the Kissimmee Chain of Lakes in central Florida, he quickly developed an interest in Florida mussels. Over the next several years, he made numerous collections throughout the state (depositing specimens in the FLMNH) and published a paper on new drainage records for Florida panhandle mussels (Butler 1990). In 1991 Butler took a position with the USFWS Ecological Services Division in Daphne, Alabama, but continued to work on Florida mussels. A year later he transferred to Jacksonville, Florida, serving as a listing and recovery biologist and assessing the conservation status of aquatic organisms throughout Florida as well as Georgia. Between 1992 and 1996, he administered USFWS funding for status surveys, life history studies, and other projects on mussels, much of it supporting University of Florida graduate students working at the USGS laboratory in Gainesville.

An updated conservation assessment of the mussel fauna of Florida was included in a comprehensive assessment of imperiled plants and animals published by the Florida Committee on Rare and Endangered Plants and Animals (Williams and Butler 1994). The printing of this publication was delayed for a period of about three years and as a result much of the information was out-of-date when it appeared in print. In 1998 Butler prepared the ESA listing documentation for seven Alabama, Florida, and Georgia mussels that occurred from Econfina Creek to the Suwannee River drainage (USFWS 1998). In 1996 he transferred to the USFWS office in Asheville, North Carolina, where he continues to work on mussel conservation projects throughout the eastern United States. In 2003 he served as senior author for the recovery plan of the seven mussels listed in 1998 that occur in Florida and adjacent states (USFWS 2003).

During the past decade, many other biologists in universities and federal agencies have contributed to our knowledge of Florida mussels. Megan M. Pilarczyk, a master's student at Troy University, Alabama, under the supervision of Paul M. Stewart, examined the biology and distribution of mussels in eastern Gulf Coast rivers (Pilarczyk et al. 2005, 2006). Another master's student, Lisa Preister, working under the direction of Carson Stringfellow at Columbus State University, Georgia, completed a thesis on *Elliptio chipolaensis* host fishes (Preister 2008). Beginning in 2005, Sandy Pursifull and Karen Herrington, USFWS, Panama City, Florida, along with Mike Gangloff, Appalachian State University, Boone, North Carolina (formerly at Auburn University, Auburn, Alabama), worked extensively on the biology and ecology of the Apalachicola River mussel community. They have also conducted surveys in the Escambia, Yellow, and Choctawhatchee Rivers in Alabama and Florida as part of an evaluation of candidate mussels considered for listing as endangered or threatened species. Pursifull compiled the documentation to propose six Florida mussels for protection under the ESA (USFWS 2012). Doug N. Shelton, from the Alabama Malacological Research Center, Mobile, Alabama, has conducted contract surveys for mussels in the panhandle portion of the state during the past two decades. His mussel collections from more than 100 sites in Florida have resulted in about 325 lots and include recent collections of rare taxa.

Contributions from previous workers provide a sound foundation for future research and conservation efforts involving Florida mussels. Progress during the past two decades in understanding mussel biology, conservation, ecology, and relationships of Florida mussels has been significant. The increasing interest by individuals from state and federal agencies, universities, and nongovernment institutions in Florida is encouraging.

Mussels in the Archaeological Record

Animal remains in the archaeological record document the vertebrates and some invertebrates used by Native Americans prior to the arrival of European settlers (Brown 1994). Animal remains in archaeological excavations provide a source of data on prehistoric species composition, distribution, and possibly relative abundance. Mussels recovered from archaeological sites are especially important in providing a record of the fauna prior to ubiquitous environmental disturbance by European settlers. Some of the oldest archaeological records of Native American settlements in Florida include the Page-Ladson site on Aucilla River, about 12,000 BP (Dunbar 2006), and the Little Salt Spring site in southwest Florida, dated occupation from 12,000 to 5,200 BP (Clausen et al. 1979). Mussels reported from these sites include *Toxolasma paulum* from the Page-Ladson site (Auffenberg et al. 2006) and *Uniomerus carolinianus* (reported as *Uniomerus obesus*) from the Little Salt Spring site (Clausen et al. 1979). The presence of *U. carolinianus* at Little Salt Creek provided evidence of the freshwater nature of the site during the period of occupation. Native Americans undoubtedly included mollusks in their diets as mussel and snail shells were present at these and other Florida archaeological sites (Brown 1994; Dunbar 2006).

Peninsular Florida has a large number of archaeological sites located along Gulf and Atlantic Coast rivers as well as inland sites located on smaller streams. Numerous sizable Native American shell middens—some up to several hundred meters long and 5 m high along St. Johns River—were described in the mid-19th century by Wyman (1868). Although predominantly composed of gastropods, principally *Pomacea paludosa* and *Viviparus georgianus*, they often included shells of *Elliptio jayensis*. Most archaeological excavations report mollusks in general terms such as freshwater or marine shell, or in some cases, mussels or snails. Some accounts of excavations in peninsular Florida identify mussels to genus, usually *Elliptio*. Shell deposits 75–85 cm thick, consisting predominantly of *E. jayensis* and *V. georgianus*, at sites such as Hontoon Island, Volusia

County, on St. Johns River, are indicative of the importance of mollusks in the lives of Native Americans (Wing and McKean 1987).

Excavated archaeological sites along the Apalachicola and Chattahoochee Rivers have revealed large quantities of shell material (Table 2.1). Three of the localities are registered archaeological sites (Percy 1976; Maymon et al. 1996). The fourth, 8Li172 known as the Otis-Hare site, is a smaller midden and included a single valve of an estuarine bivalve, *Macrocallista nimbosa*. During a Jim Woodruff Lock and Dam preimpoundment archaeological excavation along Chattahoochee River, large quantities of mussels were found (Bullen 1958). The shell deposits consisted of several slope middens, most covering hundreds of square feet. Mussels identified from the middens included *Amblema neislerii, Elliptio crassidens, Elliptio fraterna, Elliptoideus sloatianus, Hamiota subangulata, Lampsilis floridensis*, and *Quadrula infucata*. The presence of *Hamiota subangulata, Medionidus penicillatus*, and *Pleurobema pyriforme* in middens along the main channel of the Apalachicola and Chattahoochee is noteworthy since there are no recent (post-European settlement) records of these species in the main channel of the Florida reaches of these rivers. Further, most of these sites harbored *Reginaia apalachicola*, which is now extinct.

Table 2.1. Mussels recorded from archaeological sites along Apalachicola River, Florida (adjusted to taxonomy used herein). Localities are as follows: (1) 8Gd13, Sycamore site, Gadsden County, located on the east bank of Apalachicola River, near river mile 99.6, N 30.62750 W 84.89612; (2) 8Li76, Liberty County, located 0.5 kilometers east of Apalachicola River, near river mile 88, N 30.50633 W 84.98324; (3) 8Ja104, Scholz Steam Plant site, Jackson County, located near the west bank of Apalachicola River, 5.2 kilometers southeast of Sneads, near river mile 103, N 30.66855 W 84.88642; and (4) 8Li172, Otis-Hare site, Liberty County, located on the left descending bank of Apalachicola River, near river mile 73.1, about 6 kilometers south-southeast of Blountstown, N 30.39682 W 85.01517.

Species	8Gd13	8Li76	8Ja104	8Li172
Amblema neislerii	X	X	X	X
Elliptio crassidens	X	X	X	X
Elliptio fumata	X			
Elliptio pullata		X		
Elliptoideus sloatianus		X	X	X
Lampsilis floridensis	X	X		X
Lampsilis straminea		X	X	X
Medionidus penicillatus		X	X	
Megalonaias nervosa	X		X	X
Pleurobema pyriforme	X	X		
Quadrula infucata	X	X	X	X
Reginaia apalachicola	X	X	X	X
Villosa lienosa	X	X	X	
Villosa vibex	X			
Total species	**10**	**11**	**9**	**8**

Mussels in the Fossil Record

The Pleistocene Epoch is defined by fluctuating sea levels in response to glacial and interglacial periods during approximately 2.6 million years with cycles characterized by extremely low sea level stands during glacial maxima and high sea level stands during interglacial periods. During glacial cycles the peninsular region of Florida varied in width, but as recently as 18,000 years ago it was approximately twice its current width. Concurrent with this huge increase in land area was the development of large rivers draining the exposed landscape that extended considerable distances onto the Gulf of Mexico continental shelf.

Fossil unionids dating from early to late Pleistocene occur from north Florida (e.g., Suwannee River) south to the Tampa Bay area and Caloosahatchee River. The occurrence of fossil *Elliptoideus sloatianus*, *Elliptio crassidens*, and *Megalonaias nervosa* in the Suwannee River and *E. crassidens* and *M. nervosa* near Tampa suggests a large river drained much of the west coast of Florida. This distribution also indicates the presence of a drainage that connected, continuously or intermittently, peninsular Florida and the Apalachicola and/or Ochlockonee Rivers in north Florida, as these are the nearest drainage basins where living populations of these three species are found today.

Florida is one of the few southeastern states with unionids represented in its fossil record. The fossils are found primarily in the peninsular region. The first fossil unionid reported from Florida was described as a new species, *Unio caloosaensis* (= *Elliptio caloosaensis*), by Dall (1898) (Figure 2.10). The type locality was somewhat vague, but in a catalogue of type specimens of invertebrate fossils in the USNM, Schuchert (1905) noted that cotypes or syntypes of *U. caloosaensis* were from marls of the Caloosahatchee River Formation, near Fort Thompson, Florida. All that remains of Fort Thompson is a historic marker indicating its location in the present-day town of LaBelle, Hendry County, on the south side of the Caloosahatchee River. Henderson (1935) reported the Caloosahatchee fossil beds to be "chiefly on the Caloosahatchee River and Shell Creek, Florida" but did not mention any additional unionids. The age of this formation has been the subject of debate but is currently considered to be early Pleistocene Epoch (Jones 1997) and not Pliocene Epoch.

In the type lot of *Unio caloosaensis*, there are two syntypes, one measuring 45 mm and the other measuring 57 mm. In the original description, Dall (1898) recorded the length of the larger specimen, 57 mm, but he illustrated the 45 mm specimen (Figure 2.10). Dall's illustrated specimen (USNM MO 107745) is designated herein as the lectotype in order to enhance stability of the nomenclature in accordance with ICZN Article 74.7.3. The shape and umbo sculpture of the *Elliptio caloosaensis* lectotype is very similar to *Elliptio jayensis*; however, additional research is needed to determine the relationship between these two taxa. The type lot of *E. caloosaensis* was found to be a mixed lot, with the secondary specimen (57 mm) reidentified as *Uniomerus* sp. based on shell shape and umbo sculpture.

The second fossil mussel, *Elliptio pachyodon*, was described from Florida by Pilsbry (1953). It was reported from Pliocene (probably early Pleistocene) deposits in St. Petersburg, Pinellas County. The holotype, ANSP 18586 [Paleontology], was an imperfect valve measuring 52.5 mm in length. Richards (1968) did not include the museum number for *E. pachyodon* in a catalogue of fossil invertebrates in the ANSP, which possibly indicates that the specimen could not be located. The type specimen was not found during a visit to the ANSP in 2010. Pilsbry (1953) reported the valve of *E. pachyodon* to be thick and have teeth similar to those of *Elliptio crassidens*. It appears, based on the figure of the holotype and comments by Pilsbry, that it is most similar to *E. crassidens*.

Figure 2.10. *Unio caloosaensis* (= *Elliptio caloosaensis*). Lectotype (designated herein), USNM MO 107745, length 45 mm. Pliocene (now considered early Pleistocene) marls of the Caloosahatchee Formation along Caloosahatchee River, Florida. Illustration (upper) from Dall (1898) and photograph (lower) by James D. Williams.

Middle Pleistocene unionids were found in the Bermont Formation at the Leisey Shell Pits (Bogan and Portell 1995) and Caloosa Shell Pit (Portell and Kittle 2010), both located near Tampa in Hillsborough County, Florida. These fossils were estimated to be from about 1–1.6 million BP. The following taxa were recovered from the Leisey Shell Pits: *Elliptio crassidens* (reported as *Elliptoideus sloatianus), Megalonaias nervosa* (reported as *Megalonaias boykiniana*), *Utterbackia peninsularis* (reported as *Utterbackia* sp.), and *Villosa amygdalum* (reported as *Villosa* sp.) (identified by J.D. Williams). The *M. nervosa* specimen from Caloosa Shell Pit (Portell and Kittle 2010) is unusual in that both valves are from the same individual (Figure 2.11). The fossil records of *E. crassidens* and *M. nervosa* represent the most southern and eastern distribution for these species in the southeastern United States. Today their nearest living populations are found in Ochlockonee River in northern Florida.

Additional fossil unionids in the FLMNH were from Troy Spring, located adjacent to Suwannee River, about 5.5 air miles northwest of Branford, Lafayette County, Florida. They were found in chalky lime rock around the margin of the circular sink, which has a depth of approximately 30 m. Only three species, *Elliptoideus sloatianus* (Figure 2.12), *Elliptio crassidens* (Figure 2.13), and *Megalonaias nervosa*, have been identified from the site. Using radiocarbon dating, the age of the shells was determined to be 17,400 ± 100 BP. The records for these species are geographically intermediate between the Leisey Shell Pits and Ochlockonee River. The fossil unionids from Troy Spring were picked up by casual collectors, and there has been no detailed investigation of the site.

Figure 2.11. *Megalonaias nervosa*. UF 152321, length 83 mm. Caloosa Shell Pit near Tampa, Hillsborough County, Florida. Photograph by James D. Williams.

Figure 2.12. *Elliptoideus sloatianus*. UF 132094, right valve length 115 mm (left) and left valve length 118 mm (right). Troy Spring, Lafayette County, Florida. Photograph by James D. Williams.

Figure 2.13. *Elliptio crassidens*. UF 208105, lengths 90 mm (left) and 97 mm (right) of two left valves. Troy Spring, Lafayette County, Florida. © Richard T. Bryant.

There are several fossil mollusk sites in the Florida panhandle, but almost all are marine deposits. The best known freshwater locality is the Page-Ladson site on the west bank of the Aucilla River in Jefferson County. This site was periodically explored over two decades, examining both paleontological and archaeological remains (Webb 2006). The original report of mollusk remains consisted of 20 terrestrial snails, 23 freshwater snails, 2 fingernailclams, and 1 mussel, *Toxolasma paulum* (Auffenberg et al. 2006). Subsequent recovery and processing of material in the invertebrate paleontology collection in the FLMNH revealed the presence of five additional unionids from the Page-Ladson site—*Elliptio* sp., *Pyganodon grandis*, *Uniomerus* sp., *Utterbackia* sp., and

Villosa villosa. The Page-Ladson site fossil freshwater mollusk assemblage is one of the most diverse ever recorded in the eastern United States (Auffenberg et al. 2006).

There is one fossil site, reported as Pliocene age, containing freshwater mollusks on the Satilla River, about 4 miles south of Atkinson, Brantley County, Georgia, about 20 miles north of the Florida state border (Aldrich 1911). While no fossil mussels have been recovered from the site, there are several species of freshwater snails (Aldrich 1911; Thompson and Hershler 1991). The Satilla River currently supports a very limited mussel fauna—*Elliptio downiei*, *Elliptio hazlehurstianus*, and *Uniomerus carolinianus*—and only one species of snail. Today the oligotrophic waters are slightly acidic and likely limit the development of a more diverse mollusk community.

Chapter 3

Geology and Physiographic Divisions

The landscape that characterizes Florida, and to a large extent influences the composition and distribution of biological communities in the state's freshwater ecosystems, is the product of geological and climatological processes that began hundreds of million years ago. These processes continue to shape Florida's topography today. The geological structure of Florida consists of two basic components, a basement, or bottom layer of earth crust, and a platform, which was deposited over the top of the basement. As the basement and platform were being formed the process of plate tectonics moved "Florida" across the planet and, at one point, it was transferred from one megacontinent to another. The Florida basement was built by a combination of volcanic, metamorphic, and tectonic processes that occurred over a period of 750 million years (Heatherington and Mueller 1997; Smith and Lord 1997). The sand and limestone surficial landforms that constitute the platform and dominate the Florida landscape are products of depositional, erosional, and karst processes associated with sea level fluctuation and fluvial action that occurred over the past 250 million years (Randazzo 1997; Scott 1997).

Formation of the Florida Basement

Geologists estimate that the earth formed as molten rock 4.5 Ga. About 2.5 Ga the planet cooled and developed oceans and emergent landmasses on drifting tectonic plates (Dalrymple 1991). The bedrock constituting the Florida basement was formed from 1 Ga to 145 Ma as a product of undersea volcanic activity and sedimentation (Arthur et al. 1994; Smith and Lord 1997). The basement is composed of igneous rocks (cooled magma) formed from 1 Ga to 490 Ma, sedimentary rocks deposited between 490–350 Ma, and volcanic rocks (extrusive rocks) formed between 250–145 Ma. Subsurface heat and pressure resulted in the formation of metamorphic rock throughout the basement-building process (Applin 1951; Barnett 1975; Arthur et al. 1994; Heatherington and Mueller 1997; Smith and Lord 1997).

The Florida basement originated as a component of the African Plate of the megacontinent Gondwana (Smith and Lord 1997). The future continents of South America, Antarctica, and Australia were also part of Gondwana. During the late Paleozoic Era, about 350–300 Ma (Figure 3.1), Gondwana slowly collided and combined with a second megacontinent, Laurasia, which was composed of the future continents of North America and Eurasia. The collision of Gondwana and Laurasia formed the last of the supercontinents, Pangaea. While a part of Pangaea, the Florida basement separated from the African Plate and sutured to the North American portion of the former Laurasia. Pangaea then broke apart during the early Mesozoic Era between 250–200 Ma, and as the present-day continents began drifting towards their current relative locations, the Florida basement remained attached to North America.

Today the Florida basement is covered by a layered platform of sedimentary rock that was deposited beginning in the late Mesozoic Era (145 Ma). The basement is about 1 km below the surface of the platform in the north-central region of the state, more than 3.2 km beneath the surface in the panhandle, and more than 4.8 km below the surface in south Florida (Arthur et al. 1994). The Florida basement has few active faults and is seismically stable, with only six confirmed earthquakes in the period from 1879–1975 (Smith and Randazzo 1989).

Era	Age (Ma)	Period	Epoch	Age (Ma)
Cenozoic	65.5–0.01	Quarternary	Holocene	0.01
			Pleistocene	2.6
		Neogene	Pliocene	5.3
			Miocene	23.0
		Paleogene	Oligocene	33.9
			Eocene	55.8
			Paleocene	65.5
Mesozoic	251.0–65.5	Cretaceous		145.5
		Jurassic		201.6
		Triassic		251.0
Paleozoic	542.0–251.0			

Figure 3.1. Scale of geologic time expressed in millions of years (Ma) before present (modified from Walker and Geissman 2009).

Formation of the Florida Platform

The basement of igneous, metamorphic, and sedimentary rocks present after the breakup of Pangaea was the foundation upon which the sedimentary rock that formed the Florida Platform was deposited. From the mid-Jurassic Period to the mid-Oligocene Epoch (170–30 Ma), the earth was relatively warm, continental ice sheets were absent, and sea levels were extremely high (more than 91 m above present-day levels). During this interval carbonate sedimentation (deposition of calcium carbonate remains of marine organisms) built a thick limestone platform on most of the Florida basement. Structural features on the surface of the Florida Platform strongly influenced patterns of surficial sediment deposition (Scott 1997). Thinner sediment layers were deposited on higher elevations of the platform, such as the Peninsular Arch and Ocala Platform, while deep sediment layers were deposited in low areas, such as the Okeechobee basin (Figure 3.2).

In the midst of sedimentary platform building during the mid-Cretaceous Period, a channel system formed in the location of what is now northern Florida and southern Georgia. The Georgia Channel System consisted of two components (Huddlestun 1993). The first, the Suwannee Channel (Figure 3.3), existed from the mid-Cretaceous Period to the mid-Eocene Epoch (100–45 Ma) and enabled a strong oceanic flow, the Suwannee Current, to run in a northeastern direction from the Gulf of Mexico to the Atlantic Ocean (Randazzo 1997). The second component, the Gulf Trough (Figure 3.3), existed from the mid-Eocene to the mid-Miocene Epochs (45–16 Ma) and was essentially an embayment and estuary with restricted flow between the Gulf and Atlantic. The Georgia Channel System formed along the suture zone where Florida was joined to what would become the North American continent. Overall, the Georgia Channel System was biologically significant because it functioned as a barrier that limited movement of biota between the mainland North American continent and peninsular Florida as it emerged from the ocean.

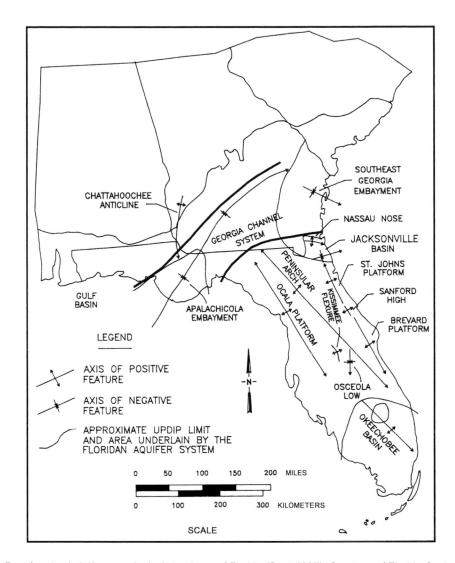

Figure 3.2. Prominent subplatform geological structures of Florida (Scott 1997). Courtesy of Florida Geological Survey.

During the Late Eocene Epoch (35 Ma), the earth entered a cooling phase and large glaciers began to form on the continental land masses. As an increasing amount of water was incorporated into glaciers, there was a corresponding decline in sea levels and Florida began to "emerge" from the sea. The first major landform to emerge on the Florida peninsula was Orange Island. It extended down the central region of the peninsula from just south of St. Marys River to northern Polk County. By the late Middle Miocene Epoch (about 12 Ma), Orange Island encompassed roughly half of the present-day peninsula, extending as far south as present-day Lake Okeechobee and was isolated to the north by the Gulf Trough (Figure 3.3). During this time period the Gulf Trough separated Florida from the mainland and likely acted as a barrier to limit colonization of Orange Island by biota from the continental landmass to the north. It is possible that Orange Island had sufficient expanse and elevation to support pond and stream systems that could also have acted as refugia for freshwater biota during the periods of high sea level that occurred intermittently from the Late Miocene Epoch through the Pliocene (Hobbs 1942; Young 1954; Clench and Turner 1956; Thompson 1968; Burgess and Franz 1978). Others have doubted the importance of Orange Island as a factor in determining the distribution of the freshwater fauna of the Florida peninsula (Johnson 1972).

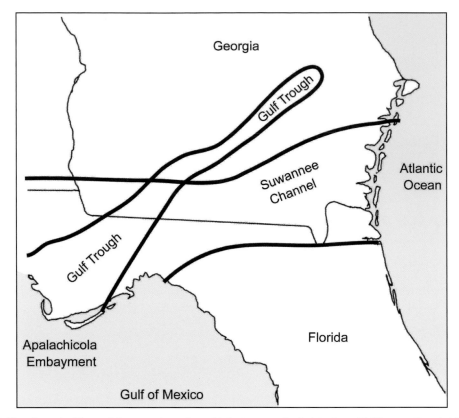

Figure 3.3. The Georgia Channel System, from the mid-Cretaceous Period to the mid-Miocene Epoch (100–12 Ma), including the Suwannee Channel and the Gulf Trough (modified from Huddlestun 1993).

Uplift and erosion of the Appalachian Mountains during the Early Oligocene Epoch (30 Ma) resulted in the movement and deposition of quartz-based sediment layers over the previously deposited carbonate sediments on much of the Florida Platform (Scott 1992). Accumulation of these quartz-based sediments was responsible, at least in part, for filling the Georgia Channel System and bridging the gap between Florida and the mainland, thus removing the channel system as a barrier to potential biological colonization of the Florida peninsula. Deposition of both quartz-based and carbonate sediments built upon the Florida Platform from the late Early Oligocene to the Pliocene Epochs (33 Ma to 5 Ma). The surface layer covering present-day Florida is a 0.3–30.5 m thick veneer of undifferentiated quartz-based sediments that were deposited from Pliocene to Holocene times (Scott 1997; Berndt et al. 1998).

Sea level fluctuations during the Miocene, Pliocene, and early Pleistocene Epochs were responsible for shaping much of present-day Florida (Scott 1997). This topography includes Hazelhurst, Coharie, Sunderland, and Wicomico Terraces, as well as several prominent surficial features, including Cody Scarp, Trail Ridge, Brooksville Ridge, and Lake Wales Ridge (Healy 1975). The Cody Scarp (Figure 3.4) is one of the more prominent of these features. It represents shorelines during the highest sea levels of Pleistocene interglacial periods between 2 Ma to 10,000 BP and is the boundary between upland areas—the Western Highlands, Marianna Lowlands, and Northern Highlands to the north and east, respectively, and the Coastal Lowlands. In some areas there is a dramatic (by Florida standards) change in elevation of more than 40 m in less than 4–8 km, which results in shoals where larger streams cross the scarp (e.g., Big Shoals on Suwannee River, Columbia and Hamilton counties).

Figure 3.4. Extent of the Cody Scarp in Florida.

Pleistocene and Holocene Topography

The Pleistocene Epoch ranged from 2.6 Ma to 10,000 BP and is often referred to as "The Great Ice Age." The Pleistocene was not a single ice age but was characterized by repeated periods of global cooling and continental glacier formation (glacials) followed by warming and glacier retreat (interglacials). Sea level fluctuations prompted by formation and recession of glaciers were responsible for the sedimentation, erosion, and karstification processes that shaped much of the present-day Florida topography that is less than 21 m in elevation (Scott 1997). These elevations include Penholoway, Talbot, Pamlico, and Silver Bluff Terraces as defined by Healy (1975). Pleistocene sea level fluctuations have played a major role in distribution of the present-day Florida biota.

Each glacial advance captured large volumes of water as individual glaciers fused into vast continental ice sheets that were up to 3.2 km thick. As a consequence of continental ice sheet formation, sea levels declined substantially. At the maximum extent of glacial coverage (21,000–12,000 BP), 30 percent of the earth's surface was covered with ice. The most southern advance of any ice sheet in North America was near the latitude of central Illinois. During the warmer interglacial periods, glaciers slowly melted and receded, causing sea levels to rise.

Until recently, geologists recognized four major glacial cycles during the Pleistocene. In North America these were named, in successive order, the Nebraskan, Kansan, Illinoisan, and Wisconsinan glacials. The three intervening periods were the Aftonian, Yarmouthian, and Sangamonian interglacials. Recently, geologists revised the naming system after determining that there were at least eleven major glacial cycles globally (Richmond and Fullerton 1986). A numbering system, the MIS, was implemented to identify glacial and interglacial cycles on a worldwide rather than regional basis (Richmond and Fullerton 1986). The largest extent of glacial coverage in North America, and hence the lowest sea levels, occurred during the period originally known as the Wisconsinan glacial, which corresponds to MIS periods two and four (period three was interglacial). MIS 2–4 extended from 110,000–10,000 BP, with the largest extent of ice sheet coverage occurring about 18,000 BP.

Glaciers never approached Florida but indirectly played a major role in shaping the state's present-day topography and determining the composition and distribution of its biota. With each succeeding rise and fall of sea level during the Pleistocene, the Florida landscape was reshaped. Shoreline wave action and sea currents acting at constantly varying elevations were responsible for the formation of a multitude of shorelines, dune fields, scarps, and terraces that would eventually contribute to formation of the present-day geomorphic districts and ecoregions of Florida (Healy 1975; Brooks 1981; Schmidt 1997).

Along with sea level fluctuation, karstification played a significant role in developing landforms during Pleistocene and Holocene times (Arthur et al. 1994; Scott 1997). Karst terrain is characterized by underground waters in carbonate sedimentary layers. Karstification occurs as acidic groundwater and/or seeping acidic rainwater dissolves the calcium carbonate (limestone) rock layers underlying quartz-based superficial layers. The process has occurred for millions of years in areas of the Florida panhandle and north-central Florida, resulting in extensive development of underground conduits and features such as springs, caverns, and sinkholes (Lane 1986). Among the most prominent of these features include Silver Springs, Wakulla Spring, Florida Caverns, and Brooks Sink. Most of Florida's 700 spring systems are located in karst districts. In some areas the carbonate dissolution process has resulted in collapse of overlying layers and lowering of the landscape, such as in the western and central valleys of the central peninsula (Arthur et al. 1994). A number of Pleistocene collapse depressions and sinkholes eventually filled with water and today are present as lakes in the central region of the state (e.g., Santa Fe and Kingsley Lakes).

At the Wisconsinan glacial maximum (18,000 BP), North American continental glaciers reached their greatest areal coverage and sea levels were at a historic low, 119 m below present-day levels. At this time the Florida Platform was completely exposed as land and, as measured above the 183 m contour level, was 595 km at its widest point and 724 km long (Figure 3.5). Today the Florida peninsula is less than half the width of the Florida Platform. Sonar studies on the 161 km wide submerged western portion of the Florida Platform have documented existence of large river channels that extended from the present-day panhandle and peninsula to the edge of the continental shelf during the late Wisconsinan glacial (Faught 2004; Adovasio and Hemmings 2008).

The Holocene Epoch began 10,000 BP when global warming caused continental ice sheets to melt and recede to near their current sizes and locations. As a consequence of ice melt, sea levels rose by over 90 m to near their present-day levels, an event termed "Holocene Transgression" (Scott 1997; Bryan et al. 2008). Biotic communities thriving in freshwater ecosystems that developed on the western and to a lesser extent eastern Florida shelves during the late Pleistocene were inundated by rising sea levels. With the absence of fossil material, these communities will largely remain unknown. Barrier islands, shorelines, and estuaries of present-day Florida are of Holocene origin and are composed of quartz-based sediments. Today most of the Florida Platform remains submerged below the Gulf of Mexico and Atlantic Ocean (Figure 3.5).

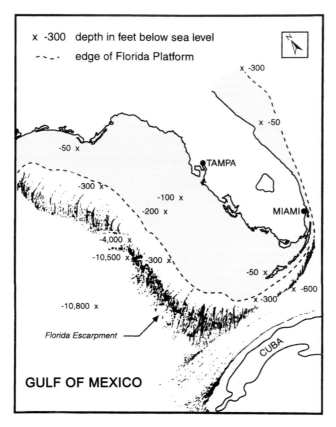

Figure 3.5. The Florida Platform from the end of the Holocene Transgression to present time (modified from Lane 1986). Blue shading represents areas of the shelf below sea level after the Holocene Transgression.

Influences of Sea Level Change on the Florida Mussel Fauna

Fossil records indicate that mussels of the family Unionidae were present in eastern North America at least as long ago as the end of the Triassic Period (200 Ma). By the beginning of the Pleistocene (2.6 Ma), unionids were distributed throughout most of the continent (Bogan and Roe 2008). Prior to the mid-Oligocene (28 Ma), sea levels were a minimum of 91.4 m above present-day levels, which would likely have precluded the existence of freshwater mollusks in any part of Florida. During the mid-Oligocene, seas receded to below present-day levels, thus exposing a large portion of Florida, but rose again and reached a maximum approximately 15 Ma in the mid-Miocene (Alt and Brooks 1965; Scott 1997). This mid-Miocene high-water event most likely eliminated most freshwater faunal assemblages established in the interim low water period, except those that could survive in refugia such as Orange Island.

The exact elevations and timing of sea level fluctuation events during the remainder of the Miocene are not well established, so the timing and mechanism of the mussel migration that produced the current Florida fauna is uncertain. However, at the location of Wicomico Terrace, a large portion of the peninsula and panhandle would have exceeded 21 m above msl (Healy 1975). At this elevation a broad connection to mainland regions of Alabama and Georgia would offer ample opportunity for mussel dispersal. According to Alt and Brooks (1965) and Laessle (1968), Wicomico Terrace was created during Pliocene times (5–2.6 Ma). As a consequence it appears that the migration of mussels into and throughout Florida was possibly well underway by the late Pliocene and early Pleistocene.

When sea levels were low during end of the Pleistocene glacial periods, west Florida rivers had much larger basins and flowed across the broad western portion of the Florida Platform. These ancient stream channels may have connected, receiving the flow of present-day large rivers, such as the Apalachicola, Ochlockonee, Suwannee, and Hillsborough, thus providing potential avenues for dispersal. Migrations along such pathways provide a possible explanation of widely disjunct distribution patterns of some mussel populations. Fossil material may corroborate this hypothesis. *Megalonaias nervosa* fossils from the Tampa area are hundreds of kilometers southeast of its current distribution in the Apalachicola and Ochlockonee basins.

The fossil record for Florida mussels is insufficient to provide a clear picture of the expansion of unionids into the state. However, current mussel distribution patterns indicate that most species had their origins in streams located to the northwest and, to a lesser extent, in streams along the southern Atlantic Coast (Clench and Turner 1956).

Physiographic Regions of Florida

The habitat characteristics of inland waters are greatly influenced by surrounding landforms. The structure of biological communities inhabiting water bodies today are a product of topography, geology, soil, and vegetation characteristics of the watershed, which are strongly influenced by past geological processes. Florida lies entirely within the Coastal Plain Physiographic Province of North America (Fenneman 1938). The Coastal Plain appears, deceptively, to be homogenous. It is characterized by low relief but has a diverse variety of terrestrial and aquatic habitats. There is an abundance of rivers, lakes, marshes, and estuaries. The surficial bedrock underlying the entire province is primarily of marine sedimentary origin and Florida follows this model closely. It is flat relative to other southeastern states, with elevations varying from sea level to just over 100 m in the Western Highlands of the panhandle and approximately 90 m in the Central Highlands of the peninsula. Nearly two-thirds of the state is less than 15 m above msl. The highest elevation in Florida is 105.2 m at Britton Hill in Walton County near the Alabama border. This is the lowest high elevation of all states.

The concept of distinct physiographic regions within the state of Florida has evolved through time and with different generations of geologists and biologists. In a series of papers between 1910 and 1927, Harper (1910, 1914, 1921, 1927) took a holistic approach and recognized divisions based upon geographical location and vegetation cover. Some of Harper's divisional classifications, such as the West Florida Pinehills and East Florida Flatwoods, are consistent with the present-day concepts of physiographic regions. A general description of the topography of the state was provided by Cooke and Mossom (1929), but there was no clear link of geology or landforms to specific geomorphic zones. Shaping of Florida landforms as a function of sea level fluctuations and currents was first described by Cooke (1939). He delineated five "topographic divisions" (Figure 3.6), which formed the basis for the present-day perception of physiographic regions in Florida (Cooke 1939, 1945; Schmidt 1997). Cooke's physiographic regions included the Western Highlands, Marianna Lowlands, Tallahassee Hills, Central Highlands, and Coastal Lowlands (Figure 3.6). Biologists in the early to mid-1900s utilized the work of Harper and Cooke in describing the distributions of some faunal groups in Florida. These included monographs on Florida damselflies and dragonflies (Byers 1930), crayfish (Hobbs 1942), mayflies (Berner 1950), and water beetles (Young 1954).

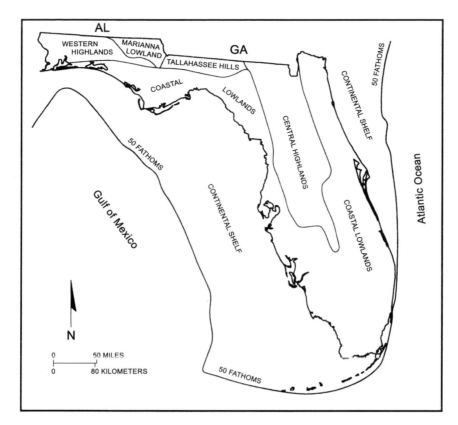

Figure 3.6. Physiographic regions of Florida as delineated by Cooke (1939).

White, Puri, and Vernon described principal geologic structures existing in Florida and accepted Cooke's (1939) concept of five major physiographic regions in the state (Puri and Vernon 1964). White (1970) drew upon the work of Cooke (1939) and Puri and Vernon (1964) to produce maps that further divided Cooke's landforms based upon distinct geomorphic features. White's map is the most detailed of all Florida physiographic maps produced to date but is not particularly useful for explaining distribution patterns of freshwater faunas. Brooks (1981) developed a map and guidebook that consolidated a number of White's (1970) landform regions, yielding a total of ten geomorphic districts in Florida.

The Florida portion of the Coastal Plain Physiographic Province is divided into ten geomorphic districts—Southern Pine Hills, Dougherty Karst Plain, Apalachicola Delta, Tifton Uplands, Ocala Karst, Okefenokee Basin, Barrier Island Sequence, Central Lakes, Sarasota River, and Everglades (Authur et al. 2007) (Figure 3.7). Geologic structure and history within each of these districts are similar; however, differences between even adjacent regions can be substantial.

Southern Pine Hills District—This district is located in the western portion of the panhandle and represents a primarily Miocene and Pliocene age sediment delta plain (Figure 3.7). The northern section is a stream-sculptured alluvial plain overlying thick, clastic sediments consisting of sand, gravel, silt, and clay. The intermediate and coastal sections were formed by marine sediments and consist of ridges, lagoons, and barrier islands. The primary terrestrial ecosystems are pine and mixed hardwood forests, sandhills, and dunes. Larger streams include Blackwater, Escambia, Perdido, Shoal, and Yellow Rivers. Small lakes and ponds are interspersed across the landscape.

Dougherty Karst Plain District—Located in the west-central panhandle, this district is characterized by karst in near-surface Eocene and Oligocene limestones (Figure 3.7). Elevations in this district are low compared to adjacent districts. Dominant terrestrial ecosystems are pine and mixed hardwood forests. Choctawhatchee and upper Chipola Rivers are the principal streams in the district, while smaller streams flow directly into Choctawhatchee Bay.

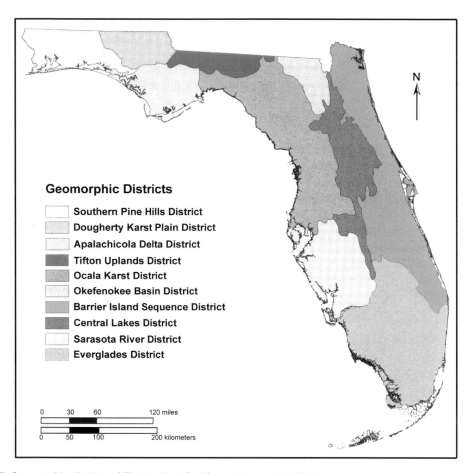

Figure 3.7. Geomorphic districts of Florida (modified from Arthur et al. 2007).

Apalachicola Delta District—This district is located in the south-central panhandle and is a Pleistocene and Holocene age delta plain (Figure 3.7). The district was formed from Apalachicola River sediments and is characterized by ridges, terraces, relic deltas, dunes, and barrier islands, without karst development. The predominant terrestrial ecosystem is pine flatwoods. Principal aquatic features include the lower reaches of Apalachicola and Chipola Rivers, as well as bottomland hardwood swamps.

Tifton Uplands District—This district is in the north-central region of the panhandle and is a Miocene and Pliocene age delta plain overlying remnants of the Gulf Trough (Figure 3.7). The sedimentary rock surface is characterized by thick layers of clay and sand sculpted by streams. Some of the hills in the district approach 90 m above msl. Principal terrestrial ecosystems include hardwood and pine forests. Reaches of Apalachicola and Ochlockonee Rivers are the largest streams in the district. St. Marks River has its headwaters in this district.

Ocala Karst District—This district is in the northwestern part of the peninsula and is geologically diverse, overlying Eocene and Oligocene limestone formations (Figure 3.7). It is characterized by a wide range of elevations and landscapes. The northern and eastern portions of the district are uplands with a karst central plain. Terrestrial ecosystems include hardwood forests, pine flatwoods, and sandhills. The Suwannee, St. Marks, Withlacoochee (southern), and Hillsborough Rivers are the major streams. Large springs in the district include Ichetucknee, Rainbow, and Silver.

There are two rivers named Withlacoochee in the Ocala Karst District. One has its headwaters in southern Georgia and flows south, entering the Suwannee River at Ellaville, Florida—herein referred to as Withlacoochee River (northern). The second is an independent river basin that enters the Gulf of Mexico south of the Suwannee River near Yankeetown, Florida—herein referred to as Withlacoochee River (southern).

Okefenokee Basin District—Located in the northeastern portion of the state, this district is a combination of terraces, ridges, and recent peat and muck deposits formed over Pliocene and Pleistocene age sediments (Figure 3.7). Elevations range from 37–46 m above msl. Dominant ecosystems include pine flatwoods, cypress swamps, and marshes. The headwaters of St. Marys River, Suwannee River, and its tributary Santa Fe River are located in this district.

Barrier Island Sequence District—Located along the eastern side of the state, this district is a series of Pliocene and Pleistocene age barrier islands and lagoons (Figure 3.7). Elevation is generally less than 27 m. Ecosystems associated with this district include pine flatwoods, prairies, cypress lakes and ponds, mangroves, and coastal dunes. Headwaters of St. Johns River are located in the southern region of this district. The St. Johns River and its tributaries, lakes, and marshes are the predominant freshwater ecosystems.

Central Lakes District—This district is located in central Florida and is a large area of mostly karst seepage lakes that developed in quartz-based sand sediments from the Miocene to Pleistocene (Figure 3.7). The major landforms are the central ridge, rolling sandhills, and solution basins. The dominant terrestrial ecosystems are sandhills and pine scrub. The permeable sands and karst terrain characterizing the region make it an important recharge area of the Floridan Aquifer.

Sarasota River District—This district is in the southwestern region of the peninsula and is low and flat, characterized by surficial sand, clay, and organic deposits overlying Miocene to Pleistocene age limestones (Figure 3.7). Dominant ecosystems are pine flatwoods, prairies, cypress domes and ponds, mangroves, and coastal dunes. Major streams include Alafia, Caloosahatchee, Little Manatee, Manatee, Myakka, and Peace Rivers.

Everglades District—Located in south Florida, this district is primarily a freshwater peat marsh that developed over Pliocene to Pleistocene age limestone (Figure 3.7). Surficial material, in addition to organic material, includes sand, marl, and limestone. Sand ridges and mangroves occur along the coastlines. Other ecosystems in this district include prairies, tree islands, and cypress domes. Aquatic systems include Lake Okeechobee, Big Cypress Swamp, Caloosahatchee River, Everglades, lower Fisheating Creek, lower Kissimmee River, and Loxahatchee River.

Chapter 4

Inland Waters

Over 1,700 streams and 7,800 lakes are distributed across the Florida landscape. These water bodies account for approximately 18.5 percent of the total surface area of the state. Florida has 170,340 km^2 of total surface area, 31,480 km^2 of which are covered with inland waters, ranking it 22nd among the 50 states in that respect. The natural inland aquatic ecosystems of Florida are geomorphically diverse and include springs, creeks, rivers, lakes, ponds, swamps, and marshes. Some of these ecosystems are among the most unique on earth. Karst geology defines many streams, while other watersheds are characterized by an abundance of lakes or swamps and marshes. Large portions of some basins, including the Apalachicola, Choctawhatchee, Escambia, Ochlockonee, St. Marys, Suwannee, and Yellow, are shared with Alabama and/or Georgia. Other basins are totally contained within Florida or have a few minor tributaries in Alabama or Georgia. All Florida basins (Figure 4.1) flow into either the Gulf of Mexico or Atlantic Ocean.

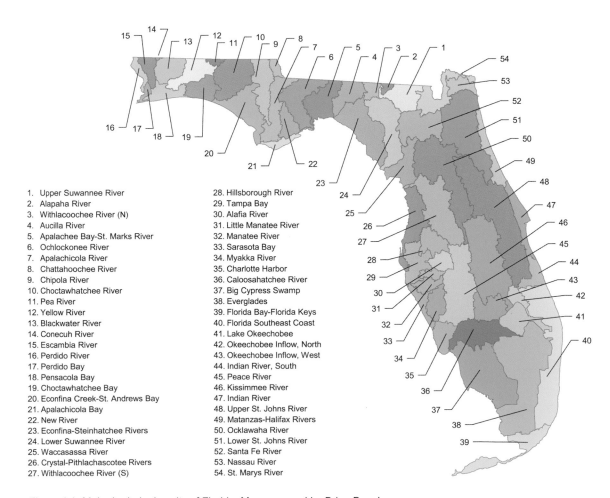

1. Upper Suwannee River
2. Alapaha River
3. Withlacoochee River (N)
4. Aucilla River
5. Apalachee Bay-St. Marks River
6. Ochlockonee River
7. Apalachicola River
8. Chattahoochee River
9. Chipola River
10. Choctawhatchee River
11. Pea River
12. Yellow River
13. Blackwater River
14. Conecuh River
15. Escambia River
16. Perdido River
17. Perdido Bay
18. Pensacola Bay
19. Choctawhatchee Bay
20. Econfina Creek-St. Andrews Bay
21. Apalachicola Bay
22. New River
23. Econfina-Steinhatchee Rivers
24. Lower Suwannee River
25. Waccasassa River
26. Crystal-Pithlachascotee Rivers
27. Withlacoochee River (S)
28. Hillsborough River
29. Tampa Bay
30. Alafia River
31. Little Manatee River
32. Manatee River
33. Sarasota Bay
34. Myakka River
35. Charlotte Harbor
36. Caloosahatchee River
37. Big Cypress Swamp
38. Everglades
39. Florida Bay-Florida Keys
40. Florida Southeast Coast
41. Lake Okeechobee
42. Okeechobee Inflow, North
43. Okeechobee Inflow, West
44. Indian River, South
45. Peace River
46. Kissimmee River
47. Indian River
48. Upper St. Johns River
49. Matanzas-Halifax Rivers
50. Ocklawaha River
51. Lower St. Johns River
52. Santa Fe River
53. Nassau River
54. St. Marys River

Figure 4.1. Major hydrologic units of Florida. Map prepared by Brian Beneke.

Management of water resources in Florida was largely accomplished by local government entities until 1972 when the Florida Legislature enacted the Florida Water Resources Act, which divided the state into five independent water management districts (Figure 4.2). Most district boundaries conform to watershed divides. These districts have taxing authority and were created for the purposes of flood control and management of regional water resources. One of the districts' most important contributions has been the purchase of sensitive lands, wetlands, and stream corridors for protection of water resources. These public lands not only provide opportunities for recreational uses, they provide protection for wetlands, floodplains, and riparian habitats. State agencies, including FDEP and FWC, are responsible for maintaining and monitoring water quality and freshwater ecosystem integrity throughout the state.

Figure 4.2. Location and boundaries of Florida's five water management districts (modified from Purdum et al. 1998).

Florida has an abundance of water resources which are maintained in part by the high average annual rainfall of about 53 inches. Within the state the highest rainfall rate is usually in the western panhandle (Escambia to Apalachicola Rivers) and the extreme southern Atlantic Coast (Henry 1998). Historically, high rainfall and plentiful groundwater has provided an ample supply for freshwater ecosystems, agriculture, industry, recreation, and personal use. However, in the past three decades, overuse and reduced rainfall have combined to make freshwater a precious resource throughout the state. Critical shortages of water available to sustain normally functioning "healthy" freshwater ecosystems have already become common in some areas of the state, such as the Apalachicola River where the deaths of thousands of mussels can be attributed to artificially manipulated water levels below Jim Woodruff Lock and Dam (USFWS 2006). The manner in which today's water managers in Florida and adjacent states allocate the use of freshwater will, to a large extent, determine the health of Florida's inland water ecosystems in the future. If the trend toward depletion of groundwater and surface water resources continues, there is great potential for widespread water shortages in the future, and eventually water managers may be forced to make difficult decisions regarding appropriation of water for human use or sustaining ecosystems (Golliday et al. 2004; Peterson et al. 2011).

Classification of Florida Waters

Water managers and scientists have developed a number of classification schemes for Florida inland waters to identify and group similar natural systems for management purposes and to explain distributions of biotic communities. Classification systems must be used with caution as every freshwater ecosystem has characteristics unique unto itself and some ecosystems (e.g., large rivers) do not fit into a single category. State and federal agencies have enacted legislation designed to protect specific ecosystems because of their unique natural attributes. The National Wild and Scenic Rivers System was created by Congress in 1968 to preserve exceptional streams. The Wekiva and Loxahatchee Rivers are the only Florida streams to be federally designated as Wild and Scenic Rivers. Portions of Myakka River in Sarasota County were designated as "Florida Wildlife and Scenic River" by the Florida Legislature in 1985. The FDEP developed an "Outstanding Florida Waters" program for water bodies worthy of protection based on their high quality. The Outstanding Florida Waters designation is significant as it limits actions that could lower ambient water quality conditions.

The first classification for Florida waters was presented by Rogers (1933). Successive workers (Carr 1940; Hobbs 1942; Berner 1950; Herring 1951; Beck 1954, 1965) built upon Rogers's system. Berner (1950) was the first to recognize the importance of vegetated habitat as a major component of stream classification and also recognized canals as a stream type. William Beck, who developed the most widely used stream classification system in Florida (Beck 1954), states that his term "stream types" pertains to "reaches of water courses" and gives the example that "the Suwannee River, in its flow across Florida, is successively a swamp-and-bog stream, a sand-bottomed stream then a calcareous stream." Similarly, the Aucilla River has a calcareous origin in springs but becomes an acidic, blackwater stream with input of waters from swamps below the headwaters (Nordlie 1990).

Lotic Waters Classification Systems

The Florida stream classifications were consolidated into a descriptive system by Beck (1965). Beck used statistical analyses of physical, chemical, and biological characteristics to demonstrate the existence of five distinct stream types—sand-bottomed streams, calcareous streams, large rivers, swamp and bog streams, and canals. The most common and widely distributed of Beck's Florida stream types is the sand-bottomed stream, which occurs statewide. These streams typically have shifting sand bottoms, limestone outcroppings, pH ranging from 5.7 to 7.4, alkalinity from 5 to 100 mg/L, hardness from 5 to 120 mg/L, moderate to high color, vegetation from sparse to dense, and moderate to swift current (Beck 1965). Examples of sand-bottomed streams include Yellow and St. Marys Rivers.

Calcareous stream classification in Beck's system included creeks and rivers having spring headwaters or substantial spring input. They are present in both highland and lowland karst areas of the panhandle and central peninsula. Calcareous streams are clear but slightly stained by tannins and have pH ranging from 7.0 to 8.2, alkalinity from 20 to 200 mg/L, and hardness from 25 to 300 mg/L. Bottom materials in areas of moderate to rapid current consist of sand, clay, limestone gravel, and/or bedrock. Large amounts of detritus may occur in low-flow areas. Beds of submerged macrophytes (e.g., *Vallisneria*) may be present. Calcareous streams range from spring runs, such as Wekiva River, to larger streams, such as Silver, St. Marks, and Wakulla Rivers.

Swamp and bog streams are acidic, have headwaters in swamps or marshes, and are referred to as "blackwater" or "softwater" systems. They occur primarily in coastal lowlands, but a few are present in the central highlands. Swamp and bog streams typically have sand bottoms and lack substantial flow. They are often clear but highly stained by tannins and carry little sediment. Concentrations of dissolved organic materials are high. The pH ranges from 3.8 to 6.5, and both alkalinity and hardness are below 40 mg/L. Examples of swamp and bog streams include the headwater reaches of Fisheating Creek and Suwannee and St. Johns Rivers.

Beck (1965) reserved the "larger rivers" category for the largest streams in the state (Figure 4.3). The physical and chemical characteristics of these rivers spanned a broad spectrum of conditions, both within and among rivers. Only four streams are included in the large river category. Three of these, Apalachicola, Choctawhatchee, and Escambia, have headwaters in other states (Figure 4.3). The fourth, St. Johns River, has its headwaters in Indian River County and flows northward to Jacksonville. It originates in marshes, has an extremely low gradient throughout its length, and is usually clear but highly stained.

Inclusion of a man-made "canal" category in Beck's classification was met with criticism as canals lack many of the physical characteristics of streams. Canals have vertical banks, flat bottoms, and lack structure and constant flow. Justification for inclusion of the category was based on the fact that canals were the only flowing water resembling streams south of Lake Okeechobee (although direction of flow was subject to change) and because canals were abundant along the southeastern coast of the state. Beck termed canals "sublotic" and further justified their inclusion as a stream type because they are a present-day substitute for the actual streams that once flowed in the southeastern region of the state.

Beyond the ecological classification of streams there is a need to standardize terms used to refer to inland lotic waters. Currently, the terms applied to streams—basins, drainages, systems, catchments, watersheds, etc.—lack precision in their definitions. To develop a hierarchal order of terms and advance a degree of consistency in usage, the following terms and definitions are adopted for use herein. Basin is used for streams that empty directly into the marine environment. This is not dependent on size of discharge, length of channel, or other measurable parameter but is based entirely on the primary connectivity to the marine environment. Drainage refers to large tributaries within basins. The term watershed is reserved for land areas within a basin or drainage and is synonymous with catchment area. This classification scheme is not perfect due to relative terms, such as large and small tributaries.

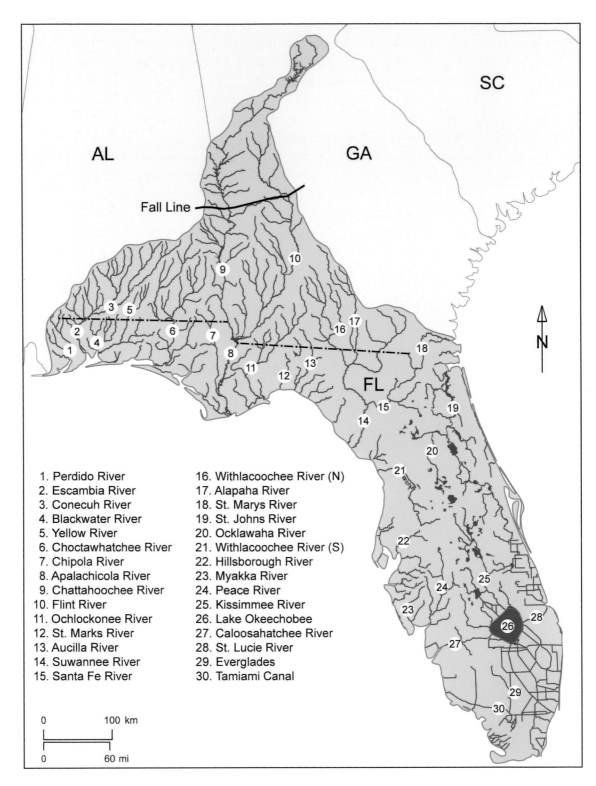

1. Perdido River
2. Escambia River
3. Conecuh River
4. Blackwater River
5. Yellow River
6. Choctawhatchee River
7. Chipola River
8. Apalachicola River
9. Chattahoochee River
10. Flint River
11. Ochlockonee River
12. St. Marks River
13. Aucilla River
14. Suwannee River
15. Santa Fe River

16. Withlacoochee River (N)
17. Alapaha River
18. St. Marys River
19. St. Johns River
20. Ocklawaha River
21. Withlacoochee River (S)
22. Hillsborough River
23. Myakka River
24. Peace River
25. Kissimmee River
26. Lake Okeechobee
27. Caloosahatchee River
28. St. Lucie River
29. Everglades
30. Tamiami Canal

Figure 4.3. Major Florida rivers. Rivers are identified from west to east from the Perdido to the Suwannee and southward from the St. Marys to Tamiami Canal.

Lentic Waters Classification Systems

There are four major categories of lentic ecosystems in Florida: lakes, marshes, ponds, and swamps. There is no universally accepted distinction between the lake and pond categories. Lakes are generally considered to be larger and deeper water bodies, usually greater than 0.4 ha in surface area. In contrast, ponds are smaller and shallower with light penetrating to the bottom allowing rooted plants to grow throughout. Some lakes and ponds support diverse and abundant mussel communities, while others have reduced diversity or are completely devoid of mussels. Marshes are defined as shallow wetlands with rooted, soft-stemmed vegetation that includes submergent and emergent plants, grasses, rushes, and sedges but not trees or other woody vegetation. Environmental conditions often include prolonged periods of low DO concentration, so only a few mussels species inhabit marshes (e.g., *Elliptio jayensis*). Swamps are forested wetlands and are dominated by hardwood trees, most commonly cypress. Mussels are usually absent from isolated swamps, but they are often present in sloughs, ponds, and lakes associated with floodplain swamps of creeks and rivers (e.g., *Uniomerus*).

Collectively, Florida lakes span a broad range of physical, chemical, and biological conditions. Most Florida lakes are natural, shallow (maximum depth less than 5 m), and were formed by the solution of the underlying lime rock. Such solution lakes are supplied by groundwater and often drain by seepage rather than overland outflow. Other Florida lakes were formed by uplifted depressions in the seabed (e.g., Lake Okeechobee) or by fluvial processes (e.g., Lake Monroe). Thousands of small lakes characterize the Florida landscape, but most are concentrated in the peninsula. The Lake Wales Ridge region of central Florida has more than half of all the lakes in the state (Brenner et al. 1990). One-quarter of the 7,800 Florida lakes are in the Kissimmee River drainage, which also includes many lakes associated with Lake Wales Ridge (Mossa 1998). Only five natural lakes in the state have surface areas exceeding 100 km². The largest, by far, is Lake Okeechobee, with a surface area of 1,770 km². Other large lakes include Lake George (190 km²), Lake Kissimmee (140 km²), and Lake Apopka (125 km²) (Brenner et al. 1990). Florida lakes are confronted by a number of problems that are the result of development. Most prominent of these problems are artificial stabilization of water levels, reduced water levels due to excessive groundwater pumping, and eutrophication from nonpoint nutrient inflows (Brenner et al. 1990).

Descriptions of River Basins

Florida stream ecosystems exhibit great diversity in physiography, watershed area, length, gradient, flow rate, and total discharge. Most Florida rivers flow to the south or west into the Gulf of Mexico, but a few (e.g., Loxahatchee, St. Johns, and St. Marys Rivers) flow east into the Atlantic Ocean. Two major rivers, St. Johns and Withlacoochee (southern), are unusual in that they generally flow from south to north. The physical attributes of selected Florida rivers are presented in Table 4.1. The lower reaches of most Florida rivers, plus numerous smaller streams, are tidal to varying degrees and have brackish waters in the lowermost reaches. The extent of brackish waters and tidal influence depends primarily on freshwater discharge and stream gradient.

Table 4.1. Physical attributes of selected Florida rivers. Data sources include Bass (1983), FDEP (2001a, 2001b, 2001c, 2002a, 2002b, 2002c, 2003a, 2003b, 2004a, 2004b, 2004c, 2005, 2006), Nordlie (1990), and Mossa (1998). Average flow of St. Johns River is average net flow corrected for tidal influence.

River	Total Length (km)	Florida Length (km)	Watershed Area (km²)	Average Gradient (m/km)	Average Discharge (m³/sec)
Alapaha	325	37	4,662	0.18	53.0
Apalachicola	805	172	56,446	0.40	702.4
Aucilla	111	98	2,201	0.55	15.6
Blackwater	99	88	2,227	0.64	30.0
Caloosahatchee	112	112	3,625	0.04	40.8
Chipola	201	136	2,657	0.17	42.8
Choctawhatchee	280	201	12,033	0.27	203.8
Escambia	386	87	10,938	0.72	185.1
Hillsborough	87	87	1,787	0.27	16.8
Kissimmee	90	90	7,614	0.07	62.0
Myakka	106	106	1,424	0.34	7.1
Ochlockonee	331	180	6,257	0.47	45.7
Ocklawaha	154	154	7,172	0.13	45.2
Peace	121	121	6,086	0.19	32.7
Perdido	101	93	2,396	0.80	21.8
Santa Fe	113	113	3,626	0.36	45.3
St. Johns	482	482	23,745	0.02	156.2
St. Marks	59	59	3,118	0.88	19.7
St. Marys	193	161	4,105	0.17	19.3
Suwannee	378	333	25,900	0.09	283.2
Waccasassa	62	62	2,424	0.38	8.3
Withlacoochee (N)	185	45	3,885	0.44	47.6
Withlacoochee (S)	253	253	5,439	0.15	32.0
Yellow	177	117	3,535	0.57	65.0

Panhandle Gulf Coast Basins: Perdido Bay to St. Andrews Bay

Basins of the western panhandle region of Florida include Perdido, Escambia, Blackwater, Yellow, and Choctawhatchee Rivers. These basins have their headwaters in the Coastal Plain Physiographic Province in south Alabama and flow south across the Southern Pine Hills, Dougherty Karst Plain, or Apalachicola Delta geomorphic districts of Florida.

The Perdido River basin is the westernmost basin in Florida and forms the border with Alabama for 93 of its 101 km total length before entering Perdido Bay. The river drains a watershed of 2,400 km². The mean annual water temperature is 19.7°C, which is cool for a Florida river (Bass 1983). Over most of its length, the river is characterized by low banks, sand and gravel bottoms, and sandbars on the river bends. The low pH and nutrient concentration precludes establishment of aquatic mollusks, including unionid mussels.

The Escambia River basin originates as the Conecuh River in southeastern Alabama. The Conecuh River becomes the Escambia River at the junction of Little Escambia Creek just below the Alabama and Florida state line. The Escambia River drains a watershed of 10,938 km². It flows for a distance of 386 km, of which only 87 km lie within Florida. The Escambia is the third longest river and has the fourth highest discharge of all Florida streams. The Escambia River in Florida has a few small tributaries (e.g., Canoe and Cotton Creeks) and is characterized by meanders, oxbow lakes, and a broad floodplain (Figure 4.4).

Figure 4.4. Escambia River, looking upstream from Chumuckla Springs boat ramp, Santa Rosa County, Florida, September 2008. Photograph by James D. Williams.

The Blackwater River has headwaters in the Conecuh National Forest in south Alabama and terminates in Blackwater Bay, Florida. The Blackwater drains a watershed of 2,227 km². For 88 of its 99 km, the Blackwater flows through the Southern Pine Hills geomorphic district. It has a moderate gradient and is characterized by swift currents, sand bottom, and large sandbars. The DO concentrations are generally high, but nutrient concentrations are low. The river is acidic with pH ranging from 4.4 to 6.3 (Bass 1983). The low pH and nutrient concentration precludes establishment of aquatic mollusks, including unionid mussels. However, the Blackwater River generally has a diverse fauna despite having low productivity (Bass 1983).

The Yellow River, with headwaters in south Alabama, is predominantly a sand-bottomed stream. The Yellow is a high-gradient stream that flows south across the Southern Pine Hills geomorphic district before entering Blackwater Bay. It drains a watershed of 3,535 km² in Alabama and Florida. The largest Florida tributary, Shoal River, joins the Yellow near Crestview, Florida. The Yellow River has a low concentration of nutrients but otherwise has physical and chemical characteristics similar to other panhandle streams (Bass 1983; Nordlie 1990). Water temperatures are cool, DO concentrations are relatively high, and hardness and conductivity levels are low. The

Yellow River is considered to be one of the highest quality ecosystems in the state (Bass 1983; Florida Department of Natural Resources 1989).

The Choctawhatchee River originates in southeast Alabama and flows south for 280 km before emptying into Choctawhatchee Bay. In Florida it flows across the Dougherty Karst Plain and Apalachicola Delta geomorphic districts. The Choctawhatchee is the fourth largest basin (12,033 km^2) and has the third highest discharge of all Florida streams (Bass 1983). It is also one of the longest undammed rivers in the southeastern United States. Its largest tributary, Pea River, is located in Alabama with the exception of a 2 km reach in northwest Holmes County, Florida. The largest Pea River tributary in Florida is Limestone Creek, a sand-bottomed stream with limestone shoals and moderate current (Figure 4.5). The Florida reach of Choctawhatchee River has several tributaries; the largest is Holmes Creek, a high-quality spring-fed stream that flows for 101 km through mostly low, swampy terrain. For much of its length the Choctawhatchee is broad and shallow, with numerous large sand bars, and occasional exposed bedrock. In addition to tributaries, the flow is supplemented by several springs (e.g., Morrison and Holmes Blue). Overall, the Choctawhatchee is a relatively high-quality stream and analyses of water quality and aquatic biodiversity are indicative of low anthropogenic impact (Livingston et al. 1991).

Figure 4.5. Limestone Creek, looking upstream, Walton County, Florida, December 2010. Photograph by Robert S. Butler.

The Econfina Creek basin is located east of Choctawhatchee River in Bay, Jackson, and Washington counties. This small coastal basin has a total watershed area of only 1,090 km^2 but has a diverse fish and mussel fauna, including three federally protected species. The relatively undisturbed watershed is characterized by a large number of springs with good water quality. The lower reach was impounded (Deer Point Lake) in 1961 just above its entry into North Bay, part of St. Andrews Bay. The reservoir serves as a water supply for Panama City.

Apalachicola Basin: Apalachicola, Chattahoochee, Flint, and Chipola Rivers

The Apalachicola River is formed by the junction of Chattahoochee and Flint Rivers near the town of Chattahoochee, Florida (Figure 4.6). The Apalachicola River drains a watershed of 56,446 km^2 in Alabama, Florida, and Georgia. The portion of the basin in Florida is about 6,200 km^2. The headwaters of Apalachicola River in Alabama and Georgia are high-gradient streams originating in the Blue Ridge and Piedmont Physiographic Provinces above the Fall Line. The Apalachicola is the second largest basin entering the eastern Gulf of Mexico (Ward et al. 2005). The Chattahoochee River is its largest tributary, with a watershed of 22,714 km^2 and flowing for a distance of 692 km in all three states. Only the lowermost reach of Chattahoochee River borders Florida. The second largest tributary to the Apalachicola is Flint River, with a watershed of 21,911 km^2. The largest Apalachicola River tributary in Florida is Chipola River, which has its headwaters in southeast Alabama. The Apalachicola, Chattahoochee, and Flint Rivers and their tributaries when considered as a whole are often referred to as the ACF basin.

Figure 4.6. Apalachicola River, looking downstream, below Jim Woodruff Lock and Dam, Chattahoochee, Gadsden County, Florida, July 2007. Photograph by James D. Williams.

Jim Woodruff Lock and Dam was constructed just below the junction of the Chattahoochee and Flint Rivers to form Lake Seminole. Following completion of the dam in 1957, the lowermost reaches of Chattahoochee and Flint Rivers were inundated and destroyed by the 150 km^2 reservoir. Apalachicola River below Jim Woodruff Lock and Dam flows south for 182 km through the Tifton Uplands and Apalachicola Delta geomorphic districts before emptying into Apalachicola Bay. Formation of an entrenched channel below the dam has resulted in negative impacts to the ecology of the river, its floodplain, and the associated mussel fauna (Brim Box and Williams 2000).

The Apalachicola has the highest discharge of all Florida rivers (Table 4.1). It is often turbid due to a high suspended load of clay and sand. The bottom is predominately sand with some gravel and occasional bedrock outcrops. The channel is characterized by continuously shifting sand bars. The river has a wide hardwood forest floodplain that is diverse and productive. In a landmark study, Light et al. (1998) demonstrated the relationship between flow and availability of aquatic habitats in the floodplain. During low flows, water withdrawals in upstream municipal areas (e.g., Atlanta) and the agricultural region of southwest Georgia result in loss of connectivity between floodplain habitat and the river. This loss of connectivity substantially impacts both the riverine and floodplain ecosystems.

The channels of the Apalachicola, Chattahoochee, and Flint Rivers have been greatly modified for flood control, hydroelectric power generation, and navigation. A total of 17 large dams have been constructed in the ACF basin. Prior to 2001 the USACE used dredging and a system of dikes to maintain a navigation channel 2.7 m deep in the Apalachicola River. The USACE's operation of Jim Woodruff Lock and Dam has periodically resulted in flow and water level reductions that have caused deaths of large numbers of federally protected mussels (e.g., *Amblema neislerii* and *Elliptoideus sloatianus*). Structural modifications have had a significant negative influence on the biological productivity in the ACF basin (Leitman et al. 1991).

The Chipola River originates in southeastern Alabama and flows south across the Dougherty Karst Plain and Apalachicola Delta geomorphic districts. It is a calcareous stream and receives flow from over 60 springs and spring runs along its length, including Jackson Blue Spring via Merritts Mill Pond and Spring Creek near Marianna. The river flows over limestone outcrops in many areas and has numerous rocky shoals. After flowing through Dead Lake, a natural feature in the lower reach of the river, it continues southward for several miles before joining Apalachicola River southeast of Wewahitchka.

Ochlockonee River to Suwannee River

The Ochlockonee River basin has a watershed of 6,257 km² in Florida and southern Georgia, with slightly over half of the basin in Florida. It originates in southern Georgia and flows for 331 km before entering Ochlockonee Bay. In Florida the lower 180 km of the river flows through the Tifton Uplands geomorphic district over the Cody Scarp and onto the Apalachicola Delta geomorphic district. The Ochlockonee drops approximately 50 m in elevation in the vicinity of the scarp. West of Tallahassee the Ochlockonee is impounded by Jackson Bluff Dam, which was built in 1929 for power generation and formed Lake Talquin, the second oldest large reservoir in the state.

The Ochlockonee River has a relatively steep gradient and is characterized by high sediment load. Numerous alluvial tributaries contribute flow to the Ochlockonee, the largest being Little River, Telogia Creek, and Sopchoppy River. There are more than 300 lakes in the Ochlockonee basin, but few of these exceed 40 ha. One of the larger, Lake Jackson, is located north of Tallahassee. During droughts it drains completely through sinkholes (FDEP 2001a).

The St. Marks River basin is located directly east of the Ochlockonee River basin, but the two differ dramatically in their geology. St. Marks River drains a watershed of 3,118 km² and is 59 km in length. It originates in the Tifton Uplands geomorphic district north of Tallahassee. The St. Marks basin has a poorly defined stream network in its headwaters where much of the flow is through swamps and underground solution channels. After flowing across the Cody Scarp, it is fed by springs and becomes clear and wide. The river disappears underground at Natural Bridge Sink near the Leon and Wakulla county line and reemerges 1.5 km downstream at St. Marks River Rise. The two river reaches differ substantially in their chemical characteristics. Wakulla River, the largest tributary of St. Marks, originates at Wakulla Spring (Figure 4.7) in Wakulla County and flows for 16 km where it joins the St. Marks just above Apalachee Bay. The lower 5 km of the St. Marks has been dredged downstream to Apalachee Bay.

Figure 4.7. Wakulla River in Wakulla Springs State Park, Wakulla County, Florida, November 1996. Photograph by James D. Williams.

The Aucilla River basin has a watershed of 2,200 km² in southern Georgia and Florida. This blackwater river flows 111 km across the Tifton Uplands and Ocala Karst geomorphic districts. The Aucilla is characterized by slow current, numerous shoals, and limestone outcroppings. The river disappears into a sinkhole and flows underground for 5 km and reemerges at Nutall Rise in Taylor County. The Aucilla's major tributary Wacissa River originates from springs in Jefferson County (Rosenau and Faulkner 1975).

Several smaller coastal basins, including Econfina, Fenholloway, and Steinhatchee Rivers, flow into the Gulf of Mexico between the Aucilla and Suwannee Rivers. The largest of these, Steinhatchee River, is 45 km long. As it approaches the coast it flows over a shallow, rocky area, locally known as "Steinhatchee Shoals," approximately 13 km upstream from the Gulf of Mexico.

The Suwannee River basin is the second largest in Florida with a watershed of 25,900 km² and has the second highest average discharge. It originates in Okefenokee Swamp in southeastern Georgia and flows 378 km across the Tifton Uplands, Okefenokee Basin, and Ocala Karst geomorphic districts to the Gulf of Mexico. In its upper reach the Suwannee is a low-nutrient, acidic, sandy stream with occasional rocky shoals (Figure 4.8). Between Ellaville and Branford, Florida, the Suwannee is characterized by karst topography overlying the Floridan Aquifer (Figure 4.9). The lower reach has few surface tributaries and is characterized by sand and gravel substrate with occasional limestone outcrops.

Figure 4.8. Suwannee River, looking upstream, at State Highway 6, Columbia and Hamilton counties, Florida, August 2002. Photograph by James D. Williams.

The largest tributaries of Suwannee River—Alapaha, Withlacoochee (northern), and Santa Fe Rivers—are located in the upper portion of the basin. In their headwaters they are blackwater rivers, but as they cross the Cody Scarp they become calcareous and are strongly influenced by groundwater and springs. The Suwannee River basin has more than 250 springs (FDEP 2001b), of which 197 are located along the river (Rosenau et al. 1977). Spring flows are critical to maintain water levels in creeks, rivers, and lakes to sustain good mussel habitat in the Suwannee basin.

Figure 4.9. Looking upstream at the confluence of Withlacoochee River (northern) (left) and Suwannee River (right) at Suwannee River State Park near Ellaville, Suwannee County, Florida, December 2012. Photograph by Gary L. Warren.

The Alapaha River drainage originates in south-central Georgia and is the largest Suwannee tributary, draining a watershed of 4,662 km² (Figure 4.10). Only the lower 37 km of the 325 km stream are in Florida. The river flows above ground year-round in Georgia, but about 7 km south of the Georgia border it often disappears underground through sinkholes in the river channel when the discharge falls below 800 cfs (Ceryak et al. 1983) (Figure 4.11). It remains underground in solution channels for approximately 30 km before reemerging via springs at Alapaha River Rise and Holton Creek Rise and within the main channel of the Suwannee River.

Figure 4.10. Alapaha River, looking upstream, 1 kilometer south of the Florida-Georgia border in Hamilton County, Florida, January 2011. Photograph by Gary L. Warren.

Figure 4.11. Alapaha River, looking downstream, near Jennings, Hamilton County, Florida—above-ground flow following a hurricane, November 2004 (left), and typical condition of dry river bed and subterranean flow, December 2010 (right). Photographs by Gary L. Warren.

The Withlacoochee River (northern), second largest tributary of Suwannee River, originates in south-central Georgia and drains a watershed of 3,885 km². The river flows across karst terrain in north Florida before joining the Suwannee River near Ellaville. It is characterized by an incised channel with limestone shoals (Figure 4.12). A first magnitude spring, Madison Blue Spring, enters Withlacoochee River (northern) in Madison County. During periods of low flow, this spring contributes up to one-quarter of the discharge of the river.

Figure 4.12. Withlacoochee River (northern), looking upstream, 5 kilometers from its confluence with Suwannee River, Hamilton County, Florida, April 2006. Photograph by Gary L. Warren.

The Santa Fe River is the third largest tributary to Suwannee River and lies entirely within the state. It drains a watershed of 3,626 km^2 and is 113 km in length. The headwaters are located at Santa Fe Lake. Two tributaries, New River and Olustee Creek, enter the upper Santa Fe from the north near Worthington Springs. Downstream from Worthington Springs, karst terrain influences the Santa Fe as it disappears into a sink at O'Leno State Park, reemerging 5 km downstream in River Rise State Preserve. After reemergence it is strongly influenced by spring flow between High Springs and Branford. One of the largest spring discharges in this reach is the Ichetucknee River (Figure 4.13).

Figure 4.13. Ichetucknee River, looking upstream from Dampier's Landing, Columbia County, Florida, March 2012. Photograph by Jennifer L. Bernatis.

Waccasassa River to Caloosahatchee River

Peninsular rivers flowing into the Gulf of Mexico south of Suwannee River are influenced by diverse landscapes and aquatic habitats. These streams range from single channels flowing a short distance to the coast to highly braided tributaries meandering for distances of over 250 km.

The Waccasassa River is a blackwater stream draining a watershed of 2,424 km² in the Ocala Karst geomorphic district. Its headwaters originate in swamps and wetlands in Gilchrist County, but it has large tributary springs (Levy Blue and Wekiva) that increase flow and alter the chemical nature of the river.

The Green Swamp is a prominent physiographic feature of the peninsular west coast region in portions of Lake, Pasco, Polk, and Sumter counties. The swamp encompasses 2,200 km² and lies on a plateau bounded by sand ridges. Among wetlands it ranks second only to the Everglades in terms of hydrologic and environmental importance in the state. Its diverse wetlands provide a mosaic of habitats for a multitude of imperiled species. Headwaters of several basins, including Withlacoochee (southern), Hillsborough, and Peace Rivers, originate in the swamp. The headwaters of Ocklawaha River, a large tributary of St. Johns River, lie adjacent to the swamp.

The Withlacoochee River (southern) basin is one of two large rivers in Florida that flow from south to north, the other being the St. Johns. The river is 253 km in length and flows across the Ocala Karst geomorphic district, draining an area of 5,439 km². It originates in northeastern Polk County and its watershed drains 80 percent of Green Swamp. From its headwaters it flows to an "overflow" area where, during periods of heavy rainfall and high water levels, the headwaters of Withlacoochee (southern) and Hillsborough Rivers intermingle.

Tsala Apopka Lake is a 77 km² lake and wetland complex encompassing three large, hydrologically distinct areas. The lake lies within the floodplain of Withlacoochee River (southern) and was formed by deposition of sand ridges. The river passes along the eastern boundary of the lake and overflows into the lake during periods of high water. The Withlacoochee flows through several small lakes in this area (e.g., Annie, Nelson, Bonnet, and Princess).

Lake Rousseau is a 17 km² reservoir formed by Inglis Lock and Dam on Withlacoochee River (southern). The reservoir is the oldest river impoundment in Florida, created by Florida Power Corporation in 1909 to generate electricity (power generation has since been discontinued). A canal that connects the reservoir with the Gulf of Mexico was built in the 1970s as the western terminus of the never-completed Cross Florida Barge Canal project.

A number of small, independent coastal basins are present along the west coast of Florida in Levy, Citrus, Hernando, Pasco, and Pinellas counties. Several of these basins contain large springs, including Crystal, Chassahowitzka, Homosassa, and Weeki Wachee, that have short runs emptying into the Gulf of Mexico. Other small coastal basins in the region include Pithlachascotee and Anclote Rivers. One large freshwater lake, Lake Tarpon , is located in Pinellas County.

There are several small basins that empty into the Gulf of Mexico via Tampa Bay. These include Hillsborough, Alafia, Little Manatee, and Manatee Rivers. The largest, Hillsborough River, flows for 87 km and drains a watershed of 1,787 km². After exiting Green Swamp, the river receives flow from Crystal Springs and several small tributaries.

A flood control structure near the confluence of Hillsborough River and Trout Creek diverts floodwaters of the river into the Tampa Bypass Canal, which empties into Tampa Bay. The lower reach of Hillsborough River flows into the City of Tampa Reservoir, a water supply impoundment behind the Hillsborough River Dam. Below the dam, Hillsborough River is brackish and receives the flow of Sulphur Spring. There are numerous lakes in the Hillsborough River basin; the largest is Lake Thonotosassa (330 ha), which is shallow and extremely eutrophic.

The Alafia River basin drains a watershed of 1,062 km^2. The river originates in western Polk and eastern Hillsborough counties where several small, westward-flowing streams join to form the North and South Prong tributaries. The prongs converge at Alderman's Ford Park to form the Alafia main stem, which empties into Hillsborough Bay, which comprises the eastern portion of Tampa Bay. The headwaters and upper reaches of the Alafia watershed flow through phosphate-rich lands that are extensively strip-mined, resulting in waters with high phosphorus concentrations. Two groups of springs, Lithia and Buckhorn, also contribute significant flows to the river. Medard Reservoir, a reclaimed phosphate pit, is the largest (312 ha) lentic water body in the Alafia basin.

The Manatee River is a blackwater stream with a watershed of 970 km^2 and a length of 96 km. It is characterized by coastal lowlands, hardwood swamps, marshes, and flatwoods. Downstream from the confluence of North and East Forks, the river is impounded by Lake Manatee Dam, forming an 809 ha reservoir that serves as a water supply for coastal communities in Sarasota and Manatee counties. Below the dam the river is narrow and meandering for about 13 km where two tributaries, Gamble and Mill Creeks, enter the river. From this point downstream to Tampa Bay, the Manatee is estuarine, braided, and characterized by emergent salt marshes and mangrove islands.

The Myakka and Peace Rivers, located south of Tampa Bay, flow across the Sarasota River geomorphic district and drain into Charlotte Harbor. The Myakka River drains a watershed of 1,424 km^2 and flows for a distance of 106 km. The Myakka is one of few remaining, relatively large and undisturbed, blackwater streams in the state. The watershed is characterized by a wide diversity of terrestrial and aquatic habitats, including extensive freshwater and estuarine wetlands. The Myakka originates in marshes in Manatee County near Myakka Head. The river flows through two natural lakes, Upper and Lower Myakka.

The Peace River basin, with a watershed of 5,900 km^2 and a length of 190 km, is one of the largest stream ecosystems lying totally within Florida (Figure 4.14). It is a blackwater stream with headwaters in the southern portion of Green Swamp in Polk County. Most of the upper watershed is highly modified where it flows through strip-mined phosphate lands, ranch pastures, and other agricultural areas. The influence of nutrient inflows from phosphate mining (since the late 1800s), as well as point source and stormwater runoff, has resulted in the highest nutrient concentrations (phosphorus and nitrogen) of all major Florida rivers (Nordlie 1990). Groundwater withdrawals from the Floridan Aquifer in the upper Peace River basin have substantially reduced flows. Kissengen Spring, which once contributed approximately 20 million gallons of water per day to the river, ceased flowing and became a sink in the 1950s.

Figure 4.14. Peace River looking downstream, about 1 mile west of Brownville, De Soto County, Florida, April 2012. Photograph by Gary L. Warren.

Prior to channelization, Caloosahatchee River was an 80 km long, shallow, meandering stream that originated near Lake Hicpochee, southwest of Lake Okeechobee. By 1883 the river was connected to Lake Okeechobee by dredging in an attempt to facilitate control of water levels in the lake. During the 1930s the Caloosahatchee, which once had significant rapids near La Belle, was channelized for much of its length to improve navigation and flood control. Today the Caloosahatchee has been reconfigured into a canal, known as C-43, and is essentially an artificial waterway with flows and levels controlled by a series of three locks. The lowermost, Franklin Lock, near Fort Myers Shores, is the upstream limit of the tidal reach of the river. The Caloosahatchee empties into San Carlos Bay in Lee County. The connection of the Caloosahatchee to Lake Okeechobee, combined with channelization and operation of water control structures, disrupted the natural seasonal pattern of freshwater flows to the Caloosahatchee estuary. Today the freshwater reach of the Caloosahatchee is a stressed ecosystem with substantial water quality problems.

Lake Wales Ridge

The Lake Wales Ridge is a sandy upland, oriented north and south in the south-central portion of peninsular Florida. The ridge is about 240 km long and consists of dunes and swales that formed during periods of high sea level fluctuations during the Pliocene and Pleistocene Epochs. Several thousand lakes are present on the Lake Wales Ridge. Most are small and characterized by sand bottoms, maximum depths of less than 3 m, chronically low DO, lack of significant coverage by rooted macrophytes, and high nutrient concentrations (Figure 4.15). Outlet streams of numerous lakes in the southern portion of the ridge flow south into Kissimmee River and ultimately into Lake Okeechobee. Lakes in the northern portion of the ridge drain into the upper Ocklawaha

River, a tributary of St. Johns River. Most ridge lakes do not have overland connections and are isolated except for those that have been connected by man-made canals.

Figure 4.15. Grassy Lake, near Lake Placid, Highlands County, Florida, April 2012. Photograph by Gary L. Warren.

Kissimmee River, Lake Okeechobee, and Everglades Basin

The greater Everglades ecosystem spans approximately 10,000 km^2 and constitutes one of the largest and most unique basins in Florida. It includes the diverse ecosystems of Kissimmee River, Lake Okeechobee, Fisheating Creek, Arbuckle Creek, Lake Istokpoga, numerous Lake Wales Ridge lakes, and the Everglades. Historically, the basin terminated in southern Florida where sheet flow from the Everglades entered Biscayne and Florida Bays and the Ten Thousand Islands region in Collier and Monroe counties. Since the 1880s, the greater Everglades ecosystem has been substantially altered by man to provide flood protection for agriculture and development. These projects continue today and include canal and levee construction designed to prevent flood and hurricane damage.

The Kissimmee River watershed is located at the northern end of the Everglades ecosystem. The Kissimmee was a shallow, meandering stream that flowed across the Osceola Plain and prairie lands for a distance of more than 160 km before emptying into Lake Okeechobee. This river was unique in being a large, alluvial, blackwater stream (Schenk et al. 2011). Between 1961 and the early 1970s, Kissimmee River was channelized by the USACE (and named C-38) to control flooding of adjacent lowlands. This project drastically altered the river by eliminating meanders, thus reducing its length from 160 to 90 km. Resulting damage to the Kissimmee River ecosystem precipitated restoration efforts—including backfilling of the canal, reopening blocked meanders, and removal of selected water control structures—in an effort to recover almost 70 km of the river and more than 10,000 ha of floodplain.

The Fisheating Creek watershed is relatively undeveloped and drains 2,200 km^2 in west-central Highlands and northern Glades counties. The creek originates as a blackwater stream in wetlands west of Lake Placid and flows south through pasture and forested lands. Near Palmdale it becomes braided as it flows through Cowbone Marsh before flowing into Lake Okeechobee near Lakeport.

Lake Okeechobee is by far the largest lake in Florida and is considered the central feature and "liquid heart" of the greater Everglades ecosystem (Figure 4.16). The lake serves as a multipurpose resource for water supply, flood control, irrigation, navigation, commercial fishing, recreation, and fish and wildlife habitat. It drains a watershed of 12,000 km², has a surface area of 1,732 km², and is shallow (mean depth = 2.7 m) and eutrophic. Approximately one-fourth of the lake's surface area is a marsh containing a diverse mosaic of vegetated habitats. Water control structures on the south end of the lake make it possible to back-pump water into the lake to control flooding. Prior to man-made alterations, Lake Okeechobee water levels were as much as 2 m higher than today (Brooks 1974; Aumen 1995). The largest tributaries to the lake are Kissimmee River and Fisheating and Taylor Creeks.

Figure 4.16. Lake Okeechobee, looking west towards Kreamer Island from Pelican Bay, Palm Beach County, Florida, August 2006. Photograph by Gary L. Warren.

Damage from hurricanes in the 1920s and 1940s prompted construction of a canal and levee system that almost completely encircles the lake (Figure 4.17). Construction of the St. Lucie Canal on the east side of the lake and dredging of the Caloosahatchee River on the southwest side created the Okeechobee Waterway, which enables boats to cross the peninsula, avoiding navigation around the southern tip of the state. Construction of five major canals—St. Lucie, West Palm Beach, Hillsboro, New River, and Miami—which began before 1910, linked Lake Okeechobee to the Atlantic Ocean, enabling flood control in the Everglades Agricultural Area and providing routes for floodwater releases.

Other prominent lakes in the Everglades basin include Lakes Istokpoga and Trafford. Lake Istokpoga, located northwest of Lake Okeechobee and east of Lake Wales Ridge, is the fifth largest lake in Florida with a surface area of 112 km² and a maximum depth of approximately 3 m. Flows from Istokpoga feed Kissimmee River and Lake Okeechobee via canals. Lake Trafford near Immokalee is the largest Florida lake south of Lake Okeechobee. It constitutes the headwaters of Corkscrew Swamp and Fakahatchee Strand.

Figure 4.17. Harney Pond Canal, west side of Lake Okeechobee, at State Route 78 bridge, near Lakeport, Glades County, Florida, March 2012. Photograph by Gary L. Warren.

The Everglades, one of the largest and most unique wetland ecosystems in existence, covers most of the southern tip of Florida (Figure 4.18). It originated as a 100 km wide sheet flow over a peat sill from the south end of Lake Okeechobee. This "River of Grass" flowed through freshwater marshes for 160 km across south Florida to the Gulf of Mexico and Atlantic Ocean. The Everglades ecosystem was characterized by low nutrient levels and slow flow through a vast mosaic of dynamic and resilient wetlands. Numerous small streams were present around the margin of the Everglades. Cory (1895) described the Miami River, which flowed from the Everglades into Biscayne Bay, as having a very rapid current and the only river in peninsular Florida having "falls or rapids worthy of the name" (Figure 4.19).

> Having completed a trip across the Everglades in January 1897, Willoughby (1898) described his descent of the rapids at the head of the Miami River, at the eastern edge of the Everglades, in Dade County, Florida, as follows:
>
> "The canoes were loaded for the last time and we shoved off into the swift current of the [Miami] river. How beautiful everything seemed on that quiet Sunday morning! No exertion was necessary but the light turn of the paddle to give the canoes their proper course. The four miles seemed like one when we reached the last bend."

The rocky rapids of the Miami River in subtropical southeast Florida may have represented the only such aquatic habitat in all of North America. The rapids occurred where the river cut through the late Pleistocene limestone of the Miami Rock Ridge, dropping over 2 m in this swift-water reach. The ridge varied from 2–4 m above msl and essentially acted as a levee on the eastern edge of the Everglades, keeping the water level two or more meters above sea level. Beginning in 1906, the New River in Fort Lauderdale was the first stream to be channelized all the way to Lake Okeechobee, followed by the Miami River in 1907. With these initial efforts to drain the Everglades, ostensibly for sugar cane production, numerous artesian freshwater springs in Biscayne

Bay began to lose their flows (Parker et al. 1955). Huge areas of the southeastern Everglades were drained, destroying thousands of hectares of wetlands.

Figure 4.18. Everglades marsh habitat, Loxahatchee National Wildlife Refuge, Palm Beach County, Florida, August 1995. Photograph by James D. Williams.

Figure 4.19. Rapids at the head of Miami River, at the edge of the Everglades, Dade County, Florida, January 1897 (Willoughby 1898).

Realization of the magnitude of harm done to the Everglades ecosystem prompted the initiation of projects intended to restore the basin. The most ambitious of these is presented in the USACE Comprehensive Everglades Restoration Plan, which includes filling canals and removing levees in some areas. A major focus of Everglades restoration is managing those areas already set aside for natural resource protection and preservation, including Everglades National Park, Big Cypress National Preserve, Fakahatchee Strand, and Corkscrew Swamp Sanctuary.

St. Johns and St. Marys River Basins

The St. Johns River is the longest river (482 km) and largest watershed (23,745 km²) lying entirely within Florida. It has qualities that place it among the most unique rivers in North America. The St. Johns was formed in the late Pleistocene Epoch when barrier islands developed along Florida's east coast, eventually separating the river from the sea and enclosing a long, shallow bay paralleling the coastline. As ocean levels dropped and freshwater accumulated, the bay was transformed from an estuary to a freshwater river with flow originating from marshes located northeast of Lake Okeechobee (Figure 4.20).

Figure 4.20. St. Johns River, Brevard and Orange counties, Florida, May 2005. © St. Johns River Water Management District.

The St. Johns River flows from south to north across a gradient that declines about 8 m from headwaters to mouth, much of it through broad, shallow, braided channels with sluggish current. Much of the river has bottom sediments composed of mud, peat, and sand. Tidal influence is evident up to 160 km from the mouth, which is considerably farther upstream than any other Florida river. The river widens at several locations along its length to form a chain of lakes. Hydrology of the St. Johns basin has been altered during the past century by draining marshes, which modified physical, chemical, and biological components of the ecosystem. Withdrawals of surface water and groundwater have influenced water availability for biological processes. Regardless of human modification, the river continues to supports diverse plant and animal communities.

The St. Johns has three distinct reaches, each with unique hydrology and morphology (Burgess and Franz 1978; Smock et al. 2005). The upper reach is approx-imately 120 km in length and originates in dense marshes near Blue Cypress Lake in Indian River County. Just upstream from Lake Harney, in the vicinity of Puzzle Lake, the St. Johns receives low-salinity (10–11 ppt) spring water from ancient marine salt deposits (DeMort 1991). However, the small volume of saline waters has little effect on the St. Johns River ecology.

The middle reach of the river extends approximately 160 km from Lake Harney to the confluence with Ocklawaha River. In this reach the river broadens to form large, shallow lakes—Harney, Monroe, Woodruff, and George. Lake George, with a surface area of 150 km², is the second largest lake in Florida. The largest tributary in this reach is the Ocklawaha River, which has headwaters in the Harris Chain of Lakes and flows 154 km before entering the St. Johns downstream of Lake George. Silver Springs, one of the largest spring systems in North America, is an Ocklawaha River tributary. Other large springs in the St. Johns include Wekiva, Volusia Blue, and Alexander. Several large springs flow directly into Lake George, including Salt, Juniper, and Silver Glen.

The lower reach of the St. Johns extends from the confluence with Ocklawaha River to its mouth. Most of the lower river flows through urbanized areas, including metro Jacksonville. Downstream from the Port of Jacksonville the river is dredged to a navigation depth of 9 m. Tributaries to the river main stem in the lower reach include Black Creek, which has headwaters near Trail Ridge and flows across the Talbot and Pamlico Terraces (Healy 1975) to the St. Johns. Julington Creek is a smaller drainage that flows through urban areas south of Jacksonville and enters the St. Johns from the east.

St. Marys River is a high-quality, 200 km long blackwater stream that forms the state boundary between Florida and Georgia for much of its length (Figure 4.21). It has a watershed of 4,105 km², with 2,400 km² in Florida, and flows into the Atlantic Ocean. The headwater reach originates in Okefenokee Swamp, which provides a large volume of water storage. The middle reach flows across Trail Ridge and is characterized by a narrow floodplain confined to the river main stem. The lower reach of the St. Marys is tidally influenced and poorly drained. Much of the river flows through undisturbed hardwood and pine forests and is lined by low sand bluffs.

Figure 4.21. St. Marys River, looking upstream, northwest of Boulogne, Nassau County, Florida, June 2010. Photograph by James D. Williams.

Interbasin Connections

In Florida there are natural and man-made interbasin connections that provide potential dispersal routes for mussels and fishes. Perhaps the most notable natural interbasin connection is Okefenokee Swamp in Georgia, which serves as the headwaters of both Suwannee and St. Marys Rivers. While this represents a permanent aquatic connection, it appears to be an inhospitable habitat for mussels and some fishes, which limits movement between basins. A similar situation exists in southeastern Pasco County, Florida, where the headwaters of Withlacoochee (southern) and Hillsborough Rivers originate in Green Swamp. In peninsular Florida there are numerous sites where two or more streams originate from a common wetland. In addition, there is the potential for adjacent low-lying headwaters to become connected during high-rainfall events (e.g., St. Johns and Lake Okeechobee watersheds).

The low relief in most of the peninsula has resulted in numerous water projects during the past 130 years. Many of these projects generate a network of canals that often connect adjacent basins. Some of these have undoubtedly resulted in the transfer of fishes with encysted glochidia. The fact that the peninsular mussel fauna has low diversity and most species are widespread makes it difficult to detect interbasin transfer via canals. The absence of early mussel studies to document natural distribution prior to canal connections further complicates efforts to evaluate dispersal routes across basin divides in peninsular Florida.

Chapter 5

Mussel Distribution

The North American freshwater mussel fauna is represented by the families Margaritiferidae, Mycetopodidae, and Unionidae, with approximately 375 mussel species (Williams et al. in review). Unionidae is the largest family with about 365 species occurring from southern Mexico to northern Canada and Alaska. The two remaining families are much smaller with Margaritiferidae represented by five species and Mycetopodidae by three species, and they have far more restricted distributions in North America. Margaritiferidae is limited to southern Canada and northern United States and a few small streams in Alabama and Louisiana, with the exception of one species that is widespread in the Mississippi basin. Mycetopodidae is limited in North America to southern Mexico but is more widespread southward into Central and South America.

The southeastern United States is the epicenter of aquatic biodiversity in North America and has been recognized as far exceeding that of other regions of the continent as well as most other temperate regions of the world. This is true for some invertebrates (e.g., mussels, snails, and crayfishes) and vertebrates (e.g., fishes and turtles) (Lydeard and Mayden 1995). Mussel diversity in this region is highest in Alabama and the Mobile basin with 182 mussel species in 44 genera (Williams et al. 2008, in review; Campbell and Lydeard 2012a). Other southeastern states have noteworthy mussel diversity: Tennessee with 129 species (Parmalee and Bogan 1998), Georgia with 120 species (J.D. Williams unpublished data), and Kentucky with 104 species (Cicerello and Schuster 2003). The exceptionally high aquatic species diversity in the Southeast owes its origin and existence to the geologically old and stable land mass, physiographic diversity, absence of glaciers, Pleistocene sea level fluctuations, high rainfall, and abundant water resources feeding numerous rivers. Of the 60 mussel species known from Florida, 58 are native and 2 (*Anodonta suborbiculata* and *Toxolasma parvum*) were introduced from other parts of the United States. Five of the native species—*Amblema plicata*, *Elliptio crassidens*, *Megalonaias nervosa*, *Pyganodon grandis*, and *Utterbackia imbecillis*—are widespread in the Mississippi basin and eastern Gulf Coast rivers to north-central Florida.

The Greater Floridan Region is defined herein as the river basins of Florida from the Perdido River east to St. Marys River along with their upstream reaches in Alabama and Georgia (Figure 5.1). There is a total of 65 species (62 native and 3 introduced) of Margaritiferidae and Unionidae (Williams et al. 2011) in the Greater Floridan Region, which represents approximately 20 percent of the mussel fauna in the United States. Of the 65 species, 5 taxa are found only in Alabama and Georgia portions of the region.

The geographic position of Florida is unique in that it encompasses several major Gulf and Atlantic Coast basins of the southeastern United States. This setting is important zoogeographically in that Florida has likely been connected to southern Atlantic Coast basins to the north (e.g., Satilla and Altamaha in Georgia) and Gulf Coast basins to the west (e.g., Mobile and Pascagoula in Alabama and Mississippi) during its geologic past. However, having access to the distinct and more diverse freshwater faunas of these two disjunct coastal areas has not had a positive effect on increasing diversity. In fact, Florida has a somewhat depauperate fauna in comparison to adjacent states. This is due in part to being farther away from those centers of diversity and having been partially inundated during interglacials. This has resulted in a corresponding reduction in number of species and genera (Table 5.1). A similar pattern is evident in the diversity of Florida fishes (Swift et al. 1986).

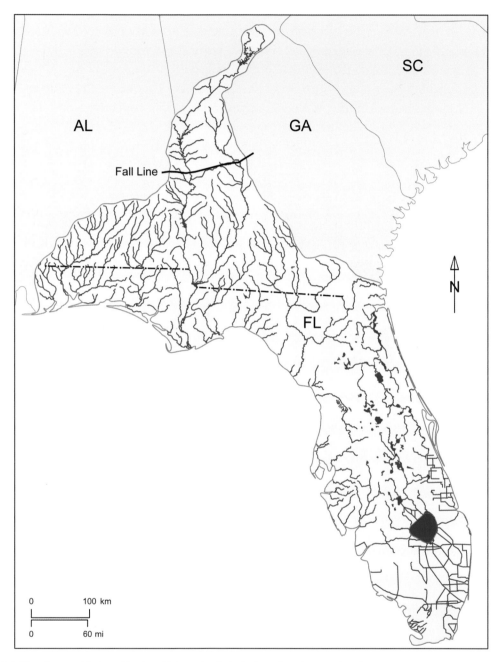

Figure 5.1. The Greater Floridan Region. This region is inclusive of all river basins, from their headwaters to the ocean, between Perdido River in Alabama and Florida and St. Marys River in Florida and Georgia.

Mussels are primarily limited in their dispersal capabilities by the fishes that serve as hosts for their glochidia. Depending on the dispersal habits of the host fish, movements may be limited to less than one hundred meters per generation (e.g., darters of the family Percidae) or several kilometers (e.g., suckers of the family Catostomidae) (Matthews 1998). Juvenile mussels move only a short distance once they drop off of the host fish and become embedded in the substrate; likewise, adults of most species have very limited movement. Most mussels are intolerant of saline environments, which further limits potential movements among basins (Johnson et al. in review). These factors combine to isolate mussels and some host fishes resulting in reduced gene flow among drainages.

Table 5.1. Distribution of 2 families and 25 genera of mussels occurring in the Greater Floridan Region. Those general that also occur in adjacent basins west and north of Florida are also indicated. Basins are labeled as follows: ACF = Apalachicola, Chattahoochee, and Flint Rivers; EYC = Escambia, Yellow, and Choctawhatchee Rivers; MOB = Mobile; OCH = Ochlockonee River; PEN = peninsular Florida rivers (Waccasassa River east to St. Marys River and south to Everglades); SAC = southern Atlantic Coast rivers in Georgia; and SUW = Suwannee River. An asterisk (*) indicates the genus occurs in the basin in Alabama and/or Georgia but not in Florida. The vertical shading represents the eastern and western boundaries of the Greater Floridan Region.

Family/Genera	MOB	EYC	ACF	OCH	SUW	PEN	SAC
Margaritiferidae							
Margaritifera	X	X*					
Unionidae							
Alasmidonta	X		X	X			X
Amblema	X	X	X				
Anodonta	X	X	X	X		X	X
Anodontoides	X	X	X				
Elliptio	X	X	X	X	X	X	X
Elliptoideus			X	X			
Fusconaia	X	X					X
Glebula	X	X	X	X			
Hamiota	X	X	X	X			
Lampsilis	X	X	X	X	X	X	X
Lasmigona	X		X*				X
Medionidus	X	X	X	X	X	X	
Megalonaias	X	X	X	X			
Obovaria	X	X					
Plectomerus	X	X					
Pleurobema	X	X	X	X	X	X	
Ptychobranchus	X	X					
Pyganodon	X	X	X	X			X
Quadrula	X	X	X	X	X		
Reginaia	X	X	X				
Toxolasma	X	X	X	X	X	X	X
Uniomerus	X	X	X	X	X	X	X
Utterbackia	X	X	X	X	X	X	X
Villosa	X	X	X	X	X	X	X
Total genera	**24**	**22**	**20**	**16**	**9**	**9**	**11**

The most important abiotic factor influencing survival, dispersal, and evolution of mussels in Florida has been fluctuating sea levels associated with major Pleistocene glacial cycles. Climate change has led to glacial cycles that raise and lower sea levels, altering the connection or isolation of adjacent basins. Lower sea levels extend rivers downstream across land formerly submerged under the ocean, which may result in the coalescence of adjacent basins. These connections result in the merging of formerly isolated aquatic faunas. Conversely, higher sea levels during interglacial periods result in the flooding of lower reaches of rivers, which may isolate portions of basins that were formerly connected or entirely inundate short coastal streams. Freshwater fauna in rivers with larger watersheds that extend farther inland usually have a better chance of survival during high sea level stands of interglacial periods due to presence of headwater refugia. Evidence of this zoogeographic pattern is found in the higher aquatic diversity in basins such as the Apalachicola compared to the Ochlockonee.

At the end of the last Pleistocene glacial cycle, about 12,000 BP, the peninsular region of Florida was approximately twice its present size with the majority of additional land area occurring in the Gulf of Mexico. Based on Florida fossil unionids, it is clear that a more diverse mussel fauna typical of large panhandle streams extended into the peninsular region during that period. The best known fossil mussel assemblages in Florida were found in Leisey Shell Pits in the Tampa Bay area and Troy Spring on the Suwannee River. These assemblages suggest that during the Pleistocene the west coast of the Florida peninsula mussel fauna was similar to those of the Apalachicola and Ochlockonee Rivers today.

Streams Naturally Devoid of Mussels

Several Florida streams naturally lack freshwater mollusks for a variety of reasons. There are entire basins in Florida and significant reaches of other streams that are unusual in being relatively undisturbed and having healthy fish communities but do not support populations of freshwater mollusks (Figure 5.2). Streams in the panhandle that are naturally devoid of freshwater mollusks include Perdido and Blackwater Rivers, some direct tributaries of Choctawhatchee Bay, and New River (east of Apalachicola River), as well as the upper reach of Shoal River (Walton County). The nonindigenous and ubiquitous *Corbicula fluminea* is also absent from these streams. Elsewhere in Florida Thompson and Hershler (1991) noted an absence of aquatic snails in the upper portions of St. Marys, Suwannee, Aucilla, and Ochlockonee Rivers, although some of these watersheds have mussels. Absence of freshwater mollusks is most likely due to some combination of low pH, lack of nutrients, and reduced levels of calcium needed to secrete shell.

Tidal reaches of many coastal streams are also devoid of freshwater mussels due to high salinity. While these rivers support healthy communities of fish and other aquatic organisms, they appear to be inhospitable to mussels. The most notable tidally influenced stream among these is the lower reach of St. Johns River from Lake George downstream to its mouth, a distance of about 160 km. Absence of mussels in the upper portions of this reach is most likely due to relatively recent penetration of saline water (Mason 1998). During the 1800s to mid-1900s, Lake George supported at least six species of mussels— *Anodonta couperiana, Elliptio ahenea, Elliptio jayensis, Elliptio occulta, Toxolasma paulum,* and *Villosa amygdalum.* These species no longer survive or they are extremely rare in Lake George. In St. Marys River, mussels are confined to the middle reach. Their absence in the upper portion of the river is most likely due to the low conductivity and pH, and in the lower main stem, salt water limits their establishment. In a detailed survey of the lower 20.4 km of Escambia River in 1952–1953, the last occurrence of unionids (*Elliptio crassidens*) was 17.1 km above the mouth of the river (ANSP 1953).

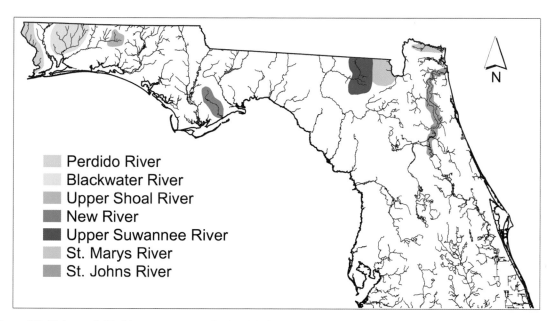

Figure 5.2. Major north Florida streams naturally devoid of mussels.

Introduced Mussels

Successful introduction of mussels outside their native ranges is unusual or rarely reported (Williams et al. 2008). The occurrence of *Toxolasma parvum* and *Anodonta suborbiculata* in Florida is known to be the result of introductions. The most likely pathway for mussel introductions is release of bait fishes or stocking of sport fishes with encysted glochidia. One or both of these were the apparent origins of *A. suborbiculata* in Escambia River and *T. parvum* in Apalachicola and Ochlockonee Rivers. Several mussel species (e.g., *T. parvum* and *Utterbackia imbecillis*) are known to inhabit commercial fish culture ponds (Williams et al. 2008). It is known that these three species were introduced into the Chattahoochee River drainage in Georgia via cultured sport fishes shipped from a fish farm in the Black Warrior River drainage, Alabama. These species were reproducing in a pond but were not found to be dispersing downstream during a short survey (R.C. Stringfellow, personal communication).

In the Blackwater River basin in west Florida, there are two man-made impoundments—Bear Lake (Santa Rosa County) and Hurricane Lake (Okaloosa County)—on small streams created to provide a recreational fishery that have established populations of the introduced mussel *Utterbackia imbecillis*. These impoundments have been regularly fertilized and treated with lime since their creation, improving nutrient and calcium levels, which enabled mussels to survive in this previously mollusk-free drainage (J.R. Knight, personal communication). *Anodonta couperiana* is native to eastern and southern peninsular Florida, but its occurrence in Lake Tarpon near Tarpon Springs, Pinellas County, is also most likely the result of an introduction.

Based on historical museum records prior to 1920, the native range of *Utterbackia imbecillis* is from the Escambia to Apalachicola basins. The earliest museum collections in the Ochlockonee and Suwannee River basins and peninsular Florida date from the mid-1950s (Clench and Turner 1956). Although not reported in an early 1960s survey of peninsular mussels (Johnson 1972), museum records from the 1970s verify its occurrence. Once established, dispersal occurred rapidly and today *U. imbecillis* is found in almost every basin in Florida.

In the Escambia River basin, the occurrence of *Anodonta suborbiculata* appears to be the result of unintentional introduction. *Anodonta* was first collected by C.A. Burke in 1917 from Escambia River at Chumuckla Springs, Santa Rosa County, Florida. It was reported as *A. suborbiculata* based on a personal communication from W.H. Heard (Butler 1990). However, it appears that Burke's specimen was actually a native *Anodonta hartfieldorum* since there are no other *A. suborbiculata* records from the drainage until the late 1900s (Williams et al. 2009). There is a large population of *A. suborbiculata* in Gantt and Point A Lakes on the Conecuh River, the largest upstream tributary of Escambia River, in southern Alabama. These reservoirs may provide the source for a downstream Escambia population in Florida, but a separate introduction cannot be ruled out. *Anodonta suborbiculata* is also introduced in the Pearl and Pascagoula River basins in Mississippi and in the Mobile basin in Alabama (Williams et al. 2008).

Pyganodon cataracta was reported from Apalachicola River in Florida by Brim Box and Williams (2000) but is not currently recognized as occurring in the state. Identification of *P. cataracta* from Florida was based on shell morphology of specimens used in a genetic analysis from reservoirs in the middle and upper portions of the basin in Alabama and Georgia. Origin of *P. cataracta* in Apalachicola basin may be the result of introduced fishes (e.g., *Morone saxatilis*) from an Atlantic Coast reservoir with encysted glochidia (Williams et al. 2008).

In addition to introduction of native mussels beyond their natural ranges, there is at least one species native to South America that has been recorded from Florida. A single live individual of *Anodontites trapesialis* (Mycetopodidae) was found in Ron's Fish Farm in southeast Florida circa 1980 (H. Rudolph, personal communication) (Figure 5.3). It was likely encysted on tropical fishes imported into the United States. While it appears to have survived to adult size in a fish culture pond, there is no evidence that *A. trapesialis* has escaped into open waters of the state. The area formerly occupied by Ron's Fish Farm has been redeveloped into a residential area.

Figure 5.3. *Anodontites trapesialis*, UF 451231, length 165 mm. Shell from Ron's Fish Farm, near [4.5 miles west-southwest of] Delray Beach, Palm Beach County, Florida, circa 1980. Photograph by Zachary Randall.

Mussel Assemblages

There are several factors that determine the number of mussel species and endemics in a particular assemblage, including habitat heterogeneity, size of the drainage basin, prehistorical relationships to other drainages, and the length of time the basin has been isolated. The mussel fauna in Florida consists of species ranging from narrowly distributed endemics (e.g., *Medionidus walkeri*) to the southeasternmost populations of widespread Mississippi basin taxa (e.g., *Megalonaias nervosa*). The general pattern is a decrease in species diversity in drainages from west to east with a distinct but depauperate fauna in peninsular Florida. This pattern reflects the recent exposure and inundation of Florida river basins during Pleistocene glacial and interglacial periods, respectively. While there are distinct mussel assemblages in Florida, there are a few species that transcend these boundaries and make them less distinct.

Early efforts to delineate mussel faunal assemblages in North America were generally confined to broadscale divisions involving large portions of the continent (Conrad 1834; Simpson 1893b; Ortmann 1913). The North American mussel fauna was divided into three faunal regions—west of the Rocky Mountains, central portion of the continent, and east of the Appalachian Mountains—by Simpson (1895). A more detailed division in the United States was delineated by van der Schalie and van der Schalie (1950), who recognized six regions based largely on endemic faunas. One of the six regions was the West Floridan or Apalachicolan Region, which extended from the Escambia River eastward to the Gulf Coast basins of peninsular Florida. The eastern coast of Florida was considered part of the Southern Division of the Atlantic Region. The West Floridan or Apalachicolan Region has generally been recognized as having a distinctive mussel assemblage, but its boundaries have been interpreted differently by various researchers (Johnson 1972; Burch 1975a; Butler 1990; Parmalee and Bogan 1998).

The relationship between mussel assemblages and river basins in Florida presented by Haag (2010) differed from those for fishes and rivers presented by Swift et al. (1986). In a phenetic clustering analysis based on presence and absence of fish species, Swift et al. (1986) recovered Apalachicola River in a clade containing Escambia, Yellow, and Choctawhatchee Rivers, and Ochlockonee and Suwannee Rivers in a clade with the St. Marys and St. Johns Rivers. Not unexpectedly, the remainder of the peninsular Florida rivers formed an independent clade sister to the latter clade. These differences, to some extent, might be explained by the different techniques utilized by Swift et al. (1986) and Haag (2010).

The most recent analysis of mussel diversity in North America north of Mexico (Rio Grande drainage) was put forth by Haag (2010). Mussel faunal similarity across the United States and Canada was analyzed, and four regions—Atlantic, Eastern Gulf, Mississippian, and Pacific—composed of 17 faunal provinces were recognized. However, the name Eastern Gulf Region proposed by Haag is somewhat misleading as it includes basins that drain into the Atlantic Ocean (e.g., Loxahatchee, St. Johns, and St. Marys). On this basis, "Greater Floridan Region" is proposed herein as a replacement name for the Eastern Gulf Region encompassing the three provinces in Florida and portions of Alabama and Georgia (Figure 5.4).

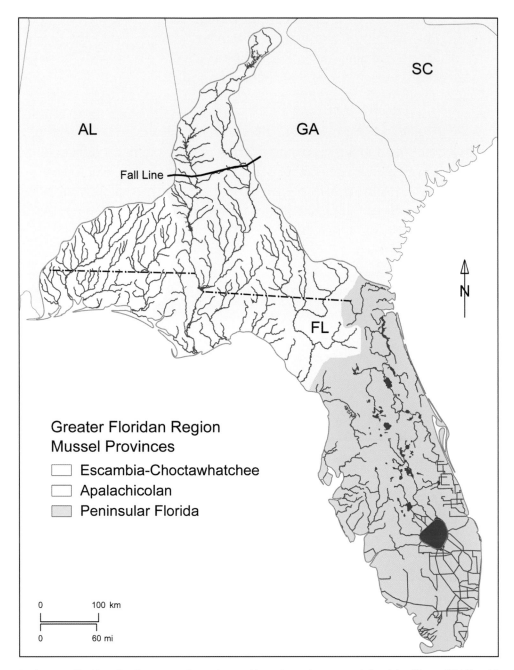

Figure 5.4. Greater Floridan Region mussel provinces. Mussel provinces as defined by Haag (2010), with minor modifications presented herein.

Three of the 17 mussel provinces proposed by Haag (2010) comprise the Greater Floridan Region. The western province is the Escambia-Choctawhatchee, which includes three major basins in west Florida and south Alabama—Escambia, Yellow, and Choctawhatchee. The Apalachicolan Province extends from the Apalachicola basin eastward to Suwannee basin. The third province, Peninsular Florida, includes all drainages south of Waccasassa River on the eastern Gulf Coast and St. Marys River on the southern Atlantic Coast (Figure 5.4).

Econfina Creek is a small Florida basin situated between the Escambia-Choctawhatchee and Apalachicolan Provinces that was not included in the analysis by

Haag (2010). It has a total of nine species, of which three—*Hamiota subangulata*, *Medionidus penicillatus*, and *Pleurobema pyriforme*—are endemic to drainages to the east (Apalachicolan Province). Further, none of the endemics found in the Escambia-Choctawhatchee Province occur in the Econfina Creek basin. Thus, the western boundary of the Apalachicolan Province as defined by Haag is extended to include Econfina Creek.

There are 62 native mussel species in the Greater Floridan Region, of which 41 (69%) are endemic to the region. Of the 58 native mussel species known from Florida, only three species (*Elliptio crassidens*, *Lampsilis floridensis*, and *Villosa vibex*) are native to all three mussel provinces (Table 5.2).

Escambia-Choctawhatchee Province—There are 35 mussel species in this province, 1 in Margaritiferidae (*Margaritifera marrianae*) and the remaining 34 in Unionidae. Two of the unionids, *Anodonta suborbiculata* and *Toxolasma parvum*, were recent introductions to the province. In the Florida portion of the province, there are 32 natives plus 1 introduced species, *A. suborbiculata*. *Margaritifera marrianae* and *T. parvum* are known only from the Alabama portion of the province.

The Escambia-Choctawhatchee is the westernmost mussel province in Florida and includes three major river basins, Escambia, Yellow, and Choctawhatchee. These basins are located entirely on the Coastal Plain Physiographic Province. They share similar mussel faunas that include seven endemic species—*Elliptio mcmichaeli*, *Hamiota australis*, *Obovaria choctawensis*, *Pleurobema strodeanum*, *Ptychobranchus jonesi*, *Quadrula succissa*, and *Toxolasma* sp. cf. *corvunculus*. Another endemic species, *Fusconaia escambia*, is shared between Escambia and Yellow Rivers. Both Escambia and Choctawhatchee Rivers have mussels that are unique to their systems—*Reginaia rotulata* in the Escambia and *Fusconaia burkei* and *Obovaria haddletoni* in the Choctawhatchee. In addition to endemics, there is a suite of mussels that occur in these three drainages that also occur in drainages to the west in Alabama, Louisiana, and Mississippi. Two smaller basins within this province, Perdido and Blackwater Rivers, are among those that do not support any native freshwater mussels.

Mobile basin fishes and mussels appear to have dispersed into the Escambia River drainage in the not-too-distant past. Five fishes (*Crystallaria asprella*, *Etheostoma histrio*, *Hybognathus hayi*, *Moxostoma carinatum*, and *Percina ouachita*) and four mussels (*Anodonta hartfieldorum*, *Lampsilis ornata*, *Margaritifera marrianae*, and *Plectomerus dombeyanus*) that occur in Escambia River basin (but not eastward) are also found westward in Mobile basin (Williams et al. 2008; Page and Burr 2011). Most of these fishes and *L. ornata* and *P. dombeyanus* typically inhabit large creeks and rivers with sand and gravel shoals, which suggests dispersal via convergent lower river main stems during a Pleistocene glacial period as opposed to a small headwater stream capture. Habitat of *A. hartfieldorum* (typically sloughs and oxbow lakes) is also associated with large rivers.

There are numerous isolated ponds and lakes in the Yellow and Choctawhatchee basins that typically lack a surface connection to streams, are shallow with sand and organic muck bottoms, and usually support emergent and submergent macrophytes. Some of these are subject to drying during droughts but nevertheless support small fish communities of native and/or introduced species. These water bodies do not appear to support native mussels, but populations of introduced species (e.g., *Utterbackia imbecillis*) may be present.

Lake Jackson is a small (200 ha) lake located on the state line in Covington County, Alabama, and Walton County, Florida, about 2 miles west of Britton Hill, the highest point (105 m) in the state of Florida. The lake was surveyed in a park in Florala, Alabama, in May 1986, and no snails, unionid mussels, or *Corbicula fluminea* were present. During an April 2003 survey at the same site, shells of two *C. fluminea* and four small (32–42 mm in length) *Utterbackia imbecillis* were found, but no live mollusks were observed. It remains to be seen if *U. imbecillis* and *C. fluminea* become established in Lake Jackson.

Apalachicolan Province—There are 40 species (Unionidae), including 38 native and 2 introduced, *Toxolasma parvum* and *Pyganodon cataracta*, in the Apalachicolan Province in Alabama, Florida, and Georgia. In the Florida portion, there are 39 species, including 38 natives and *T. parvum*. Fifteen of the native mussels are endemic to the province. *Utterbackia imbecillis* is native to the province but appears to have been introduced beyond its native range in some areas. Major basins having diverse mussel faunas in this province include (from west to east) Econfina Creek and Apalachicola, Ochlockonee, and Suwannee Rivers.

The Apalachicola is the largest river in the Apalachicolan Province and the Greater Floridan Region. It is also the only river in Florida with its headwaters located above the Fall Line in the geologically older Piedmont and Blue Ridge Physiographic Provinces. Apalachicola River (and its primary tributaries, Chattahoochee and Flint Rivers) has a distinctive mussel fauna and shares only a few wide-ranging species with the basins to the east and west. It has a total of 33 native mussel species, of which 7—*Alasmidonta triangulata*, *Amblema neislerii*, *Elliptio chipolaensis*, *Elliptio fumata*, *Elliptio nigella*, *Lampsilis binominata*, and *Reginaia apalachicola*—are endemic to the basin. Two of the endemics, *E. nigella* and *L. binominata*, are not presently known to have ranged downstream into Florida. In addition to these seven species, there are three species—*Anodonta heardi*, *Elliptoideus sloatianus*, and *Hamiota subangulata*—endemic to the province that occur in Apalachicola River and eastward in Ochlockonee River. There are two mussels, *Elliptio fraterna* and *Lasmigona subviridis*, in the Apalachicola basin that also occur in Atlantic Coast basins. These two species are not known to occur in southern Atlantic Coast basins south of Savannah River or in eastern Gulf Coast rivers east of the Apalachicola basin. The occurrence of these species in Gulf and Atlantic Coast basins has been attributed to headwater stream capture between Chattahoochee and Savannah Rivers (Johnson 1970); however, the Apalachicola population may represent an undescribed species. An additional species, *Pyganodon cataracta*, is known from Chattahoochee River impoundments and their tributaries in Alabama and Georgia, but its presence there is most likely the result of a recent introduction. Although Brim Box and Williams (2000) reported *P. cataracta* from the Apalachicola River in Florida, further research is needed to verify its occurrence. *Strophitus subvexus* also reported from the basin by Brim Box and Williams (2000) was reidentified as *Anodontoides radiatus* based on genetic analyses.

Ochlockonee River, located east of Apalachicola River, is a smaller basin which has its headwaters in southwest Georgia. It has two endemic species, *Alasmidonta wrightiana* and *Medionidus simpsonianus*, as well as the three provincial endemics shared with Apalachicola River, *Anodonta heardi*, *Elliptoideus sloatianus*, and *Hamiota subangulata*. For some species (e.g., *Toxolasma paulum* and *Pleurobema pyriforme*), Ochlockonee

River supports populations that share haplotypes with the Apalachicola basin to the west and Suwannee River basin to the east. This suggests that during one or more of the four Pleistocene glacial periods these drainages were connected to Ochlockonee River, which allowed movement of mussels and host fishes among these three basins.

Between Ochlockonee and Suwannee Rivers, there are several small independent eastern Gulf Coast basins with depauperate mussel faunas. Higher sea levels during the most recent interglacial period reduced available habitat for mussels and their host fishes and is likely the overriding factor responsible for their depauperate faunas. After the sea level dropped, there was no opportunity for recolonization from adjacent basins with higher mussel diversity.

Peninsular Florida Province—There are 14 native and 1 introduced species, *Utterbackia imbecillis*, in the Peninsular Florida Province. Considering the land area and number of lakes, streams, and springs, the mussel fauna is depauperate when compared to the other two mussel provinces in Florida. Two species, *Elliptio monroensis* and *Villosa amygdalum*, are endemic to this province. There is some separation of species between Gulf and southern Atlantic Coast basins of peninsular Florida, but four species— *Elliptio jayensis*, *Toxolasma paulum*, *Uniomerus carolinianus*, and *V. amygdalum*—are widespread on both sides of the Gulf and Atlantic Coasts divide. A fifth species, *Villosa vibex*, occurs on both sides of the divide but is not found in the southern portion of the peninsula south of Tampa Bay and the St. Johns River basin. A contrasting pattern is presented by *Anodonta couperiana*, which occurs in southern Atlantic Coast basins and the southern portion of the peninsula, including Myakka and Peace Rivers on the eastern Gulf Coast.

Lakes are one of the dominant aquatic features of the central portion of the Florida peninsula. While they vary in size, shape, depth, substrate, and water quality, most are characterized by a sand bottom with emergent and submergent macrophytes along the shore line. Some of the smaller sandhill lakes on Lake Wales Ridge support no native mussels. Low mussel diversity in lakes on Lake Wales Ridge is surprising given the presence of a relatively diverse fish community of 25–30 species. Mussel diversity varies among large peninsular lakes, but most support a suite of species that includes *Anodonta couperiana*, *Elliptio ahenea*, *Elliptio jayensis*, *Toxolasma paulum*, *Uniomerus carolinianus*, *Utterbackia imbecillis* (introduced), and *Villosa amygdalum*. Another species, *Elliptio monroensis*, is absent on the ridge but occurs in lakes and riverine habitats along the main channel of St. Johns River and lower reaches of some of its large tributaries.

Table 5.2. Distribution of mussels within the boundaries of Florida by major basin. Abbreviations are as follows: APC = Apalachicola and Chattahoochee Rivers; CHO = Choctawhatchee River; ECF = Econfina Creek; ESC = Escambia River; MGS = minor Gulf Coast streams; MYP = Myakka and Peace Rivers; OCH = Ochlockonee River; SFL = Caloosahatchee, Kissimmee, and St. Lucie Rivers, Lake Okeechobee, and the Everglades; STJ = St. Johns River; STM = St. Marys River; SUW = Suwannee River; TMB = Tampa Bay tributaries, including Hillsborough River; WTH = Waccasassa and Withlacoochee (southern) Rivers; and YEL = Yellow River. The vertical shading delimits the three mussel provinces of Florida (Escambia-Choctawhatchee, Apalachicolan, and Peninsular Florida). An "X" indicates a native occurrence in Florida, and "I" is an introduction (nonnative occurrence).

Species	ESC	YEL	CHO	ECF	APC	OCH	MGS	SUW	WTH	TMB	MYP	SFL	STJ	STM
Alasmidonta triangulata					X									
Alasmidonta wrightiana						X								
Amblema neislerii					X									
Amblema plicata	X		X											
Anodonta couperiana										I	X	X	X	X
Anodonta hartfieldorum	X													
Anodonta heardi					X	X								
Anodonta suborbiculata	I													
Anodontoides radiatus			X		X									
Elliptio ahenea								X				X	X	X
Elliptio arctata					X									
Elliptio chipolaensis					X									
Elliptio crassidens	X		X		X	X								X
Elliptio fraterna					X									
Elliptio fumata					X									
Elliptio jayensis					I	X	X	X	X	X	X	X	X	X
Elliptio mcmichaeli		X	X											
Elliptio monroensis													X	X
Elliptio occulta							X	X					X	X
Elliptio pullata	X	X	X	X	X									
Elliptio purpurella			X		X	X								
Elliptoideus sloatianus					X	X								
Fusconaia burkei			X											
Fusconaia escambia	X	X												
Glebula rotundata	X		X		X	X								
Hamiota australis		X	X											
Hamiota subangulata				X	X									

Species	ESC	YEL	CHO	ECF	APC	OCH	MGS	SUW	WTH	TMB	MYP	SFL	STJ	STM
Lampsilis floridensis	X		X		X	X		X	X	X				
Lampsilis ornata	X													
Lampsilis straminea	X	X	X		X	X		X						
Medionidus acutissimus		X	X											
Medionidus penicillatus				X	X									
Medionidus simpsonianus						X								
Medionidus walkeri								X		X				
Megalonaias nervosa	X				X	X								
Obovaria choctawensis	X	X	X											
Obovaria haddletoni			X											
Plectomerus dombeyanus	X													
Pleurobema pyriforme				X				X		X				
Pleurobema strodeanum	X	X	X		X	X								
Ptychobranchus jonesi			X											
Pyganodon grandis	X		X		X	X								
Quadrula infucata					X	X								
Quadrula kleiniana								X						
Quadrula succissa	X	X	X											
Reginaia apalachicola					X									
Reginaia rotulata	X													
Toxolasma parvum					–	–								
Toxolasma paulum				X	X	X	X	X	X	X	X	X	X	
Toxolasma sp. cf. corvunculus	X	X	X											
Uniomerus carolinianus				X		X	X	X	X	X	X	X	X	X
Uniomerus columbensis				X	X									
Uniomerus tetralasmus	X	X												
Utterbackia imbecillis	X			X	X	X	–	–	–	–	–	–	–	
Utterbackia peggyae	X			X	X	X	X							
Utterbackia peninsularis									X	X				
Villosa amygdalum									X	X	X	X	X	
Villosa lienosa	X	X	X	X	X	X		X						
Villosa vibex	X	X	X	X	X	X	X	X	X	X	X	X	X	X
Villosa villosa			X		X	X	X	X						
Total species	23	16	25	9	31	23	8	15	8	11	6	7	10	8

Chapter 6

Biology and Ecology

Unionid mussels are fascinating organisms for scientific study or casual observation. They have intrigued mankind for millennia, primarily due to their ability to form pearls, and have served as a food and tool source throughout human history (de Lozoya and Araujo 2011). Mussels are largely sedentary and spend their lives inconspicuously in the substrates of inland waters. They have a unique life history in which their larvae, called glochidia, typically undergo a parasitic period on the gills or fins of fishes. This relationship ties mussels intimately to the fish community, and an amazing array of host-infecting strategies has evolved to maximize the probability of glochidial attachment (Haag 2012). The origin of parasitism in mussels is unknown, but it may have evolved to aid in dispersal on more mobile fishes (Haag 2012). Mussels can dominate the benthic biomass in streams and lakes (Hanson et al. 1988; Layzer et al. 1993; Howard and Cuffey 2006) (Figure 6.1), and they may be keystone members of aquatic communities that have important effects on a wide variety of other organisms through various ecological services (Vaughn 2010).

Figure 6.1. Mussel bed in Chipola River, Calhoun County, Florida, June 1991. Photograph by Jayne Brim Box.

Physiology and Feeding Ecology

Mussels obtain food and oxygen by taking in water through the incurrent aperture. Water flows through the gill filaments where food particles are collected and oxygen diffuses into the hemolymph and is transported throughout the body (McMahon and Bogan 2001; Cummings and Graf 2010). Filtered water passes into the suprabranchial chamber where waste products are excreted before exiting the mussel via the excurrent aperture. During spawning season sperm are also released through the excurrent aperture. Mussels have variable oxygen requirements, and species tolerant of low oxygen concentration are potentially capable of anaerobic metabolism (Holwerda and Veenhof 1984; Massabuau et al. 1991). Metabolic rate varies among species and is largely dictated by temperature (Allen and Vaughn 2009). Some species are tolerant of wide temperature ranges, but others are thermally sensitive, with optimal performance only occurring in a narrow range of conditions (Spooner and Vaughn 2008, 2009). Given Florida's warm climate, species are expected to have relatively high thermal tolerance, especially in lentic peninsular environments. Some species adapted to fluctuating environments potentially regulate their respiration rates under unfavorable conditions (Sheldon and Walker 1989). *Uniomerus* can survive desiccation for many months by aestivating, a superb adaptation for waters susceptible to drying completely (Simpson 1899; Frierson 1903a; Isely 1914).

Adult mussels are typically considered suspension feeders (McMahon and Bogan 2001), but recent evidence suggests they may also deposit feed, taking organic sediments in through their shell gape (Nichols et al. 2005). Patterns of food availability and hydrodynamic conditions may result in seasonal or spatial shifts between suspension- and deposit-feeding strategies (Ostrovsky et al. 1993; Vaughn and Hakenkamp 2001). Mussels are omnivores and feed on items typically less than 20 μm in size, such as phytoplankton, bacteria, fungal spores, protozoans, zooplankton, and detritus (Strayer 2008). Newly metamorphosed juveniles may have different nutritional needs than adults and may depend more heavily on deposit feeding (Yeager et al. 1994; Bolden and Brown 2002). In small streams and low-productivity lakes, mussels tend to feed mostly on bacteria and detritus rather than phytoplankton, but phytoplankton and other suspended material may be important food sources in productive lentic habitats and large rivers (Raikow and Hamilton 2001; Griffiths and Cyr 2006). Some lentic *Elliptio* populations potentially depend on deposit feeding in rich organic substrates (Cahoon and Owen 1996).

Filtering capacity and gaping behavior in mussels has both environmental and biological determinants. The volume of water pumped has been directly correlated with mussel size and gill surface area (Price and Schiebe 1978; Kryger and Riisgård 1988). Filtration rates may range from approximately 1 liter per day to approximately 1 2 liters per hour (Helfrich et al. 1995; Bärlocker and Brendelberger 2004; Vaughn et al. 2008). Gaping behavior is manifest in rhythms of activity for feeding and respiring, and it may be influenced by factors such as shell morphology, photoperiod, temperature, and food supply (Imlay 1968; McCorkle et al. 1979; Englund and Heino 1994a). Diel gaping patterns vary among species as a function of temperature and metabolic rate (Chen 1998; Rodland et al. 2009). Some authors have suggested that variable productivity and food levels in lakes influence the degree of gaping (Englund and Heino 1994b; Patzner and Müller 2001). Chen (1998) found that *Pyganodon grandis* gaped mostly nocturnally and speculated that this pattern may be expected for lake species where downward nocturnal migration of phytoplankton occurs; however, lotic species depending less on phytoplankton may not display diurnal gaping rhythmicity.

Differences in food availability among habitats likely have a large influence on the occurrence of mussels in Florida. Mussels in low-productivity panhandle streams can apparently subsist largely on bacteria and detritus, whereas mussels in lakes are probably dependent on richer food sources. In the peninsula, lake outlets within chains-of-lakes may represent particularly important mussel habitats because of their high food availability and quality. Higher mussel abundance and growth in European lake outlet streams compared to lakes or nonoutlet streams have been similarly attributed to food quality and abundance (Brönmark and Malmqvist 1982; Lewandowski 1990). In Florida Kunz (1898) noted anecdotally that mussel growth was "best…in lakes with outlets, the water pure and fresh." In upper St. Johns River, mussel densities were highest in a lake outlet near the banks where the substrate was perceived to be rich in organic matter (Belanger et al. 1990). In lab experiments *Elliptio complanata* and *Pyganodon grandis* preferred mud over sand substrates (Downing et al. 2000). These studies collectively suggest that substrates rich in organic matter may be important to some Florida species, especially considering the prevalence of these sediments in the state.

Burrowing and Movement

Shell and foot morphology, reproductive activities, and environmental conditions influence anchoring and burrowing behavior (Allen and Vaughn 2009; Wilson et al. 2010). Mussels anchor themselves in the stream or lake bottom with their muscular foot. Various sculptured shell morphologies may assist with anchoring (Watters 1993, 1994a). Early stage juveniles of many species produce byssal threads with which they secure themselves to hard objects (Sterki 1891a, 1891b; Howard 1922; Smith 2000). Some species (e.g., *Lampsilis* and *Medionidus*) may retain the byssus into adulthood (Coker et al. 1921; Haag 2012). Frierson (1903b) suggested that the byssus aids in dispersal, having noted groups of little mussels (mostly *Toxolasma*) suspended in the water column by "little raft[s]." Mussels are mostly sedentary but are capable of significant movements under some conditions. They burrow or move by probing the foot into the substrate, which is facilitated by a series of muscular contractions and expansion of the foot through hemolymph hydraulics (Lewis and Riebel 1984). Mussels burrow or move to prevent dislodgement by floods or wave action (Schwalb and Pusch 2007; Cyr 2008), avoid stranding during droughts (Watters et al. 2009; Newton et al. 2011), and elude predators (Watters 1993).

Sculpture may aid in reducing dislodgement, but smooth-shelled species may be able to burrow more readily once dislodged (Watters 1993, 1994a). Burrowing is more rapid in soft sediments than in firmly packed gravel (Lewis and Riebel 1984). Smaller and younger mussels appear to be buried more often than adults (Schwalb and Pusch 2007; Negishi et al. 2011). Mussels typically burrow no deeper than 15–20 cm and rarely to 30–40 cm (Simpson 1899; Amyot and Downing 1991; Nichols and Wilcox 1997). Sediment weight and DO probably limit burrowing depth (Watters 1993; Schwalb and Pusch 2007). Seasonal burrowing behavior may be influenced by water temperature or day length, and mussels typically burrow more deeply in winter but emerge in warmer seasons (Nichols and Wilcox 1997; Amyot and Downing 1998; Perles et al. 2003). High flows may prompt mussels to burrow more deeply to avoid dislodgement (Balfour and Smock 1995; Allen and Vaughn 2009). Movement to the substrate surface facilitates efficient sperm dispersal or acquisition and glochidial discharge (Balfour and Smock 1995; Watters et al. 2001; Schwalb and Pusch 2007; Galbraith 2009). Emergence for species with mantle lures (e.g., *Lampsilis*, *Medionidus*, and *Villosa*) increases visibility and the probability of host fish contact (Barnhart et al. 2008) while potentially exposing them to increased predation risks (Watters 1993).

In addition to vertical movements within the substrate, mussels may move laterally for substantial distances, especially in response to changing water levels (Coker et al. 1921; Gough et al. 2012). *Amblema neislerii* is thought to track water level changes in Apalachicola River, while *Amblema plicata* was reported to burrow with dropping levels in upper Mississippi River (Gangloff 2011; Newton et al. 2011). Trails from crawling mussels are often evident on recently exposed shorelines or in very shallow waters

(Samad and Stanley 1986; Schwalb and Pusch 2007) (Figure 6.2). During a 1987 draw-down of Lake Tohopekaliga, Osceola County, dozens of *Elliptio jayensis* were observed at the ends of long haphazard trails. *Toxolasma* is well known for tracking changing water levels along stream and lake margins, as do some *Anodonta* and *Lampsilis* (Evermann and Clark 1917; Clench and Turner 1956; Valentine and Stansbery 1971; Newton et al. 2011). Large-river species appear less able than small-stream species to adapt to varying water levels (Grier 1922; Tudorancea 1972; Brown et al. 2011). Authors have hypothesized that mussels may move laterally in attempts to find deeper, cooler water in summer, to seek more stable substrates, and in conjunction with deposit feeding (Cahoon and Owen 1996; Waller et al. 1999; Allen and Vaughn 2009); less mobile species are more susceptible to drought, while species that tend to burrow are more resistant (Gough et al. 2012).

Figure 6.2. Trail created by *Elliptio crassidens* in shallow water of Escambia River, Santa Rosa County, Florida, June 1988. Photograph by Robert S. Butler.

Habitat

Mussels in Florida occupy a broad range of habitats, but the greatest numbers of species are found in streams. Mussels in lotic habitats can occur in nearly any size stream, ranging from small headwaters to large rivers such as the Apalachicola. Florida has many large springs, but conditions in these habitats are usually not conducive to mussels due to low DO and low food concentration. However, large mussel populations can occur in spring outlet streams (e.g., Alexander Spring, Lake County, and Jackson Blue Spring, Jackson County). Mussels are often found in tidally influenced streams (Strayer et al. 1994), but only a few species (e.g., *Glebula rotundata*, *Plectomerus dombeyanus*, *Quadrula apiculata*) have a tolerance to low levels of salinity, permitting co-occurrence with brackish-water species such as *Rangia cuneata* (Haag 2012).

In streams, mussels can be found in riffles, runs, and pools and in many substrate types. Several habitat associations appear consistent, but these vary among species (Haag 2012). Florida panhandle species that are restricted to streams are typically found in gravel or sand substrates, or in soft substrates adjacent to rock ledges (e.g., *Elliptio crassidens*, *Elliptoideus sloatianus*, *Hamiota subangulata*, *Megalonaias nervosa*, and

Quadrula infucata) (Gagnon et al. 2006; Gangloff and Hartfield 2009; Gangloff 2011). Species that occur in Florida lakes, as well as some stream species, are concentrated in silty sand or sandy mud substrates along the banks (Clench and Turner 1956; Brim Box et al. 2002); in particular, *Toxolasma* often occurs nearly at the shoreline. Throughout the state *Elliptio* and *Uniomerus* may be found abundantly in rootwads, undercut banks, and among cypress trees and stumps. Habitats with dense aquatic vegetation have been considered inhospitable to mussels (Cvancara 1972; Burlakova and Karatayev 2007), but this perception may be an artifact of the difficulty of sampling in these habitats. In Florida, beds of submergent and emergent vegetation often support sizable mussel populations, particularly in the peninsula. A cow lily (*Nuphar*) bed in the tidally influenced South Prong Black Creek, Clay County, harbored thousands of mussels (predominantly *Elliptio*) highly clumped in rich muddy ooze around the rhizomes. Mussels appeared suspended in the thick, gelatinouslike substrate that created a "false bottom," and mussels inhabit similar habitats elsewhere in the state (e.g., lower Chipola River).

A primary determinant of mussel occurrence in streams appears to be substrate stability (Strayer 2008). Bank habitats in larger streams often have high mussel abundance. These areas typically have a more stable substrate, lower current velocity, and structure—such as rootwads, cypress knees, and aquatic vegetation—that provides refuge from scouring flows that may dislodge buried mussels (Belanger et al. 1990; Brim Box et al. 2002). To some extent the perceived association of mussels with stream margin habitats may be an artifact of human-induced habitat alterations that destabilized main channel habitats, resulting in the loss of most mussels. This is particularly evident in Florida rivers characterized by large bed loads of shifting sand substrates.

Lakes, wetlands, reservoirs, and other lentic waters are prevalent in the state and provide important habitats for many mussel species. Large accumulations of decaying detritus and runoff from agricultural and developed areas with adverse levels of nutrients and pesticides may prevent mussel persistence, particularly in altered habitats. Chains of subtropical lakes in the St. Johns, upper Kissimmee, and Myakka River drainages represent unique aquatic environments in the United States. Mussel assemblages in lentic habitats may be limited by periodic low DO and seasonal drying of the habitat. Consequently, lentic habitat assemblages throughout the state are generally restricted to mussels that readily adapt to these conditions, namely *Anodonta* and a few species of *Elliptio*, *Toxolasma*, *Uniomerus*, *Utterbackia*, and *Villosa*. As in streams, several consistent mussel-habitat associations are seen in Florida lakes. In peninsular lakes like Kissimmee, dense mussel aggregations are often found in beds of emergent vegetation. Permanent wetlands may have mussel faunas similar to lakes, but ephemeral wetlands are often inhabited only by *Uniomerus*. Old phosphate mining ponds in the upper Peace River drainage have species typical of peninsular lakes. Borrow pits and floodplain lakes in the panhandle may have large populations of *Anodonta hartfieldorum*, *Pyganodon grandis*, and *Utterbackia imbecillis*. Isolated sand-bottomed lakes in the western panhandle are generally devoid of mussels, except for possibly *U. imbecillis*, perhaps a result of stocked bait or sport fishes. Numerous *Utterbackia peggyae* were found in a clear water sinkhole that is occasionally flooded by an adjacent spring run in Florida Caverns State Park, Jackson County. Reservoirs and canals may harbor lentic mussel populations unless DO levels are low and pollutant levels are excessively high. In lakes in other regions, mussel distribution may be limited by cold temperatures, low DO, and perhaps pressure at great depths (Headlee 1906; Cvancara and Freeman 1978; Ostrovsky et al. 1993; Burlakova and Karatayev 2007). Because most Florida lakes are shallow, depth is probably not an important limiting factor. An *Elliptio* shell was found on the bottom of Lake Annie, the deepest lake (approximately 20 m) on the Lake Wales Ridge.

Apart from habitat characteristics specific to rivers or lakes, water chemistry has a major influence on the occurrence of mussels in Florida. Mussels are dependent on sufficient calcium content (primarily the bicarbonate form) in the water to produce shells, and poorly buffered or low pH waters can erode shells rapidly (Mackie and Flippance 1983; McMahon and Bogan 2001). Many Florida streams and lakes are darkly stained with tannins and humic acids but are otherwise clear. Although some blackwater streams have water quality that is conducive for abundant mussel populations (e.g., Black Creek, Clay County), others may have low pH and buffering capacities because of a lack of carbonate-bearing rocks in the watershed. Consequently, they may have very low mussel diversity and abundance (e.g., Aucilla and Steinhatchee Rivers) or lack mussels entirely (e.g., Perdido, Blackwater, upper Suwannee, and upper St. Marys Rivers). Diverse and abundant mussel faunas in Florida are generally limited to regions with carbonate-bearing rocks, such as large panhandle river drainages and most of the peninsula.

Growth and Longevity

Mussels grow throughout their lives, but growth is fastest in juveniles and slows dramatically when resources are diverted to gamete production at sexual maturity. Shell growth is achieved by deposition of calcium carbonate by the mantle, increasing both shell length and thickness over time. Shell growth slows in cold temperatures, resulting in distinct annual growth rings in the shell—similar to tree rings—especially in temperate latitudes (Neves and Moyer 1988). Growth rings may also be produced in response to daily cycles or disturbance and stress (Haag and Commens-Carson 2008). Annual rings produced during winter are evident even at southerly latitudes in the United States (Rypel et al. 2008), but mussel growth has not been studied in subtropical south Florida. It is possible that the long growing season and mild winters in this area allow mussels to grow throughout the year and may reduce the conspicuousness of annual rings (Simpson 1899).

Mussels are often characterized as slow growing and long lived, but growth and longevity may be variable among species by two orders of magnitude, showing this generalization to be of limited usefulness (Haag and Rypel 2011). In general, growth rate is inversely correlated with life span such that fast-growing species are shorter lived than those in which growth is slower (Haag and Rypel 2011; Ridgway et al. 2011). Consequently, fast-growing anodontines and some lampsilines typically live less than 10 years, while slower-growing species often live more than 30 years and some can exceed 100 years. Because mussel assemblages in peninsular Florida are composed predominantly of anodontines and lampsilines (with the exception of *Elliptio* and *Uniomerus*), most species in this part of the state are probably fast growing and short lived. The most long-lived species are found in panhandle rivers, especially among *Amblema*, *Megalonaias*, *Fusconaia*, and *Quadrula*. Maximum life spans of Florida mussel populations have not been determined. It is likely that many of these species, particularly those that reach large size (e.g., *Elliptoideus sloatianus* and *Megalonaias nervosa*), live for many decades. The greatest life span reported for any freshwater mussel is nearly 200 years for *Margaritifera margaritifera*, a Holarctic species (Ziuganov et al. 2000).

Growth rate and longevity vary considerably within species relative to several factors. Annual growth is typically higher during years of low stream flow potentially due to greater food concentration, lesser energy demands to maintain position, and increased filtering efficiency during those periods of low turbidity (Rypel et al. 2009). Growth appears higher in more productive waters and in warmer climates (Arter 1989; Morris and Corkum 1999; Schöne et al. 2007). Conversely, food limitation or suppressed water temperatures (e.g., from hypolimnetic dam release) may decrease growth rates but result in greatly increased life spans (Kesler et al. 2007; Moles and Layzer 2008). In addition to limiting mussel occurrence, low calcium concentration may result in smaller and thinner shells and lower growth rates compared to mussels in calcium-rich, well-buffered waters (Mackie and Flippance 1983). Depth, wind, and wave action may also influence growth rates in lakes (Ghent et al. 1978; Cyr 2008). Males may reach larger size and more advanced ages in some species with shell sexual dimorphism, but growth and longevity are usually similar between sexes in other species (Jones et al. 2009; Haag and Rypel 2011).

Environmental effects likely result in considerable differences in growth and longevity among mussel populations in Florida. Populations in subtropical south Florida should be expected to grow faster and have shorter lives than populations of the same species in the panhandle. Similarly, populations in eutrophic and well-buffered lakes are expected to have faster growth and shorter lives than those in poorly buffered, low productivity blackwater streams. However, excessive human-caused eutrophication and turbidity may be detrimental to growth and survival (Valdovinos and Pedreros 2007). Mussel populations in large dam tailwaters (e.g., Jim Woodruff Lock and Dam on the Apalachicola River) may have reduced growth and increased life spans because of lowered food concentration and water temperature, but smaller dams may increase growth because of phytoplankton production in the impounded area above the dam (Moles and Layzer 2008; Singer and Gangloff 2011).

Reproductive Biology

Unionid mussels are unique in possessing an obligate parasitic larval stage, typically (about 98 percent) requiring a host to complete their life cycle (Figure 6.3). The typical pattern consists of a gravid female mussel releasing glochidia that attach to the gills or fins of a host, usually a fish (Figure 6.4). After dropping off of its host, the juvenile mussel begins a period of rapid growth (Figure 6.5). A few species, however, including *Utterbackia imbecillis*, have been reported to develop directly from larvae to free-living juveniles (Howard 1914a; Heard 1975b; Watters and O'Dee 1998; Dickinson and Sietman 2008). Knowledge of reproductive biology remains incomplete for the vast majority of mussel species but has expanded dramatically in recent decades (Watters 1994a; Williams et al. 2008; Cummings and Graf 2010).

Genetics and the Sexes

Unionid mussels are diploid with 38 chromosomes (Jenkinson 1976). Nearly all mussels exhibit DUI of mtDNA, known only from several bivalve groups within the entire animal kingdom (Theologidis et al. 2008). The gender associations of DUI include two distinct mitotypes, with the female (F) type occupying gonads and somatic tissues and passed on to all their offspring, and the male (M) type being restricted to gonads and sperm and passed on only to male offspring (Hoeh 2010). Hermaphroditic species lack paternally transmitted mtDNA (Breton et al. 2011), but most species have separate sexes. Either nuclear-coded proteins or possibly the unique divergent F-type and M-type mtDNA genomes may determine sex (Hoeh 2010; Breton et al. 2011). Sex ratios are typically even despite numerous reported deviations (Cummings and Graf 2010).

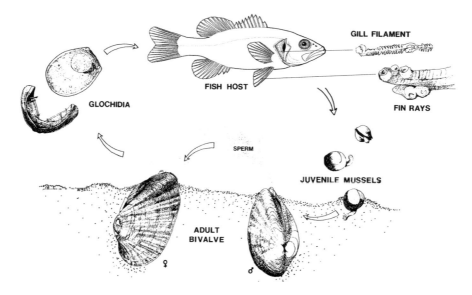

Figure 6.3. Generalized mussel life cycle.

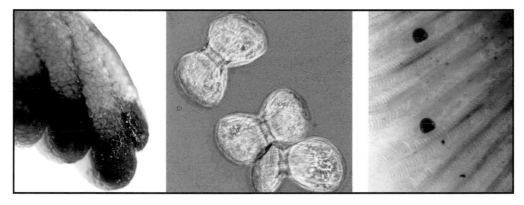

Figure 6.4. Gravid gill of *Hamiota subangulata* (left), glochidia free from gill (center), and two glochidia on gills of host fish *Micropterus salmoides* (right). Photographs by Christine O'Brien.

Hermaphroditism is exhibited in approximately 10 percent of the North American mussel fauna (Henley 2002); however, most of those species are not functionally hermaphroditic. Two types of hermaphroditism occur, sequential and simultaneous. Sequential hermaphroditism, in which sex changes at some point in life, has been reported in a few species (Downing et al. 1989) but is poorly documented and appears to be rare in mussels (Haag 2012). Simultaneous hermaphroditism, in which individuals possess gonads of both sexes at the same time, is better documented but occurs commonly in a minority of species. *Toxolasma parvum* and *Utterbackia imbecillis* are the only Florida species in which a significant proportion of the population is known to be simultaneous hermaphrodites (Sterki 1898a, 1898b; Heard 1975b; Hoeh et al. 1995). This condition is hypothesized to be beneficial in low density populations or in lentic or stressful habitats where chances of cross fertilization are low (Ghiselin 1969; Galbraith 2009; Şereflişan et al. 2009). Populations of many other Florida species occasionally have a low proportion of simultaneous hermaphrodites, including *Anodonta couperiana*, *Elliptio jayensis*, *Elliptio fumata/pullata*, *Megalonaias nervosa*, *Pyganodon grandis*, *Uniomerus tetralasmus*, and *Utterbackia peggyae* (van der Schalie 1970; Heard 1975b, 1979b; Johnston et al. 1998; Heinricher and Layzer 1999).

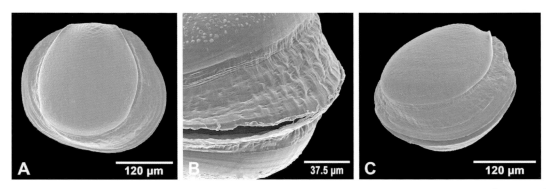

Figure 6.5. Scanning electron micrographs of juvenile *Hamiota australis* showing growth after 5 days of culture—(A) glochidial shell and growth along the margins; (B) lateral view; and (C) ventral lateral view.

Gametogenesis and Spawning

Gametogenesis, the sequential formation of gametes, appears to be regulated primarily by temperature but may also be dependent on having sufficient energy reserves (Bauer 1998; Galbraith and Vaughn 2009). The period of gametogenesis is highly variable among species (Zale and Neves 1982a; Garner et al. 1999). Mussels are sperm casters in which spawning males release sperm into the water while females retain eggs, making fertilization internal. The spawning act appears to be controlled largely by temperature and other environmental factors. Spawning is synchronous between the sexes for most species regardless of seasonality and often occurs in the fall when stream flows are relatively low (Zale and Neves 1982a; Cummings and Graf 2010). Species that spawn asynchronously have extended gamete production periods that may indicate single or multiple spawnings (Heard 1998; Henley 2002). The warm Florida climate may facilitate gametogenesis for some species through most of the year.

Most mussel species appear to release sperm in spermatozeugmata (sperm balls). Spermatozeugmata are spherical or bowl-shaped bodies up to 80 μm in diameter containing hundreds or thousands of sperm embedded in a semipermeable membrane (Waller and Lassee 1997; Ishibashi et al. 2000). These aggregations extend dispersal distance by creating a more favorable osmotic environment, enhance unidirectional movement, and potentially respond to chemical cues associated with female mussels (Barnhart and Roberts 1997; Ishibashi et al. 2000). Consequently, spermatozeugmata may facilitate fertilization of females over long distances or in sparse populations (Moles and Layzer 2008; Ferguson 2009). Spermatozeugmata are filtered from the water by female mussels where the structure then disintegrates, possibly due to osmotic pressure or a secreted compound, and individual sperm are liberated (Watters and Wolfe 2011).

Fertilization apparently occurs in the suprabranchial chamber as eggs are passed from the gonad to the marsupia. The fertilized eggs reside in the marsupial gills where they develop into glochidia (Richard et al. 1991). The portion of the gills used for brooding is modified into a marsupium, which varies among taxa. Anodontines brood glochidia in the entire outer gills, amblemines use all four gills or just the outer pair, and most lampsilines use the entire outer gills or only a portion of the outer gills. Modifications to the marsupial portion of the gill include distensible water tubes and pliable gill septa, which appear to increase tensile strength permitting larger broods (Smith et al. 2003). Anodontines develop secondary water tubes, which may lessen reduction in respiratory efficiency associated with brooding (Richard et al. 1991; Tankersley and Dimock 1992, 1993).

Most short-term brooders spawn during spring and discharge glochidia the same summer; however, some species may begin as early as late winter and last until early summer depending on environmental conditions. In long-term brooders spawning occurs during late summer or early fall and glochidia are discharged during the following winter, spring, or early summer. However, the reproductive strategy of some species is more complex. Northern populations have been reported as being either short-term or long-term brooders in both *Megalonaias nervosa* (Howard 1914b; Utterback 1916a; Woody and Holland-Bartels 1993) and *Pyganodon grandis* (Ortmann 1919; Lewis 1985). *Megalonaias nervosa* in northern Alabama is a winter short-term brooder (Haggerty et al. 2005). *Anodonta couperiana* is also a winter short-term brooder in Florida (Heard 1998).

In many mussel species, fertilization success is high even at low mussel density and in the absence of flow and is likely facilitated by spermatozeugmata (Haag and Staton 2003; Moles and Layzer 2008; Mosley et al. 2011). Fertilization or egg development may be decreased or curtailed by adverse environmental conditions, such as abnormally low temperature or eutrophication (Heinricher and Layzer 1999; Galbraith and Vaughn 2009; Gascho Landis et al. 2011). Low fertilization rates appear to be typical in species that produce conglutinates with nonfertile, structural eggs (Barnhart et al. 2008). Broods of individual females may be fertilized by multiple males (Christian et al. 2007; Ferguson 2009). Because many species spawn simultaneously, there is a possibility of hybridization between closely related species (Mock et al. 2004; Cyr et al. 2007), but the extent of hybridization is unknown. Hybridization could contribute to the difficulty of identifying some Florida mussel taxa, particularly within *Elliptio* (Williams et al. 2011). Difficulty in identifying individuals could also be the result of ecophenotypic variation (Clench and Turner 1956).

Most unionids typically produce a single brood annually (Haag and Staton 2003); however, production of multiple broods has been reported, primarily from southern species (Heard 1975b, 1998; Smith et al. 2003). Based on studies of subsets of populations, *Utterbackia peggyae* had two broods in Lake Talquin and only one brood in Holmes Creek (both sites having similar latitudes), while *Villosa villosa* sometimes had multiple broods in Lake Talquin (Heard 1975b, 1998). Most Florida species appear to discharge glochidia in summer and winter, and *Elliptio* are generally short-term brooders with multiple broods. *Elliptio jayensis* in southwestern Florida were asynchronically gravid in all months except July, which was suggestive of multiple broods produced via consecutive short-term brooding (Heard 1998). Conversely, three *Elliptio* species in South Carolina held consecutive broods less than 40 days apart (Price and Eads 2011). *Glebula rotundata* in Louisiana was reported to produce at least three broods annually (Parker et al. 1984). Species having a protracted asynchronous spawning period, characterized by the extended presence of gravid individuals, may be confused with multiple broods (Heard 1998; Smith et al. 2003). *Elliptio pullata* from Escambia River in late May had highly distended marsupial gills concurrent with abundant egg development in the gonad, which suggests multiple broods.

Glochidia and Fecundity

Centuries ago there was controversy over whether glochidia were mussel larvae or unrelated parasites, and they were formally described as a distinct species, *Glochidium parasiticum* (Heard 2000; Watters et al. 2009). According to Heard (2000), Houghton (1862) appears to be the first to experimentally confirm glochidia as parasites on fishes, and Forel (1866) provided the first description of glochidial attachment and encystment.

Unionids are larviparous, a form of ovoviviparity where partially developed young (glochidia) are discharged (Heard and Dinesen 1999). A glochidium is comprised of two chitinous, hinged valves connected by an adductor muscle, as well as mantle tissue, primordial gills, sensory hairs, and, in some taxa, a larval thread. Glochidia come in three basic forms—hooked, hookless, and ax-head-shaped (the latter only in *Potamilus*) (Lefevre and Curtis 1910). In general, hooked glochidia attach to fins of the host, while hookless glochidia attach to gills. Hooked glochidia are generally larger (approximately 300–450 μm length) than hookless glochidia (approximately 50–250 μm) (Hoggarth 1999).

After fertilization embryos rapidly develop into glochidia at warm temperatures (2–3 weeks), but maturity may be delayed at colder temperatures (Matteson 1948; Wood 1974; Bruenderman and Neves 1993). During development glochidia are nourished by energy storage cells and by contributions of calcium and other nutrients from the brooding female (Silverman et al. 1987; Schwartz and Dimock 2001; Wächtler et al. 2001). The length of time glochidia are brooded to maturity can generally be categorized as short or long term but varies among species.

Characteristics of fecundity were reviewed by Haag (2012). Annual reproductive output of glochidia varies tremendously among species (from a few thousand to over eight million). Fecundity is not strongly related to phylogenetic group, glochidia size, or life span but rather is determined primarily by body size. Mature mussels produce glochidia throughout their lives, but glochidial production may decline in very old individuals of certain species (Haag and Staton 2003). The age at reproductive maturity varies among species and is negatively correlated with growth rate and life span. In Florida relatively short-lived species such as *Anodonta*, *Pyganodon*, *Toxolasma*, and *Utterbackia* may mature as early as their first year of life, but species that live longer such as *Amblema*, *Fusconaia*, and *Quadrula* may not begin to produce glochidia until 5–8 years of age. Little is known about variation in fecundity among populations within species. However, more consistent annual glochidial production has been hypothesized for lake species relative to those that live in streams, the latter experiencing greater environmental variability (Jokela et al. 1991).

In analyzing variation in mussel life history patterns, Haag (2012) categorized three strategies—opportunistic, equilibrium, and periodic—that represent groups of species sharing similar traits (e.g., longevity, age at maturity, fecundity, size, and glochidial release strategy). Most of Florida's mussel species fall with these categories. Mussels that represent the opportunistic strategy are characterized by short life span, early maturity, high fecundity, high growth rate, generally moderate to large body size, and primarily long-term brooding. This strategy, which occurs only in anodontines and some lampsilines (e.g., *Anodonta*, *Pyganodon*, *Utterbackia*, and *Villosa*), makes excellent colonizers of new and disturbed or stressed lentic habitats typical of peninsular Florida. Conversely, species of the equilibrium strategy are characterized by long life span, late maturity, low fecundity (except for broadcasters), moderate to low growth rate, variable size except for an absence of very small species, and most (especially those species with low fecundity) are short-term brooders. Mostly amblemines are characterized by this strategy (e.g., *Fusconaia*, *Pleurobema*, and *Quadrula*) but also a few lampsiline genera (e.g., *Ptychobranchus*). These species fare well in stable, productive habitats, such as channels of larger panhandle streams. Mussels that are periodic strategists share intermediate life span, maturity and fecundity, moderate to high growth rate, generally smaller body size than opportunistic strategists, and nearly all are long-term brooders. This strategy is utilized by several anodontines and lampsilines (e.g., *Alasmidonta*, *Hamiota*, and *Villosa*) but may include amblemines (e.g., some *Elliptio*). These strategies may be adapted for smaller, generally unproductive or periodically stressed stream habitats.

Glochidial Discharge and Host Fish Attraction Strategies

Glochidial discharge appears to be triggered primarily by water temperature (Watters and O'Dee 2000), but stream discharge, host fish availability, and photoperiod may also play a role (Yokley 1972; Jokela et al. 1991; Kondo 1993). Winter and summer dischargers expel their glochidia based on temperature thresholds—winter discharge is triggered by falling temperatures and summer discharge is triggered by rising temperatures (Watters and O'Dee 2000). Some *Pyganodon grandis* populations are winter dischargers with glochidia remaining encysted for extended periods, sometimes for most of the season. This is in contrast with parent-overwintering summer dischargers, typically associated with long-term and some short-term brooders (Watters and O'Dee 2000). After discharge from the female, glochidia may live from a few hours to weeks depending upon species, temperature, and water quality (O'Brien and Williams 2002; Zimmerman and Neves 2002), though they may be physiologically able to infect hosts for only 1–2 days (Bringolf et al. 2011). Glochidial discharge and host fish attraction strategies are more variable between long-term versus short-term brooders (Culp et al. 2011). Short-term brooders' discharge periods generally last less than 12 weeks (Surber 1912; Hove and Neves 1994). Nearly continuous discharge is reported in the northernmost species of *Medionidus* (*M. conradicus*) except during gametogenesis (Zale and Neves 1982b). Release mode has been hypothesized as a major determinant of glochidial production (Culp et al. 2011). Physical disturbance, low DO, and thermal shocks may cause females to prematurely abort glochidia (Lefevre and Curtis 1910; Aldridge and McIvor 2003).

Glochidial discharge and host fish attraction strategies take several fascinating forms in two broad categories, passive and active (Barnhart et al. 2008). Passive discharge involves simple expulsion of glochidia; four general categories of active dischargers are known and most of the described methods can be found among Florida taxa.

Most anodontines and several amblemines (e.g., *Amblema*, *Elliptoideus*, and *Megalonaias*) broadcast masses of glochidia, either enmeshed in mucus strands or freely discharged, that rely on host fish encounters (Dartnall and Walkey 1979; O'Brien and Williams 2002; Haag and Warren 2003). Broadcasters typically have glochidia with larval threads that increase drag and buoyancy, thus enhancing probability of pelagic host contact (Haag and Warren 1997; Roberts and Barnhart 1999; Wächtler et al. 2001).

Some amblemines (e.g., *Elliptio*, *Fusconaia*, and *Pleurobema*) and lampsilines (e.g., *Ptychobranchus*) package their glochidia in conglutinates. These structures are formed in the water tubes of the gills and expelled from the excurrent aperture, and the glochidia parasitize gill tissues of the host (Ortmann 1910b, 1911; Watters 2008). Conglutinate color and morphology varies tremendously among taxonomic groups; however, all variations are actively foraged by fish (Haag and Warren 2003; Watters 2008). *Ptychobranchus* expel highly complex, adhesive conglutinates that resemble fish fry, insect larvae, or pupae (Hartfield and Hartfield 1996; Watters 1999). Species that discharge conglutinates tend to produce considerably fewer glochidia than broadcasters or those that employ a mantle lure. It has been hypothesized that conglutinates increase host infection probability by prolonging glochidial activity (Neves and Widlak 1988; Haag and Staton 2003; Barnhart et al. 2008).

Some *Quadrula* have swollen mantle margins associated with the excurrent aperture that serve as temporary storage compartments for conglutinates (Kurth and Hove 1997; Heath et al. 1998). Storage of glochidia in these "mantle magazines" allows rapid discharge in response to host fish (usually catfishes) attack (Barnhart et al. 2008). Other *Quadrula* have colorful, anatomically complex excurrent apertures that appear to be visual lures for attracting cyprinid hosts (Butler 2005; Fobian 2007). Both groups are reflexive releasers that rapidly discharge conglutinates, conglutinate fragments, or masses of glochidia in response to shadows, vibrations, tactile stimulation, or possibly chemical cues (Lefevre and Curtis 1910; Barnhart et al. 2008).

Species of the recently described genus *Hamiota* produce superconglutinates. This mode of glochidial discharge was discovered by R.S. Butler and W.R. Hoeh in late June 1988, being utilized by *Hamiota australis* in Limestone Creek, Walton County, Florida (Roe and Hartfield 2005). The term "superconglutinate" was coined by Williams and Butler (1994). A superconglutinate is composed of individual conglutinates (*sensu* Watters 2008) of both marsupialized outer gills tethered by two fused transparent mucilaginous tubes (which may reach approximately 2.5 m in length) apparently secreted from the suprabranchial chambers (Haag et al. 1995; Hartfield and Butler 1997). The superconglutinate resembles a minnow in size, shape, and coloration, complete with lateral stripes and eyespots (Figure 6.6). This means of host attraction is analogous to anglers using an artificial lure on monofilament line, but *Hamiota* (of the feminine gender and translating from the Greek to "hook") have been using this method for millions of years to "catch" fish. Interestingly, *Hamiota* may employ a bimodal infestation strategy, also exhibiting mantle flaps, but these structures are typically reduced (Hartfield and Butler 1997; Haag et al. 1999).

Some lampsiline genera have mantle modifications exhibiting rhythmic contractions that facilitate host fish infection (Barnhart et al. 2008). This display behavior is only exhibited periodically during their brooding period. Mantle lures include flaps (e.g., *Lampsilis*), villi (e.g., *Villosa*), caruncles (e.g., *Toxolasma*), and other modifications. Lure display appears to be influenced by temperature and day length (Gascho Landis et al. 2012). Besides movements, these structures may exhibit bright colors or actual mimicry of various fish prey items that serve to attract host fishes (Barnhart and Roberts 1997; Haag and Warren 1999). Mantle flaps in *Lampsilis* resemble fishes in shape, color, and rhythmic movements to attract large predatory fishes, such as basses and sunfishes (Barnhart et al. 2008). Some investigators maintained that glochidia are released by the bursting of the extruded marsupium when the female snaps her valves shut in response to physical stimuli (Simpson 1899; Watters and O'Dee 1996; Watters et al. 2009). Observations originally by Grier (1926) and subsequently by Morrison (1973) and Barnhart et al. (2008) demonstrated that host fish attack and rupture the marsupium, thereby releasing glochidia. To attract darters *Medionidus* lie on the substrate, sometimes upside-down, displaying contrasting colors between the mantle and marsupia or variably colored mantle margins that pulsate rhythmically (Zale and Neves 1982b; Brim Box and Williams 2000; O'Brien and Williams 2002; Haag and Warren 2003). The thumblike caruncles of *Toxolasma* exhibit a "twiddling" motion while the ventral mantle margin ripples rhythmically (Howells et al. 1996; Schwegman 1998). *Villosa* display moving villi up to 1.5 cm in length (Haag et al. 1999). *Obovaria* have reduced pulsating mantle lures (C.M. Barnhart, personal communication).

Figure 6.6. *Hamiota subangulata* in an artificial stream, discharging a superconglutinate. Individual from Flint River drainage, Georgia, May 1995. Photograph by Noel M. Burkhead.

Host capture is another highly unique host attraction behavior recently discovered in the genus *Epioblasma*, which are absent from Florida. Gravid females lure foraging darters into their shell gape, then they clamp their valves closed, holding the fish while glochidia are discharged onto its gills (Jones et al. 2006a). Unrelated to host capture, incidental fish capture has been reported in *Elliptio* (Cordeiro 2007).

Mussels have several other traits that promote glochidial attachment. Researchers have suggested that some female mussels may sense chemical cues when host fish are near (Jokela and Palokangas 1993; Pepi and Hove 1997; Haag and Warren 2000). Glochidial discharge in *Anodonta* increased in laboratory tests when host fish were present versus absent (Jansen et al. 2001). Similar results were reported for *Villosa nebulosa* and *Villosa vibex*, but lures were displayed even in the absence of fish (Haag and Warren 2000). *Villosa vibex* physically contacted by suitable host fish discharged 20–80 times more glochidia and with greater frequency than with nonhosts. Most species display diurnally, but some are known to display nocturnally (e.g., *Lampsilis floridensis*) (Brim Box and Williams 2000).

Host attraction strategies among unionids were speculated to have evolved to exploit feeding guilds and predator-prey relationships among fishes (Haag and Warren 1999). Active fish-attracting lures, particularly mantle flaps, arose early in mussel evolution (Zanatta and Murphy 2006). Strategies of some species shifted to complex conglutinates (*Ptychobranchus*), superconglutinates (*Hamiota*), and host capture (*Epioblasma*), which appears to be the most evolutionarily recent strategy in the group.

Host Fishes and the Parasitic Stage

Most North American mussel species are host specialists to some degree, typically using one to a few host fishes (Haag 2012). Cyprinidae (shiners and minnows), Ictaluridae (catfishes), Centrarchidae (sunfishes and black basses), Percidae (darters and perches), Cottidae (sculpins), and Sciaenidae (drums) represent major fish families for host specialists (Fuller 1974; Hoggarth 1992). These groups of fishes are abundantly represented in Florida except for cottids and the molluscivorous *Aplodinotus grunniens* (Freshwater Drum). Panhandle streams with diverse riverine fish faunas are dominated by host specialists (e.g., *Fusconaia*, *Pleurobema*, and *Quadrula*), whereas host generalists (e.g., *Anodonta*) are predominant in peninsular waters with its depauperate fish fauna and in lentic habitats in the panhandle. Typically, amblemine specialists use cyprinids and ictalurids (O'Brien and Williams 2002; Steingraber et al. 2007; White et al. 2008), while lampsilines use centrarchids and percids (Zale and Neves 1982b; O'Brien and Brim Box 1999). Among *Lampsilis* and *Villosa* using centrarchids, *Micropterus* tend to be primary hosts and *Lepomis* secondary, or marginal, hosts (Haag et al. 1999). Because they depend on their glochidia encountering a particular fish species, specialists typically have more highly developed host attraction strategies (e.g., *Hamiota*) compared to generalists, many of which are broadcasters (e.g., *Pyganodon*).

Host specificity is considered immunologically based and categorized as either natural or acquired (Arey 1932a; Watters and O'Dee 1996; Kirk and Layzer 1997). In natural immunity the nonhost fish's immune system causes rejection and sloughing of glochidia within days (Yeager and Neves 1986; Waller and Mitchell 1989), while in acquired immunity previously infected fish produce antiglochidial antibodies that reduce or prevent subsequent infections (Bauer and Vogel 1987; O'Connell and Neves 1999). Developing acquired immunity to one species appears to make fish immune to glochidia of other species, particularly among gill parasites (Reuling 1919; Dodd et al. 2005). The ability of mussel glochidia to circumvent the host immune response is species dependent, varying from specialists that transform on only one or a few fish species (usually within a genus or family of fishes) to generalists that can use many species. For example, glochidia of *Elliptio crassidens* are known to transform only on *Alosa chrysochloris* (Skipjack Herring), but *Pyganodon grandis* appears able to transform on nearly any species. Amphibians may serve as hosts in the laboratory in some cases but are likely unimportant, if used at all, in the wild (Watters and O'Dee 1998).

Host fishes are typically determined by conducting laboratory trials to assess the ability of glochidia to develop on various fishes or by examining natural infestations among wild fishes. However, the latter method is hampered by difficulty identifying many glochidia to species and the uncertainty about whether or not glochidia will complete development on the fish with which it was observed. Recent advances in DNA barcoding methods have allowed researchers to identify unionid larvae to the species level (Boyer et al. 2011). Transformation of glochidia is most robust and consistent on primary host species, but secondary hosts may allow some glochidial transformation (Haag and Warren 1997; O'Brien and Williams 2002).

Infestation rates under natural conditions are typically low (Strayer 2008). The number of glochidia on individual fish is highly variable but often less than 25 glochidia attach per host (Stern and Felder 1978; Neves and Widlak 1988). Infestation rates may be extremely low (1–5 per fish) in riverine species (Bruenderman and Neves 1993). Hundreds or thousands of glochidia per individual fish may occur on large fishes, especially at high mussel densities or during laboratory infestations (Surber 1913; Stern and Felder 1978; Blažek and Gelnar 2006). Juvenile fishes may represent the most naive hosts for some mussels since they presumably lack acquired immunity; however, their relatively small size, reduced gill surface (allowing them to carry fewer glochidia), and their occurrence not coinciding with the timing of glochidial release may limit their success as hosts (Jokela et al. 1991; Bauer 1994; McLain and Ross 2005).

The mechanics of glochidial attachment were reviewed in Arey (1924) and Hoggarth and Gaunt (1988). Clamping is posited to be triggered by mechanical stimulation in hooked glochidia and chemical cues in hookless glochidia (Lefevre and Curtis 1910, 1912; Shadoan and Dimock 2000). The host fish tissue clasped between the valves of the glochidium is assimilated and nutrients are likely obtained (Arey 1932b, 1932c; Fisher and Dimock 2002a, 2002b). Energy gained from carbon uptake, host nutrients, and digestion of larval organs is needed to complete metamorphosis (Jansen et al. 2001). Most encysted glochidia generally do not grow, but species with small glochidia (less than 100 μm) may double their size (Barnhart et al. 2008). It has been suggested that glochidia attached to gills are more dependent upon host nutrients than are skin parasites (Blystad 1923).

The period of encystment is temperature dependent and lasts a few weeks but possibly months, particularly in overwintering species (Watters and O'Dee 2000). Developmental stages of encysted glochidia progress from absorption of larval adductor muscle, "mushroom body" formation, digestion of host tissues, and development of other structures characteristic of adults (Blystad 1923; Fisher and Dimock 2002b). Excystment occurs when temperatures are favorable and their foot movements rupture thinning capsule walls (Jansen et al. 2001).

Factors Influencing Mussel Population Structure

The relationship between mussels and their hosts appears to play a larger role in structuring mussel assemblages than does habitat or foods. This is due to the high degree of host specialization (Haag 2012). Mussel-host relationships are very important in mussel community structuring at various levels (e.g., mussel species richness is related to host diversity and availability, mussel abundance is related to host abundance, certain host infection strategies are more effective than others across habitats). Dynamics of glochidia and host fish populations may be the driving force affecting recruitment in lakes (Jansen and Hanson 1991; Jokela et al. 1991). Common, widespread mussel species typically use common and widespread fish as hosts (Vaughn 1997). Host mobility logically tracks mussel dispersal capability, especially upstream (McLain and Ross 2005;

Woolnough 2006). Dispersal capability is reduced for mussels using small, low-mobility fishes like darters and sculpins, which may restrict colonization potential, impede recovery, and increase extinction risk (Zale and Neves 1982a, 1982b; Haag and Warren 1997; Schwalb 2009). Jokela et al. (1991) hypothesized that where host-fish communities change seasonally, selection should favor mussels that synchronize glochidial discharge with host availability. Spatial and seasonal differences in glochidial infestation patterns potentially limit competition for host fishes (Rashleigh and DeAngelis 2007; Haag 2012). Strayer (2008) speculated that fish immunity may result in intraspecific competition and that the immunological status of host fishes may range from entire populations of immunologically naive species to populations of older individuals having some level of acquired resistance.

Several factors besides host fish characteristics affect mussel demographics (e.g., hydrologic characteristics, hydraulic variables, habitat type and quality, temperature, fertilization and glochidial survival rates, longevity, and predation) (Vaughn and Taylor 1999; Payne and Miller 2000; Haag 2002; Meador et al. 2011). Flow conditions during critical early life history stages may be the most important physical factor affecting recruitment in streams (Johnson and Brown 1998; Peterson et al. 2011). Low discharge periods may enhance juvenile settlement and survival (Payne and Miller 2000; Jones and Neves 2011); however, low flow conditions may also be detrimental to recruitment due to high temperatures, low DO, and stranding (Johnson 2001; Golladay et al. 2004; Haag and Warren 2008; Galbraith 2009). Floods may result in significant downstream dispersal (Imlay 1982; Bauer 2001). Sperm probably travels very short distances, limiting dispersal capability, especially in lentic systems (Lefevre and Curtis 1910; Bauer 1987; Downing and Downing 1993). Extended glochidial viability promotes dispersal, especially in lotic systems (Zimmerman and Neves 2002). Although modeling simulations predict long-distance juvenile transport (Morales et al. 2006; Daraio et al. 2010), dispersal distances from lab and field experiments are equivocal (Ferguson 2009; Schwalb 2009).

Recruitment rates display two primary patterns in mussels: high and variable, and low and constant (Haag 2012). Recruitment also appears to be negatively related to longevity and positively related to fecundity. High recruitment rates may be expected in fast-growing, short-lived, and high-fecundity species, with the potential for dramatic short-term swings in age structure (Haag and Warren 2007, 2010). Recruitment variability appears normal for long-lived, lotic species, but strong cohorts may dominate age-class structure for years (Hove and Neves 1994; Payne and Miller 2000; Negishi et al. 2011). Mussel recruitment increased with host abundance in ponds for both a glochidia broadcaster and lure displayer (Haag and Stoeckel 2011).

Intermittent recruitment may characterize mussel populations affected by pollutants and altered habitats (Patzner and Müller 2001; Haag and Warren 2010; Wang et al. 2010; Gascho Landis et al. 2011). Drastically altered density and diversity of host fishes and/ or loss of migratory species above dams may result in recruitment failure (Smith 1985; Kelner and Sietman 2000; Locke et al. 2003). It is possible that the Ochlockonee River endemic *Alasmidonta wrightiana* became extinct due to construction of Jackson Bluff Dam in 1927. It may have been dependent on a migratory fish host that was blocked by the dam. The last collections of this presumably short-lived anodontine were made in 1931 directly upstream of Lake Talquin (Williams and Butler 1994). Disruption of any life stage could constrain recruitment, potentially leading to recruitment failure and gradually expiring populations (Saha and Layzer 2008; Strayer 2008).

Mortality appears to be very high for early life stages of all species and adults of short-lived species (Bauer 2001; Haag 2012). Mortality rates approaching 100 percent may occur for newly discharged glochidia (Jansen et al. 2001). Prodigious numbers of glochidia are therefore produced to offset natural mortality. A Canadian lake *Pyganodon grandis* population exhibited an encystment rate of only 0.007 percent (Jansen and Hanson 1991), and the survival rate for a glochidium to free-living juvenile was an estimated 0.000001. Glochidial survival rates were 0.00001–0.00004 in several southern amblemines (Haag 2002). Low rates of recruitment may be partially compensated for by longevity, high fecundity, and a lifetime of reproductive potential (Jokela et al. 1991; Akiyama and Iwakuma 2007). Dislodgement of adult mussels during stream floods and lake storms may result in high mortality (Brown et al. 1938; Hastie et al. 2001). Females may experience increased predation risk during lure display (Jansen et al. 2001; Crabtree and Smith 2009). Models suggested a strong negative relationship between high spring and summer flows and survival in *Hamiota subangulata*, *Medionidus penicillatus*, and *Pleurobema pyriforme* in Georgia due to their susceptibility to dislodgement and mortality when individuals are at the surface for spawning and glochidial discharge (Peterson et al. 2011).

Some patterns of mussel assemblage structure and abundance are discernible across Florida waters and reflect similarities in life history strategies and habitat. Streams in the Greater Floridan Region (slightly modified from the Eastern Gulf Region as defined by Haag 2010), especially the Florida panhandle, are dominated by a suite of species similar to that found in small to midsize Mississippian Region streams (e.g., *Amblema* spp., *Elliptio* spp., *Pleurobema* spp., and *Quadrula* spp.) (Haag 2012). These species are mostly periodic strategists with equilibrium life history strategies, including many specialists with limited host breadth (except for the generalist *Amblema plicata*). Periodic strategists tend to be more adapted to small streams where mussel populations are unstable due in part to susceptibility to intermittent water level changes and variable host fish abundance. Equilibrium strategists are more prevalent in larger streams having relatively stable habitats and higher host fish diversity and abundance.

There are other similarities between the mussel faunas of the Greater Floridan and Mississippian Regions. Relatively few species tend to dominate in both regions and species evenness in populations is generally high. However, mussel diversity and abundance increases sharply with stream size in the Mississippian Region, but this pattern is not obvious in the Greater Floridan Region (Haag 2012). Large panhandle rivers tend to have slightly higher species richness than smaller streams because of the higher diversity of habitat types and the presence of large-river specialists, such as *Amblema neislerii*, *Elliptoideus sloatianus*, *Reginaia rotulata*, and *Megalonaias nervosa*. However, mussel richness and abundance in small streams is often comparable to midsize streams across much of the state. This is potentially due to the presence of large headwater wetlands in many watersheds that may result in relatively high, stabilized flows in otherwise small stream channels.

In lentic habitats that are prevalent in the peninsular landscape and lentic habitats in panhandle streams (e.g., stream margins and backwaters), opportunistic life history strategists dominate (primarily anodontines) (Haag 2012). Most of these species appear to be host generalists, but these assemblages may also include some host specialists that use lures to attract hosts (e.g., *Toxolasma* and *Villosa*). Assemblages comprised of lentic-adapted species may be constrained by limited habitat types, absence of riverine host fishes, and restricted dispersal opportunities relative to lotic species. Consequently, lentic mussel assemblages throughout the state are generally restricted to some combination of *Anodonta*, *Elliptio*, *Toxolasma*, *Uniomerus*, *Utterbackia*, and *Villosa*. These species, particularly the anodontines, have several adaptations that allow them to colonize and persist in lentic habitats, including high individual and population growth rates, tolerance to seasonal hypoxia (prevalent in warm, shallow lakes), and use of lentic-adapted host fishes or a wide spectrum of hosts (Chen et al. 2001; Crail et al. 2011; Haag 2012). Many of these species adapt to disturbed and stressed sites and readily invade newly created habitats such as reservoirs, channelized streams, and canals. Fast growth rates and high energy requirements of anodontines may play a role in restricting them to depositional habitats along stream banks (Haag 2012). Even though species richness is typically highest in lotic situations, peninsular lakes may have species richness comparable to small and midsize streams in the same area. *Elliptio jayensis* is clearly the dominant species in many peninsular lakes. Therefore, these waters will have an expected reduction of evenness scores relative to the panhandle but more similar to Atlantic Slope Region streams where *Elliptio complanata* easily dominates mussel assemblages (Haag 2012).

Predation

The role of predators in regulating mussel populations is poorly studied but may have significant ramifications in some circumstances (Haag 2012). The Muskrat is the most conspicuous predator of mussels in North America but is absent from Florida. Florida vertebrates that feed on mussels include mammals (e.g., Raccoon, Mink, River Otter, Striped Skunk, feral hogs, and rats), birds (e.g., Limpkin, Boat-tailed Grackle, crows, and waterfowl), reptiles (e.g., American Alligator and turtles), amphibians (e.g., amphiumas, sirens, salamanders, and frogs), and fishes (e.g., Redear Sunfish, catfishes, and bullheads) (Fuller 1974; McDiarmid 1978; Trautman 1981; Rice 2004). Among invertebrates, flatworms, hydra, chironomid and dragonfly larvae, and crayfish are predators on juvenile mussels (Kunz 1898; Klocker and Strayer 2004). In some situations fishes may consume large numbers of mussels, particularly thin-shelled species and juvenile individuals (Haag 2012). Fish may also represent a major source of predation on glochidia (Jansen et al. 2001). Chironomid and dragonfly larvae are known to ingest gill tissues, which may affect respiration and impact the reproductive cycle when marsupial structures are damaged (Gordon et al. 1978; Levine et al. 2009).

The Limpkin is an example of a predator that may influence mussel populations in some situations. During several weeks of abnormally low water in May and June 1987, a group of five Limpkins was frequently observed probing for mussels with their long bills in emergent grasses along the sandy shallows of Lake Kissimmee, Osceola County (Figure 6.7). When birds encountered a mussel, they ran with it to feeding sites, opened the shell by delivering a series of blows to the posterior end, snipped the adductor muscle with their mandible, and then consumed the soft parts. The birds moved on after several weeks, leaving paired valves of three species (950 *Elliptio jayensis*, 63 *Villosa amygdalum*, and 8 *Utterbackia imbecillis*) at their feeding sites along 50 m of shoreline.

Similar evidence of Limpkin predation on mussels has been observed in other parts of peninsular Florida.

Figure 6.7. Limpkin feeding on *Elliptio jayensis* along the shore of Lake Kissimmee, Osceola County, Florida, June 1988. Photograph by Robert S. Butler.

Parasites and Diseases

Mussel parasites include bacteria, bryozoans, protozoans, roundworms, flatworms, trematodes, oligochaetes, leeches, copepods, water mites, and insect larvae (Beedham 1971; Fuller 1974; Gordon et al. 1978; Grizzle and Brunner 2007). Some invertebrates (e.g., bryozoans, oligochaetes, leeches, water mites, and insects) may actually be commensal rather than parasitic (Heard and Dinesen 1999; Cummings and Graf 2010). Monogenic trematodes are common mussel parasites that cause sterility and shell deformations (Hendrix et al. 1985; Jokela et al. 1993; Benz and Curren 1997). Trematode infection varies among species but normally affects a small percentage of individuals in a population (Haag and Staton 2003). Numerous unionicolid mite species parasitize mussel gill and mantle tissues; some are specific to certain mussel species, but others are host generalists (Vidrine 1989, 1996). The density of mite infestations and prevalence across species appears to be relatively high in Florida mussels. Parasitism by mites can lower reproductive output and physiological condition (Gangloff et al. 2008). Parasitic copepods may infect mussel gills (Taskinen and Saarinen 1999). Chironomids are known from Florida *Utterbackia*, but their effects on mussels are unknown (Roback et al. 1979).

The occurrence of diseases in mussels is virtually unknown (Grizzle and Brunner 2007; Cummings and Graf 2010). Spongy disease, potentially caused by protozoans, was reported from *Margaritifera falcata* and produced lesions on the foot that prevented normal burrowing (Pauley 1968). Mass mussel die-offs have been documented in many parts of the United States, but it is unknown if these were caused by disease (Neves 1986). Introduction of parasites and diseases with alien species is poorly understood but may potentially have devastating effects on native mussel populations (Strayer 1999a, 2010; Poulin et al. 2011).

Ecological Role of Mussels

Where abundant (Figure 6.1), mussels may represent the largest invertebrate biomass component of the aquatic ecosystem, sometimes by an order of magnitude (Layzer et al. 1993; Howard and Cuffey 2006). Total mussel production may equal that of the remainder of the combined macrobenthic community (Strayer et al. 1994). In some situations large mussel populations can daily filter nearly the entire water volume of streams or lakes, especially during low-water conditions, making them integral in nutrient cycling (Strayer et al. 1994; Vaughn et al. 2004). This tremendous filtering potential can increase water clarity, which aids sight-feeding fishes and increases visibility of host-attracting structures such as mantle lures and conglutinates (Nezlin et al. 1994). Mussel density and biomass is often positively correlated with species richness and abundance of other aquatic organisms (Vaughn and Spooner 2006a; Aldridge et al. 2007). The many important roles that mussels play suggest that they are keystone species that greatly influence aquatic ecosystems and improve environmental conditions.

Mussels influence nutrient cycling and energy transfer at several levels. Filter-feeding mussels remove algae and suspended organic matter from the water column and convert it to feces and pseudofeces, which are easily uptaken by primary producers (Ostrovsky et al. 1993; Hauswald 1997; Vaughn et al. 2004). Dense populations of large, long-lived species (e.g., *Elliptio crassidens*, circa 70 years, in several panhandle rivers) represent significant stores of nutrients eventually released from mussel tissue decomposition and shell dissolution (Strayer and Malcom 2007; Atkinson et al. 2010). In addition to having positive impacts on other organisms, nutrient cycling by mussels may promote the establishment and maintenance of healthy, diverse mussel assemblages. Mussels are also important in bioturbation as burrowing mixes and releases nutrients from the substrate to the water column (Vaughn and Hakencamp 2001; Meysman et al. 2006). Within an assemblage the combined feeding and bioturbation activities of different species with different physiological optima may facilitate higher mussel growth and survival under a wider range of environmental conditions (Vaughn et al. 2008; Spooner and Vaughn 2009). These activities stimulate primary and secondary production across trophic levels and may be especially important where nutrients are limiting (Waller et al. 1999; Vaughn et al. 2007).

Mussels also serve as ecosystem engineers by creating, modifying, or maintaining physical habitat (Gutiérrez et al. 2003; Spooner and Vaughn 2006; Vaughn et al. 2008). Shell production per unit area by abundant populations in hard-water streams and lakes may rival that of wood production by temperate forest trees (Gutiérrez et al. 2003; Strayer and Malcom 2007). Shell material and buried live mussels may stabilize substrates and provide important substrate and shelter for a wide variety of macroinvertebrates and other organisms (Vaughn et al. 2002; Strayer and Malcom 2007; Zimmerman and de Szalay 2007). Increased habitat heterogeneity provided by live and dead mussel shells may be particularly important in regions like peninsular Florida that lack hard substrates (Beckett et al. 1996). Recreational anglers in Florida, particularly those targeting Redear Sunfish (locally known as "shellcrackers"), learned long ago that "shell beds" in peninsular Florida lakes are especially productive fishing spots.

Clearly, large, healthy mussel populations greatly influence aquatic ecosystems. The widespread decline in mussel populations over the past century has likely been accompanied by profound loss of ecosystem services, which has affected humans in ways we are only beginning to understand (Vaughn and Hakenkamp 2001; Vaughn 2010). It will take an extraordinary effort to restore mussel populations to a point where ecosystem services are even a fraction of what they were historically (Vaughn et al. 2004; Haag 2012).

Chapter 7

Mussel Conservation

The continental United States has the highest diversity of freshwater mussels in the world, with approximately 300 species in two families, Margaritiferidae and Unionidae. The southeastern United States is a mussel biodiversity hot spot harboring 94 percent of the 300 species, including 98 percent of the taxa listed as federally endangered or threatened (Williams et al. 2008). About 30 southeastern species are now considered extinct (Haag 2009). Decline of mussels over the past century and their level of imperilment have been greater than that of any other wide-ranging faunal group in North America (Strayer et al. 2004; Strayer and Dudgeon 2010). The loss of mussel diversity reflects a global pattern of freshwater mollusk imperilment (Abell et al. 2000). Freshwater mollusks currently have the highest global extinction rate of any group of organisms, far exceeding that of terrestrial faunas (Ricciardi and Rasmussen 1999; Régnier et al. 2009). Mussels have extinction rates three times higher than estimations for other species over geological time (Nott et al. 1995). The level of mussel imperilment continues to increase due to habitat alteration and population fragmentation (Haag 2009; Williams et al. in review).

The complex life history of mussels has served them well for millions of years but recently has become a handicap for many species, limiting successful reproduction and recruitment. Many mussels are host specific and have narrow windows of opportunity for successful reproduction (Barnhart et al. 2008; Haag 2012). Strayer (2008) surmised that mussel size, number of host fishes, and their dispersal capability are significant predictors of conservation status. Increased sensitivity to anthropogenic stressors, particularly at early life stages, represents a major limitation to recruitment (Cope et al. 2008). The limited ability of mussels to seek refuge from threats and their susceptibility to habitat alterations at all life stages also play key roles in their imperilment (Galbraith and Vaughn 2011). The multitude of complex stressors in aquatic environments may combine to compromise reproduction and recruitment, reduce population viability, and drive populations to extirpation and ultimately species to extinction (Ormerod et al. 2010).

Concern for mussel populations due to anthropogenic impacts began over 150 years ago (Higgins 1858). Habitat alteration from disposal of municipal and industrial wastes was a major contributing factor to early mussel declines (Simpson 1899; Ortmann 1909a, 1924b, 1925). Harvest of pearls and shell for pearl buttons was also cited as a causal factor for the decline of mussels (Coker 1919; Williams et al. 2008). In the early 1900s, biologists at the U.S. Bureau of Fisheries became alarmed at the depletion of mussel resources (Ellis 1931a, 1931b). A search for management options to reverse the decline of mussel populations prompted years of research on life history and culture at the U.S. Bureau of Fisheries laboratory in Fairport, Iowa. Initial goals of the Fairport lab were to replenish depleted mussel stocks and propagate fishes (Coker 1914). Seminal works of Lefevre and Curtis (1912) and Coker et al. (1921) contributed to the foundation of modern mussel conservation biology. Research by many scientists continued until the lab closed in 1933 (Pritchard 2001).

Over a century of mussel decline occurred before the issue of imperiled mollusks received significant attention. In 1968 the American Malacological Union hosted a symposium on rare and endangered aquatic mollusks, which resulted in a series of papers published in the scientific journal *Malacologia*, including the first list of extinct and imperiled mussels (Table 7.1). The following year USFWS convened a meeting to discuss management and protection options for imperiled species (Stansbery 1971). The renewed interest in conservation during the late 1960s and early 1970s resulted in national legislation culminating, in part, with the Endangered Species Act of 1973. This act directed the USFWS to identify, list, and recover endangered and threatened terrestrial and freshwater species. As of 2012 a total of 87 species of mussels (29 percent of the fauna) have been added to the federal list, of which 15 (25 percent) occur in Florida.

In 1993 the American Fisheries Society Endangered Species Committee reviewed the conservation status of mussels in the United States and Canada (Williams et al. 1993). The review determined that the imperilment rate for the Florida fauna exceeded 50 percent. The committee recently reconvened and generated a draft list updating the status assessment (Williams et al. in review). Of the 42 Florida taxa recognized in both assessments (after taxonomic and nomenclatural changes were considered), the status of 9 (21 percent) declined over the nearly two-decade period. This increased the imperilment rate of the 60 currently recognized Florida taxa to 60 percent.

Mussel Conservation in Florida

The first conservation status assessment of Florida mussels was produced by William H. Heard under contract with USFWS (Heard 1975a). Approximately 50 Coastal Plain sites were sampled from Alabama to South Carolina, including 22 sites in all major drainages of north Florida. In the late 1970s the Florida Committee on Rare and Endangered Plants and Animals assessed the conservation status of Florida biota (Franz 1982). Only three mussels were recognized—*Obovaria choctawensis* as threatened and *Obovaria haddletoni* and *Ptychobranchus jonesi* as special concern (Thompson 1982). An updated second edition in the early 1990s provided an annotated list of 28 mussel taxa, including 5 endangered, 14 threatened, 5 special concern, and 4 undetermined status (Williams and Butler 1994). The increase in number of species between the two assessments primarily reflected the increased knowledge of distribution and threats in Florida streams. During the past four decades, there have been several conservation status reviews that included Florida mussels (Table 7.1). These status reviews have been instrumental in providing guidance in the management and conservation of Florida mussels.

Increasing interest in the imperiled mussel fauna in Florida prompted FWC in 1995 to establish a rule to protect mussels. Under the rule, commercial harvest of unionids, either living or dead, is prohibited. The rule also established a daily "bag limit" and a "possession limit" of unionid mussels for personal use (e.g., fish bait). The one-day bag limit is 10 live unionid mussels or 20 half shells (or 20 valves). A person may not be in possession of more than two-day's bag limit (20 live mussels or 40 half shells). Freshwater bivalves not belonging to Margaritiferidae or Unionidae (e.g., Cyrenidae and Sphaeriidae) are unregulated and may be taken for personal use. Mussels may only be collected by hand since mechanical methods are prohibited. Permits for taking unionid mussels in excess of the numbers specified may be issued for scientific purposes only.

Table 7.1. Conservation status of Florida mussels as reported in publications between the years 1970 and 2012. Status categories are as follows: E = Endangered, PX = Possibly Extinct, R = Rare, SC = Special Concern, T = Threatened, U = Undetermined, V = Vulnerable, and X = Extinct. Years correspond to the following publications: Athearn (1970), Stansbery (1971), Thompson (1982), Williams et al. (1993), Williams and Butler (1994), and USFWS (1998, 2012).

Taxa	1970	1971	1982	1993	1994	1998	2012
Alasmidonta triangulata	R&E	R&E		SC	E[1]		
Alasmidonta wrightiana				E-PX	U		
Amblema neislerii		R&E		E	T	E	
Amblema plicata					U[2]		
Anodonta hartfieldorum							
Anodonta heardi					E[3]		
Anodonta suborbiculata					T		
Anodontoides radiatus				SC	T		
Elliptio ahenea				SC			
Elliptio arctata				SC			
Elliptio chipolaensis				T	T	T	
Elliptio crassidens							
Elliptio fraterna				E			
Elliptio jayensis				U			
Elliptio mcmichaeli		R&E		SC			
Elliptio monroensis							
Elliptoideus sloatianus	R&E	R&E		T	T	T	
Fusconaia burkei	R&E[4]	R&E[4]		T[4]	T[4]		T
Fusconaia escambia				T	T		T
Glebula rotundata					SC		
Hamiota australis	R&E[5]	R&E[5]		T[5]	T[5]		T
Hamiota subangulata				T[6]	SC[6]	E	
Lampsilis ornata				SC	T		
Medionidus acutissimus				T		T	
Medionidus penicillatus	R&E	R&E		E	T	E	
Medionidus simpsonianus				E	E	E	
Medionidus walkeri				T	T		
Megalonaias nervosa					SC[7]		
Obovaria choctawensis			T[8]	T[8]	T[8]		E[8]
Obovaria haddletoni			SC[9]				
Plectomerus dombeyanus					U		
Pleurobema pyriforme	R&E	R&E		E	T	E	
Pleurobema strodeanum					T		T
Ptychobranchus jonesi	R&E	R&E	SC	T	U		E
Quadrula infucata				SC			
Quadrula kleiniana							
Quadrula succissa				SC[10]	SC[10]		
Reginaia apalachicola							
Reginaia rotulata		R&E[11]		E[11]	E[11]		E[11]
Utterbackia peninsularis							
Villosa villosa				SC			

[1]reported as *Alasmidonta undulata*; [2]reported as *Amblema perplicata*; [3]reported as *Anodonta* sp. (undescribed); [4]reported as *Quincuncina burkei*; [5]reported as *Villosa australis*; [6]reported as *Villosa subangulata*; [7]reported as *Megalonaias boykiniana*; [8]reported as *Villosa choctawensis*; [9]reported as *Lampsilis haddletoni*; [10]reported as *Fusconaia succissa*; [11]reported as *Fusconaia rotulata*.

The USFWS office in Panama City, Florida, has the lead for 15 federally protected endangered and threatened mussels. These 15 species occur in one or more of the major Gulf Coast basins in Florida between the Escambia and Hillsborough Rivers (Table 7.2). In 1998 the USFWS listed seven endangered and threatened mussels that were known to occur in the Apalachicola, Ochlockonee, and Suwannee River basins (USFWS 1998). Critical habitat was not included in the original listing but was subsequently designated in 2007 (USFWS 2007). One threatened mussel, *Medionidus acutissimus*, was listed in 1993, but Florida was not originally included in its range (USFWS 1993). Reevaluation of the distribution of *Medionidus* in Alabama, Florida, and Georgia by Williams et al. (2008) concluded that *Medionidus* from the Escambia, Yellow, and Choctawhatchee basins were *M. acutissimus*, not *M. penicillatus*. A recovery plan for the seven endangered and threatened mussels listed in 1998 was finalized in 2003 (USFWS 2003). This plan outlines the species' status, extant and extirpated populations, past and current threats, and conservation measures needed to move them to recovery and eventual delisting from the ESA. Additional Florida mussels, three endangered and four threatened, with designated critical habitat, were added to the federal list in 2012 (USFWS 2012). These seven mussels occur in basins between the Escambia and Choctawhatchee Rivers.

Table 7.2. Florida mussel taxa included on the federal endangered species list, including status, year listed, and year critical habitat was designated (USFWS 1993, 1998, 2004, 2007, 2012). Status categories are endangered (E) and threatened (T).

Taxa	Status	Year Listed	Critical Habitat
Amblema neislerii	E	1998	2007
Elliptio chipolaensis	T	1998	2007
Elliptoideus sloatianus	T	1998	2007
Fusconaia burkei	T	2012	2012
Fusconaia escambia	T	2012	2012
Hamiota australis	T	2012	2012
Hamiota subangulata	E	1998	2007
Medionidus acutissimus	T	1993	2004
Medionidus penicillatus	E	1998	2007
Medionidus simpsonianus	E	1998	2007
Obovaria choctawensis	E	2012	2012
Pleurobema pyriforme	E	1998	2007
Pleurobema strodeanum	T	2012	2012
Ptychobranchus jonesi	E	2012	2012
Reginaia rotulata	E	2012	2012

Of the 15 federally protected mussels that occur in Florida, all but one (*Medionidus acutissimus*) have designated critical habitat in state streams. A total of 1,909 km of streams were designated as critical habitat for seven protected species in 2007 (USFWS 2007). Of the 1,909 km of critical habitat, approximately one-fourth is located in the Florida portions of Apalachicola, Chipola, Ochlockonee, and Suwannee Rivers, and Econfina Creek. More recently the USFWS (2012) designated an additional 2,404 km of streams as critical habitat for eight protected mussels that occur in the Escambia, Yellow, and Choctawhatchee Rivers and their tributaries in southern Alabama and western Florida. Constituent elements of mussel critical habitat include a geomorphically stable stream channel; a predominantly sand, gravel, and/or cobble stream substrate with low to moderate amounts of silt and clay; permanently flowing water; water quality (including temperature, turbidity, DO, and chemical constituents) that meets or exceeds the current aquatic life criteria established under the Clean Water Act; and fish hosts that support the larval life stage (USFWS 2007).

In 1998 NatureServe published an assessment of imperiled mussels in the United States by faunal region and river (Master et al. 1998). In the Florida Gulf Region, Escambia and Apalachicola Rivers were tied for fourth nationwide in the number of at-risk mussels, with the remainder of Florida and southern Atlantic Coast basins in Georgia ranked tenth. Over the past several years, The Nature Conservancy has supported projects that directly benefit mussels in the Apalachicola River basin. These include working with various partners to restore natural conditions and ecological functions to streams, restoring riparian habitats, and promoting management and stewardship of riparian lands. Large-scale habitat restoration of streams and lakes continues to be an important tool in conserving mussel resources (Bernhardt et al. 2005; Cooke et al. 2005).

Anthropogenic Threats to Mussels

Aquatic habitats and mussel communities throughout the continent have been profoundly affected by the cumulative effects of anthropogenic activities (Strayer 2006, 2008). Humankind is placing enormous pressure on mussel populations by increasing demands on freshwater ecosystems (Strayer et al. 2004; Strayer 2006). Diffuse and chronic threats that act insidiously to harm mussel populations combined with nonindigenous species may now present the greatest challenge to their conservation and management (Strayer et al. 2004). There are four broad categories of threats to mussels—exploitation, habitat alteration (chemical and physical), introduction of nonindigenous species, and population fragmentation. The relative impact of these threats varies among basins. Without changes in water conservation and management practices, additional species of mussels may be at risk of extinction (Golliday et al. 2004; Peterson et al. 2011).

Exploitation

Utilization of mussel resources has a long history in the eastern United States. For at least 10,000 years, Native Americans consumed untold numbers of mussels and utilized shells for ornaments, ceremonial purposes, tools, and utensils (Parmalee and Bogan 1998). Individual kitchen middens that occurred along southeastern rivers often extended a kilometer or more and contained millions of mussel valves (Williams et al. 2008; Watters et al. 2009). Middens up to hundreds of meters long and more than 4 m high were described along the middle and lower St. Johns River. Although typically containing snails (e.g., *Pomacea paludosa* and *Viviparus georgianus*), they also included valves of *Elliptio jayensis* (Wyman 1868). There are numerous archaeological sites in the vicinity of Apalachicola River that contain large numbers of unionids (Percy 1976; White 1994). The largest archaeological site in Apalachicola River basin in Florida containing extensive mounds of unionid shells was along the Chattahoochee River but was inundated by Jim Woodruff Lock and Dam (Bullen 1958).

In the late 1800s, German immigrant John F. Boepple started the first pearl button factory in Muscatine, Iowa (Claassen 1994). Most factories were located along the middle and upper Mississippi River and its tributaries. Mussels were harvested in midwestern and southeastern rivers for production of pearl buttons for half a century (Figure 7.1). The intense commercial harvest ultimately contributed to conservation laws and regulations that broadly benefited mussel resources. Pearls continued to be valued as a byproduct of mussels harvested for button manufacturers. The advent of plastic buttons in the 1940s brought a gradual end to the pearl button industry, with the last factories closing in the 1960s.

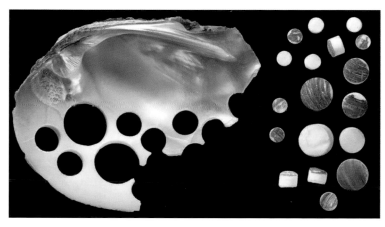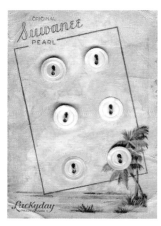

Figure 7.1. Shell of *Megalonaias nervosa* used in production of blanks for buttons (left). © Richard T. Bryant. Despite the name and palm trees on the card of buttons (right), the shell stock for these most likely came from Mississippi River basin (Alexander 2007). Photograph by James D. Williams.

Mussel harvest continued in the 1950s with an export market for the cultured pearl industry (Isom 1969). No substance has proven more reliable than freshwater mussel shell for a nucleus or "seed" in production of cultured pearls (Fassler 1997). Shells are diced and tumbled, producing a spherical bead for insertion into pearl oysters. Mussel harvest for the cultured pearl industry continues, but harvest rates fluctuate widely and are currently low after peaking in the 1990s (Hubbs 2009). Human exploitation of unionids for shells and pearls in the post-European era has been implicated in their widespread decline (Kunz 1898; Anthony and Downing 2001) despite the absence of data to substantiate that these activities actually resulted in any extinctions. The consensus among conservationists is that exploitation has been a relatively minor threat to the mussel fauna compared to the ubiquitous and profound effects of habitat alteration (Stansbery 1971; Williams et al. 1993; Richter et al. 1997; Downing et al. 2010).

Exploitation of Florida mussels has been limited and highly localized. There was interest in commercial mussel harvest for pearl buttons in the early 1900s and the state of Florida apparently leased sections of the lower Chipola River bottom to the Southern Pearl Corporation (Figure 7.2). Florida shells were of poor quality and generally too small for buttons, which ultimately precluded even small-scale commercial harvest. However, Florida mussels have been harvested for their soft parts. *Elliptio jayensis* has been harvested by trawl in Lake Monroe for baiting trot lines, and anglers occasionally use mussels and *Corbicula fluminea* as bait for catfishes and other species. There are also anecdotal reports of individuals harvesting *C. fluminea* for food.

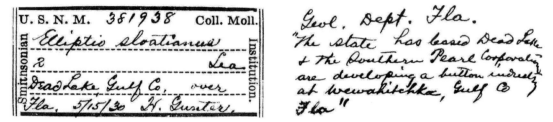

Figure 7.2. U.S. National Museum of Natural History, Smithsonian Institution, collection label (front and back) noting the lease of the bottom of Dead Lake on the Chipola River for commercial shell harvest in May 1930. Label scanned courtesy of Robert Hershler.

Habitat Alteration

Some of the more important habitat modifications contributing to mussel decline in Florida and elsewhere include impoundments, channelization, sedimentation, mining, chemical contamination, and development activities (Downing et al. 2010). Historical mussel decline was generally the result of acute impacts (e.g., dams). Diffuse and chronic threats (e.g., contaminants) that act insidiously to impact mussel populations represent the greatest current conservation challenge (Strayer et al. 2004). Causative agents responsible for the declines are difficult to identify due to insufficient empirical evidence (Downing et al. 2010; Haag 2012). Nevertheless, it is clear that anthropomorphic factors in Coastal Plain streams have altered substrate stability and reduced mussel diversity and abundance (Haag and Warren 2007).

Dams and their impounded waters are the most egregious example of habitat alteration in riverine ecosystems (Yeager 1993; Poff et al. 2007). Habitat losses from dams and impoundments are the leading cause of decline, imperilment, and extinction of mussels in the eastern United States (Haag 2009). Dams disrupt most of a river's ecological processes and faunal communities by converting lotic habitat to lentic habitat (Williams et al. 1992; Yeager 1993; Watters 2000, 2001). Even low mill dams can have effects on mussels (Watters 1996; Brainwood et al. 2008). Dams also alter downstream water quality and riverine habitats, thus negatively affecting tailwater mussel populations (Vaughn and Taylor 1999; Poff et al. 2007). In the mid-1990s the substrate of the entire channel of the lower Ochlockonee River below Jackson Bluff Dam was unstable detritus and muck, apparently due to the lack of flushing flows from Lake Talquin, and many areas were devoid of mussels (J. Brim Box, personal communication). Physical changes to tailwater habitat include altered thermal and flow regimes, sedimentation, channel entrenchment, and altered channel morphology from scouring and bed load movement. While very few large dams are currently being built, legacy effects continue to adversely impact mussel communities and limit their recovery (Williams et al. 1992; Haag 2012).

Although dams are more prevalent in other southeastern states, several have altered habitats in Florida. Jim Woodruff Lock and Dam (Figure 7.3), the largest in the state, was constructed by USACE on Apalachicola River where Chattahoochee and Flint Rivers converge. The purpose of this dam was to facilitate navigation along with the production of a small amount of hydropower. Although barge traffic is now virtually nonexistent on this waterway, channel maintenance operations, such as dredging and spoil disposal, have had a negative legacy effect on mussel habitat (USFWS 2006) (Figure 7.4). Fluctuating flows associated with the small amount of hydropower generation also continue to impact riverine habitat.

Figure 7.3. Jim Woodruff Lock and Dam on the Apalachicola River, just upstream of U.S. Highway 90, near Chattahoochee, Gadsden and Jackson counties, Florida, June 1993. Photograph by James D. Williams.

Figure 7.4. Apalachicola River, spoil from navigation channel maintenance dredging, locally known as Sand Mountain, located near river mile 36, Gulf County, Florida, June 1998. Photograph by James D. Williams.

Other sizable dams on Florida rivers include Jackson Bluff Dam (Ochlockonee), Rodman Lock and Dam (Ocklawaha) (Figure 7.5), and Inglis Lock and Dam (Withlacoochee [southern]). Completed in 1928, Jackson Bluff Dam (18 m high) and its associated reservoir Lake Talquin were constructed for hydropower production and recreation. Rodman and Inglis Lock and Dams are part of the unfinished and now defunct Cross Florida Barge Canal project, which was initiated in 1964. A few dams are located on smaller streams (e.g., Lake Manatee, Manatee River, Manatee County) and spring runs (e.g., Merritts Mill Pond, Spring Creek, Jackson County).

Figure 7.5. Rodman Lock and Dam on Ocklawaha River, southwest of Palatka, Putnam County, Florida, August 2002. Photograph by James D. Williams.

The USACE constructed dams on the outlets of many natural lakes in central Florida (e.g., upper Kissimmee Chain of Lakes). These dams serve as flood control structures but have fostered lakeside development with their stringently managed, stabilized water levels. Secondary impacts to mussels have resulted from lost lakeshore habitat and runoff from lawns and roads. The dam on Lake Tohopekaliga in Osceola County disrupted a seasonal water level regime that historically fluctuated several meters, reducing fluctuations to a meter and facilitating massive buildups of live and decomposing vegetation up to 0.6 m thick in wide bands parallel to the shoreline (Moyer et al. 1992). Muck accumulation, exacerbated by nutrient loading from several wastewater treatment plants, and the resultant nuisance plant growth formed extensive berms that smothered benthic habitats and limited fish movements. Drawdowns to consolidate lake bottoms and remove excessive decaying vegetation are management tools used routinely to restore habitat quality for benthic organisms in some lakes (Butler et al. 1992).

Most mussels are adapted to free-flowing streams whose levels vary seasonally and with climatic conditions, making natural flow variability crucial for their recruitment (Vaughn and Taylor 1999; Galbraith and Vaughn 2011). Increases in dam discharge during spring and early summer (when many juvenile mussels are excysting from their hosts) may prevent successful juvenile settlement into suitable habitat. Conversely, minimum flow releases from dams are often detrimental since mussels apparently require sufficient low-flow thresholds (Steuer et al. 2008). Pervasive decline in water levels due to management of Jim Woodruff Lock and Dam and Apalachicola River channel maintenance activities (Light et al. 2005) may severely affect highly aggregated *Amblema neislerii* populations along shorelines (Gangloff 2011). However, some Florida species that inhabit streams are reservoir tolerant (e.g., *Anodonta couperiana*, *Elliptio jayensis*, *Pyganodon grandis*, and *Utterbackia imbecillis*).

In recent years dam removal has come to be recognized as necessary for restoring ecological functions to streams (Bednarek 2001; Hart and Poff 2002). Impacts on mussels from dam deconstruction activities should be outweighed by the long-term benefits of restored flows and habitat conditions (Sethi et al. 2004). In the mid-1950s Clench and Turner (1956) noted a substantial population of *Amblema neislerii* that was "amazingly abundant in Dead Lake" before impoundment of the lower Chipola River in 1960 altered mussel habitat. Dead Lake is a natural feature that is characterized by a broad expanse of water with a slow current. Removal of the dam in 1987 has allowed natural habitat restoration by reestablishing flow and removing the barrier to fish movement, which permitted recolonization of this reach by imperiled riverine species.

Channelization, dredging streams for navigation, and ditching wetlands for drainage and flood control have profoundly altered surface waters nationwide. Channelization alters many of the physical and biological functions of streams (Watters 2000). Channelized streams transport high sediment levels that may smother juvenile mussels and create conditions of perpetual turbidity. Mussels that produce mantle lures and depend upon sight-feeding fishes as hosts may also be adversely affected by high turbidity levels, ultimately reducing recruitment (Österling et al. 2010). Alteration of watersheds by stream channelization has a long history in Florida. The first major efforts were by Hamilton Disston in the 1880s, ostensibly to drain portions of the Everglades basin for cattle grazing and development. These efforts included channelization of Kissimmee and Caloosahatchee Rivers, with the latter being connected to Lake Okeechobee via a canal. This allowed steamboats to transport goods between the town of Kissimmee and the Gulf of Mexico while draining thousands of hectares of wetlands.

While mussel populations in Florida undoubtedly have been reduced or lost due to dredging activities and canal construction, some peninsular species appear to be tolerant of canal conditions. However, in channelized streams and canals where extreme environmental conditions exist (e.g., low DO, chemical runoff, excessive organic buildup, and eutrophication), colonization by mussels is unlikely. Based on distribution patterns of introduced *Utterbackia imbecillis*, it appears that canals have potentially served as conduits for interbasin transfers in peninsular Florida.

Excessive sedimentation is pervasive in U.S. streams and has long been implicated in decline of mussel populations (Brim Box and Mossa 1999), although precise causative factors remain elusive (Haag 2012). Biological impacts may include altered feeding and respiration, reduced growth rates, and limited burrowing activity (Ellis 1936; Waters 1995). Reduced primary productivity can potentially limit mussel foods (Kanehl and Lyons 1992). Physical effects alter mussel habitat by affecting flows, bed loads, and channel stability (Vannote and Minshall 1982; Geist and Auerswald 2007). Sedimentation reduces crucial habitats for juvenile mussels by clogging substrates and lowering DO (Sparks and Strayer 1998). Sediment runoff may also act as a vector for delivering nutrients, pesticides, and other contaminants to streams (Salomons et al. 1987). Ingestion of contaminated silt particles by juveniles may increase their exposure to toxicants (Yeager et al. 1994).

Prodigious amounts of sediment can clearly impact benthic habitats and mussel populations through smothering (Figures 7.6 and 7.7). The channels of some Florida streams now have copious amounts of unstable sandy sediments. In northern peninsular and panhandle streams, several species typically found in channels represent some of the most imperiled mussels in Florida (e.g., *Pleurobema pyriforme* and *Medionidus penicillatus*). The Ochlockonee River channel above Lake Talquin harbored abundant mussel populations through the late 1980s. The channel is now characterized by beds of unstable sand deposits, and the vastly reduced mussel fauna is now found only in narrow strips along the banks where the substrate is more stable. Based on museum records, the last verified *Medionidus simpsonianus* from this reach was collected in 1993.

Figure 7.6. Sediment plume following a major rainfall event in Lake Tohopekaliga at mouth of St. Cloud Canal from East Lake Tohopekaliga, Osceola County, Florida, May 1987. Photograph courtesy of Florida Fish and Wildlife Conservation Commission.

Agricultural activities are a major source of sedimentation and chemical runoff in much of the United States (Waters 1995; Richter et al. 1997). Cultivation and live-stock in riparian areas threaten mussels by eliminating vegetation, increasing water temperatures, reducing filtering capacity, and increasing bank erosion (Trimble and Mendel 1995). Riparian timber harvest and access road construction contribute to stream sedimentation and channel alterations (Brim Box and Mossa 1999; Henley et al. 2000). Mussel occurrence has been correlated to forested riparian zones (Poole and Downing 2004; Hopkins 2009). Buffer zones filter sediments, reduce nutrient loading, and regulate hydrological responses, while ameliorating temperatures, particularly in small streams (Naiman and Decamps 1997).

Wastewater and sediment associated with mining of minerals and fossil fuels have adversely affected many mussel populations (Neves 1993; Williams et al. 1993). In Florida mining activities are primarily for phosphate, lime rock, gravel, and sand. Phosphate and aggregate mining can affect mussel habitat by reducing groundwater base flows, destabilizing stream channels, and degrading water quality (Kanehl and Lyons 1992). These factors may also reduce fish populations and alter their behavior, potentially compromising their ability to serve as glochidial hosts. Phosphate mining has drastically affected drainages in localized portions of northeast Florida and the southwestern peninsula. Streams have been rerouted, drainage patterns disrupted, and water quality reduced, particularly in the upper Peace River basin. Phosphate mining has lowered groundwater levels resulting in dewatering of portions of nearby stream channels (Livingston 1990). In-stream and alluvial gravel mining in north Florida have affected some stream reaches. One of the most extreme cases occurred along the lower reaches of

Big Escambia Creek, Escambia County, which not only destroyed habitat in the creek but affected Escambia River downstream of its mouth. A restoration project was completed in 2005 that reestablished the creek to its original channel, and preliminary observations indicate the creek is recovering.

Figure 7.7. Sediment plume, Santa Fe River, looking downstream, County Route 241, Alachua and Union counties, Florida, June 1987. Photograph by Robert S. Butler.

Contaminants in point and nonpoint discharges can degrade water and substrate quality and adversely impact mussel populations. Environmental contaminants potentially influence all mussel life stages which have common and unique characteristics that affect exposure and sensitivity to toxicants in four habitat exposure routes—surface water, pore water, sediments, and diet (Cope et al. 2008). Potential exposure times may vary from days (sperm and free glochidia) to years (juveniles and adults). Encysted glochidia may potentially be exposed to contaminants in host fish tissues.

Episodic chemical spills and point sources may result in mussel mortality, but the subtle effects of chronic, low-level contamination may also affect mussels (Naimo 1995). Metals and ammonia are among the most toxic compounds to mussels, and pedal-feeding juveniles may readily ingest contaminants adsorbed to sediment particles (Augspurger et al. 2003; Newton 2003; Wang et al. 2010). Heavy metals may reduce glochidial viability or inhibit attachment to hosts and, similar to pesticides, may affect calcium metabolism (Keller and Zam 1991; Huebner and Pynnönen 1992; Pynnönen 1995; Frank and Gerstmann 2007). Sediment samples from various Apalachicola basin streams tested for heavy metals (e.g., cadmium and copper) had concentrations markedly above background levels (Frick et al. 1998). In Chipola River, elevated heavy metal concentrations in

Corbicula fluminea and sediment samples included levels known to be toxic to mussels (Winger et al. 1985; Hemming et al. 2008).

The pervasiveness of ammonia in aquatic habitats warrants priority attention for its toxic effects on mussels (Newton and Bartsch 2007; Haag 2012). The nonionized form (NH_3) is considered the most toxic, although the ammonium (NH_4^+) may be toxic under certain conditions (Newton 2003; Strayer 2008). These compounds alter metabolic processes and reduces mussel survival, growth, and reproduction, potentially representing a limiting factor for populations (Chetty and Indira 1994, 1995; Augspurger et al. 2003). Ammonia may be at its highest concentration in stream interstitial habitats, resulting in juvenile mortality at concentrations as low as 0.7 parts per million. Toxicity to juveniles increases during periods of low flow and high temperature (Newton 2003; Cherry et al. 2005). Common sources of ammonia include animal feedlots, nitrogenous fertilizers, effluents of wastewater treatment plants, and natural decomposition of organic nitrogen. Hemming et al. (2008) found pore water ammonia levels at several Chipola River sites exceeding water quality criteria recommended for mussel protection by Augspurger et al. (2003).

Pharmaceuticals and personal care products typically are not filtered out at wastewater treatment plants and are increasingly appearing in surface waters (Daughton and Ternes 1999; Bringolf et al. 2010). It has been demonstrated that some compounds adversely affect mussel immune systems (Gagné et al. 2006). Certain compounds may mimic estrogen and cause feminization of male mussels (McMaster 2001; Gagné et al. 2011). The antidepressant fluoxetine caused untimely release of *Elliptio* sperm and glochidia downstream from a wastewater treatment plant (Bringolf et al. 2010).

Agricultural sources of chemical contaminants include two broad categories, nutrients and pesticides (Frick et al. 1998). Aquatic ecosystems in general and mussels in particular are negatively affected when nutrients (e.g., nitrogen and phosphorus) enter streams at concentrations that cannot be assimilated, resulting in overenrichment, a condition exacerbated by low-flow conditions (Patzner and Müller 2001). Primary sources of nutrients are runoff from livestock feedlots and fertilizers used in row cropping (Peterjohn and Correll 1984). Nutrient levels are often higher in streams with grassy versus wooded riparian buffers, stressing the importance of forested buffer zones for filtering nutrients and reducing sediment loading (Morris and Corkum 1996; Hopkins 2009). Nutrient overenrichment may result in lower benthic DO levels (Strayer 2008). Since DO levels in substrates are a small fraction of that found in the water column, these conditions are capable of affecting mussels, particularly juveniles (Buddensiek et al. 1993; Dimock and Wright 1993; Tankersley and Dimock 1993). Eutrophication has been correlated with mussel declines in lakes, probably due to DO sags (Arter 1989; Nalepa et al. 1991; Kołodziejczyk et al. 2009). Severe eutrophication is widespread in Florida, manifest in concentrations of filamentous algae in many rivers and in large accumulations of both native and nonnative macrophytes in numerous peninsular lakes (Engstrom et al. 2006).

Pesticide runoff may represent one of the greatest threats to the mussel fauna (Haag 2012). Commonly used pesticides have been implicated in a mussel die-off in North Carolina (Fleming et al. 1995). Organochlorine pesticides are still detected in streams and aquatic organisms decades after being banned, often at levels exceeding chronic exposure criteria for protection of aquatic life (Buell and Couch 1995; Frick et al. 1998). Pesticide registration is based on toxicity of the active ingredient alone. Other chemicals used in pesticide applications, including surfactants and other ingredients, may be even more toxic to mussels but remain unregulated (Folmar et al. 1979; Bringolf et al. 2007a, 2007b). Atrazine is the most widely used herbicide nationally and among the most commonly detected in stream and groundwater surveys (Kolpin et al. 1998). Although not toxic to *Lampsilis* glochidia and juveniles in acute tests, the surfactants used with Atrazine were toxic to juveniles in chronic tests (Bringolf et al. 2007a). Mussels disappeared from some areas in Lake Okeechobee where there were repeated applications of herbicide, but populations remained extant in untreated areas. These chemicals are likely responsible for reduced mussel populations in other waters of the state.

Nonindigenous Species

Invasion of nonindigenous species is a complex and perplexing issue (Strayer 2010). The proliferation of species introductions and the profound consequences they often trigger has led to coining of the terms "Homogocene" and "New Pangaea" to describe an era of increasing homogenization of biotic diversity (Rosenzweig 2001; Strayer 2010). Highly altered and currently stressed ecosystems appear acutely susceptible to invasions (Strayer 1999b; Kaushall et al. 2010). Many plant and animal groups have nonindigenous representatives in Florida waters. Nearly 1,600 nonindigenous species are found in south Florida; scores of them are aquatic-dependent species that comprise a considerable percentage of the current biota (Rodgers et al. 2011). Floating, submergent, and emergent plant species that reach high concentrations may have significant effects on ecosystem functions that negatively affect mussels (Strayer 2010). Some highly piscivorous fishes introduced beyond their native ranges (e.g., *Pylodictis olivaris* in Apalachicola River) may considerably reduce the number of some host fishes for native mussels. Among mollusks, several grazing snails that influence primary production are firmly established in Florida (e.g., *Melanoides*, *Pomacea*, and *Tarebia*). Suspension-feeding bivalves, such as *Corbicula fluminea*, represent a group of invaders that are most likely to displace native mussels in some systems. Nonindigenous species potentially carry diseases and parasites that could be devastating to the native mussel fauna (Strayer 2010).

Corbicula is the most widespread and visible aquatic nonindigenous mollusk in North America. *Corbicula* is a primary consumer that disrupts food webs and competes with native mussels, especially juveniles, for food, nutrients, and space (Leff et al. 1990; Atkinson et al. 2010, 2011). They may ingest unionid sperm, glochidia, and newly metamorphosed juveniles, potentially reducing fertilization and recruitment rates, and impacting juvenile habitats via sediment disturbance (Strayer 1999b). In laboratory experiments there was a positive correlation between *Corbicula fluminea* density and juvenile mussel mortality, and juvenile growth rates were also reduced (Yeager et al. 2000). In an artificial stream, an induced *C. fluminea* die-off resulted in the release of ammonia at concentrations lethal to juvenile mussels (Cherry et al. 2005). Periodically during summer months there is a *C. fluminea* die-off in Apalachicola River below Jim Woodruff Lock and Dam and dead individuals with soft parts accumulate along the eastern bank. In late summer of 2006 and 2007, droughts and river drawdowns caused a die-off of *C. fluminea* and native mussels (e.g., *Amblema neislerii* and *Elliptoideus sloatianus*) stranded in isolated pools in slough channels. Mussel mortality apparently occurred only after the death and decomposition of *C. fluminea*, suggesting ammonia toxicity as a potential cause.

Corbicula fluminea densities vary widely in habitat patches with sparse mussel concentrations but appear to preferentially invade sites where mussels are in decline (Strayer 1999a). *Corbicula fluminea* numbers are rarely high in densely populated mussel beds, suggesting that it is a poor competitor at the localized habitat scale, but where dense it may increase unionid imperilment and impede recovery (Vaughn and Spooner 2006b). Field experiments in St. Johns River appeared to indicate that growth of *Elliptio* (probably *E. jayensis*) declined somewhat with increasing *C. fluminea* density (Belanger et al. 1990), which reached 2,700 per square meter in a Lake Harney study site in 1984. In the Santa Fe River drainage, densities exceeded 2,000 per square meter in 1973 (Bass and Hitt 1974) and 1998 (Blalock and Herod 1999).

Dreissena polymorpha is a major threat to mussels in the Midwest (Strayer 1999a), but the species is not known from Florida. It is not known if the species could become established and what effects *D. polymorpha* may have on native mussels. Based on a model presented by Hayward and Estevez (1997), some Florida freshwater habitats, including such diverse water bodies as springs and lakes, are suitable for colonization by *D. polymorpha*.

Population Fragmentation

Many disjunct mussel populations are geographically limited and isolated due to artificial and natural barriers, making them more susceptible to declines and extirpation due to stochastic environmental and demographic variables. Fragmentation results from habitat reductions, and barriers impede natural mussel dispersal (Vaughn 1997). Therefore, recolonization is unlikely in fragmented populations without human intervention through reintroductions (Davis and Howe 1992; Strayer et al. 2004). Increasing fragmentation promotes genetic and demographic constraints that inhibit the ability of populations to cope with natural and anthropogenic changes (Murphy et al. 1990). Small populations may experience declines in fertility and survival (Allee effects), reductions in fitness from inbreeding depression, and loss of genetic diversity that compromises adaptability to environmental conditions (Traill et al. 2010).

Fragmentation increases the patchiness of mussel populations with effects being exacerbated for rare species and taxa exhibiting narrow host specificity (Vaughn 1993). Mussel dispersal among patches is primarily dependent upon host fishes transporting encysted glochidia (Vaughn and Taylor 2000). Colonization rates are usually very low due to low juvenile survival rates, generally low densities, and other life history attributes (Neves 1993). Isolated mussels in small tributary streams may be sink populations that rely on source populations in larger streams for sustainability. Due to longevity of individuals in some species, extirpation of populations may take decades (Haag 2009; Jones and Neves 2011). As a result of legacy habitat alteration, the mussel fauna has incurred a substantial extinction debt that is manifest in continuing declines of small, isolated populations for decades to come (Tilman et al. 1994). This phenomenon led Haag (2009) to predict that many more mussel species experiencing population fragmentation may soon become extinct. Mussel rarity makes maintaining adequate genetic heterogeneity crucial for imperiled species recovery efforts (Jones et al. 2006b). Minimum viable population size is critical for species survival. Traill et al. (2010)

advocated a population size of 5,000 individuals to ensure adequate genetic diversity in adapting to environmental change, withstanding catastrophic events, and persisting long-term.

Other Anthropogenic Activities

Watershed disturbance associated with development and urbanization is detrimental to stream habitats (Paul and Meyer 2001), often affecting mussel habitats due to increased erosion and contaminant runoff. These threats are increased where adequate buffers are not maintained. Several peninsular Florida counties with high numbers of natural lakes have become increasingly urbanized (Figure 7.8). Development often destroys shoreline habitat by wetland conversion and bulkhead construction.

Figure 7.8. Residential lakeshore development near Orlando, Florida, 1990. Photograph by Robert S. Butler.

Development has resulted in a vast increase in impervious surfaces, which alter rainfall runoff and groundwater replenishment. High impervious surface percentages in a watershed can cause multiple deleterious effects (Doyle et al. 2000). These effects exacerbate flooding by accelerating stormwater runoff, triggering a series of actions that destabilize in-stream and riparian habitats, while impacting mussel and host fish populations (Brim Box and Mossa 1999). High peak flows and shear stress result in substrate movement that may reduce juvenile abundance and increase displacement to unsuitable habitats (Layzer and Madison 1995; Gangloff and Feminella 2007).

Withdrawal for municipal, agricultural, and industrial water supplies lowers water tables, subjecting mussel populations to low flow levels in streams (Haag and Warren 2008). Increased frequency of low-flow periods in southwest Georgia streams in recent decades are due to increasing usage of center pivot irrigation (Stamey 1996; Albertson and Torak 2002). Although there have been no significant changes in annual rainfall levels in recent years, seasonality of rain events has been altered (Rugel et al. 2011). Exacerbated by periodic droughts, this coupling of threats has resulted in mussel population declines in some streams (Johnson et al. 2001; Gagnon et al. 2004; Golladay et al. 2004). Mussel strandings are becoming commonplace for several imperiled species. Modeled simulations of the effects of pre- and post-irrigation hydrologic regimes on three federally protected mussels in a southwest Georgia stream indicated a 20-fold increase in extinction risk (Peterson et al. 2011).

Beginning a century ago in Florida, altered coastal hydrology has resulted in salt water intrusion, detrimentally affecting the freshwater biota (Sonenshein 1996). After canals were constructed (beginning in early 1900s) to drain portions of the Everglades, wells in Miami soon yielded salt water, forcing the drilling of replacement wells further inland (Parker et al. 1955). Former freshwater habitats were eliminated in large areas. The lowered waters of the Everglades coupled with the extensive canal system has allowed both subsurface and surface salinity levels to advance westward, especially during droughts, increasingly impacting aquatic ecosystems. Portions of the Everglades and other near-coastal areas in Florida may no longer support mussels or *Corbicula fluminea* due to salt water intrusion. Museum records indicate Lake George historically supported populations of at least six unionids. However, recent survey efforts failed to locate freshwater mussels or *C. fluminea*. These bivalves have potentially been extirpated from Lake George also due to salt water intrusion.

Climate Change

Accelerating climate change resulting from the cumulative effects of humankind on our planet is well accepted within the scientific community. Species have evolved within a matrix of environmental conditions and some will perish if they are unable to adapt to altered climatic conditions (Galbraith et al. 2010). Major effects on aquatic ecosystems include alterations in temperature and precipitation patterns, primary and secondary productivity, species interactions, and timing of life history events (Parmesan 2006; Strayer and Dudgeon 2010). Effects of altered conditions on mussels may be exacerbated by increasing levels of urbanization and susceptibility to nonindigenous species invasions (DeWalle et al. 2000; Kaushall et al. 2010; Strayer 2010). Changes to the mussel fauna will likely have negative cascading effects throughout aquatic communities given their importance in driving ecosystem processes (Galbraith et al. 2010; Vaughn 2010). Stream mussel imperilment will increase since they are unable to migrate to suitable waters (Strayer and Dudgeon 2010). Modeled simulations demonstrate that mussels are more susceptible to flow alterations than fishes (Spooner et al. 2011). Further, continental patterns of extirpations were strongest in the southeastern United States and considered severe given current climate change scenarios. Mussels inhabiting lakes may be able to move to greater depths as temperatures increase, but their functional role in the littoral zone may change (Cyr 2008). The shallowness of most Florida lakes will likely preclude mussels from escaping rising temperatures. Other factors (e.g., lake size, water color, and transparency) may influence changes in lake environments (Fee et al. 1996).

The importance of climate change will vary since mussels differ in their physiological responses to environmental stimuli (Spooner and Vaughn 2008). Species with narrow temperature tolerance ranges may be ill-adapted to adjust to the effects of climate change (Poff et al. 2001). A rise in temperature may result in a higher metabolic energy demand while altering fitness, behavior, and reproduction (Pandolfo et al. 2010). Increases in temperature have been correlated with decreased glochidial survival times and increased antibody production in fishes (Plumb et al. 1986; Jansen et al. 2001). Both of these scenarios negatively affect recruitment by reducing rates of glochidial survival and encystment. Drastically altered flows and temperatures could lead to altered genetic structure and reproductive failure (Galbraith 2009). In southeastern Oklahoma a shift in dominance from thermally sensitive to thermally tolerant species was noted over a 15-year period (Galbraith et al. 2010). Rare headwater stream mussels had a higher risk of extirpation during an Alabama drought since survivability was primarily a function of predrought abundance (Haag and Warren 2008). Increasing global temperatures will lead to higher sea levels and result in greater incidences of salt water intrusion in coastal Florida (South Florida Water Management District 2009). This scenario will be exacerbated by groundwater pumping. Reedy Creek, an upper Kissimmee River tributary, was among 20 of 40 streams nationwide (and the only one studied in Florida) that had warmed significantly in recent decades (Kaushall et al. 2010). Based on conditions associated with global climate change and physiological constraints on mussels, it would appear that some populations may disappear from Florida.

As serious as the effects of climate change may be, the effects of humanity's response to these changes may be worse. Remedial engineering solutions in response to water shortages (e.g., flood control structures, dredging, and water diversions) and even preventative measures to reduce greenhouse gas emissions (e.g., more renewable energy from hydroelectric facilities) may have dramatic environmental consequences that exacerbate the effects of climate change (Hastie et al. 2003; Strayer and Dudgeon 2010). These complex and sometimes emotionally driven issues pose a profound challenge to conservationists and water managers alike when attempting to balance the environmental requirements of aquatic organisms and the increasing demand for freshwater (Richter et al. 2003; Vorosmarty et al. 2010).

Emerging Conservation Challenges

Losses of imperiled mussels dictate that population reintroductions must be initiated to recover endangered and threatened species (USFWS 2003). Attempts to establish new populations of aquatic organisms demand a long-term commitment of capital and human resources, have potential pitfalls (Shcchan et al. 1989; Dodd and Seigel 1991; Jones et al. 2006b), and many efforts will fail (Morell 2008). Factors limiting population reestablishment success include insufficient biological information, risks associated with diseases and parasites, deleterious effects of reduced genetic stocks, and potentially suites of ecological and economic problems (Strayer and Dudgeon 2010). Stressors that initially drove species extirpations must be ameliorated for reintroductions to be successful. An additional limitation on reestablishing mussels in historical habitats is the scarcity of either natural or restored habitat in which to put them. Several mussel culture facilities operate in the southeastern United States. However, none are located in Florida and no facilities currently focus specifically on species that occur in the state. The Alabama Department of Conservation and Natural Resources culture facility—Alabama Aquatic Biodiversity Center—is capable of targeting highly imperiled west Florida species.

Consideration of population genetics for imperiled species is crucial for maintaining healthy, genetically diverse populations that are able to adapt to anthropogenic changes (Frankham et al. 2002). Guidelines have been developed for managing genetic stocks of propagated mussels to decrease the potential for harming both recipient and donor populations (Jones et al. 2006b). Several genetic tools have been applied to mussel conservation, including identifying evolutionarily significant units, resolving taxonomic ambiguities, and describing population genetic structure and reduced gene flow among populations (Mulvey et al. 1998).

Managing streams to maintain biodiversity and accommodate water for humans is a major conservation challenge (Strayer and Dudgeon 2010). Apalachicola basin was used as a case study for ecologically sustainable water management. Concerns regarding flows in the early 1990s among the three states sharing the watershed—Alabama, Florida, and Georgia—led to a comprehensive study attempting to determine ecosystem flow requirements for aquatic organisms and evaluating human impacts on riverine resources. It also promoted a collaborative approach in searching for solutions to problematic issues, such as conducting water experiments to resolve issues that frustrate efforts at integrating human and ecosystem needs. Most streams can compatibly serve in the dual role of fulfilling ecosystem and human needs, but plan implementation is urgently needed before water becomes an even more overappropriated resource (Richter et al. 2003).

Continuing research indicates that the ecological roles mussels play are integral in providing vital ecosystem functions, highlighting the importance of maintaining diverse and abundant populations of both rare and common mussel species (Galbraith and Vaughn 2009; Vaughn 2010). Services provided by healthy mussel populations enhance water quality, providing a direct and profound benefit to humankind. Fortunately, research over the past several decades has significantly advanced the science of mussel conservation (Strayer and Dudgeon 2010). Through the actions of federal and state agencies and nongovernmental organizations, mussels are increasingly being considered resources crucial for the restoration and maintenance of healthy surface waters. Protection and management efforts could eventually curtail, if not reverse, the mussel imperilment crisis in Florida and beyond.

Chapter 8

Shell Morphology and Soft Anatomy

Freshwater mussels are bivalve mollusks characterized by two calcareous shells that protect the soft-bodied animal within. The shells, or valves, are connected dorsally by a proteinaceous hinge ligament. Mussels vary greatly in shell morphology among and within species. Variation of shell morphology is determined by the interaction of inherited traits and environmental factors. This variation is expressed in shell size, shape, color, inflation, and degree of surface sculpture development.

Shell Characteristics

Shell size, mass, and shape are very important characteristics used for identification of mussels. Maximum sizes vary widely, from about 30 mm in species of *Toxolasma* to more than 275 mm in *Megalonaias*. Florida populations of some species are often smaller than those in other areas. This may be attributed in part to differences in productivity of aquatic habitats. Shell mass varies greatly among taxa and depends in part on water chemistry. Concentrations of calcium and bicarbonate in streams may be proportional to shell thickness (Tevesz and Carter 1980). Other factors affecting shell size include stream productivity and food availability, as well as hydraulic conditions and substrate (Bailey and Green 1988; Morris and Corkum 1999; Kesler et al. 2007). Differences in shell mass among species are far greater than those of shell size. Shells of *Megalonaias* are several orders of magnitude heavier than those of *Toxolasma*. Shell thickness varies from less than 1 mm in *Utterbackia* to more than 20 mm in large *Megalonaias*. Within species, lake populations may have thinner valves than those in streams (Surber 1912), except those in highly turbulent lakeshores, which may be thicker (Bailey and Green 1988; Hinch and Bailey 1988).

Shell margins are often the most distinguishing characteristic used for species identification (Figure 8.1). Mussel outlines vary but can generally be described as oval, elliptical, lanceolate, quadrate, rhomboidal, trapezoidal, or triangular. Most Florida mussels tend to be oval or elliptical, but some panhandle species are quadrate or triangular. Some species are lanceolate (e.g., *Elliptio ahenea*), with fewer approaching round (e.g., *Reginaia rotulata*). Measurement of shell length is taken as the distance between the anterior and posterior margins, measured approximately parallel to the hinge ligament. Shell height is the greatest distance between dorsal and ventral margins, measured perpendicular to the length measurement. Shell width is the greatest distance between the outer surfaces of the valves. Width relative to height is referred to as shell inflation and species range from being compressed (e.g., *Elliptio arctata*) to highly inflated (e.g., *Pyganodon grandis*). Juveniles of some species are compressed but become highly inflated with age (e.g., *Glebula rotundata*). Of the 23 genera of mussels that occur in Florida, species of the genus *Elliptio* have the most variable shell morphology.

The umbo, located dorsally and anterior to the hinge ligament, represents the oldest part of the valve and originates from the juvenile shell (Figure 8.2). Relative umbo position may be a distinguishing shell feature among species. The umbo is typically anterior to the center of the shell and may or may not be elevated above the hinge line (shell margin) (Figure 8.2). Sculpture is present on the umbo in the form of raised ridges that vary in pattern and thickness. It is usually consistent within genera and can be important in identifying small individuals (Marshall 1890; Van Hyning 1917). Umbo sculpture seldom persists in adults, depending on degree of shell erosion, and has never been observed on some rare species (e.g., *Reginaia rotulata*). Individuals in low pH systems or habitats with swift water and large amounts of sediment are often badly eroded around the umbo. When the two valves are articulated, the small crescent-shaped depression between the valves anterior to the umbo is referred to as the lunule.

Figure 8.1. Margins of a freshwater mussel shell, *Anodontoides radiatus*. © Richard T. Bryant.

Figure 8.2. Elevation of umbo relative to dorsal margin–well above in *Pyganodon grandis* (left) and even with in *Utterbackia imbecillis* (right). © Richard T. Bryant.

Shell shape is usually the same in both males and females. However, sexually dimorphic shells are typical of long-term brooders and a few other taxa. It is typically expressed as posterioventral inflation, or swelling, in the shells of females (Figure 8.3). The post-basal swelling accommodates the host-attracting lures and most of the gravid gills (Jones and Neves 2011). The extent of dimorphism varies but can be very subtle (e.g., *Glebula*). Sexually dimorphic Florida genera include *Glebula, Hamiota, Lampsilis, Medionidus, Obovaria, Toxolasma*, and *Villosa*.

External shell characters are important in species identification. The portion of the valve below the umbo is referred to as the shell disk. The disk is bounded posteriorly by the posterior ridge, which extends obliquely from the umbo to posterioventral margin (Figure 8.4). The posterior ridge may be sharp (e.g., *Alasmidonta triangulata*), rounded (e.g., *Lampsilis straminea*), biangulate (e.g., *Elliptio pullata*), or triangulate (e.g.,

Ptychobranchus jonesi) (Figure 8.5). The shell disk posterior to the ridge is referred to as the posterior slope (Figure 8.4). In some species a shallow sulcus may be present immediately anterior to the ridge. Shells of most Florida species are smooth, but some have varying degrees of sculpture.

Figure 8.3. Shell sexual dimorphism in *Villosa lienosa* female (left) and male (right). © Richard T. Bryant.

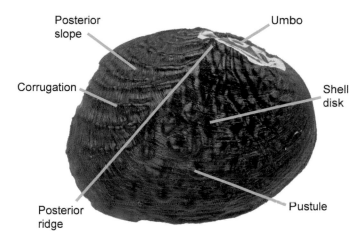

Figure 8.4. External morphology of a freshwater mussel shell, *Quadrula kleiniana*. © Richard T. Bryant.

The various layers of the shell are continually produced as the animal grows. Periods of slow growth produce growth rests, which externally appear as concentric rings or annuli and may be darker in color (Figure 8.5). External growth annuli result from thickening of periostracum (outer covering of the shell). Some peninsular Florida populations may grow more or less continuously, even during winter (Simpson 1899). Although used for estimating age, counting external rings generally results in underestimation, particularly in older animals (Neves and Moyer 1988).

Various forms of ornamentation or sculpture, including plications, corrugations, wrinkles, knobs, tubercles, pustules, and spines, may be present on the shell disks. Sculpture presence or absence is usually consistent within a given species but may vary greatly in its development among individuals, especially those from different drainages. Definitions and usage of terms describing sculpture are not always consistent among malacologists or conchologists. Sculpture on Florida mussels is somewhat limited in its development and is restricted, with the exceptions of *Elliptio monroensis* and *Medionidus walkeri*, to species in the panhandle. Sculpture on panhandle shells consists of variously developed corrugations, plications, wrinkles, and pustules. These terms are defined herein and used consistently. Corrugations are small ridges, generally oriented parallel to each other (Figure 8.6). If present they are usually found on the posterior slope. Plications are larger, deeper ridges and are usually found on the disk and may or may not be parallel. In some species they may extend across the posterior ridge onto the posterior slope (Figure 8.6). Wrinkles are small ridges or grooves characterized by their irregular shape. A pustule is a small prominence that may be round or oval to oblong in shape. They can be irregularly arranged or evenly spaced on the disk.

Figure 8.5. Posterior ridge variation—sharply angled in *Alasmidonta triangulata* (upper left), rounded in *Lampsilis straminea* (upper right), biangulate in *Elliptio pullata* (lower left), and triangulate in *Ptychobranchus jonesi* (lower right). © Richard T. Bryant.

Species that are typically smooth may occasionally have sculpture. The shell disk anterior to the posterior ridge of *Elliptio* is usually smooth, but some large individuals may have shallow vertical plications near the center of the disk giving it a ribbed appearance. *Elliptio jayensis* typically has a smooth shell, but occasionally individuals may exhibit a small patch of evenly spaced pustules on the lower portion of its disk.

The periostracum varies in texture, ranging from smooth (e.g., *Villosa vibex*) to clothlike (e.g., *Uniomerus carolinianus*). Texture may influence luster of the periostracum, which varies from shiny, dull, to clothlike.

The periostracum color ranges from uniformly dark brown to vivid shades of yellow and green with variable rays. The striking appearance of *Hamiota subangulata* was described as "one of the most beautiful of our North American freshwater [mussels]" by Clench and Turner (1956). Color may be influenced by several environmental factors, including turbidity, substrate, chemical composition of suspended solids, and pollutants (Grier 1920a, 1920b; Hanlon and Smith 1999). An example of environmental influence on periostracum color in Florida waters is spring populations of *Elliptio*, which often have a golden to coppery tint, where as those from streams and lakes are dark olive or brown to black.

Figure 8.6. Corrugations on the posterior slope of *Medionidus walkeri* (left) and plications on the shell disk of *Amblema neislerii* (right). © Richard T. Bryant.

Shell background colors of most species are usually uniform throughout their ranges, but some vary within and among populations. While periostracum color is generally uniform, the shell often becomes darker marginally and with size. Descriptions of periostracum colors are subjective. Grier (1920a) used a scientific color nomenclature guide of hundreds of colors developed by Ridgeway (1912) in an attempt to standardize usage that was mostly ignored by later investigators. Colors ascribed by conchologists in the 1800s to characterize periostracum remain in general usage today (e.g., brownish, greenish, olive green, and blackish).

Many species have secondary color patterns that break up the background color of the periostracum. These patterns typically consist of rays of various greenish or brownish hues that radiate from the umbos to all margins of the shell (Figure 8.7). Presence or absence of rays is relatively consistent within species, but their intensity and size are variable among individuals. Rays often vary in width and coverage of the shell and may be interrupted, wavy, or coalesced into regular or irregular patterns. Ray patterns are distinct and useful taxonomically in some Florida species (e.g., *Anodonta couperiana* and *Villosa vibex*). In some species they may be present but obscured in dark and particularly large specimens. The dark rays of *Villosa villosa* may be difficult to discern except when transmitting light through the valves (Figure 8.7). Secondary color patterns may represent the byproducts of metabolic processes sequestered in the periostracum (Watters et al. 2009).

Figure 8.7. Dark green rays on a light background in *Hamiota subangulata* (left). Photograph by James D. Williams. Dark, somewhat obscured rays in *Villosa villosa* (photographed with a light behind the shell) (right). © Richard T. Bryant.

It was noted during examination of thousands of museum lots that the periostracum of some specimens collected during the 1800s and early 1900s are brilliant shades of green and yellow. In these same streams today, shells of these species are usually shades of brown. A similar observation was made by Calvin Goodrich in correspondence with Grier (1920a). Prior to widespread deforestation, streams flowed clear in much of the eastern United States. Recent tendency towards darker shell color may be the result of high levels of turbidity now commonplace in streams. Similarly, brighter colors were noted when comparing mussels in relatively clear Lake Erie with those from sediment-laden streams (Walker 1913; Grier 1920a). Clench and Turner (1956) reported *Toxolasma paulum* and other species from the highly spring-influenced Santa Fe River to have a bright green periostracum. Populations of *T. paulum* from streams with no spring influence are typically brownish in color.

Internally along the dorsal shell margin of most mussels is a thickened structure called the hinge plate. Located on the hinge plate are the pseudocardinal and lateral teeth (Figure 8.8), usually the most prominent features on the inner shell surface. The pseudocardinal and lateral teeth in opposing valves interlock to provide stability and resist shear stress. Pseudocardinal teeth are usually jagged projections beneath the umbo near the anterior end of the valves, typically two in the left valve and one in the right. A small accessory denticle may be present anterior and/or posterior to pseudocardinal teeth. Pseudocardinal teeth are variable, ranging from thick and triangular to thin and bladelike structures. Lateral teeth are elongate ridges on the hinge plate, usually two in the left valve and one in the right (Figure 8.8). Variability of lateral teeth among species is primarily manifest in length, thickness, and degree of curvature. The relatively flat area on the hinge plate between the pseudocardinal and lateral teeth is the interdentum. The umbo cavity is located beneath the interdentum and varies from being very shallow (e.g., *Anodontoides*) to very deep (e.g., *Alasmidonta*). The thickened hinge plate and teeth are reduced or absent in anodontines (e.g., *Alasmidonta*, *Anodonta*, and *Utterbackia*).

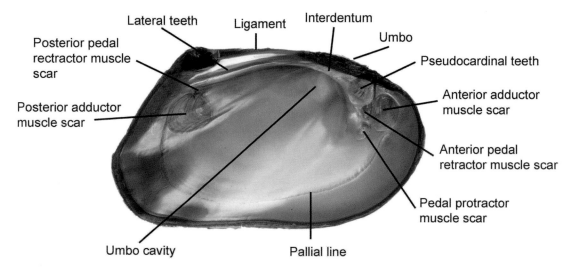

Figure 8.8. Internal morphology of a freshwater mussel shell, *Elliptoideus sloatianus*. © Richard T. Bryant.

There are other prominent internal shell features that are formed by the attachment of musculature. Anterior to the pseudocardinal teeth is the deep, generally roughened attachment scar of the anterior adductor muscle (Figure 8.8). The posterior adductor muscle scar, located ventral to the distal end of the lateral teeth, is shallow and usually smooth. The other paired muscle scars are the anterior pedal retractor, located between the pseudocardinal teeth and lateral teeth, and the posterior pedal retractor, located between the lateral teeth and posterior adductor muscle scar. The pedal protractor is an unpaired muscle whose scar is located ventral to the anterior adductor and pedal retractor muscle scars. The pedal retractors and protractor may be immediately adjacent to, and bundled with, their associated adductor muscle scars, making them difficult to distinguish. The pallial line is a scar formed by the mantle attachment and roughly parallels the ventral shell margin extends between the two adductor muscle scars. Other muscle scars may be present in the cavity of the shell but are generally small and inconspicuous.

Shell material is produced by the mantle, a thin, sheetlike organ that envelops the soft parts of the mussel. Valves have three basic layers of material, with a fourth layer strictly associated with muscle attachments scars. The outermost layer is the periostracum, a thin, proteinaceous layer that protects the shell from the hypoosmotic medium of freshwater. Underlying the periostracum is the prismatic layer, composed primarily of minute calcium carbonate crystals, oriented perpendicularly to the valve surface. The innermost layer is the nacre, which is primarily aragonite, a lustrous form of calcium carbonate crystals arranged parallel to the valve surface. The prismatic layer and nacre also have small quantities of organic protein components, including conchiolin. The organic matrix of these layers protects the shell from dissolution in acidic waters and the abrasive effects of sediment and current. The microstructure of conchiolin, which binds with aragonite in the nacre to strengthen the shell, is composed of three distinct layers and has been suggested as a taxonomic character (Kat 1983a). Bivalve nacre is a complex conglomeration of mineral and biological compounds with mechanical properties that far surpass those of its component parts, making it an intriguing compound for engineers to study in the new field of biomimetics when fabricating products for commercial applications (Jackson et al. 1988; Sanchez et al. 2005; Cartwright and Checa 2007). A fourth layer, the hypostracum, formed by fibers of the anterior and posterior adductor muscles and is found only beneath the attachments.

The nacreous layer forms the thickest and heaviest part of the valve. Only the margin of the mantle produces the periostracum and prismatic layers, whereas the entire distal surface of the mantle produces nacre. Mussels can repair shell dissolution and erosion only by producing more nacre. Mussels form pearls around foreign objects that lodge between the mantle and nacre in the same manner as they form new shell (Cartwright and Checa 2007). This ability led to the naming of both families of North American mussels, the Margaritiferidae and Unionidae. The terms *margaritos* and *unio* are Greek and Latin, respectively, for pearl (Martin 1997).

Nacre color is variable, but in most mussels it is lustrous white. Others are various hues of cream, pink, salmon, purple, blue, or greenish gray. Within species, nacre color is generally consistent throughout their ranges, but in some taxa there appears be a gradient in nacre color from white in the north to purple in the south (e.g., *Villosa lienosa*). Some species may also have a bicolor nacre with a darker color marginally and a lighter shell cavity. In examining a large number (about 100 individuals) of *Glebula rotundata* from one site in the Apalachicola River, approximately half of the individuals had white nacre and the remainder had shades of pink or purple. Nacre color within *Elliptio jayensis* varies from purple to pink, peach, cream, or white within a single lake. Nacre is often iridescent posteriorly, which is most apparent in shells with bluish-white nacre (e.g., *Villosa villosa*). The mechanisms that dictate nacre color have not been critically studied despite having been observed for centuries.

Variation in Shell Morphology

The wide range of shapes and ornamentation among mussels has long intrigued malacologists and conchologists. Hundreds of species were described in the 1800s and early 1900s based on nuances in shell shape. This was particularly true in peninsular Florida where dozens of nominal species of *Elliptio* were described (Williams et al. 2011). Shell shape is influenced by various genetic and environmental factors (Stanley 1970, 1975; Watters 1994b). Environmental factors include hydrology, stream gradient, substrate characteristics, and food availability (Eager 1950; Tevesz and Carter 1980). The range of shapes in *Elliptio* is a classic example of environmentally influenced shell morphology (Clench and Turner 1956). Understanding the taxonomic implications of ecophenotypic variation, Ball (1922) observed that "many forms now described as distinct species are but variations connected with the environmental conditions." Ecophenotypic variation has confounded malacologists for over a century and continues to obscure species boundaries and relationships. Smooth-shelled forms, the majority of Florida mussels, are more easily misidentified than sculptured species (Shea et al. 2011).

The greatest force influencing shell inflation and sculpture in lotic mussels is unidirectional current (Watters 1994b). The same species may appear quite different when comparing lotic and lentic individuals as well as along lotic gradients (Surber 1912). The concept of shell shape changing along the gradient from headwaters to larger rivers was studied by Ortmann (1920) and subsequently referred to as Ortmann's rule or law of stream position (Watters et al. 2009; Hornbach et al. 2010). According to this law of clinal variation, within a species, the headwater forms tend to have thinner shells and are more compressed than large-river forms, which tend to have thicker shells and be more inflated. The umbo is often higher in downstream or large-river populations.

To address buoyancy in the soft substrates of lakes and lentic habitats associated with streams, mussels have adapted by reducing weight or increasing inflatedness by enlarging the percentage of shell surface area in contact with the substrate (Watters 1994b). Buoyancy and shell inflatedness may be directly proportional (Anderson and Ingham 1978; Ghent et al. 1978). Relatively few species, however, have combined inflatedness with thin shells (e.g., *Pyganodon*) to further enhance buoyancy (Watters 1994b). Species with thin shells are at greater risk of predation, abrasion, and desiccation (Gagnon et al. 2006).

Some lentic mussels appear to have proportionally higher shells than stream species (Grier 1918; Ortmann 1919; Roper and Hickey 1994). Mussels in northern lakes (e.g., *Elliptio complanata*, *Lampsilis siliquoidea*, and *Pyganodon grandis*) were less obese with increased depth (Hinch et al. 1989; Amyot and Downing 1991). Turbulence in wave-washed littoral zones of lakes may affect shape (Green et al. 1989), generally manifest in greater inflatedness (Roper and Hickey 1994). Inherent differences between sand and mud habitats may also account for variable shell morphologies (Bailey 1989).

One of the most common shell defects is umbonal erosion, due primarily to physical abrasion and dissolution from acidic waters (Tevesz and Carter 1980; Hinch and Green 1988; Roper and Hickey 1994). Erosion of shells in Florida was noted by Simpson (1899), who observed "in Florida many Unios are destroyed by having holes actually eaten through to the animal." There is a direct relationship with age and degree of erosion (Roper and Hickey 1994). Many mussels exhibit brass-colored splotches on the nacre, which are often referred to as oil or grease spots (Clench and Turner 1956; Martin 1997). These spots appear to represent adventitious conchiolin material and may be a response to reinforce shell material since they are often found opposite sites of erosion (Tevesz and Carter 1980).

Deformed shells are not an uncommon occurrence and can be caused by a number of environmental and possibly genetic factors. Bubbling or blistering of the nacre can occur in one or both valves and in one instance, in *Pyganodon grandis*, was reported to be caused by a trematode parasite (Simpson 1899). Commensal larval chironomids along the growing edge of the shell may actually become incorporated into capsules during shell formation (Forsyth and McCallum 1978; Roper and Hickey 1994). Accumulation of lime, algal growth, and sediments trapped between the mantle and shell may result in deformation (Tevesz and Carter 1980). Shell deformities are not uncommon among mussels that become sufficiently lodged between rocks, roots, or debris so as to prevent space for normal growth and development. A collection of *Elliptio pullata* from Holmes Creek, Florida, had many individuals whose shells were twisted to the left or right, but no explanation was offered (Clench and Turner 1956). A more unusual deformity is the heterodont condition, where the typical development of two pseudocardinals and two laterals are found in the right valve instead of the left.

Internal Anatomy

Inside the shell, a mussel body appears as an amorphous mass but displays bilateral symmetry and provides all functions needed for life. Color of the body mass is typically creamy white or light brown but is orange to reddish orange in some species. The most prominent mussel body feature is the foot, which extends anterioventrally from the center of the animal (Figure 8.9). Internal organs (e.g., heart, stomach, and gonad) are housed within the visceral mass at the proximal end of the muscular foot.

Immediately inside of the valves is the mantle, which envelops the other soft tissues (Figure 8.9). The mantle is a thin layer of tissue that lies against both valves, fused dorsally where adjacent to the visceral mass but unfused posteriodorsally. Posteriorly, the unfused edges of the mantle margins are modified into incurrent and excurrent apertures. These differ from true siphons, which are retractable, fused tubular structures, found in most marine and some freshwater bivalves (excluding Margaritiferidae and Unionidae). The incurrent aperture is located at the posterior margin of the shell and the excurrent aperture is just dorsal to the incurrent aperture. A third aperture, the supra-anal, is directly dorsal of the excurrent aperture. A mantle bridge separates the excurrent aperture from the supra-anal aperture in most unionids but is variable among and within species and absent in margaritiferids. The mantle adheres to the valves in a long arc of pallial muscles that lie roughly parallel to the anterior, ventral, and posterior margins of the shell. The dorsal muscles, located in the umbo cavity, attach the body of the mussel to the valves.

Between the mantle and visceral mass are the inner and outer gills, which are attached dorsally and extend most of the length of the shell.

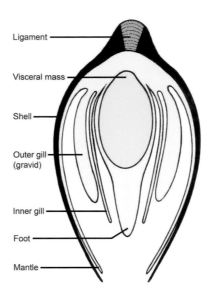

Figure 8.9. Diagrammatic cross section of a freshwater mussel. Illustration by Susan Trammell.

A pair of adductor muscles, located near the anteriodorsal and posteriodorsal shell margins, contract to close the valves (Figure 8.10). A single pedal protractor and pair of pedal retractor muscles for controlling foot movements are directly adjacent to the two adductor muscles.

Water containing food and oxygen is brought into the mantle cavity through the incurrent aperture by actions of cilia located on the surface of the gills. Food particles are trapped by the gill cilia, passed ventrally to the food groove located on the gill distal margin, and then anteriorly to the labial palps, which funnel the food into the mouth. Nonfood particles are passed to the suprabranchial chamber, which is a narrow compartment located dorsal to the gills. Those particles are then expelled with water currents through the excurrent aperture. The suprabranchial chamber is separated from the mantle cavity by an incomplete diaphragm composed of the dorsal portions of the gills in unionids and the dorsal portions of the gills, as well as by diaphragmatic septa of the mantle, in margaritiferids.

Mussel gills are multifunctional, serving as organs for respiration and food capture, as well as marsupia in brooding females. One pair of sheetlike gills, individually referred to as demibranchs or ctenidia, is located on each side of the body (Figure 8.11). Gills are composed of fused filaments that form water tubes and are aligned perpendicular to the long axis of each demibranch. They are usually semi-lunar in shape with variations depending on the species and position of the marsupia.

The pair of roughly triangular labial palps, which are fused basally and may be partially fused dorsally, is located at the anterioventral end of the gills. They serve to sort food particles and funnel them into the mouth. A short esophagus leads from the mouth to a well-defined stomach that is entirely enveloped within the digestive gland. A detailed description of the anatomy and functions of mussel organ systems was presented by Cummings and Graf (2010).

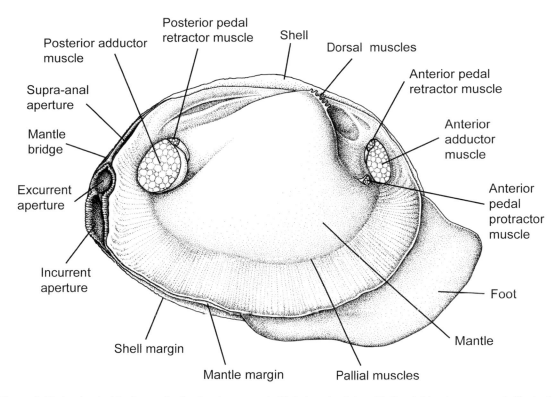

Figure 8.10. Anatomical features of a freshwater mussel, *Glebula rotundata*, with the right valve removed. Illustration by Susan Trammell.

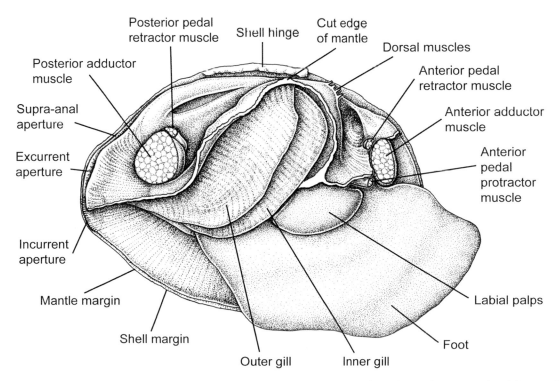

Figure 8.11. Anatomical features of a freshwater mussel, *Glebula rotundata*, with the right valve removed and the mantle cut away. Illustration by Susan Trammell.

Soft parts of mussels are often considered of little taxonomic value, however, there are exceptions. The genus *Alasmidonta* has square to rectangular blotches of dark pigment along the mantle margin. In some genera (e.g., *Elliptoideus* and *Quadrula*), the papillae associated with the incurrent aperture are dendritic as opposed to the typical condition of simple (no branching) or bifid (split distally). This condition is observed in live individuals and those that are relaxed prior to preservation. Marsupial gill margins in lampsilines are often pigmented (gray or black), while others are white. In some species the band of dark pigment is continuous in nongravid individuals but becomes interrupted at the margins of the water tubes as the gill expands when gravid (e.g., *Lampsilis* spp. and *Villosa* spp.). Marsupial regions of gills may appear differently than the remainder depending on the color of glochidia being brooded. In lampsilines the mantle margin and its associated structures (e.g., flaps and papillae) are highly variable in color patterns and shape in most species and are best observed when the animal is alive and in its natural habitat.

Chapter 9

Format of Accounts for Genera and Species

Genus Account

Accounts are included for all 23 genera of unionid mussels. Genera are arranged in alphabetical order followed by their assigned species. Each account contains the number of species included in the genus, distribution, taxonomic history, and general information.

Type Species

The genus, species, author, and year from the original description, and its current binomial are given in each account.

Diagnosis

Shell morphology, color, and soft anatomy characters that are diagnostic for all species included in the genus are provided.

Species Account

Accounts are included for all 60 species of Florida unionid mussels. The accounts are arranged in alphabetical order by genus and species. Information for each species is arranged in the same sequence to standardize accounts.

Scientific and Common Names

Scientific names follow those recognized in Turgeon et al. (1998) or proposed by Williams et al. (2008, 2011, in review), and Campbell and Lydeard (2012a). Taxonomic changes that are not addressed or deviate from Turgeon et al. (1998) and adopted herein are presented in Table 9.1. These changes were usually based on analysis of genetic data, shell morphology, and soft anatomy published in peer-reviewed literature. Author and year of publication follow the scientific name in the heading of each account.

Common names are those given in Turgeon et al. (1998) and recent updates provided by Williams et al. (2008) and herein. All common names utilized herein are capitalized, as is the practice for many other groups of animals (e.g., birds, reptiles, amphibians, and fishes). The standard taxonomic practice of placing parentheses around author and date of species names was followed when they were described in a genus different from the one currently recognized.

Illustrations

Color photographs of the exterior of the right valve and the interior of the left valve are provided for each species unless otherwise noted. Shells are horizontally aligned by the dorsal margins of anterior and posterior adductor muscle scars. This positioning resulted in some species appearing misaligned as the dorsal and/or ventral margins are not parallel to the top and bottom of the page. For species with pronounced sexual dimorphism, both males and females are included. For some species, additional photographs were included to illustrate variation in shell morphology or juvenile form. The length, museum catalogue number or field collection number, most complete locality data available, and collection date (if known) are included for each figured specimen.

Distance information (usually given in miles) recorded on the museum label is repeated without conversion to the metric system. Interpolated data for figured specimens are enclosed in brackets.

In addition to the introductory photographs of the species, some accounts include illustrations of other shells (e.g., archaeological), various morphological features (e.g., umbo sculpture and mantle flaps), and habitats.

Table 9.1. Scientific names of mussels that are not included in, or deviate from, those used in Turgeon et al. (1998).

Current Taxonomy	Turgeon et al. (1998)
Anodonta hartfieldorum	recently described species
Elliptio fumata	not recognized
Elliptio occulta	not recognized
Elliptio pullata	not recognized
Elliptio purpurella	not recognized
Fusconaia burkei	*Quincuncina burkei*
Hamiota australis	*Lampsilis australis*
Hamiota subangulata	*Lampsilis subangulata*
Lampsilis floridensis	not recognized
Obovaria choctawensis	*Villosa choctawensis*
Obovaria haddletoni	*Lampsilis haddletoni*
Quadrula infucata	*Quincuncina infucata*
Quadrula kleiniana	not recognized
Quadrula succissa	*Fusconaia succissa*
Reginaia apalachicola	recently described genus and species
Reginaia rotulata	*Obovaria rotulata*
Uniomerus carolinianus	not recognized

Description

Shell descriptions are based on examination of specimens from Florida and adjacent drainages in Alabama and Georgia. Descriptions are presented in telegraphic style in the following sequence: size, shell, teeth, nacre, and periostracum. The size information is based on the largest reported shell for the species and the largest known for Florida. If the largest known shell is from Florida, only one length is given.

Glochidium Description

Descriptions and scanning electron micrographs of glochidia, if available, are presented for each species. Glochidium descriptions and measurements are based on direct observations or published information. Most of the measurement data for glochidia reported from published accounts are based on specimens collected from basins outside the state of Florida. Terminology used to describe glochidial shape is based primarily on Hoggarth (1999) (Figure 9.1).

Similar Sympatric Species

A comparison of conchologically similar species is included in each account. Similar species are limited to those that are sympatric regardless of taxonomic affiliation. Distinguishing features are established principally on shell characteristics. For some taxa, obvious differences in easily observed soft anatomy characteristics are also provided.

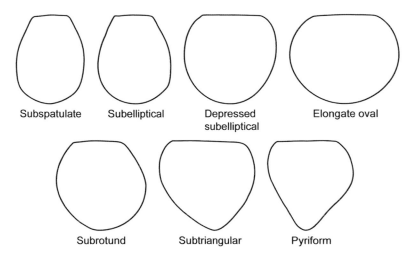

Figure 9.1. Glochidial shapes and terminology based on Hoggarth (1999). Illustration by S. Trammell.

Distribution in Florida

Information for Florida mussel distribution is based primarily on museum records, published and unpublished documents, and recent field work. Distribution for species in the panhandle is presented from west to east and in the peninsula from north to south. With the exception of *Reginaia apalachicola*, species recorded from archaeological and fossil sites are not plotted on the distribution maps.

Distribution Map

A dot distribution map is presented for each species. Locality data depicted on the map is from museum records, published and unpublished documents, and recent field work. Some of the early collections from peninsular Florida with specific locality data were difficult to plot as many of the landmarks of the late 1800s to the mid-1900s no longer exist. Other localities with imprecise or vague data were evaluated on a case-by-case basis before plotting. There are some instances where a single dot may represent multiple collections made in close proximity or collections from a given site on multiple dates. An inset map depicting the approximate range of the species in the United States is also provided.

Ecology and Biology

Information on ecology and biology is taken from published and unpublished data and field observations of the authors. Very little information is available on the biology of Florida mussels. For these taxa, inferences are made based on knowledge of related species. For rare or extirpated species, inferences are made based on what is known about their biology in adjacent states.

There is very little information available on host fishes for Florida mussels. For those species that have host fish data, we included information from published reports and the OSU Museum of Biological Diversity mussel/host database. Specific information on host fishes is limited to those species, native and introduced, that are known to occur in Florida.

Conservation Status

Range-wide and Florida conservation statuses are presented for all species that have been evaluated. Federal conservation status is presented for species protected under the federal ESA. There are no mussels on the official state list of endangered and threatened species. Current status information on occurrence throughout the species' Florida range and its relative abundance are also presented.

Remarks

A variety of topics such as taxonomy, phylogenetic relationships, morphological variation, distributional changes, introduction beyond native range, erroneous distribution records, and archaeological records are covered in this section. An explanation or the origin of the species name is presented for some taxa.

Synonymy

The genus, species, author, year, type designation, museum number, length, and the type locality are presented in the synonymies of all 60 taxa. Type species' designations (e.g., holotype, holotype by monotypy, and lectotype) are those reported by Johnson (1965, 1967a, 1970, 1974), Johnson and Baker (1973), and Williams et al. (2011). An illustration of the type specimen from the original description or a recent photograph (by James D. Williams unless otherwise noted) is also included if available. A partial synonymy for each species is included but limited to those described from Florida and contiguous drainages in Alabama and Georgia. A more extensive discussion of type material and associated information for mussel taxa from Florida was provided by Williams et al. (2011).

Chapter 10

Accounts for Genera and Species

Family Unionidae

The family Unionidae is represented by more than 140 genera and 674 species worldwide (Graf and Cummings 2007). In North America there are approximately 365 species and subspecies of freshwater mussels and approximately 300 are known to occur in the United States (Williams et al. in review). Most of the mussel diversity in the United States is centered in the southeastern states (Haag 2010). Of currently recognized United States mussel taxa, 60 (approximately 20 percent) occur in Florida, which is considerably less than those in Alabama (182) and Georgia (120) (Williams et al. in review). The Greater Floridan Region harbors a large number of endemic taxa, with 40 species (approximately 65 percent) confined to the region.

Within the Unionidae in North America, there are three recognizable groups based on conchological characters, soft anatomy, and life history traits. In some classification schemes, the three groups are recognized as subfamilies Lampsilinae, Anodontinae, and Ambleminae (Ortmann 1912). This breakdown of the family is still useful for purposes of subdividing the North American mussel fauna into three groups that share common conchological characters. However, when examined in the larger worldwide Unionidae context using molecular analyses, this subfamily classification is not supported. Instead, two subfamilies and five tribes are recognized in North America (Unioninae: Anodontini; and Ambleminae: Amblemini, Lampsilini, Pleurobemini, and Quadrulini) (Campbell and Lydeard 2012a).

Diagnosis

The two shells are approximately equal in size and bound together by a hinge ligament located along the dorsal margin. The shell is smooth or sculptured with corrugations, plications, wrinkles, and/or pustules. The periostracum may be shiny, dull, cloth-like, variously colored, and with or without rays. The umbo is typically sculptured and variable in elevation, ranging from even with the hinge line to elevated well above it. It may be narrow or broad and vary from slightly to highly inflated.

Internally, there is a hinge plate along the dorsal margin, typically with two pseudocardinal and two lateral teeth in the left valve, and one pseudocardinal and one lateral tooth in the right valve, though the hinge plate and teeth are reduced or absent in some species. The umbo cavity, located anteriorly under the hinge plate, may be very shallow to very deep. The two adductor muscle scars are variously developed; the anterior scar is most prominent and is represented by a well-defined oval to circular depression, while the posterior scar is often not well defined and difficult to delineate.

The inner valve surface is covered by the mantle, a thin layer of soft tissue that deposits shell nacre. The mantle margins are joined dorsally as well as along a mantle bridge separating the supra-anal and excurrent apertures. Posteriorly, the mantle is differentiated into incurrent and excurrent apertures, but the mantle separating the two is not fused into a mantle bridge. Mantle margins anterior to the incurrent aperture may be smooth, papillate, or developed into elaborate mantle lures in some species.

There are two pairs of gills, one on each side of the body. The gills are comprised of water tubes formed by interlamellar connections developed as continuous septa oriented perpendicular to the dorsal margin. In females the gills function as marsupia for developing glochidia, occupying all four gills, only the outer gills, or only a portion of the outer gills depending on the genus. Mature glochidia vary in size and shape among species, and some have styliform hooks on their ventral margins.

Distribution

The Unionidae occurs on most major land masses with freshwater habitats except Australia, Greenland, New Guinea, New Zealand, and South America (Figure 10.1). They inhabit most of Europe, Asia, North America, portions of northern and sub-Saharan Africa, and northwestern Madagascar. In Asia they occur in India, Japan, Philippines, Borneo, Sumatra, and Java (Bogan 2008).

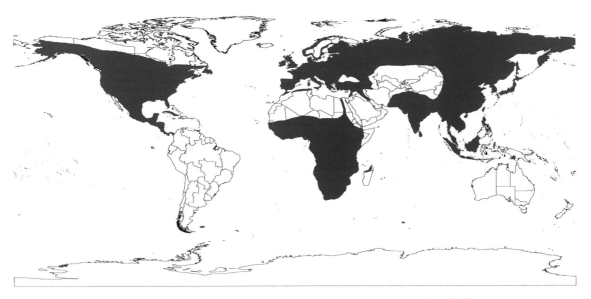

Figure 10.1. Worldwide distribution of the family Unionidae (modified from Bogan and Roe 2008).

Genus *Alasmidonta* Say 1818

Twelve species of *Alasmidonta* are recognized, all from North America (Turgeon et al. 1998). *Alasmidonta* occurs in the Great Lakes and Mississippi basins, Atlantic Coast rivers from Canada to Georgia, and eastern Gulf Coast rivers in Alabama, Florida, and Georgia. Two described species of *Alasmidonta* are known from Florida rivers—*A. triangulata* endemic to the Apalachicola basin and *A. wrightiana* endemic to the Ochlockonee basin. Only two other *Alasmidonta* species occur in eastern Gulf Coast rivers, *A. mccordi* and an undescribed species, both from the Mobile basin (Williams et al. 2008). There are no *Alasmidonta* in eastern Gulf Coast rivers east of Ochlockonee River and southern Atlantic Coast rivers south of Altamaha River.

Alasmidonta was included in a monograph of the tribe Alasmidontini by Clarke (1981), who recognized four subgenera. The two species occurring in the eastern Gulf Coast rivers of Florida, *Alasmidonta triangulata* and *Alasmidonta wrightiana*, were assigned to the subgenus *Alasmidonta*. The only other species assigned to this subgenus, *Alasmidonta arcula* and *Alasmidonta undulata*, are endemic to Atlantic Coast rivers.

Type Species
Monodonta undulata Say 1817 = *Alasmidonta undulata* (Say 1817)

Diagnosis
Shell thin to moderately thick; smooth; inflated; outline elliptical to rhomboidal; posterior ridge sharp, well developed; posterior slope with rudimentary to well-developed sculpture consisting of fine corrugations; umbo sculpture moderately thick, single-looped ridges, usually nodulous where they cross posterior ridge, nodules may extend onto shell disk; periostracum typically shiny, occasionally dull; pseudocardinal teeth poorly to moderately developed, usually 2 in left valve, 1 in right valve; interdental projection on left valve with corresponding depression in hinge of right valve; lateral teeth rudimentary to absent.

Soft tissues usually some shade of orange; posterior mantle margin with irregularly spaced, square to rectangular, dark brown to black blotches, which may be visible through the shell margin; inner lamellae of inner gills completely or partially attached to visceral mass; incurrent and excurrent apertures separate; outer gills marsupial, with secondary water tubes when gravid; marsupium extended beyond original gill margin when gravid; glochidium outline subtriangular to pyriform, with styliform hooks (Ortmann 1912; Simpson 1914; Clarke 1981).

Alasmidonta triangulata (Lea 1858)
Southern Elktoe

Alasmidonta triangulata – Upper image: length 55 mm, MCZ 190391. Chipola River, Dead Lake, Chipola Park, 20 miles south of Blountstown, Calhoun County, Florida, 3 September 1954. © Richard T. Bryant. Lower image: length 50 mm, UF455465. Flint River on the upper reach of backwaters of Seminole Reservoir [Lake Seminole], 7.7 air miles north-northeast of Bainbridge, Decatur County, Georgia, 9 August 2012. Photograph by Zachary Randall.

Description

Size: length to 70 mm, in Florida to 64 mm.

Shell: moderately thin, thickened anteriorly; smooth; moderately to highly inflated, width 40%–65% of length, small individuals usually more compressed; outline typically subtriangular, occasionally oval; anterior margin rounded; posterior margin obliquely truncate, bluntly point posterioventrally; dorsal margin slightly convex; ventral margin slightly convex to broadly rounded; posterior ridge sharp, weakly biangulate posterio-ventrally, especially in small individuals; posterior slope steep, flat, occasionally slightly concave dorsally, often with very thin corrugations from posterior ridge to posteriodor-sal margin; umbo broad, triangular, highly inflated, elevated well above hinge line; umbo sculpture 3–5 thick, single-looped ridges, nodulous where they cross posterior ridge; umbo cavity wide, very deep.

Teeth: pseudocardinal teeth thin in small individuals, moderately thick in large individuals, 2 teeth in left valve, anterior tooth low, rudimentary, posterior tooth high, triangular, typically 1 subtriangular tooth in right valve, with flat crest, considerably larger than those in left valve; interdental projection on left valve low, wide; lateral teeth absent, dorsal margin thickened.

Nacre: white, pinkish purple to purple.

Periostracum: somewhat shiny; small individuals yellowish brown to olive, occasionally with dark green rays, large individuals dark olive brown to black, usually obscuring rays.

Glochidium Description

The glochidium of *Alasmidonta triangulata* is unknown.

Similar Sympatric Species

Individuals of *Alasmidonta triangulata* may superficially resemble some small *Elliptio crassidens*. Small (length to 35 mm), highly inflated *E. crassidens* differ from *A. triangulata* in having a thicker shell, less inflated umbo, more rounded posterior ridge, and well-developed pseudocardinal and lateral teeth.

Alasmidonta triangulata can be distinguished from other mussel taxa in Apalachicola River by having a sharp posterior ridge, steep posterior slope with corrugations, rudimentary pseudocardinal teeth, and no lateral teeth.

Distribution in Florida

Alasmidonta triangulata occurs in Apalachicola and Chipola Rivers (Figure 10.2).

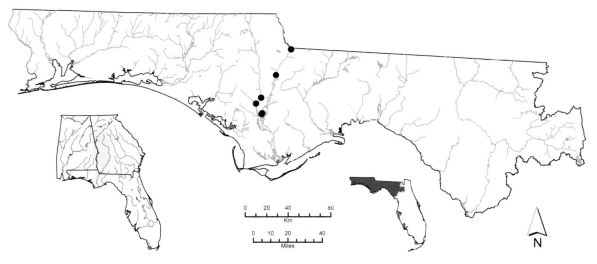

Figure 10.2. Distribution of *Alasmidonta triangulata*.

Ecology and Biology

Alasmidonta triangulata inhabits large creeks and rivers in substrates ranging from mixed sand and gravel to sandy mud. It may be found in backwater channels behind islands in slow current or pools. Its continued presence in the middle reach of the Apalachicola River is noteworthy. This reach of the river (river mile 30–60) has a bed elevation similar to historical conditions and has a more stable substrate and higher connectivity to the floodplain. The Apalachicola River bed is entrenched from Jim Woodruff Lock and Dam downstream to the vicinity of river mile 50.

Alasmidonta triangulata also inhabits the Flint River arm of Lake Seminole in southwestern Georgia. It occurs in the upper reach of the reservoir adjacent to the river channel where there is still some current over isolated patches of sandy mud habitat. The

Chattahoochee River arm of Lake Seminole in Florida has similar isolated patches of sandy mud habitat but not adjacent to a river channel with current. It is unlikely that *A. triangulata* inhabits the Florida portion of Lake Seminole.

Alasmidonta triangulata is presumably a long-term brooder, gravid from late summer or autumn to the following summer. Glochidial hosts for *A. triangulata* are unknown.

Conservation Status

Most conservation status reviews have reported *Alasmidonta triangulata* as endangered throughout its range (Athearn 1970; Stansbery 1971, 1976; Williams et al. 1993, in review; Brim Box and Williams 2000). *Alasmidonta triangulata* (as *Alasmidonta undulata*) was considered to be endangered by Williams and Butler (1994). The USFWS was petitioned in 2010 to list this species for protection under provisions of the federal ESA.

Alasmidonta triangulata is extant in the Chipola and Apalachicola Rivers in Florida but is very rare.

Remarks

Alasmidonta triangulata has been treated as a synonym of *Alasmidonta undulata* by Clarke (1981) and Williams and Butler (1994). Clench and Turner (1956), Heard (1979a), Brim Box and Williams (2000), and Williams et al. (2008) recognized it as a valid species endemic to the Apalachicola River basin. Simpson (1892b) did not include *A. triangulata* in his review of Florida Unionidae. Johnson (1970) recognized *A. triangulata* but gave its distribution as the Apalachicola River basin and southern Atlantic Coast rivers. Some specimens from Savannah and Ogeechee Rivers are conchologically similar to *A. triangulata*. However, based on preliminary genetic analyses, *A. triangulata* from the Savannah and Ogeechee Rivers are aligned with *Alasmidonta arcula* (A.E. Bogan, personal communication), not *A. undulata*. Additional research is needed to determine species and distributional boundaries of *Alasmidonta* in the southeastern United States.

Synonymy

Margaritana triangulata Lea 1858 (Lea 1858a:138). Lectotype (Johnson 1970), USNM 86249, length 56 mm. Type locality: published as upper Chattahoochee [River], Georgia, [Chattahoochee River drainage,] restricted by Clench and Turner (1956) to Chattahoochee River, Columbus, [Muscogee County,] Georgia.

Alasmidonta wrightiana (Walker 1901)
Ochlockonee Arcmussel

Alasmidonta wrightiana – Upper image: length 51 mm, UF 4167. Ochlockonee River on highway from Jacksonville to Pensacola [U.S. Highway 90], Gadsden County, Florida, 8 June 1930. © Richard T. Bryant. Lower image: length 41 mm, UF 4167. Ochlockonee River on highway from Jacksonville to Pensacola [U.S. Highway 90], Gadsden County, Florida, 8 June 1930. © Richard T. Bryant.

Description

Size: length to 55 mm.

Shell: thin to moderately thick, thicker anteriorly; smooth; highly inflated, width 45%–55% of length; outline typically oval, occasionally subtriangular; anterior margin rounded; posterior margin rounded; dorsal margin slightly convex; ventral margin slightly convex; posterior ridge sharp dorsally, rounded posterioventrally; posterior slope steep, flat, occasionally slightly concave dorsally, with fine corrugations from posterior ridge to posteriodorsal margin; umbo broad, highly inflated, elevated well above hinge line; umbo sculpture thick, nodulous ridges parallel to annuli; umbo cavity wide, very deep.

Teeth: pseudocardinal teeth moderately thick, usually 2 teeth in left valve, anterior tooth small, rudimentary, occasionally absent, posterior tooth triangular, typically 1 subtriangular tooth in right valve; interdental projection on left valve low, wide; lateral teeth absent, dorsal margin thickened.

Nacre: white to bluish white, occasionally with yellowish stain in cavity of shell.

Periostracum: dark olive brown to black, with broad dark green rays intermingled with narrow dark green rays, may be obscure in darker individuals.

Glochidium Description

The glochidium of *Alasmidonta wrightiana* is unknown.

Similar Sympatric Species

Individuals of *Alasmidonta wrightiana* may superficially resemble some small *Elliptio crassidens*. Small (length to 35 mm), highly inflated *E. crassidens* differ from *A. wrightiana* in having a thicker shell, less inflated umbo, more rounded posterior ridge, and well-developed pseudocardinal and lateral teeth.

Alasmidonta wrightiana can be distinguished from other mussel taxa in Ochlockonee River by having a sharp posterior ridge, steep posterior slope with corrugations, rudimentary pseudocardinal teeth, and no lateral teeth.

Distribution in Florida

Alasmidonta wrightiana is known only from Ochlockonee River (Figure 10.3).

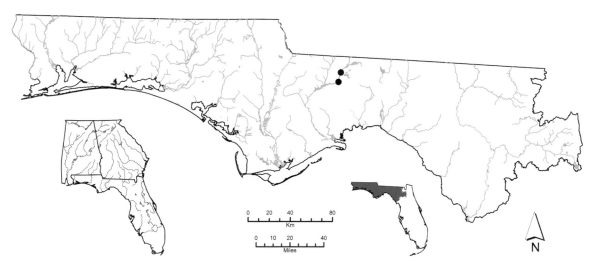

Figure 10.3. Distribution of *Alasmidonta wrightiana*.

Ecology and Biology

Alasmidonta wrightiana likely inhabited side channels and backwater riverine habitats. Ochlockonee River upstream of Lake Talquin, where the species was last collected, has predominantly sand to sandy mud substrate, with pools and runs of slow to moderate current. Much of the *A. wrightiana* habitat was likely inundated by Jackson Bluff Dam, which forms Lake Talquin. Substrate of Ochlockonee River downstream of Lake Talquin is highly disturbed for several miles due to channel entrenchment following construction of the dam.

Alasmidonta wrightiana was presumably a long-term brooder, gravid late summer or autumn to the following summer. Glochidial hosts for *A. wrightiana* are unknown. It is possible that this species had an anadromous fish host (e.g., *Alosa alabamae*) whose migration up Ochlockonee River was blocked by Jackson Bluff Dam, thus contributing to its demise (Williams and Butler 1994).

Conservation Status

Most conservation status reviews have reported *Alasmidonta wrightiana* as extinct or possibly extinct (Athearn 1970; Stansbery 1971, 1976; Williams et al. 1993, in review). It was listed as a candidate for protection under provisions of the federal ESA (USFWS 1984, 1989, 1991, 1994).

Alasmidonta wrightiana was listed as status undetermined by Williams and Butler (1994) based on the absence of a thorough Ochlockonee River survey. The fact that it has not been found since 1931 is potentially due to the lack of sampling effort in appropriate quieter backwater habitats. However, since the species is not known to occur in Georgia and has not been seen in Florida since 1931, it may be extinct.

Remarks

The last collection of *Alasmidonta wrightiana* was from Ochlockonee River, west of Tallahassee, Gadsden and Leon counties, Florida, in November 1931. During the past two decades, numerous collections have been made in Ochlockonee River in Florida and Georgia. While the river still supports a good mussel assemblage, no *A. wrightiana* have been found.

Based on a paucity of museum records, *Alasmidonta wrightiana* appears to have been a rare mussel and was known only from the Florida portion of Ochlockonee River. In his monograph on *Alasmidonta* and related genera, Clarke (1981) found 4 individuals in natural history museum collections but overlooked 11 specimens in FLMNH.

The distinctive shell morphology of *Alasmidonta wrightiana* led Simpson (1914) to recognize it as a species of *Strophitus* and placed it in a new subgenus, *Jugosus*. Clench and Turner (1956) considered *Alasmidonta wrightiana* to be a synonym of *Alasmidonta triangulata*, but Johnson (1967b) recognized its distinctiveness and treated it as a valid species. *Alasmidonta wrightiana* was reported by Heard (1979a) to be endemic to the Ochlockonee River basin. It was erroneously reported from the Choctawhatchee River basin in Alabama by Stansbery (1976).

Synonymy

Strophitus wrightianus Walker 1901 (Walker 1901:65). Holotype by monotypy, UMMZ 74938, length 54 mm. Type locality: Ochlockonee River, Florida, [Ochlockonee River basin]. In the original description the type locality was erroneously published as "tributaries of the Flint River, Baker County, Georgia," but was subsequently corrected by Walker (*in* Simpson 1914).

Genus *Amblema* Rafinesque 1820

There are three species in *Amblema* (Turgeon et al. 1998), which is confined to lakes and streams of eastern North America. One species, *Amblema plicata*, is wide-ranging, occurring throughout the Mississippi and Great Lakes basins and Gulf Coast rivers from the Guadalupe River, Texas, east to Choctawhatchee River, Florida. A second species, *Amblema neislerii*, is endemic to Apalachicola and Flint Rivers in Florida and Georgia, and the third species, *Amblema elliottii*, is endemic to Mobile basin, Alabama and Georgia.

Type Species
Amblema costata Rafinesque 1820 = *Amblema plicata* (Say 1817)

Diagnosis
Shell large; sculpture usually present, variable plications on posterior two-thirds of disk; corrugations present on the posterior slope; outline round to trapezoidal; umbo positioned anteriorly; umbo sculpture thick ridges, nodulous where they cross posterior ridge; periostracum dull to clothlike; pseudocardinal and lateral teeth thick, 2 in left valve, 1 in right valve.

Supra-anal aperture separated from excurrent aperture by very short mantle bridge, absent in some individuals; inner lamellae of inner gills connected to visceral mass only anteriorly; all 4 gills marsupial; marsupium not extended beyond original gill margin when gravid; glochidium outline depressed subelliptical to subrotund, without styliform hooks (Ortmann 1912; Utterback 1915; Baker 1928).

Amblema neislerii (Lea 1858)
Fat Threeridge

Amblema neislerii – Upper image: length 70 mm, UF 369. Chipola River, Dead Lake at Chipola Park, 20 miles south of Blountstown, Calhoun County, Florida, 13 September 1954. © Richard T. Bryant. Lower image: length 41 mm, UF 449287. Apalachicola River at river mile 46.8, Gulf County, Florida, 7 August 2006. © Richard T. Bryant.

Description

Size: length to 115 mm.

Shell: moderately thick; with 7–11 slightly oblique, parallel plications, extending from anterior portion of shell to posterior margin, forming undulations at posterior edge; moderately to highly inflated, width 35%–60% of length; outline oval to quadrate; anterior margin typically rounded, occasionally truncate; posterior margin rounded to obliquely truncate; dorsal margin straight to slightly convex; ventral margin straight to convex; posterior ridge rounded, obscured by plications; posterior slope flat to slightly concave; umbo broad, moderately inflated, elevated slightly above hinge line, occasionally well above; umbo sculpture 3–4 thick ridges, nodulous where they cross posterior ridge; umbo cavity wide, moderately deep.

Teeth: pseudocardinal teeth large, thick, rough, triangular, 2 divergent teeth in left valve, 1 tooth in right valve, occasionally with accessory denticle anteriorly and/or posteriorly; lateral teeth moderately long, moderately thin to thick, straight to slightly curved, 2 in left valve, 1 in right valve; interdentum short to moderately long, narrow to moderately wide.

Nacre: white to bluish white.

Periostracum: small individuals dull to somewhat shiny, large individuals cloth-like; dark olive brown to black.

Glochidium Description

Outline depressed subelliptical; length 180–220 μm; height 200–219 μm; dorsal margin 135–149 μm; ventral margin with internal lanceolate micropoints (O'Brien and Williams 2002) (Figure 10.4).

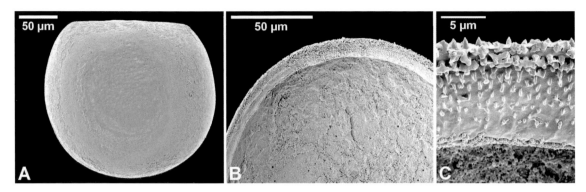

Figure 10.4. Scanning electron micrographs of glochidial valves of *Amblema neislerii* from Apalachicola River at Brickyard Island, Franklin County, Florida, 5 June 1997—(A) external view; (B) internal view of ventral margin and micropoints; and (C) close-up of micropoints (modified from O'Brien and Williams 2002).

Similar Sympatric Species

Amblema neislerii occurs sympatrically with two large, thick-shelled species, *Megalonaias nervosa* and *Elliptoideus sloatianus*, which also have highly sculptured shell disks. *Amblema neislerii* differs from these species by having plications on the shell disk that are roughly parallel to the ventral margin and fewer wrinkles on the umbo.

Distribution in Florida

Amblema neislerii occurs in Apalachicola and Chipola Rivers (Figure 10.5).

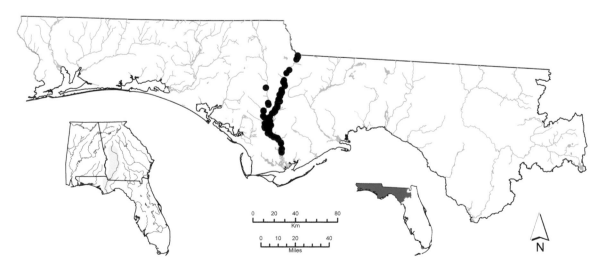

Figure 10.5. Distribution of *Amblema neislerii*.

Ecology and Biology

Amblema neislerii inhabits large rivers and floodplain distributaries in slow to moderate current (Figure 10.6). They are typically found in a substrate of stable sand or sandy mud along stream banks and side channels behind small islands. They have also been found in sand mixed with fine gravel and rocky rubble. *Amblema neislerii* was reported as "amazingly abundant" in Dead Lake, Chipola River near Wewahitchka, by Clench and Turner (1956) where they found 10–15 individuals per square meter along 200 m of shoreline. Van Hyning (1925) reported that a "nest" of *A. neislerii* had been located by a fisherman in Chipola River, Florida, but no specific locality was given.

Amblema neislerii is gravid with developing glochidia in early May and mature glochidia in early June when water temperature warms to approximately 24°C. In captivity *A. neislerii* glochidia were broadcast in a white, adhesive, weblike mass of larval threads, suggesting that it is a host generalist. Currents in the holding tank would cause the mass to expand and wrap around the body of any fish that swam into it, affording glochidia a chance for attachment. Glochidia remained viable for two days after being released from the mussel (O'Brien and Williams 2002). In laboratory trials *Amblema neislerii* glochidia metamorphosed on five fishes that occur in Florida—Centrarchidae: *Lepomis macrochirus* (Bluegill), *Lepomis microlophus* (Redear Sunfish), and *Micropterus salmoides* (Largemouth Bass); Cyprinidae: *Notropis texanus* (Weed Shiner); and Percidae: *Percina nigrofasciata* (Blackbanded Darter). These fishes are considered to be potential hosts as they occur in the same habitat as *A. neislerii*. *Notropis texanus* transformed the smallest number (mean = 3) of juveniles per fish, and *L. macrochirus* transformed the largest number (mean = 27). Glochidia transformation occurred 10–14 days in 23°C well water. Two additional fishes that occur in Florida—Poeciliidae: *Gambusia holbrooki* (Eastern Mosquitofish) and Ictaluridae: *Noturus leptacanthus* (Speckled Madtom)—were exposed to glochidia, but no juvenile mussels were recovered (O'Brien and Williams 2002).

Conservation Status

Amblema neislerii was considered to be endangered in several rangewide status assessments (Athearn 1970; Stansbery 1971; Williams et al. 1993, in review; Brim Box and Williams 2000). The species was deemed to be endangered in Florida by Williams and Butler (1994). In 1998 *A. neislerii* was added to the list of endangered species under the authority of the federal ESA (USFWS 1998). Critical habitat was designated in 2007 and included a total of 786.6 km in Florida and Georgia. In Florida, critical habitat was designated primarily in the main channels of the Apalachicola and Chipola Rivers (USFWS 2007). It also included Swift Slough (Figure 10.6), a distributary located in the middle reach of the Apalachicola River near river mile 40. A finalized recovery plan was approved in 2003 (USFWS 2003).

Amblema neislerii is extant throughout Apalachicola River and some distributaries, as well as lower Chipola River. It is most common in the middle reach of Apalachicola River and less so elsewhere. *Amblema neislerii* appears to be reproducing and recruiting at most localities in the Apalachicola and Chipola Rivers.

Figure 10.6. Swift Slough, a distributary of Apalachicola River, Liberty County, Florida, 10 October 2008. View from the dewatered Swift Slough during a drought and river drawdown (looking west-southwest with Apalachicola River in the background). Photograph by James D. Williams.

Remarks

There are two published records of *Amblema neislerii* from Escambia River in Florida (B.H. Wright 1898a; Heard 1979a), which are most likely based on *Amblema plicata*. All examined museum specimens of *Amblema* from Escambia River basin in Florida and Alabama have been *A. plicata*. *Amblema neislerii* was reported from an archaeological site on Chattahoochee River just upstream of Jim Woodruff Lock and Dam (Bullen 1958).

Synonymy

Unio neislerii Lea 1858 (Lea 1858b:165). Lectotype (Johnson 1974), USNM 83993, length 70 mm. Type locality: Flint River at Lanier, [about 6 miles north of Oglethorpe, Macon County,] Georgia, [Flint River drainage].

Amblema plicata (Say 1817)
Threeridge

Amblema plicata – Upper image: length 96 mm, UF 3233. Choctawhatchee River, lake at Oakey Bend, Walton County, Florida, 1933. Photograph by Zachary Randall. Middle image: length 70 mm, UF 47024. Conecuh River, 1 mile southeast of Brewton, Escambia County, Alabama, 14 August 1976. © Richard T. Bryant. Lower image: length 51 mm, UF 3233. Choctawhatchee River, lake at Oakey Bend, Walton County, Florida, 1933. Photograph by Zachary Randall.

Description

Size: length to 178 mm, in Florida to 113 mm.

Shell: thick; with large, deep, oblique plications, forming undulations at posterior edge; moderately compressed to inflated, width 35%–60% of length; outline round (Escambia River basin) to quadrate (Choctawhatchee River basin); anterior margin typically rounded, occasionally truncate; posterior margin rounded to obliquely truncate; dorsal margin straight to slightly convex; ventral margin straight to rounded; posterior ridge rounded, typically obscured by plications; posterior slope flat to slightly concave, usually with corrugations curving upward from posterior ridge to dorsal margin; umbo broad, elevated slightly above hinge line; umbo sculpture irregular ridges, nodulous where they cross posterior ridge; umbo cavity moderately wide, deep.

Teeth: pseudocardinal teeth large, thick, rough, triangular, 2 divergent teeth in left valve, 1 tooth in right valve, occasionally with accessory denticle anteriorly; lateral teeth moderately long, straight to slightly curved, 2 in left valve, 1 in right valve; interdentum short, wide.

Nacre: white to bluish white, occasionally with pink or purple tint posteriorly.

Periostracum: small individuals dull to somewhat shiny, large individuals cloth-like; small individuals yellowish green to olive, large individuals dark brown or black.

Glochidium Description

Outline subrotund; length 177–224 μm; height 191–242 μm; dorsal margin 92–165 μm, based on specimens from areas outside eastern Gulf Coast basins (Lefevre and Curtis 1910; Ortmann 1912, 1914; Surber 1912; Howard 1914b; Utterback 1915, 1916a; Coker et al. 1921; Kennedy and Haag 2005).

Similar Sympatric Species

Amblema plicata may resemble *Megalonaias nervosa* but differs in having deeper plications on the shell disk, fewer wrinkles, and a smooth umbo.

Distribution in Florida

Amblema plicata occurs in Escambia and Choctawhatchee Rivers (Figure 10.7).

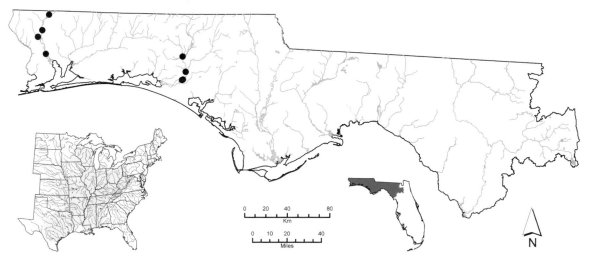

Figure 10.7. Distribution of *Amblema plicata*.

Ecology and Biology

Amblema plicata occurs in large creeks to rivers, natural lakes, and some impoundments. It can be found in swift waters as well as pools. In riverine reaches it inhabits a variety of substrates. In Escambia River it inhabits deeper water in substrates of sand mixed with clay or fine gravel. No information is available on habitat conditions associated with the Choctawhatchee River population.

Amblema plicata is a short-term brooder, gravid from May or June to August (Frierson 1904; Ortmann 1909b, 1912, 1914; Surber 1912; Utterback 1915, 1916a; Coker et al. 1921; Holland-Bartels and Kammer 1989; Howells 2000) and attains sexual maturity at four years of age (Stein 1969). A population of mature females in Sipsey River, Alabama, was reported to have a gravidity rate of 71 percent and fecundity averaged 229,738 glochidia per individual (Haag and Staton 2003). Conglutinates in northern populations are reported to be white, compressed and lanceolate, and discharged in broken or loose, irregular masses (Ortmann 1912; Utterback 1915).

Host fish investigations for glochidia of *Amblema plicata* include natural infestations and laboratory infestations and transformations. These investigations included 25 species of fishes belonging to 10 families (OSU Museum of Biological Diversity 2012), suggesting *A. plicata* is a host generalist. Of the potential host fishes, 11 occur in Florida. Natural infestations on potential hosts include three fishes—Centrarchidae: *Lepomis gulosus* (Warmouth); Ictaluridae: *Ictalurus punctatus* (Channel Catfish); and Moronidae: *Morone chrysops* (White Bass). Laboratory infestations have been documented for two fishes—Ictaluridae: *Pylodictis olivaris* (Flathead Catfish) and Percidae: *Perca flavescens* (Yellow Perch)—but transformation success was not reported. Successful laboratory transformations were reported for seven fishes—Centrarchidae: *Lepomis cyanellus* (Green Sunfish), *Lepomis macrochirus* (Bluegill), *Lepomis megalotis* (Longear Sunfish), *Micropterus salmoides* (Largemouth Bass), *Pomoxis annularis* (White Crappie), and *Pomoxis nigromaculatus* (Black Crappie); and Percidae: *Perca flavescens* (Yellow Perch) (OSU Museum of Biological Diversity 2012).

Conservation Status

Amblema plicata was reported as currently stable throughout its range by Williams et al. (1993, in review). In Florida its conservation status (as *Amblema perplicata*) was undetermined (Williams and Butler 1994).

Status of *Amblema plicata* in Florida is somewhat tenuous. It has not been found in Choctawhatchee River since the early 1900s in spite of concerted efforts during the past 20 years. It is extant in Escambia River in isolated patches of habitat but in low numbers.

Remarks

Shell morphology and growth rates of *Amblema plicata* are highly variable among individuals from different habitats as well as among populations in different basins (Ortmann 1920; Ball 1922; Stein 1969). During the 1800s several *A. plicata* synonyms were described based on shell morphology variation. In eastern Gulf Coast rivers, the name *perplicata* has been recognized as both a species and subspecies by various authors. Based on genetic analyses, Mulvey et al. (1997) found *Amblema perplicata* to be conspecific with *A. plicata*. *Amblema plicata* was reported from Escambia River by Heard (1979a) as *A. perplicata*. Clench and Turner (1956) did not report *A. plicata* from Florida. *Amblema plicata* was erroneously reported from Yellow River, Florida, by Burch (1975a), but the record was actually from Choctawhatchee River, Florida. *Amblema plicata* is not known from reaches of Choctawhatchee River in Alabama (Williams et al. 2008).

Synonymy

Unio plicata Say 1817 (Say 1817:[unpaginated, pages 11–12]). Syntype(s) lost; neotype (Haas 1930), SMF 4305, length 133 mm, not figured. Type locality: Ohio River, [Ohio River drainage]. The neotype designated by Haas (1930) may not be valid (Watters et al. 2009).

Genus *Anodonta* Lamarck 1799

Ten *Anodonta* species were recognized in the most recent AFS list of mollusk names (Turgeon et al. 1998). One new species, *Anodonta hartfieldorum*, was recently described from eastern Gulf Coast rivers of Louisiana, Mississippi, Alabama, and Florida (Williams et al. 2009). Three additional species—*Anodonta couperiana*, *Anodonta heardi*, and *Anodonta suborbiculata*—occur in eastern Gulf Coast rivers. As currently recognized, *Anodonta* is Holarctic in its distribution and in North America occurs from Alaska and Canada south to Mexico.

Taxonomy of *Anodonta* is currently in a state of flux and relationships of North American and European species are being examined. A phylogeny of *Anodonta*, which included the type species *A. cygnea* from Western Europe, was based on morphological and allozymic data and resulted in dividing North American members of the genus into three clades. The clades *Anodonta*, *Pyganodon*, and *Utterbackia* were recognized as genera by Hoeh (1990). However, a recent genetic analysis suggests that Asian, European, and North American species cannot be resolved as a monophyletic group. This study suggests that eastern North American species currently assigned to *Anodonta* be placed in the next available genus, *Utterbackiana* (Zanatta et al. 2007; A.E. Bogan, personal communication).

Type Species
Mytilus cygnea Linnaeus 1758 = *Anodonta cygnea* (Linnaeus 1758)

Diagnosis
Shell thin; smooth; compressed to inflated; outline elliptical to oval; umbo sculpture thin, undulating ridges; periostracum shiny, occasionally dull; without pseudocardinal and lateral teeth.

Inner lamellae of inner gills connected to visceral mass only anteriorly; supra-anal aperture usually small, separated from excurrent aperture by wide mantle bridge (may be as long as or longer than either of the apertures); outer gills marsupial, with secondary water tubes when gravid; marsupium extended ventrally beyond original gill margin when gravid; glochidium outline subtriangular to pyriform, with styliform hooks (Ortmann 1912; Simpson 1914).

Anodonta couperiana Lea 1840
Barrel Floater

Anodonta couperiana – Upper image: length 78 mm, UMMZ 103955. St. Marys River, Florida. Photograph by James D. Williams. Lower image: length 65 mm, UF 3699. Newnans Lake, 6 miles east of Gainesville, Alachua County, Florida, 27 June 1933. Photograph by Zachary Randall.

Description

Size: length to 125 mm.

Shell: thin; smooth; moderately compressed to inflated, width 25%–55% of length; outline oval; anterior margin broadly rounded; posterior margin narrowly rounded, angle of posterior and dorsal margin usually 145°–160°; dorsal margin straight; ventral margin broadly rounded; posterior ridge rounded, indistinct; posterior slope moderately steep, flat to slightly convex; umbo broad, inflated, elevated slightly above hinge line; umbo sculpture thin, undulating ridges; umbo cavity wide, shallow.

Teeth: pseudocardinal and lateral teeth absent.

Nacre: white to bluish white, usually iridescent.

Periostracum: usually shiny; greenish yellow, olive to brown, with very fine green rays radiating from umbo, increasing in width distally, often obscure in large individuals.

Glochidium Description

Outline pyriform; length 252–259 µm; height 275–282 µm; width 164–171 µm; with styliform hooks (Figure 10.8).

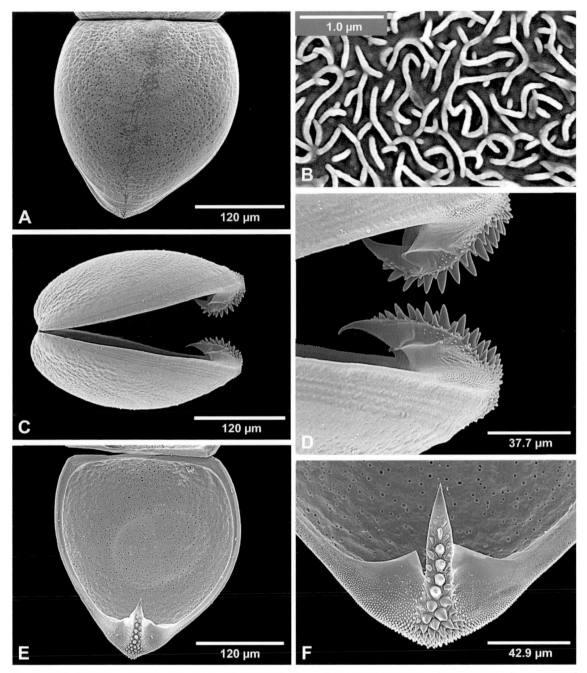

Figure 10.8. Scanning electron micrographs of glochidial valves of *Anodonta couperiana* from Arbuckle Creek, east of Sebring, Highlands County, Florida, 7 December 2007—(A) external view; (B) surface sculpture; (C) lateral view; (D) lateral view of styliform hooks; (E) internal view of styliform hook on ventral margin; and (F) close-up of styliform hook.

Similar Sympatric Species

Anodonta couperiana closely resembles *Utterbackia imbecillis* and *Utterbackia peninsularis* but attains a larger size, has a more rounded ventral margin, an umbo elevated very slightly above hinge line, and periostracum that is typically yellowish with numerous fine green rays, which may be obscure in large individuals.

Anodonta couperiana superficially resembles *Villosa amygdalum* and *Villosa vibex*, which have thin shells with well-developed dark green rays of various widths. However, *A. couperiana* lacks lateral and pseudocardinal teeth, which are always present in *V. amygdalum* and *V. vibex*.

Distribution in Florida

Anodonta couperiana occurs in peninsular Florida in Myakka and Peace River basins on the Gulf Coast and St. Marys and St. Johns River basins on the Atlantic Coast. It also occurs in Everglades basin and in some isolated coastal basins along the southern Atlantic Coast. There is a population that appears to be introduced in Lake Tarpon, just north of Tampa Bay in Pinellas County (Figure 10.9).

Figure 10.9. Distribution of *Anodonta couperiana*.

Ecology and Biology

Anodonta couperiana is found in a variety of habitats including streams and lakes. It is found in areas with little or no current in substrates ranging from firm sand, sandy mud, and mud, often containing fine particulate organic matter. It is also tolerant of altered environments such as reservoirs, phosphate pits, and canals and their adjacent wetlands. In canal habitats it may be found in substrates of mixed sand and lime rock rubble.

Anodonta couperiana is gravid from October to March. Females from the middle reach of the St. Johns River were checked in February (water temperature 72°F); some individuals had mature glochidia and others were partially spent. The gravid gill was yellowish to golden in color. Glochidial hosts for *A. couperiana* are unknown. Other species of *Anodonta* and several other anodontines (e.g., *Anodontoides*, *Pyganodon*, and *Utterbackia*) broadcast glochidia in loose masses in mucus webs and appear to be host generalists (Williams et al. 2008; Watters et al. 2009).

Anodonta couperiana has been observed to be commonly infested with mites (Unionicolidae). Of six individuals examined from two localities in St. Johns River basin, all were parasitized by *Unionicola formosa* (Vidrine 1996). Casual observations of preserved *A. couperiana* from peninsular Florida revealed a high percentage of individuals infested with mites.

Conservation Status

Anodonta couperiana was reported as currently stable throughout its range by Williams et al. (1993, in review).

Anodonta couperiana is extant in eastern and southern portions of peninsular Florida. It remains common throughout its range.

Remarks

Anodonta couperiana was reported from Florida by Simpson (1892b). It was reported from peninsular basins by Clench and Turner (1956), who erroneously included a record from Blackwater Creek, tributary to Hillsborough River, Hillsborough County. This represents a misidentified specimen of *Utterbackia peninsularis*, which was not described until 1995. *Anodonta couperiana* was reported from Apalachicola and Ochlockonee Rivers by Heard (1975a, 1977, 1979a). These records are most likely based on *Anodonta heardi*, which resembles *A. couperiana* but was not described until 1995.

Anodonta couperiana was named for J.H. Couper, who collected the shell used in the original description. The specific name was misspelled as "*cowperiana*" in the original description but was subsequently corrected to "*couperiana*" as a justified emendation since it was a patronym for J.H. Couper (Lea 1842).

Anodonta hartfieldorum Williams, Bogan, and Garner 2009
Cypress Floater

Anodonta hartfieldorum – Upper image: length 108 mm, UF455466. Gravel pit ponds at Mystic Springs boat ramp on Escambia River, Escambia County, Florida, 15 August 2012. Photograph by Zachary Randall. Middle image: length 89 mm, UF 428537. Escambia River at Bluff Springs boat ramp, Escambia County, Florida, 19 September 2007. Photograph by Zachary Randall. Lower image: length 36 mm, UF455466. Gravel pit ponds at Mystic Springs boat ramp on Escambia River, Escambia County, 15 August 2012. Photograph by Zachary Randall.

Description
Size: length to 120 mm.

Shell: thin; smooth; moderately inflated, width 35%–55% of length; outline oval; anterior margin broadly rounded; posterior margin narrowly rounded to bluntly pointed, angle of posterior and dorsal margins usually 140°–155°; dorsal margin straight, occasionally with very low dorsal wing; ventral margin broadly rounded; posterior ridge moderately sharp dorsally, rounded posterioventrally; posterior slope moderately steep, flat to slightly concave; umbo broad, moderately inflated, elevated slightly above hinge line; umbo sculpture thin, undulating ridges, often somewhat nodulous; umbo cavity wide, shallow.

Teeth: pseudocardinal and lateral teeth absent.

Nacre: white to bluish white, occasionally with pink tint in umbo cavity.

Periostracum: usually shiny; yellowish, olive to brown, typically with very thin dark green rays, often obscure in large individuals.

Glochidium Description

Outline subtriangular; length 290–315 μm; height 310–325 μm; with styliform hooks (Williams et al. 2008).

Similar Sympatric Species

Anodonta hartfieldorum resembles *Anodonta suborbiculata* but is more elongate and has a more inflated shell disk and umbo. The angle between dorsal and posterior shell margins of *A. hartfieldorum* (mean = 146°, N = 68) is greater than that of *A. suborbiculata* (mean = 129°, N = 127) (Williams et al. 2009).

Anodonta hartfieldorum may also resemble *Pyganodon grandis*, but that species has a much more inflated umbo that is elevated well above the hinge line.

Anodonta hartfieldorum may vaguely resemble *Utterbackia imbecillis* and *Utterbackia peggyae*, but those species are more elongate, usually more compressed, and their umbos are even with the hinge line.

Distribution in Florida

Anodonta hartfieldorum occurs in Escambia River (Figure 10.10).

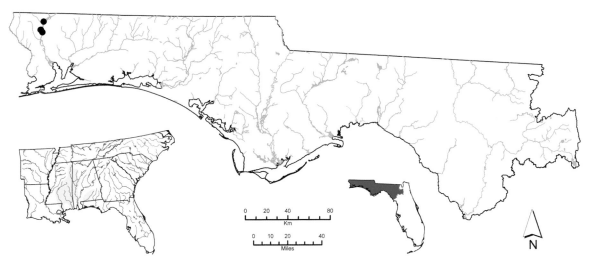

Figure 10.10. Distribution of *Anodonta hartfieldorum*.

Ecology and Biology

Anodonta hartfieldorum typically inhabits backwater sloughs and oxbow lakes. It is also found in abandoned, periodically flooded, gravel mining pits adjacent to the river and in reservoirs (e.g., Gantt and Point A Lakes on Conecuh River, Alabama). Substrates in these habitats range from mud to muddy sand often mixed with fine detritus.

Anodonta hartfieldorum is a long-term brooder, gravid in autumn presumably to the following summer. Gravid individuals brooding mature glochidia have been observed in late October in Pascagoula River, Mississippi, and in early November in Gantt Lake,

Alabama. Glochidial hosts for *A. hartfieldorum* are unknown. Other species of *Anodonta* and several other anodontines (e.g., *Anodontoides*, *Pyganodon*, and *Utterbackia*) broadcast glochidia in loose masses in mucus webs and appear to be host generalists (Williams et al. 2008; Watters et al. 2009).

Conservation Status

Anodonta hartfieldorum was not included in earlier conservation status reviews because it was not widely recognized as a distinct species until recently. It was considered to be vulnerable throughout its range by Williams et al. (in review). A survey of floodplain lakes and sloughs in Florida is needed before its conservation status can be accurately determined.

Anodonta hartfieldorum is extant but uncommon in backwater habitats along Escambia River. It may be common in some gravel mining pits in the floodplain of the Escambia River.

Remarks

There has been a significant reduction in oxbow lake and backwater slough habitats due to channelization and impoundment of large rivers. These habitats are often overlooked or avoided during mussel surveys due to sampling difficulties (e.g., not easy to access, very soft substrate). These factors may account for the paucity of *Anodonta hartfieldorum* specimens in museum collections.

The earliest known Escambia River basin specimen of *Anodonta* was collected in 1917 by C.A. Burke from Chumuckla Springs in Santa Rosa County, Florida. This specimen (UMMZ 101375) was reported as *Anodonta suborbiculata* by Butler (1990) based on a personal communication with W.H. Heard. The specimen could not be located to confirm identification, but it was likely *Anodonta hartfieldorum* since it appears to predate the introduction of *A. suborbiculata* into the Escambia River basin. The *A. suborbiculata* (MCZ 267518) reported from Gantt Lake, Covington County, Alabama, by Johnson (1969a) was *A. hartfieldorum*.

Anodonta hartfieldorum is named in honor of Paul D. and Elizabeth A. Hartfield in recognition of their contributions to conservation and natural history in the southeastern United States (Williams et al. 2009).

Synonymy

Anodonta hartfieldorum Williams, Bogan, and Garner 2009 (Williams, Bogan, and Garner 2009:26). Holotype, UF 375595, length 112 mm. Type locality: Fish Lake, oxbow of Pascagoula River, 1 air mile, southeast of Highway 614 bridge, southwest of Wade, Jackson County, Mississippi, [Pascagoula River basin,] 27 October 2000. © Richard T. Bryant.

Anodonta heardi Gordon and Hoeh 1995
Apalachicola Floater

Anodonta heardi – Upper image: length 113 mm, UF 358656. Harrison Creek, north side of bend, Franklin County, Florida, 7 September 1991. © Richard T. Bryant. Lower image: length 101 mm, UF 381286. Florida River, from downstream near southwest edge of Acorn Lake to point downstream of head of feeder slough into Everett Slough, Liberty County, Florida, 4 June 2002. Photograph by Zachary Randall.

Description

Size: length to 141 mm.

Shell: thin; smooth; moderately inflated, width 30%–45% of length; outline oval; anterior margin broadly rounded; posterior margin narrowly rounded, angle of posterior and dorsal margins 145°–160°; dorsal margin straight; ventral margin broadly rounded; posterior ridge rounded, indistinct; posterior slope moderately steep, flat to slightly convex; umbo broad, inflated, elevated slightly above hinge line; umbo sculpture thin, undulating ridges; umbo cavity wide, shallow.

Teeth: pseudocardinal and lateral teeth absent.

Nacre: white to bluish white, usually iridescent.

Periostracum: usually shiny; greenish yellow to yellowish brown, usually with very fine green rays, increasing in width distally, often obscure in large individuals.

Glochidium Description

The glochidium of *Anodonta heardi* is unknown.

Similar Sympatric Species

Anodonta heardi may superficially resemble *Pyganodon grandis* as well as *Utterbackia imbecillis* and *Utterbackia peggyae*. *Anodonta heardi* has the umbo elevated slightly above the hinge line in contrast to *Pyganodon*, which has a highly inflated umbo that is elevated well above the hinge line. *Anodonta heardi* has a more rounded ventral margin, very fine green rays radiating from the umbo (may be absent in large individuals), and is usually more inflated than *U. imbecillis* and *U. peggyae*. The population of *U. peggyae* in Lake Talquin is more inflated than most stream populations and can be difficult to distinguish from *A. heardi*. However, *U. peggyae* observed in Lake Talquin did not have the fine green rays radiating from the umbo.

Distribution in Florida

Anodonta heardi occurs in Apalachicola and Ochlockonee Rivers (Figure 10.11).

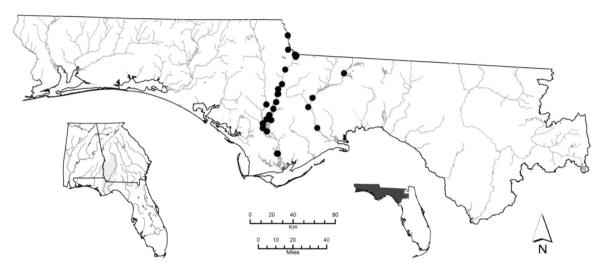

Figure 10.11. Distribution of *Anodonta heardi*.

Ecology and Biology

Anodonta heardi inhabits floodplain lakes and backwater sloughs, as well as some reservoirs. These habitats typically have little or no current and substrates composed of mud, sandy mud, and sand, often with detritus.

Anodonta heardi is presumably a long-term brooder, gravid from autumn to the following summer. A single hermaphroditic *A. heardi* was reported (as *Anodonta couperiana*) from an Apalachicola population by Heard (1975b). Glochidial hosts for *A. heardi* are unknown. Other species of *Anodonta* and several other anodontines (e.g., *Anodontoides*, *Pyganodon*, and *Utterbackia*) broadcast glochidia in loose masses in mucus webs and appear to be host generalists (Williams et al. 2008; Watters et al. 2009).

Conservation Status

Anodonta heardi was considered currently stable throughout its range by Williams et al. (in review). Williams and Butler (1994) assigned it a status of endangered in Florida. This species appears to be more stable than previously believed, primarily due to its presence in appropriate habitat along most of the Apalachicola River and tolerance of reservoir habitats. It is less common in the Ochlockonee River where there is less backwater and floodplain habitat.

Anodonta heardi is extant throughout its range in a Florida.

Remarks

Prior to its recognition as a distinct species in 1995, *Anodonta heardi* was reported under the name *Pyganodon gibbosa* by Clench and Turner (1956) and *Anodonta couperiana* by Johnson (1969a) and Heard (1975a, 1977, 1979a). *Pyganodon gibbosa* is endemic to the Altamaha River basin, Georgia, and *A. couperiana* is confined to southern Atlantic Coast rivers and peninsular Florida.

The single record of *Anodonta heardi* reported from the Chattahoochee River drainage of Alabama (Brim Box and Williams 2000) was reexamined and found to be a conchologically atypical specimen of *Utterbackia imbecillis* (Williams et al. 2008). However, it has been found at several sites in impounded waters of Chattahoochee River in Barbour, Henry, and Russell counties, Alabama, and Chattahoochee and Clay counties, Georgia (The Catena Group 2010; J.D. Williams, personal observation).

Anodonta heardi was named for William H. Heard, Florida State University (retired), in recognition of his service and contributions to malacology. The original description of *A. heardi* by Gordon and Hoeh (1995) was published in the journal *Walkerana* and dated 1993–1994 but not copyrighted, printed, and distributed until 1995. The recognized date of authorship is 1995 (ICZN 1999: Articles 21 and 22).

Synonymy

Anodonta heardi Gordon and Hoeh 1995 (Gordon and Hoeh 1995:265). Holotype, UMMZ 250516, length 89 mm. Type locality: Apalachicola River, approximately 9.7 kilometers north of Blountstown at Ocheesee Landing, Calhoun County, Florida, [Apalachicola River basin,] 4 August 1968. In the original description the shell below is indicated as being the holotype in the figure legend; however, the number is referenced in the text as being a paratype (Gordon and Hoeh 1995).

Anodonta suborbiculata Say 1831
Flat Floater

Anodonta suborbiculata – Upper image: length 124 mm, UF 376151. Coosa River, Weiss Reservoir [Lake], mouth of Big Cedar Creek, about 2 air miles east of Alabama and Georgia state line, river mile 258, Floyd County, Georgia, 20 August 1997. © Richard T. Bryant. Lower image: length 49 mm, UF 134930. Escambia River, State Highway 4 crossing, 2.8 miles east of Century, 13 kilometers north-northeast of McDavid, 7.8 kilometers west of Jay, Santa Rosa County, Florida, 3 June 1988. © Richard T. Bryant.

Description

Size: length to 200 mm, in Florida to 49 mm (the only known specimen).

Shell: thin; smooth; compressed, width 25%–50% of length; outline round to oval; anterior margin broadly rounded; posterior margin broadly rounded, angle of posterior and dorsal margins usually 120°–140°, often with slight dorsal wing posterior of umbo, wing generally more pronounced in small individuals; dorsal margin straight; ventral margin broadly rounded; posterior ridge low, rounded; posterior slope moderately steep, slightly concave; umbo broad, moderately inflated, typically even with hinge line, sometimes elevated slightly above; umbo sculpture thin, undulating, irregular ridges, often nodulous; umbo cavity wide, shallow.

Teeth: pseudocardinal and lateral teeth absent.

Nacre: white to bluish white, large individuals occasionally with pink tint.

Periostracum: usually shiny; light yellow, greenish yellow to dark brown, typically with fine green rays on umbo, often obscure in large individuals.

Glochidium Description

Outline subtriangular; length 323–328 µm; height 320–328 µm; dorsal margin 231–237 µm; with styliform hooks (Surber 1915; Utterback 1916b; Hoggarth 1999) (Figure 10.12).

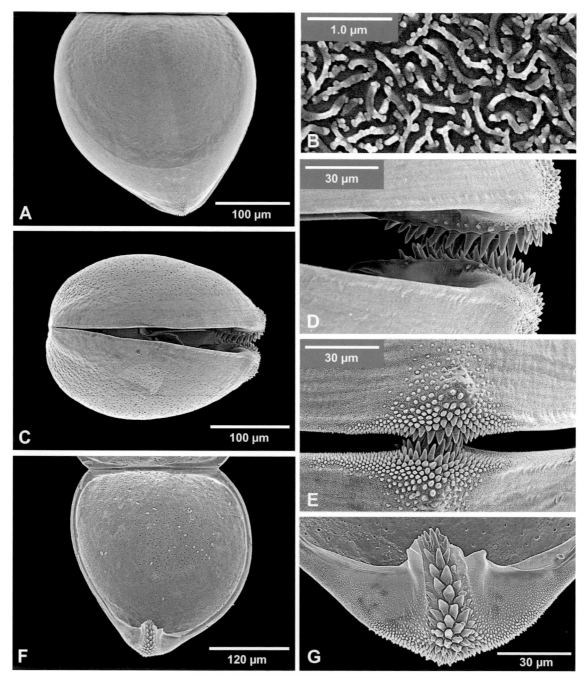

Figure 10.12. Scanning electron micrographs of glochidial valves of *Anodonta suborbiculata* from Angelina River, Sam Rayburn Reservoir, San Augustine and Angelina counties, Texas, 12 December 1995—(A) external view; (B) surface sculpture; (C) lateral view; (D) lateral view of styliform hooks; (E) ventral view of styliform hooks; (F) internal view of styliform hook on ventral margin; and (G) close-up of styliform hook.

Similar Sympatric Species

Anodonta suborbiculata resembles *Anodonta hartfieldorum* but is less elongate and more compressed. The angle between dorsal and posterior shell margins of *A. suborbiculata* (mean = 129°, N = 127) is less than that of *A. hartfieldorum* (mean = 146°, N = 68) (Williams et al. 2009).

Anodonta suborbiculata also resembles *Utterbackia imbecillis* and *Utterbackia peggyae*, but those species are more elongate and more inflated.

Small *Anodonta suborbiculata* may superficially resemble small *Pyganodon grandis*, but that species has a much more inflated umbo that is elevated well above the hinge line.

Distribution in Florida

Anodonta suborbiculata is nonindigenous in Florida and found only in Escambia River (Figure 10.13).

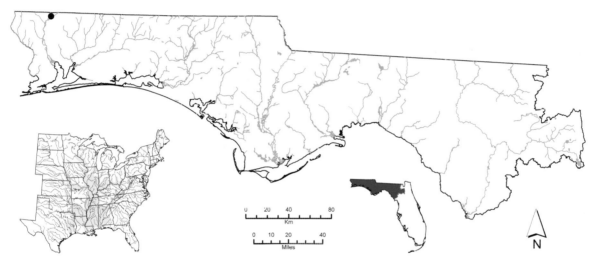

Figure 10.13. Distribution of *Anodonta suborbiculata*.

Ecology and Biology

Anodonta suborbiculata inhabits large creeks, rivers, ponds, floodplain lakes, sloughs, and oxbows in areas with little or no current and mud or muddy sand substrate. It is also found in reservoirs and canals.

Anodonta suborbiculata is a long-term brooder, with mature glochidia present between October and December and brooded until March (Utterback 1915; Barnhart and Roberts 1997). Watters et al. (2009) stated that this species broadcasts glochidia in an extensive web of mucus that entangles host fishes. The mucus-web strategy is often used by host generalists such as *A. suborbiculata*.

Host fish investigations for glochidia of *Anodonta suborbiculata* included laboratory transformations on seven fish species belonging to three families (OSU Museum of Biological Diversity 2012). All seven potential hosts occur in Florida—Centrarchidae: *Lepomis cyanellus* (Green Sunfish), *Lepomis gulosus* (Warmouth), *Lepomis megalotis* (Longear Sunfish), *Micropterus salmoides* (Largemouth Bass), and *Pomoxis annularis* (White Crappie); Cyprinidae: *Notemigonus crysoleucas* (Golden Shiner); and Ictaluridae: *Ictalurus punctatus* (Channel Catfish) (OSU Museum of Biological Diversity 2012).

Conservation Status

Anodonta suborbiculata was assigned a conservation status of currently stable throughout its range by Williams et al. (1993, in review). *Anodonta suborbiculata* was considered to be threatened in Florida (Williams and Butler 1994) based on the assumption that it was native to Escambia River.

Anodonta suborbiculata is nonindigenous in Florida and rare in Escambia River.

Remarks

The earliest known Escambia River basin specimen of *Anodonta* was collected in 1917 from Chumuckla Springs in Santa Rosa County, Florida. This specimen (UMMZ 101375) was reported as *Anodonta suborbiculata* by Butler (1990) based on a personal communication with W.H. Heard. The specimen could not be located to confirm identification, but it was likely *Anodonta hartfieldorum* since it appears to predate the introduction of *A. suborbiculata* into the Escambia River basin. Williams et al. (2009) hypothesized that the UMMZ specimen was actually *A. hartfieldorum*. Heard (1979a) illustrated what clearly is *A. suborbiculata* in his Florida mussel guide but did not provide locality data for the shell. Butler (1990) reported *A. suborbiculata* from Escambia River in Florida based on a specimen (UF 134930) collected in 1988. The specimen, 49 mm in length, is the only known record of *A. suborbiculata* from Florida.

Anodonta suborbiculata has expanded its range in recent decades by invading impounded waters and likely by the introduction of host fishes with glochidia attached (Williams et al. 2008). A record of *A. suborbiculata* (MCZ 267518) reported from Gantt Lake on Conecuh River in southern Alabama (Johnson 1969a) was reidentified as *Anodonta hartfieldorum* Williams et al. (2009). The populations of *A. suborbiculata* in Gantt and Point A Lakes, on the Conecuh River in Alabama, are likely unintentional introductions (Williams et al. 2008) and may be the source of *A. suborbiculata* in Florida.

Synonymy

Anodonta suborbiculata Say 1831 (Say 1831:[no pagination]). Syntype(s) lost. Type locality: ponds near the Wabash River, Indiana, [Ohio River drainage].

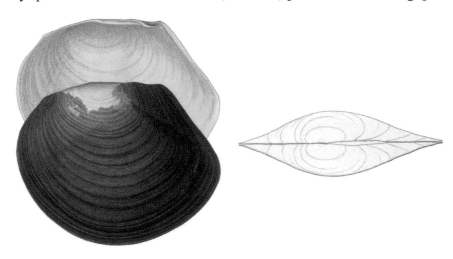

Genus *Anodontoides* Simpson 1898

Turgeon et al. (1998) recognized two species in *Anodontoides*. A third species, *Anodontoides denigrata*, was subsequently elevated from synonymy by Cicerello and Schuster (2003) and occurs in the Cumberland River upstream of Cumberland Falls in Kentucky and Tennessee. *Anodontoides* occurs in Hudson Bay, Great Lakes, Mississippi, and Susquehanna basins and eastern Gulf Coast basins from Louisiana to Apalachicola basin in north Florida.

Type Species
Anodonta ferussacianus Lea 1834 = *Anodontoides ferussacianus* (Lea 1834)

Diagnosis
Shell thin; smooth; moderately inflated; outline elliptical; posterior ridge rounded; umbo elevated slightly above hinge line; umbo sculpture thin to moderately thick ridges; periostracum shiny to dull; dorsal margin slightly curved; pseudocardinal and lateral teeth rudimentary or absent.

Inner lamellae of inner gills only connected to visceral mass anteriorly; excurrent aperture with well-developed papillae; outer gills marsupial, with secondary water tubes when gravid; marsupium not extended beyond original gill margin when gravid; glochidium outline subtriangular, with styliform hooks (Ortmann 1912; Simpson 1914; Baker 1928).

Anodontoides radiatus (Conrad 1834)
Rayed Creekshell

Anodontoides radiatus – Upper image: length 56 mm, UF 358658. Big Sandy Creek, County Road 14, 12 air miles southeast of Union Springs, Bullock County, Alabama, 10 May 2000. © Richard T. Bryant. Middle image: length 48 mm, UF 64086. Sandy Creek near Evergreen, Conecuh County, Alabama, May 1910. © Richard T. Bryant. Lower image: length 41 mm, UF 449430. Bruce Creek on Walton Bridge Road, about 14 kilometers (air) south-southwest of Ponce de Leon, Walton County, Florida, 22 April 2011. © Richard T. Bryant.

Description

Size: length to 78 mm, in Florida to 61 mm.

Shell: thin to moderately thick; smooth; moderately inflated, width 30%–45% of length; outline oval to elliptical; anterior margin rounded; posterior margin narrowly rounded to bluntly pointed; dorsal margin straight to slightly convex; ventral margin convex to broadly rounded; posterior ridge rounded, occasionally weakly biangulate distally; posterior slope moderately steep, flat to slightly convex; umbo broad, slightly inflated, elevated slightly above hinge line; umbo sculpture moderately thick ridges; umbo cavity wide, very shallow.

Teeth: pseudocardinal teeth small, thin, compressed, 1 tooth in left valve, 1 tooth in right valve, crests almost parallel to shell margin; lateral teeth absent.

Nacre: bluish white.

Periostracum: typically shiny; small individuals yellowish brown to dark olive, large individuals brownish black, with dark green rays of varying width and intensity.

Glochidium Description

Outline subtriangular; length 220–313 µm; height 241–280 µm; width 175–183 µm; with styliform hooks (Williams et al. 2008) (Figure 10.14).

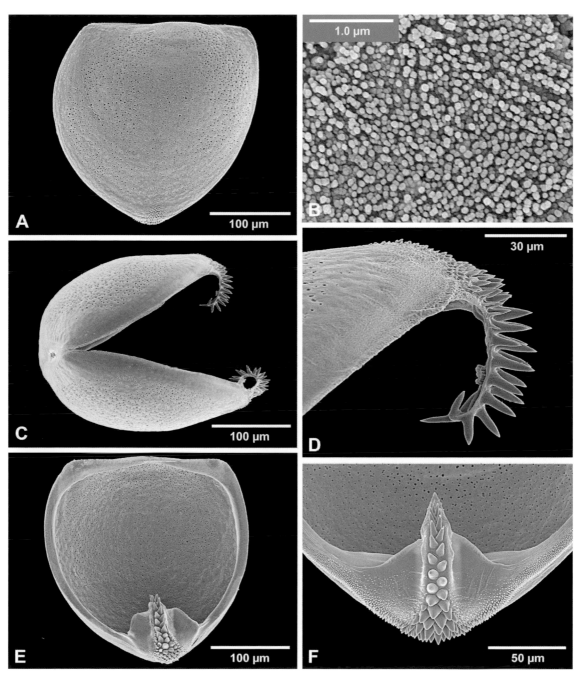

Figure 10.14. Scanning electron micrographs of glochidial valves of *Anodontoides radiatus* from Uchee Creek, Russell County, Alabama, 16 March 2007—(A) external view; (B) surface sculpture; (C) lateral view; (D) lateral view of styliform hook; (E) internal view of styliform hook on ventral margin; and (F) close-up of styliform hook.

Similar Sympatric Species

Anodontoides radiatus superficially resembles *Villosa vibex*, especially small individuals. Shells of *V. vibex* have well-developed pseudocardinal and lateral teeth as opposed to *A. radiatus*, which lack lateral teeth and only have one rudimentary pseudocardinal tooth in each valve.

Anodontoides radiatus may resemble some small *Pyganodon grandis* but are less inflated, have pseudocardinal teeth, and rays.

Distribution in Florida

Anodontoides radiatus occurs in Choctawhatchee and Apalachicola River basins (Figure 10.15).

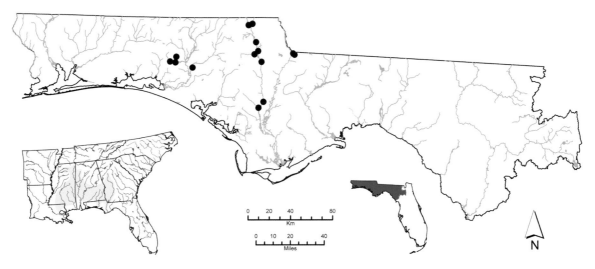

Figure 10.15. Distribution of *Anodontoides radiatus*.

Ecology and Biology

Anodontoides radiatus inhabits small creeks with slow to moderate current and substrates composed of sand, mixtures of sand and gravel, and occasionally sandy mud. It is known to have occurred in the main channel of rivers but is rarely encountered in riverine habitats today.

Anodontoides radiatus is a long-term brooder, gravid from autumn to the following summer. Gravid females have been collected from Apalachicola basin in March, August, and September and Escambia basin in December (Brim Box and Williams 2000). Glochidial hosts for *A. radiatus* are unknown. Watters et al. (2009) stated that a related species, *Anodontoides ferussacianus*, broadcasts glochidia in mucus strands and appears to be a host generalist.

Conservation Status

Anodontoides radiatus was reported as a species of special concern throughout its range by Williams et al. (1993) and assigned the equivalent status of vulnerable by Williams et al. (in review). In Florida its conservation status was undetermined by Williams and Butler (1994).

Anodontoides radiatus is extant in tributaries of the Choctawhatchee River and upper reaches of the Chipola River but is uncommon.

Remarks

Anodontoides radiatus (as *Anodontoides elliottii*) was reported from Apalachicola and Chipola Rivers in Florida by Clench and Turner (1956). *Strophitus spillmanii* reported from the Chipola River by van der Schalie (1940) was based on *A. radiatus*. The species was likely more widespread in the Apalachicola River main stem prior to habitat degradation from construction of Jim Woodruff Lock and Dam and subsequent channel entrenchment and navigation maintenance activities. Heard (1979a) reported *A. radiatus* from Escambia River basin in Alabama, but there are no known records from the Florida portion of the basin. Reports of *Strophitus subvexus* from the Chipola and Apalachicola Rivers by Brim Box and Williams (2000) are based on *A. radiatus*.

The uncertain phylogenetic relationship of *Anodontoides radiatus* has resulted in its previous alignment within five genera—*Alasmidonta*, *Anodon*, *Anodonta*, *Margaritana*, and *Strophitus* (Williams et al. 2008). However, based on preliminary genetic analysis, it appears that the genus *Anodontoides* is not monophyletic.

Synonymy

Alasmodonta radiata Conrad 1834 (Conrad 1834:341). Lectotype (Johnson 1967b), ANSP 41147, length 64 mm. Type locality: small streams in south Alabama.

Margaritana elliottii Lea 1858. Lea 1858b:138. Type locality: Chattahoochee River, [downstream of Uchee Bar,] near Columbus, [Muscogee County,] Georgia, [Chattahoochee River drainage].

Genus *Elliptio* Rafinesque 1819

Elliptio is the largest mussel genus in North America, with 36 species recognized by Turgeon et al. (1998). Four additional species—*Elliptio fumata*, *Elliptio occulta*, *Elliptio pullata*, and *Elliptio purpurella*—were subsequently recognized (Brim Box and Williams 2000; Williams et al. 2008, 2011). However, in a review of taxa described from Florida and contiguous drainages in Alabama and Georgia (Williams et al. 2011), *Elliptio buckleyi* was synonymized with *Elliptio jayensis* and *Elliptio waltoni* with *Elliptio ahenea*. There are currently 38 recognized species of *Elliptio* in the United States.

Elliptio is widespread in eastern North America, occurring in Hudson Bay, Great Lakes, and Mississippi basins southward to Gulf Coast basins from Louisiana to Florida. *Elliptio* also occurs in Atlantic Coast rivers from Nova Scotia south through peninsular Florida. Most of the *Elliptio* diversity is found in basins along the southern Atlantic and eastern Gulf Coasts. There are only 2 species, *Elliptio crassidens* and *Elliptio dilatata*, in the entire Mississippi basin compared to 12 recognized in Florida (Williams et al. 2011). Based on molecular genetic analysis, *E. dilatata* may not belong to *Elliptio* (Campbell et al. 2005; Campbell and Lydeard 2012a).

Within *Elliptio* there appears to be two well-defined groups based on glochidial and adult shell morphology (corrugate and noncorrugate). Shells of the "corrugate" group are characterized by corrugations on the posterior slope. Glochidia of the corrugate group are subtriangular in outline and have a broad shelf covered with micropoints on the inside of the ventral margin (Surber 1915; O'Brien et al. 2003). Glochidial hosts of the corrugate group appear to be the highly migratory shad of the genus *Alosa* (Clupeidae) (Howard 1914b; Williams et al. 2008; J. Stoeckel, personal communication). There are four species of "corrugate" *Elliptio* in Florida—*E. crassidens*, *E. fraterna*, *E. mcmichaeli*, and *E. monroensis*—and two that occur outside Florida—*E. congaraea* and *E. dariensis*; the remaining *Elliptio* species are noncorrugate. Shells of the "noncorrugate" group are characterized by a smooth posterior slope. Glochidia of the noncorrugate group are subelliptical or depressed subelliptical in outline and lack the broad micropoint-covered shelf on the inside of the ventral margin. Glochidial hosts for the noncorrugate group include species of six different families—Centrarchidae, Cyprinidae, Fundulidae, Lepisosteidae, Percidae, and Poeciliidae (OSU Museum of Biological Diversity 2012). The presence of two distinct morphology-based lineages within *Elliptio* raises the possibility that the genus may be polyphyletic. However, molecular genetic studies have failed to recover any phylogenetic structure consistent with grouping species based on these conchological characters (Campbell et al. 2005, 2008).

Elliptio species are by far the most common mussels in Florida, and one or more occur in most of the state's basins. Their ability to survive, grow, and reproduce in a wide range of habitats has produced a plethora of shell morphologies, making it extremely difficult to impossible to delineate species based on shell morphology in many cases. When sorting and identifying *Elliptio*, it is often more helpful to examine smaller size classes rather than looking at the largest individuals. The older (larger) individuals that have been exposed to environmental variables often display unusual or "atypical" shapes due to ecophenotypic variation. The highly plastic shell shape often makes it difficult to reliably identify some sympatric species (e.g., *Elliptio fumata* versus *Elliptio pullata*; *Elliptio jayensis* versus *Elliptio occulta*).

New tools are needed to accurately identify many species with certainty. This is reflected in comments by Simpson (1892b) when he expressed his doubts about the number of valid *Elliptio* species in peninsular Florida.

Species in the genus *Elliptio* in peninsular Florida have always been difficult to identify. This was as perplexing to earlier workers as it is today. In his 1892 publication *Notes on the Unionidae of Florida and the Southeastern States*, Charles T. Simpson wrote:

"It is my desire in this list to considerably reduce the number of so-called species [of *Elliptio*] by showing that they vary into each other, or that many of them have without proper study simply been assumed to be new and renamed, and to give such descriptions and notes as will materially assist in properly determining material."

Clench and Turner (1956) also commented on the difficulty in delineating distributional boundaries of *Elliptio* taxa. Like others before them, they expressed their frustration over the paucity of morphological characters to distinguish species. They also commented on the ability of *Elliptio* species to persist under extremely adverse conditions. Some populations of *Elliptio* rival *Uniomerus* species in their ability to survive without water for prolonged periods of time.

Clench and Turner (1956) noted the continuing problem of delineating species' boundaries and providing precise distribution of species in the genus *Elliptio* in North America.

"This genus is probably more confused than any other in the Unionidae of North America. The limits of distribution of the many species are still unknown, and it is exceedingly difficult to assign taxonomic limits to any one species. Most of the difficulty, besides the lack of marked differential characters to separate the various species, lies in the fact that they are able to live in environments which are impossible for many other genera. They may exist under the most adverse conditions, withstanding pollution, silting, and other seemingly unfavorable conditions. The result is that local populations may show marked differences in size, color, thickness, and shape of shell, as a direct result of the environment. This, unfortunately, has led to a large number of names which have been based upon these ecological forms."

Elliptio species inhabit a wide range of aquatic habitats in Florida, especially in the peninsula. Nowhere is this better demonstrated than the St. Johns River basin, which originates in Blue Cypress Lake, a large water body surrounded by St. Johns Marsh. The river exits the lake and flows northward through a series of braided channels of varying sizes and lakes that vary greatly in physical and chemical characteristics. Along its course there are numerous spring systems, some of which have saline waters. The largest St. Johns tributary, Ocklawaha River, has its headwaters in the central Florida ridge lakes. Additionally, Ocklawaha River enters lower St. Johns River in the tidally influenced mussel-less reach, so freshwaters of the two streams are isolated except for possible interchange of host fishes. Other large tributaries in the tidal reach of St. Johns River (e.g., Black Creek) are likewise isolated. The habitat complexity and periodic isolation during several Pleistocene interglacial periods have combined to produce extreme divergence of shell morphology in peninsular Florida *Elliptio* (Figure 10.16).

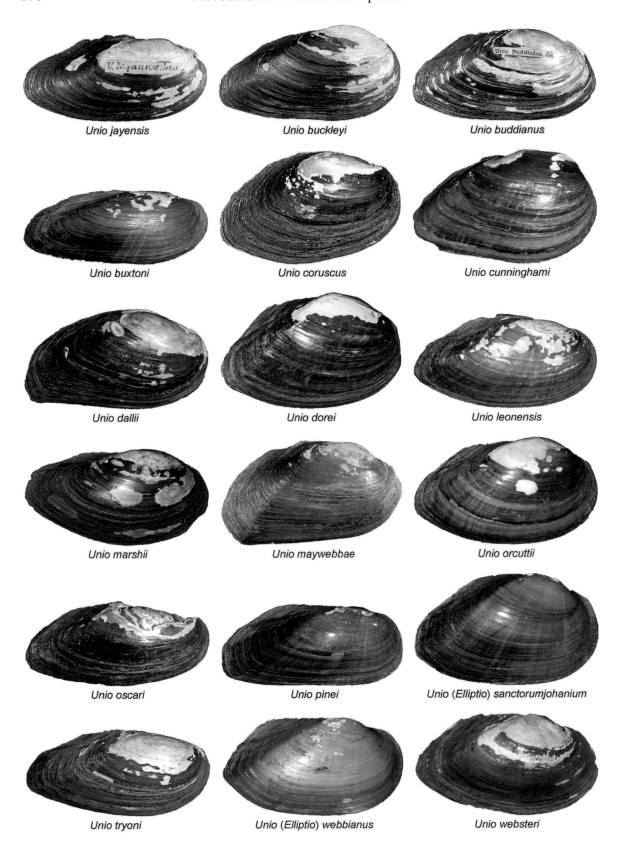

Figure 10.16. Variation in shell morphology of *Elliptio jayensis* and 17 (of 26) junior synonyms from peninsular Florida. Photographs by James D. Williams.

The smooth and nondescript shells of most Florida mussels can make identification challenging. However, in some groups there are soft anatomy characters (e.g., papillate mantle margin in lampsilines) that will often narrow the possibilities down to genus level. In the absence of soft part characters, subtle differences in shell shape, shell thickness, development of teeth, umbo sculpture, and color of periostracum often make for reliable landmarks for identification of species. However, extremely variable shell morphology in *Elliptio* creates a difficult to impossible task when trying to sort out sympatric species (e.g., *Elliptio fumata* versus *Elliptio pullata*; *Elliptio jayensis* versus *Elliptio occulta*).

The state of Florida has been home to species of *Elliptio* since at least the Pliocene. *Elliptio crassidens* is present in the Pleistocene fossil record from Suwannee River at Troy Spring, Lafayette County, Florida, that was determined to be about 18,000 BP with radiocarbon dating. *Elliptio pachyodon* was described from early Pleistocene (reported as Pliocene) deposits in the Tampa area (Pilsbry 1953). *Elliptio pachyodon* is a thick-shelled species and appears to be related to *E. crassidens*. *Elliptio caloosaensis* was described from Pliocene deposits along Caloosahatchee River (Dall 1898). Conchologically this species is similar to *Elliptio jayensis*.

Type Species

Unio (*Elliptio*) *nigra* Rafinesque 1820 = *Elliptio crassidens* (Lamarck 1819)

Diagnosis

Shell thin to thick; smooth, except for spines or weak corrugations on a few species; outline elongate, varying from lanceolate to rectangular; umbo even with or elevated slightly above hinge line, located near anterior end; umbo sculpture variously developed, typically consisting of thin ridges, which may be somewhat angular and nodulous where they cross posterior ridge, often with a slight indentation ventrally; periostracum shiny to dull, occasionally clothlike; pseudocardinal and lateral teeth well developed, 2 in left valve, 1 in right valve.

Incurrent aperture with simple, occasionally bifurcate papillae; excurrent aperture with simple papillae; mantle bridge separating excurrent and supra-anal apertures short; inner lamellae of inner gills only connected to visceral mass anteriorly; outer gills marsupial; glochidia held across entire gill; marsupium smooth and thickened, not extended beyond original gill margin when gravid; glochidium outline variable, without well-developed styliform hooks, may have modified flange with micropoints (Ortmann 1912; Oesch 1995; O'Brien et al. 2003).

Elliptio ahenea (Lea 1843)
Southern Lance

Elliptio ahenea – Upper image: length 74 mm, UF 455455. St. Johns River at outlet of Lake Monroe, under Interstate 4 bridge, about 9 miles north of Sanford, Seminole County, Florida, 6 September 1986. © Richard T. Bryant. Middle image: length 67 mm, UMMZ 58692. Wekiwa [Wekiva] River, Lake County, Florida. Photograph by Zachary Randall. Lower image: length 38 mm, UF 233175. Blue Cypress Lake, west shore, junction canal, Indian River County, Florida, 21 February 1971. Photograph by Zachary Randall.

Description

Size: length to 101 mm.

Shell: thin to moderately thick; smooth; moderately compressed to highly compressed, width 15%–30% of length; outline elongate, oval to elliptical; anterior margin typically rounded, occasionally truncate; posterior margin narrowly rounded to bluntly pointed; dorsal margin long, straight, rarely slightly convex; ventral margin straight to slightly convex; posterior ridge low, moderately sharp dorsally, rounded posteriorly, typically biangulate distally; posterior slope low, flat to slightly concave; umbo broad, slightly inflated, even with hinge line or elevated slightly above; umbo sculpture moderately thick, parallel ridges; umbo cavity wide, shallow.

Teeth: pseudocardinal teeth triangular, 2 small teeth in left valve, may be joined at base, 1 large tooth in right valve, occasionally with accessory denticle anteriorly; lateral teeth moderately long, straight to slightly curved, 2 in left valve, 1 in right valve; interdentum short to moderately long, narrow.

Nacre: white, with pale pink to purple tint.

Periostracum: shiny to clothlike; small individuals brown to olive green, with green rays of varying width and intensity, large individuals dark olive to black, usually obscuring rays.

Glochidium Description

Outline depressed subelliptical; length 180–215 μm; height 230–264 μm; width 120 μm; ventral margin with internal and external irregularly arranged lanceolate micropoints (Figure 10.17).

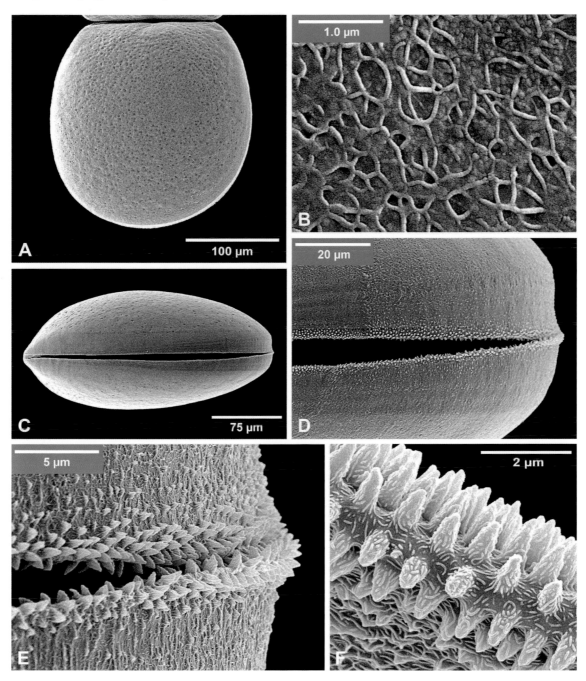

Figure 10.17. Scanning electron micrographs of glochidial valves of *Elliptio ahenea* from St. Marys River, Nassau County, Florida, 22 May 2008—(A) external view; (B) surface sculpture; (C) lateral view; (D) lateral view of micropoints along margin; (E) lateral view of ventral margin and micropoints; and (F) close-up of micropoints.

Similar Sympatric Species

Elliptio ahenea may resemble some elongate individuals of *Elliptio jayensis*. However, typical *E. ahenea* have thinner shells and teeth, are less inflated, and the umbo is elevated only slightly, or not at all, above the hinge line.

Distribution in Florida

Elliptio ahenea occurs in Suwannee, St. Marys, and St. Johns River basins, and Kissimmee River and Lake Okeechobee drainages (Figure 10.18).

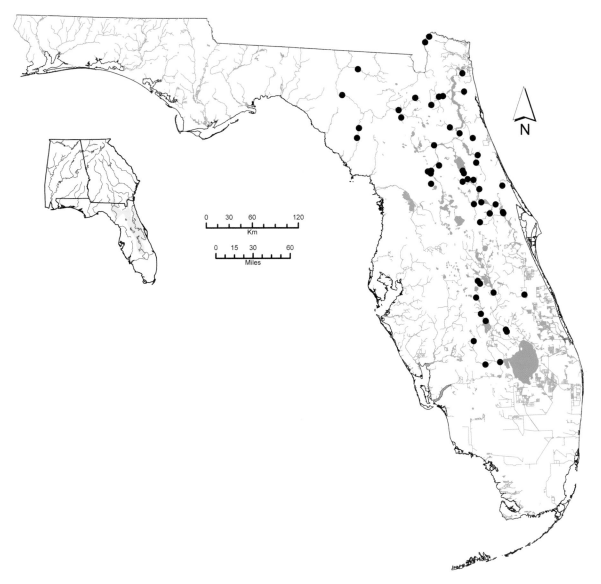

Figure 10.18. Distribution of *Elliptio ahenea*.

Ecology and Biology

Elliptio ahenea occurs in medium to large creeks, rivers, and some lakes. It is found in stable sand, usually associated with structure such as woody debris and roots.

Elliptio ahenea is gravid from spring into early summer. Glochidial hosts for *E. ahenea* are unknown.

Conservation Status

The first conservation status review to include *Elliptio ahenea* was by Williams et al. (1993), who listed it as a species of special concern. It was assigned a rangewide conservation status of threatened by Williams et al. (in review). *Elliptio ahenea* has never appeared on the USFWS candidate species list. However, its synonym *Elliptio waltoni* was a candidate species (USFWS 1994).

Elliptio ahenea is extant in the St. Marys and St. Johns River basins and upper reaches of Suwannee River basin. Its status in the Kissimmee River and Lake Okeechobee is unknown. The species is typically rare throughout its range.

Remarks

In the past *Elliptio ahenea* has been variously treated as a valid species (Simpson 1892b; Frierson 1927; Turgeon et al. 1988, 1998; Williams et al. 1993) or a synonym of *Elliptio jayensis* (Johnson 1972; Burch 1975a). *Elliptio waltoni*, considered valid by Williams et al. (1993) and Turgeon et al. (1998), is treated herein as a synonym of *E. ahenea* (Williams et al. 2011).

Synonymy

Unio aheneus Lea 1843 (Lea 1843:[one page privately published]). Lectotype, USNM 86030, length 54 mm. Type locality: Black Creek, Florida, [St. Johns River basin].

Unio waltoni B.H. Wright 1888 (B.H. Wright 1888:114). Type locality: Lake Woodruff, Volusia County, Florida, [St. Johns River basin].

Elliptio arctata (Conrad 1834)
Delicate Spike

Elliptio arctata – Upper image: length 66 mm, UF 369597. Apalachicola River just south of Rock Bluff Landing, Liberty County, Florida, 18 June 1991. © Richard T. Bryant. Lower image: length 60 mm, UF 370415. Sepulga River, Flat Rock Ford, Conecuh County, Alabama, 30 August 2000. © Richard T. Bryant.

Description

Size: length to 100 mm.

Shell: thin; smooth; compressed to slightly inflated, width 20%–35% of length; outline elongate, oval to elliptical, large individuals often arcuate; anterior margin typically rounded, occasionally truncate; posterior margin obliquely truncate; dorsal margin straight to slightly convex; ventral margin straight to slightly concave; posterior ridge low, rounded, broadly biangulate posterioventrally involving most of posterior margin of shell; posterior slope low, flat to slightly concave; umbo broad, slightly inflated, even with hinge line; umbo sculpture unknown; umbo cavity wide, shallow.

Teeth: pseudocardinal teeth small, low, triangular, 2 widely separated teeth in left valve, anterior tooth typically smaller, 1 tooth in right valve; lateral teeth long, thin, straight to slightly curved, 2 in left valve, 1 in right valve; interdentum moderately long, narrow.

Nacre: white to bluish white, occasionally purplish.

Periostracum: shiny to dull, occasionally clothlike; small individuals dark olive or brown, with dark green rays of varying width and intensity, large individuals dark brown to black, usually obscuring rays.

Glochidium Description

Outline subelliptical; length 187–200 μm; height 200–237 μm (Williams et al. 2008).

Similar Sympatric Species

Elliptio arctata resembles *Elliptio pullata* but has a thinner, more compressed and often arcuate shell, smaller pseudocardinal teeth, and thinner lateral teeth.

Elliptio arctata also resembles *Elliptio purpurella* but is more compressed and usually has white to bluish-white nacre, whereas that of *E. purpurella* is usually purple.

Elliptio arctata superficially resembles *Elliptio fumata* but typically has a thinner shell and is more elongate and arcuate.

Distribution in Florida

Elliptio arctata occurs in Apalachicola River basin (Figure 10.19).

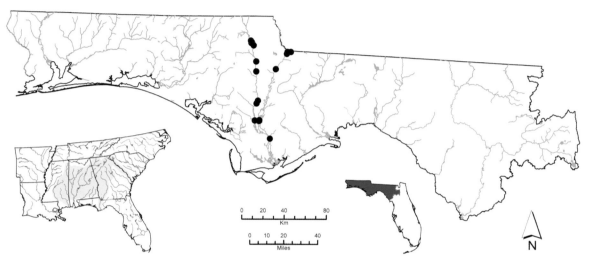

Figure 10.19. Distribution of *Elliptio arctata*.

Ecology and Biology

Elliptio arctata occurs in large creeks and rivers where it is often associated with sandy and rocky substrates in slow to moderate current. It occurs less commonly in root mats or other structures.

Elliptio arctata is presumably a short-term brooder. The sexes are usually separate, but Heard (1979b) found 4 of 126 individuals from Apalachicola basin to be hermaphroditic. Glochidial hosts for *E. arctata* are unknown.

Conservation Status

Elliptio arctata was considered a species of special concern throughout its range by Williams et al. (1993) and threatened by Williams et al. (in review).

Elliptio arctata is rare in Florida and is extant in isolated reaches of Apalachicola River and its floodplain distributaries.

Remarks

Phylogeographic analysis of *Elliptio arctata* populations throughout its range may reveal additional species. Conchological and molecular genetic characters will be required to detect cryptic species.

Synonymy

Unio arctatus Conrad 1834 (Conrad 1834:340). Lectotype (Johnson 1970), ANSP 41356, length 57 mm. Type locality: Alabama River, [Alabama, Alabama River drainage].

Unio strigosus Lea 1840 (Lea 1840:287). Type locality: Chattahoochee River, Columbus, [Muscogee County,] Georgia, [Chattahoochee River drainage].

Unio tortivus Lea 1840 (Lea 1840:287). Type locality: Chattahoochee River, Columbus, [Muscogee County,] Georgia, [Chattahoochee River drainage].

Unio viridans Lea 1859 (Lea 1859a:170). Type locality: near Columbus, [Muscogee County,] Georgia, [Chattahoochee River drainage].

Elliptio chipolaensis (Walker 1905)
Chipola Slabshell

Elliptio chipolaensis – Upper image: length 62 mm, UMMZ 138388. Chipola River, Merritts bridge, Marianna, Jackson County, Florida, July 1918. © Richard T. Bryant. Lower image: length 53 mm, UMMZ 138409. Chipola River, Peacock Bridge, near Sink Creek, Jackson County, Florida, 1918. © Richard T. Bryant.

Description

Size: length to 85 mm.

Shell: moderately thick; smooth; moderately inflated, width 30%–45% of length; outline elliptical to oval; anterior margin rounded; posterior margin bluntly pointed to narrowly rounded; dorsal margin slightly convex; ventral margin straight to convex, occasional large individuals slightly arcuate; posterior ridge high, rounded, typically biangulate posterioventrally; posterior slope moderately steep, flat to slightly concave; umbo broad, inflated, elevated slightly above hinge line; umbo sculpture moderately thick ridges; umbo cavity wide, moderately shallow.

Teeth: pseudocardinal teeth triangular, 2 teeth in left valve, occasionally joined at base, crests parallel to shell margin, oriented anterioventrally, anterior tooth somewhat compressed, 1 tooth in right valve; lateral teeth long, straight to slightly curved, 2 in left valve, 1 in right valve; interdentum short to moderately long, narrow.

Nacre: white to bluish white, often with salmon tint in umbo cavity.

Periostracum: shiny to dull; yellowish brown to brown, typically with dark annuli, large individuals often darker.

Glochidium Description

Outline depressed subelliptical; length 180–203 μm; height 174–180 μm; width 99 μm; ventral margin with internal and external irregularly arranged lanceolate micropoints (Figure 10.20).

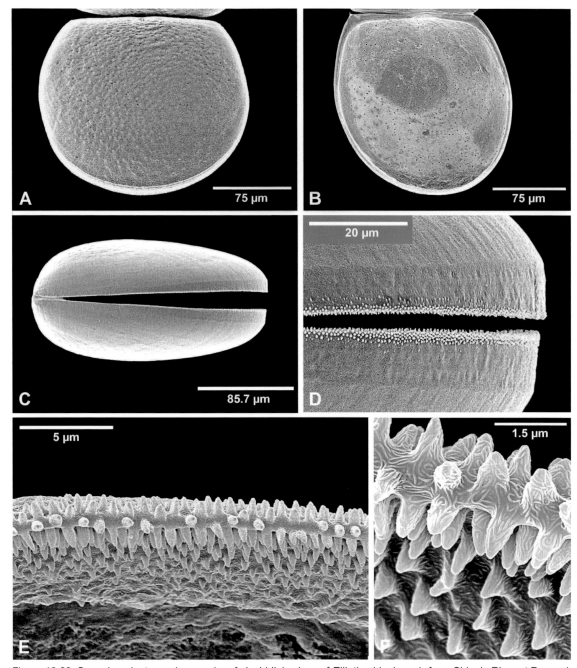

Figure 10.20. Scanning electron micrographs of glochidial valves of *Elliptio chipolaensis* from Chipola River at Peacock Bridge, Jackson County, Florida—(A) external view; (B) internal view; (C) lateral view; (D) lateral view of micropoints along margin; (E) internal view of micropoints on ventral margin; and (F) close-up of micropoints.

Similar Sympatric Species

Elliptio chipolaensis resembles some *Elliptio* in the Apalachicola basin but can be distinguished by its oval shape, yellowish-brown periostracum with dark annuli, umbo elevated above the hinge line, and anterior pseudocardinal tooth somewhat compressed and oriented anterioventrally.

Distribution in Florida

Elliptio chipolaensis occurs in Chipola River drainage and Apalachicola River proper near the confluence of Chipola River (Figure 10.21).

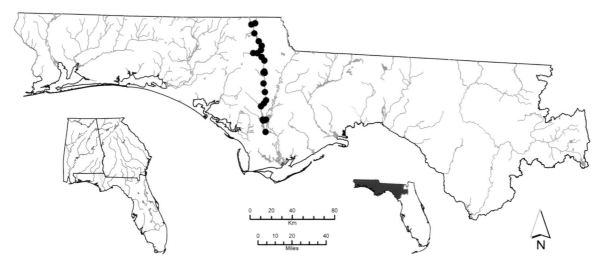

Figure 10.21. Distribution of *Elliptio chipolaensis*.

Ecology and Biology

Elliptio chipolaensis inhabits large creeks to rivers in slow to moderate current and in silty sand substrates.

Elliptio chipolaensis is a short-term brooder, gravid from early June to early July. Conglutinates released naturally were approximately 13 mm long, 3 mm wide, and white in color (Preister 2008).

In laboratory trials *Elliptio chipolaensis* glochidia metamorphosed on two fishes that occur in Florida—Centrarchidae: *Lepomis auritus* (Redbreast Sunfish) and *Lepomis macrochirus* (Bluegill). Glochidia successfully transformed on 60 percent of *L. macrochirus* and 80 percent of *L. auritus* individuals. Transformation occurred three to four weeks after infestation. Four additional fishes that occur in Florida were exposed, but no metamorphosed glochidia were recovered—Cyprinidae: *Hybopsis* sp. (undescribed chub) and *Notropis texanus* (Weed Shiner); Ictaluridae: *Ictalurus punctatus* (Channel Catfish); and Percidae: *Percina nigrofasciata* (Blackbanded Darter) (Preister 2008).

Conservation Status

Elliptio chipolaensis was assigned a conservation status of threatened throughout its range by Williams et al. (1993, in review). In Florida it was considered to be threatened by Williams and Butler (1994). In 1998 *E. chipolaensis* was listed as threatened under the authority of the federal ESA (USFWS 1998). Critical habitat was designated in 2007 and included a total of 228.7 km in the Chipola River and selected tributaries in Alabama and Florida (USFWS 2007). A final recovery plan was approved in 2003 (USFWS 2003).

Elliptio chipolaensis was reported to be abundant at selected localities in Chipola River during the early 1970s (Heard 1975a). *Elliptio chipolaensis* is extant through most of its historical range and is common at some localities.

Remarks

Since its description, *Elliptio chipolaensis* was generally considered to be a Chipola River endemic, but Brim Box and Williams (2000) reported it from Howard's Mill Creek, a Chattahoochee River tributary, Houston County, Alabama. It was recently found at a single site on the Apalachicola River near the junction of the Chipola.

The specific epithet *"chipolaensis"* was taken from Chipola River. The origin of the word chipola is from the Native American Chatot dialect and translated means "sweet" (Morris 1995).

Synonymy

Unio chipolaensis Walker 1905 (Walker 1905a:135). Holotype, UMMZ 96363, length
57 mm. Type locality: published as Chipola River, Florida, incorrectly and unnec-
essarily restricted by Clench and Turner (1956) to Chipola River, 1 mile north of
Marianna, Jackson County, Florida, [Apalachicola River basin]. They give no evidence that the holotype came from the site they chose, merely noting that it was
"a locality from which we have a good series." The restriction of type locality by
Clench and Turner (1956) is invalid.

Elliptio crassidens (Lamarck 1819)
Elephantear

Elliptio crassidens – Upper image: length 111 mm, UF 37822. Apalachicola River below Jim Woodruff Lock and Dam at Chattahoochee, Gadsden County, Florida, 20 May 1981. © Richard T. Bryant. Middle image: length 87 mm, UF 371. Escambia River, 3 miles southeast of Century, Escambia County, Florida, 27 August 1954. © Richard T. Bryant. Lower image: length 34 mm, UF 449431. Apalachicola River, backwater pool, river mile 46.8, Gulf County, Florida, 7 August 2006. © Richard T. Bryant.

Description

Size: length to 150 mm, in Florida to 115 mm.

Shell: moderately to very thick; smooth; moderately inflated, width 30%–60% of length; outline variable, ranging from rhomboid to elliptical or subtriangular, large individuals often arcuate; anterior margin rounded; posterior margin bluntly pointed posterioventrally; dorsal margin slightly convex to broadly rounded; ventral margin straight to slightly convex, large individuals often concave; posterior ridge sharp to rounded, typically biangulate posterioventrally; posterior slope steep dorsally, flat to slightly concave posterioventrally, typically with thin corrugations extending from posterior ridge to posteriodorsal margin; umbo broad, moderately inflated, elevated slightly above hinge line; umbo sculpture 3–5 moderately thick, nodulous ridges, angular on posterior ridge; umbo cavity wide, shallow.

Teeth: pseudocardinal teeth large, triangular, 2 divergent teeth in left valve, anterior tooth usually smaller, 1 tooth in right valve, occasionally with accessory denticle anteriorly; lateral teeth moderately long, thick, straight to slightly curved, 2 in left valve, 1 in right valve; interdentum short to moderately long, moderately narrow to wide.

Nacre: variable, white to salmon, pink, or purple.

Periostracum: shiny to dull, occasionally clothlike; small individuals olive to brown, occasionally with faint green rays, large individuals dark brown to black.

Glochidium Description

Outline subtriangular; length 130–150 µm; height 141–160 µm; ventral margin with broad shelf covered with micropoints (Surber 1915; O'Brien et al. 2003) (Figure 10.22). A small, straplike hook was reported on glochidia of *Elliptio crassidens* by Howard (1914b).

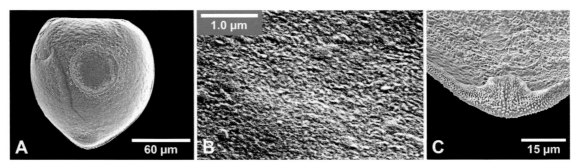

Figure 10.22. Scanning electron micrographs of glochidial valves of *Elliptio crassidens* from Coolewahee Creek, north of Newton, Baker County, Georgia, 2 April 1997—(A) external view; (B) surface sculpture; and (C) internal view of modified flange with micropoints (modified from O'Brien et al. 2003).

Similar Sympatric Species

Elliptio crassidens resembles *Elliptio mcmichaeli* in that both species have a well-developed posterior ridge and sculpture on the posterior slope. *Elliptio crassidens* is typically less elongate, more triangular in outline, has thicker pseudocardinal teeth (especially the anterior tooth), and the ventral margin of the shell is usually less rounded than that of *E. mcmichaeli*.

Elliptio fraterna is the only other *Elliptio* in Apalachicola River with sculpture on the posterior slope. Compared to *Elliptio crassidens*, it typically has a thinner shell and is elliptical to oval in outline.

Small (length up to 40 mm) *Elliptio crassidens* may resemble *Fusconaia burkei* but differ in having a thicker shell with sculpture restricted to the posterior slope and thicker pseudocardinal teeth.

Elliptio crassidens superficially resembles *Alasmidonta triangulata* and *Alasmidonta wrightiana* in having a sharp posterior ridge with sculpture on the posterior slope. However, *A. triangulata* and *A. wrightiana* have a more inflated umbo, thinner shell, and lack well-developed pseudocardinal and lateral teeth.

Distribution in Florida

Elliptio crassidens occurs in Escambia, Yellow, Choctawhatchee, Apalachicola, and Ochlockonee River basins. In Ochlockonee River it is confined to the reach of the river downstream of Jackson Bluff Dam. There is also a small population in St. Marys River (Figure 10.23).

Figure 10.23. Distribution of *Elliptio crassidens*.

Ecology and Biology

Elliptio crassidens inhabits rivers and lower reaches of large creeks. It occurs in substrates of sandy mud, sand, gravel, and rubble in slow to moderate current.

Elliptio crassidens is apparently a short-term brooder that is gravid from April to August in Apalachicola basin (Brim Box and Williams 2000; O'Brien et al. 2003). More than 500 individuals collected from the Apalachicola basin between June and September were checked, but only 12 gravid females were found, suggesting a spring and early summer period of gravidity. One small female, 51 mm in length, was reported to be gravid (Brim Box and Williams 2000). Conglutinates of *E. crassidens* from Mississippi basin were described as leaf-shaped and white in color (Ortmann 1912; Utterback 1915). Based on natural infestations, only one fish that occurs in Florida has been identified as a potential glochidial host for *E. crassidens*—Clupeidae: *Alosa chrysochloris* (Skipjack Herring) (Howard 1914b).

Conservation Status

Elliptio crassidens was considered to be currently stable throughout its range by Williams et al. (1993). Declines in northern populations have led Williams et al. (in review) to assign it a rangewide status of vulnerable.

Elliptio crassidens remains common and its populations are apparently stable in Apalachicola and Chipola Rivers. The species is extant in isolated reaches of Escambia, Yellow, Choctawhatchee, and St. Marys Rivers but is uncommon.

Remarks

Unio incrassatus was described from the Chattahoochee River near Columbus, [Muscogee County,] Georgia. Since its description, it has been variously treated as a valid species, a synonym of *Elliptio crassidens*, or a southeastern subspecies, *Elliptio crassidens incrassatus* (Clench and Turner 1956). This may well represent a valid species, but it is not possible to make that determination using morphology or current molecular genetic techniques.

Elliptio crassidens is present in most archaeological sites in Apalachicola basin. Fossil *E. crassidens* have been recovered from Troy Spring on Suwannee River and Leisey Shell Pits near Tampa Bay (Bogan and Portell 1995). These fossils indicate that it had a wider distribution in prehistoric times. The fossil record from Suwannee River helps explain distribution of the species in the adjacent St. Marys River, its only occurrence on the southern Atlantic Coast.

Synonymy

Unio crassidens Lamarck 1819 (Lamarck 1819:71). Lectotype (Johnson 1969b), MNHN uncatalogued, length 66 mm. Type locality: published as Ohio, restricted by Johnson (1969b) to Ohio River, Cincinnati, Ohio, [Ohio River drainage].

Unio incrassatus Lea 1840 (Lea 1840:286). Type locality: Chattahoochee River near Columbus, [Muscogee County,] Georgia, [Chattahoochee River drainage].
Unio danielsi B.H. Wright 1899 (B.H. Wright 1899:31). Type locality: Spring Creek, [tributary to Flint River,] Decatur County, Georgia, [Flint River drainage].

Elliptio fraterna (Lea 1852)
Brother Spike

Elliptio fraterna – Length 70 mm, UF 441283. Apalachicola River near river mile 46.7, just north of Iola Landing, Gulf County, Florida, 18 August 2011. Photograph by Zachary Randall.

Description

Size: length to 80 mm, in Florida to 70 mm.

Shell: moderately thin; smooth; compressed, width 25%–35% of length; outline elliptical to subrhomboidal; anterior margin rounded; posterior margin narrowly rounded to obliquely truncate; dorsal margin straight to slightly convex; ventral margin straight to slightly convex; posterior ridge moderately sharp to rounded, biangulate posterioventrally; posterior slope low, flat to slightly concave, typically with weak corrugations extending from posterior ridge to dorsal and posterior margins, usually absent distally; umbo broad, slightly inflated, elevated slightly above hinge line; umbo sculpture unknown; umbo cavity wide, shallow.

Teeth: pseudocardinal teeth moderately thin to thick, triangular, finely striate, 2 teeth in left valve, anterior tooth usually smaller, 1 tooth in right valve, occasionally with accessory denticle anteriorly; lateral teeth moderately short to moderately long, straight to slightly curved, 2 in left valve, 1 in right valve; interdentum short to moderately long, narrow to moderately wide.

Nacre: white to purple.

Periostracum: dull; brown to brownish black, small individuals occasionally with green rays.

Glochidium Description

The glochidium of *Elliptio fraterna* is unknown.

Similar Sympatric Species

Elliptio fraterna has corrugations across the posterior slope that distinguishes it from all other *Elliptio* in Apalachicola basin except *E. crassidens*. *Elliptio fraterna* differs from *E. crassidens* in being more elongate, having a thinner, more compressed shell, and smaller pseudocardinal teeth.

Distribution in Florida

    Elliptio fraterna occurs in Apalachicola River (Figure 10.24).

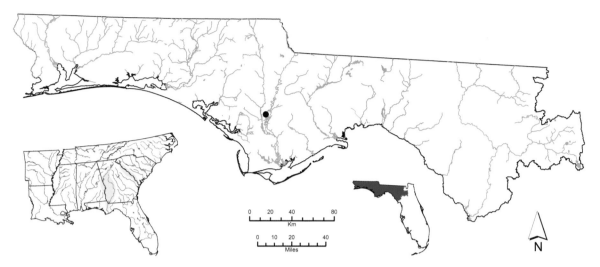

Figure 10.24. Distribution of *Elliptio fraterna*.

Ecology and Biology

    Elliptio fraterna inhabits large rivers in sandy mud to sand substrates. Individuals collected in Florida came from depositional areas with very slow current along the shore of Apalachicola River. At least one sample of *E. fraterna* from the Savannah River came from a similar habitat.

    Elliptio fraterna is presumably a short-term brooder, graved in spring and early summer. The only gravid individual observed was collected in the Apalachicola River in June. Glochidial hosts for *E. fraterna* are unknown.

Conservation Status

    Elliptio fraterna was considered endangered throughout its range by Williams et al. (1993, in review).

    Elliptio fraterna is extant but appears to be rare in Apalachicola River.

Remarks

    The original description of *Elliptio fraterna* was based on specimens from Chattahoochee River, Columbus, Georgia, and Abbeville District, Savannah River basin, South Carolina. The Abbeville District specimen was designated as the lectotype, restricting the type locality to the Savannah River basin (Johnson 1970, 1974).

    Available information indicates that *Elliptio fraterna* has always been rare. The Savannah River was sampled regularly for decades with only a few individuals found (Britton and Fuller 1980). While the Apalachicola was sampled less thoroughly, none were found by Brim Box and Williams (2000). It has been found in both Savannah and Apalachicola Rivers in the past decade.

Synonymy

Unio fraternus Lea 1852 (Lea 1852b:263). Lectotype (Johnson 1970), USNM 85396, length 61 mm. Type locality: Abbeville District, South Carolina, [Savannah River basin].

Elliptio fumata (Lea 1857)
Gulf Slabshell

Elliptio fumata – Upper image: length 74 mm, UF 2875. Blue Springs, 3 miles east of Marianna, Jackson County, Florida, 8 June 1930. Photograph by Zachary Randall. Middle image: length 69 mm, UF 388490. Spring Creek, 1 river mile above Smith Landing, 9 air miles west of Bainbridge, Decatur County, Georgia, 25 September 1992. © Richard T. Bryant. Lower image: length 54 mm, UF 2875. Blue Springs, 3 miles east of Marianna, Jackson County, Florida, 8 June 1930. Photograph by Zachary Randall.

Description

Size: length to 145 mm.

Shell: moderately thin; smooth; moderately inflated, width 25%–40% of length; outline elliptical to subrhomboidal; anterior margin rounded; posterior margin bluntly pointed to narrowly rounded or truncate; dorsal margin straight to slightly convex; ventral margin straight to slightly convex, occasional large individuals slightly concave posteriorly; posterior ridge low, broadly rounded, occasionally biangulate posteriorly; posterior slope low, flat to slightly convex; umbo broad, slightly inflated, may be even with hinge line or elevated slightly above; umbo sculpture low ridges; umbo cavity wide, shallow.

Teeth: pseudocardinal teeth triangular, 2 teeth in left valve, anterior tooth usually smaller, oriented ventrally to anterioventrally, 1 tooth in right valve, occasionally with accessory denticle anteriorly; lateral teeth long, straight to slightly curved, 2 in left valve, 1 in right valve; interdentum long, narrow.

Nacre: white, often with purple or salmon tint.

Periostracum: dull to clothlike; greenish brown to dark brown or black, occasionally with green rays, usually obscure in large individuals.

Glochidium Description

The glochidium of *Elliptio fumata* is unknown.

Similar Sympatric Species

Elliptio fumata most closely resembles *Elliptio pullata* and *Elliptio fraterna*. *Elliptio fumata* is typically higher in relation to length than *E. pullata*, and *E. fraterna* has corrugations on the posterior slope.

Elliptio fumata typically has a thicker shell, is higher in relation to length, and less elongate and arcuate than *Elliptio arctata*.

Elliptio fumata is less elongate and inflated than *Elliptio purpurella*. The shell nacre is generally white in *E. fumata* and purple in *E. purpurella*.

Distribution in Florida

Elliptio fumata occurs in Apalachicola and Chipola Rivers (Figure 10.25).

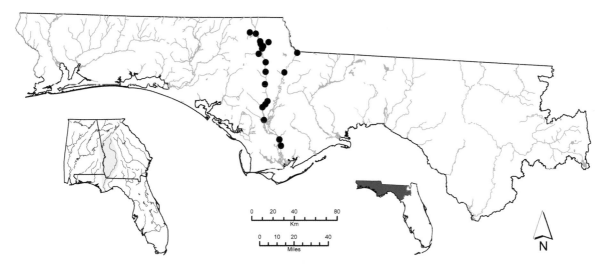

Figure 10.25. Distribution of *Elliptio fumata*.

Ecology and Biology

Elliptio fumata inhabits small creeks to rivers. It typically occurs in slow to moderate current in substrates of stable sand to sandy mud, gravel, and crevices of bedrock.

Elliptio fumata is a short-term brooder, with gravid females reported from June to August (Brim Box and Williams 2000). It is typically dioecious; however, occasional hermaphroditic individuals may be encountered (Heard 1979b). Glochidial hosts for *E. fumata* are unknown.

Conservation Status

Elliptio fumata was only recently recognized as a valid taxon (Williams et al. 2008) and thus has never been assigned conservation status rangewide. The species was considered currently stable by Williams et al. (in review). *Elliptio fumata* is extant throughout its range in Florida and is relatively common.

Remarks

Elliptio fumata is a highly variable species endemic to Apalachicola basin. Historical records of the Atlantic Coast endemic *Elliptio complanata* in Apalachicola basin generally refer to *E. fumata*. While this taxon was recognized as a distinct species by Williams et al. (2008), it needs to be thoroughly studied to understand its relationships with other *Elliptio* in eastern Gulf and southern Atlantic Coast rivers.

Elliptio fumata is common in archaeological samples from Apalachicola basin, indicating widespread harvest by Native Americans (Williams et al. 2008).

Synonymy

Unio fumatus Lea 1857 (Lea 1857a:171). Lectotype (Johnson 1970), USNM 85552, length 59 mm. Type locality: Chattahoochee River near Columbus, [Muscogee County,] Georgia, [Chattahoochee River drainage]. © Richard T. Bryant.

Unio plantii Lea 1857 (Lea 1857a:171). Type locality: Flint River near Macon [County], Georgia, [Flint River drainage].

Unio subniger Lea 1857 (Lea 1857a:172). Type locality: Flint River near Macon [County], Georgia, [Flint River drainage].

Unio hallenbeckii Lea 1859 (Lea 1859a:170). Type locality: Flat Rock [Flatrock] Creek near Columbus, [Muscogee County,] Georgia, [Chattahoochee River drainage].

Unio salebrosus Lea 1859 (Lea 1859a:170). Type locality: Flat Rock [Flatrock] Creek, [Muscogee County,] Georgia, [Chattahoochee River drainage].

Unio basalis Lea 1872 (Lea 1872:161). Type locality: Carter's Creek near Columbus, [Muscogee County,] Georgia, [Chattahoochee River drainage].

Unio corneus Lea 1874 (Lea 1874b:423). Type locality: [Chattahoochee River,] Columbus, [Muscogee County,] Georgia, [Chattahoochee River drainage].

Unio dooleyensis Lea 1874 (Lea 1874c:424). Type locality: Dooley [Dooly] County, Georgia, [Flint River drainage].

Unio gesnerii Lea 1874 (Lea 1874c:424). Type locality: Uchee River [Creek, Lee and Russell counties, Alabama,] near Columbus, Georgia, [Chattahoochee River drainage].

Unio invenustus Lea 1874 (Lea 1874c:424). Type locality: [Chattahoochee River,] Columbus, [Muscogee County,] Georgia, [Chattahoochee River drainage].

Elliptio jayensis (Lea 1838)
Florida Spike

Elliptio jayensis – First image: length 79 mm, ANSP 73861. Lake in Withlacoochee River [southern] basin, Hernando County, Florida. © Richard T. Bryant. Second image: length 58 mm, UF 381374. North Canal, tributary of St. Johns River, on Duda Ranch, about 5 miles west of junction of Interstate 95 and Route 502, Brevard County, Florida, 9 February 2001. © Richard T. Bryant. Third image: length 55 mm, UF 449293. Lake Jesup, Seminole County, Florida, 9 April 1996. © Richard T. Bryant. Fourth image: length 24 mm, UF 449432. Lake Weir at Hampton Beach Park, Marion County, Florida, 1 May 2008. © Richard T. Bryant.

Description
   Size: length to 100 mm.

Shell: thin to moderately thick; smooth; moderately compressed to inflated, width 25%–55% of length; outline elongate, oval to elliptical, occasionally subtriangular; anterior margin rounded; posterior margin obliquely truncate; dorsal margin straight to slightly convex; ventral margin straight, occasionally slightly convex; posterior ridge sharp to broadly rounded, biangulate posterioventrally; posterior slope low to moderately steep, flat to slightly concave; umbo broad, inflated to highly inflated, elevated well above hinge line; umbo sculpture moderately thick ridges, angular on posterior ridge, with shallow ventral indentation (Figure 10.26); umbo cavity wide, shallow.

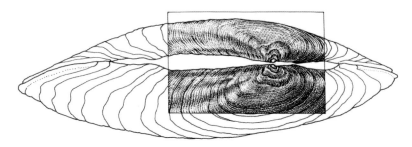

Figure 10.26. Umbo sculpture of *Elliptio jayensis,* length 37 mm, UF 449116. Outlet River of Lake Panasoffkee at boat ramp in park near County Route 470 bridge, Sumter County, Florida, 24 May 2006. Illustration by Susan Trammell.

Teeth: pseudocardinal teeth large, compressed, 2 teeth in left valve, occasionally joined at base, approximately equal in size or anterior tooth slightly smaller, anterior tooth crest oriented anterioventrally, 1 tooth in right valve, with or without accessory denticle anteriorly; lateral teeth variable, usually long, moderately thin to thick, straight to slightly curved, 2 in left valve, 1 in right valve; interdentum short to moderately long, usually narrow, occasionally moderately wide.

Nacre: variable, white to orange, pink, or purple.

Periostracum: usually shiny, occasionally clothlike; yellowish green, dark olive to brown, occasionally black, typically with green rays varying in width and intensity, often obscure in large individuals.

Glochidium Description

Outline depressed subelliptical; length 180–188 µm; height 186–195 µm; width 109 µm; ventral margin with internal and external irregularly arranged lanceolate micropoints (Figure 10.27).

Similar Sympatric Species

Elliptio jayensis most closely resembles *Elliptio occulta,* but *E. jayensis* typically has thicker teeth, the umbo is elevated farther above the hinge line, and the junction of the posterior and dorsal margins are usually angular. The outline of *E. occulta* is typically more oval than that of *E. jayensis.*

Elongate individuals of *Elliptio jayensis* may resemble some *Elliptio ahenea.* However, *E. jayensis* usually have thicker shells and teeth, are more inflated, and the umbo is elevated farther above the hinge line than typical *E. ahenea.*

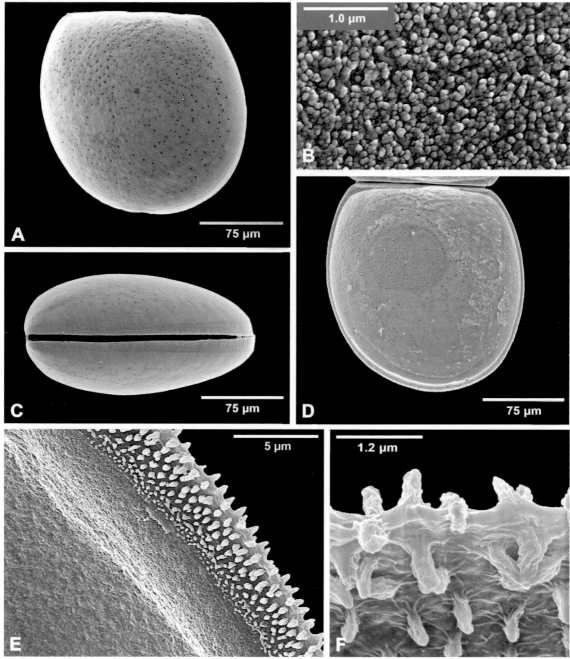

Figure 10.27. Scanning electron micrographs of glochidial valves of *Elliptio jayensis* from Hitchcock Lake (Slough), about 0.25 miles upstream of junction with Ochlockonee River, Liberty County, Florida, 30 July 2008—(A) external view; (B) surface sculpture; (C) lateral view; (D) internal view; (E) internal view of ventral margin and micropoints; and (F) close-up of micropoints.

Distribution in Florida

Elliptio jayensis occurs from Ochlockonee River to St. Marys River and throughout peninsular Florida including the Everglades. Populations in Mosquito Creek (tributary to Apalachicola River) and Spring Creek (tributary to Chipola River) appear to be introduced (Figure 10.28).

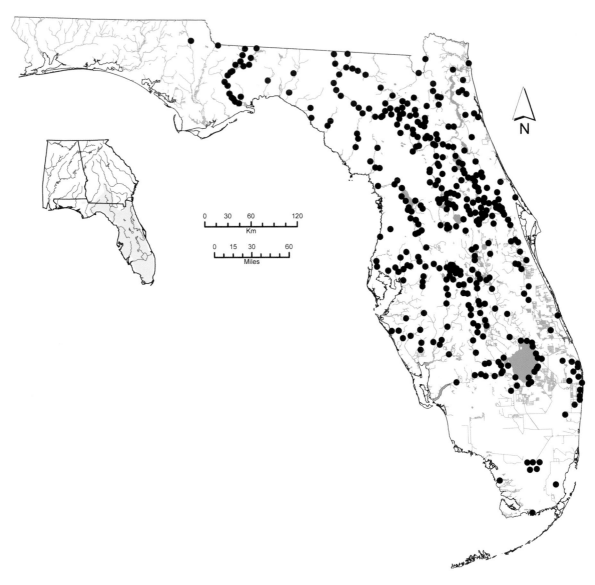

Figure 10.28. Distribution of *Elliptio jayensis*.

Ecology and Biology

Elliptio jayensis inhabits almost every type of water body in peninsular Florida, including creeks, rivers, marshes, ponds, and lakes. It also inhabits man-made water bodies such as canals, wastewater and aquaculture ponds, and reservoirs. It is found in mud, sand, gravel, and rubble substrates.

Elliptio jayensis is gravid from winter to early summer. Based on laboratory infestations, three host fishes that occur in Florida have been identified—Centrarchidae: *Lepomis macrochirus* (Bluegill) and *Micropterus salmoides* (Largemouth Bass); and Lepisosteidae: *Lepisosteus platyrhincus* (Florida Gar) (Keller and Ruessler 1997a).

Elliptio jayensis has been observed to be infested with mites (Unionicolidae). Of 370 individuals examined from seven localities in peninsular Florida, 146 were parasitized by one or more of the following species: *Unionicola abnormipes*, *Unionicola aculeata*, *Unionicola alleni*, and *Unionicola causeyae* (Vidrine 1996).

Conservation Status

Elliptio jayensis was listed as undetermined throughout its range by Williams et al. (1993) due to its taxonomic uncertainty. It was considered to be currently stable by Williams et al. (in review).

Elliptio jayensis is the most widespread and common mussel in peninsular Florida.

Remarks

Throughout the past century *Elliptio buckleyi* was the name most commonly applied to the species of *Elliptio* that dominated aquatic habitats throughout peninsular Florida. The correct name for this taxon is *Elliptio jayensis*, which has priority over *E. buckleyi* (Williams et al. 2011). Common names applied to *E. buckleyi* and *E. jayensis* by Turgeon et al. (1988, 1998) were Florida Shiny Spike and Flat Spike, respectively. Since *E. jayensis* is virtually endemic to Florida and is neither consistently shiny nor flat, we propose a new common name, Florida Spike.

The wide variety of habitats occupied by *Elliptio jayensis* has produced extreme variation in shell morphology (Figure 10.16). This has resulted in 26 nominal taxa being described based on minor differences in shell shape (Williams et al. 2011).

The species name *jayensis* is a patronym for John Clarkson Jay, a conchologist who was active between 1830 and 1860. He provided numerous specimens of unionids to colleagues who were actively describing new species.

Synonymy

Unio jayensis Lea 1838 (Lea 1838a:28). Lectotype (Johnson 1974), USNM 86031 (right valve only), length 64 mm. Type locality: published as Florida, restricted by Williams et al. (2011) to Lake Monroe, Seminole and Volusia counties, Florida, St. Johns River basin. This restriction was in error as further review of the USNM ledger indicated that the specimen came from Lake George, [Volusia County,] Florida, [St. Johns River basin].

Unio buckleyi Lea 1843 (Lea 1843:[one page privately published]). Type locality: Lake George, [Putnam and Volusia counties,] Florida, [St. Johns River basin].

Unio buddianus Lea 1843 (Lea 1843:[one page privately published]). Type locality: Lake George, [Putnam and Volusia counties,] Florida, [St. Johns River basin].

Unio coruscus Gould 1856 (Gould 1856:15). Type locality: River St. Johns near Lake Beresford, [Volusia County,] Florida, [St. Johns River basin].

Unio prasinatus Conrad 1866 (Conrad 1866:279). Type locality: published as Florida, restricted by Williams et al. (2011) to Lake Monroe, Seminole and Volusia counties, Florida, St. Johns River basin. This action was in error as no evidence was provided (as required by the ICZN) to justify restricting the locality to Lake Monroe.

Unio cunninghami B.H. Wright 1883 (B.H. Wright 1883:58). Type locality: published as lakes of Sumter County, Florida, restricted by Johnson (1967a) to Lake Harris, Yalaha, Lake County, Florida, [St. Johns River basin].

Unio dallii B.H. Wright 1888 (B.H. Wright 1888:119). Type locality: Lake Beresford, Volusia County, Florida, [St. Johns River basin].

Unio dorei B.H. Wright 1888 (B.H. Wright 1888:115). Type locality: Lake Monroe, [Volusia and Seminole counties,] Florida, [St. Johns River basin].

Unio hinkleyi B.H. Wright 1888 (B.H. Wright 1888:117). Type locality: Lake Monroe, [Volusia and Seminole counties,] Florida, [St. Johns River basin].

Unio marshii B.H. Wright 1888 (B.H. Wright 1888:118). Type locality: Lake Woodruff, Volusia County, Florida, [St. Johns River basin].

Unio nolani B.H. Wright 1888 (B.H. Wright 1888:116). Type locality: a creek flowing into St. Johns River, near Palatka, [Putnam County,] Florida, [St. Johns River basin].

Unio orcuttii S.H. Wright 1888 (S.H. Wright 1888:60). Type locality: Lake Miakka [Myakka, Sarasota County], Florida, restricted by Johnson (1972) to Upper Miakka [Myakka] Lake, Sarasota County, Florida, [Myakka River basin].

Unio simpsoni B.H. Wright 1888 (B.H. Wright 1888:117). Type locality: Lake Woodruff, Volusia County, Florida, [St. Johns River basin].

Unio tryoni B.H. Wright 1888 (B.H. Wright 1888:120). Type locality: Lake Woodruff [Spring Garden Lake] near De Leon Springs, Volusia County, Florida, [St. Johns River basin].

Unio ferrissii Marsh 1891 (Marsh 1891:30). Type locality: a small creek near Pilatka [Palatka, Putnam County,] Florida, [St. Johns River basin].

Unio leonensis B.H. Wright 1892 (B.H. Wright *in* Simpson 1892b:419). Type locality: Lake Woodruff [Spring Garden Lake] near De Leon Springs, [Volusia County, Florida, St. Johns River basin].

Unio oscari B.H. Wright 1892 (B.H. Wright 1892:124). Type locality: a creek from Lake Osceola at Winter Park, [Orange County,] Florida, [St. Johns River basin].

Unio subluridus Simpson 1892 (Simpson 1892b:432). Type locality: Orange Springs, Volusia [Marion] County, Florida, [St. Johns River basin].

Unio buxtoni B.H. Wright 1897 (B.H. Wright 1897a:55). Type locality: lakelets of Marion County, Florida, [St. Johns River basin or Withlacoochee River (southern) basin].

Unio diazensis S.H. Wright 1897 (S.H. Wright 1897:5). Type locality: Lake Diaz [Dias], Volusia County, Florida, [St. Johns River basin].

Unio pinei B.H. Wright 1897 (B.H. Wright 1897b:40). Type locality: unnamed lake in Witthacoochee [Withlacoochee] River region of Hernando County, Florida, [Withlacoochee River (southern) basin].

Unio suttoni B.H. Wright 1897 (B.H. Wright 1897a:56). Type locality: lake near Candler, Marion County, Florida, [St. Johns River basin].

Unio tenuisculus Frierson 1911 (Frierson 1911:29). Type locality: Reedy Lake, [Frostproof,] Polk County, Florida, in the Gulf [Coast] drainage, [Kissimmee River drainage].

Unio (Elliptio) sanctorumjohanium B.H. Wright 1933 (B.H. Wright 1933:17). Type locality: Lake Druid near Floral City, [Citrus County,] Florida, [Withlacoochee River (southern) basin].

Elliptio maywebbae B.H. Wright 1934 (B.H. Wright 1934a:28). Type locality: near Seminole Springs, 15 miles southeast of Eustis, [Lake County,] Florida, [St. Johns River basin].

Unio (*Elliptio*) *webbianus* B.H. Wright 1934 (B.H. Wright 1934b:94). Type locality: Lake Consuelo near Floral City, Citrus County, Florida, [Withlacoochee River (southern) basin].

Unio (*Elliptio*) *webbianus hartii* B.H. Wright 1934 (B.H. Wright 1934b:95). Type locality: Lake Consuelo near Floral City, Citrus County, Florida, [Withlacoochee River (southern) basin].

Elliptio mcmichaeli Clench and Turner 1956
Fluted Elephantear

Elliptio mcmichaeli – Upper image: length 83 mm, UF 388513. Pea River, 0.8 miles west of Antioch Church on County Road 14, Bullock County, Alabama, 15 June 1998. © Richard T. Bryant. Middle image: length 72 mm, USNM 710723 (paratype). Choctawhatchee River, 8 miles west of Miller Crossroads, State Route 2, Holmes County, Florida. © Richard T. Bryant. Lower image: length 35 mm, UMMZ 163341. Pea River, Andrew's Fish Trap, Barbour County, Alabama, November 1915. © Richard T. Bryant.

Description

Size: length to 130 mm, in Florida to 116 mm.

Shell: moderately thin; typically smooth anterior to posterior ridge; compressed to moderately inflated, width 20%–35% of length; outline elliptical to subrhomboidal; anterior margin rounded; posterior margin obliquely truncate to bluntly pointed; dorsal margin straight to slightly convex; ventral margin straight to convex, occasional large individuals concave; posterior ridge sharp dorsally, rounded posterioventrally, typically biangulate posterioventrally; posterior slope steep dorsally, flat to slightly concave posterioventrally, typically with thin corrugations extending from posterior ridge to posteriodorsal margin; umbo broad, slightly inflated, elevated slightly above hinge line; umbo sculpture unknown; umbo cavity wide, shallow.

Teeth: pseudocardinal teeth low, triangular, 2 divergent teeth in left valve, anterior tooth typically smaller, 1 tooth in right valve, occasionally with accessory denticle anteriorly; lateral teeth long, straight to slightly curved, 2 in left valve, 1 in right valve; interdentum short to moderately long, narrow to moderately wide.

Nacre: bluish white to purplish, occasionally with salmon tint in umbo cavity.

Periostracum: small individuals shiny, large individuals dull or clothlike; small individuals greenish brown, with dark green rays of varying width and intensity, large individuals dark brown to black, often obscuring rays.

Glochidium Description

Outline subtriangular; length 130–157 μm; height 149–161 μm; width 106–116 μm; ventral margin with broad shelf covered with micropoints; small, straplike hook present near midline of ventral margin (Figure 10.29).

Similar Sympatric Species

Elliptio mcmichaeli may resemble *Elliptio crassidens* but differs in having a thinner shell and less angular posterior ridge. The outline of *E. mcmichaeli* is more elongate than *E. crassidens*, which is typically more triangular.

Small individuals of *Elliptio mcmichaeli* can be confused with *Fusconaia burkei*; however, *E. mcmichaeli* is typically more compressed and the pseudocardinal and lateral teeth are thicker than those of *F. burkei*.

Distribution in Florida

Elliptio mcmichaeli occurs in Yellow and Choctawhatchee Rivers and large tributaries of Choctawhatchee River (Figure 10.30).

Ecology and Biology

Elliptio mcmichaeli occurs in large creeks and rivers in sand and sandy clay substrates in slow to moderate current.

Elliptio mcmichaeli is a short-term brooder. Mature glochidia have been reported in *E. mcmichaeli* in May in the Choctawhatchee River (O'Brien et al. 2003).

Glochidial hosts for *Elliptio mcmichaeli* are unknown. It has disappeared from Pea River upstream of Elba Dam, Alabama, which was built in the early 1900s (Hall and Hall 1916) and impedes fish migration above that point. This suggests that an anadromous fish is used as a host. There are three anadromous fishes known to make spawning runs up Pea River—Acipenseridae: *Acipenser oxyrinchus desotoi* (Gulf Sturgeon); and Clupeidae: *Alosa alabamae* (Alabama Shad) and *Alosa chrysochloris* (Skipjack Herring).

Elliptio mcmichaeli has been observed to be infested with mites (Unionicolidae). Of 76 individuals examined from four localities on the Choctawhatchee River, one was parasitized by *Unionicola alleni*, one by *Unionicola sakantaka*, and one by *Unionicola vamana* (Vidrine 1996).

Conservation Status

Elliptio mcmichaeli was reported by Williams et al. (1993) as a species of special concern throughout its range. It was assigned the status of currently stable by Williams et al. (in review).

Elliptio mcmichaeli is extant in Choctawhatchee River and lower reaches of many of its tributaries and Yellow River upstream of its confluence with Shoal River. It remains common through most of its Florida range.

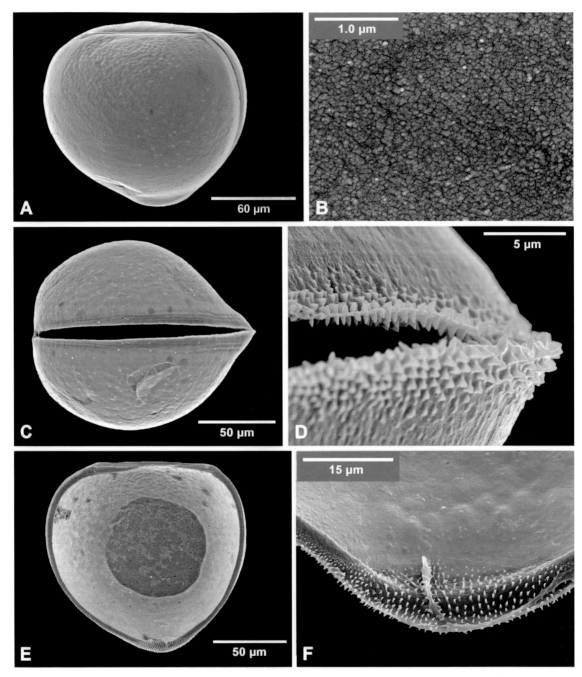

Figure 10.29. Scanning electron micrographs of glochidial valves of *Elliptio mcmichaeli* from West Fork Choctawhatchee River near Blue Springs, Dale County, Alabama, 5 May 1986—(A) external view; (B) surface sculpture; (C) lateral view; (D) lateral view of ventral margin and micropoints; (E) internal view; and (F) internal view of modified flange with micropoints and a small, straplike hook.

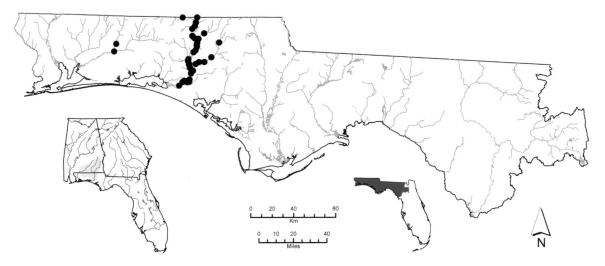

Figure 10.30. Distribution of *Elliptio mcmichaeli*.

Remarks

Elliptio mcmichaeli was described from the Choctawhatchee River basin by Clench and Turner (1956). It was subsequently placed in the synonymy of *Elliptio fraterna* by Johnson (1970), but Fuller and Bereza (1973, 1974) resurrected *E. mcmichaeli* based on mantle margin pigmentation and the presence of arborescent papillae on the incurrent aperture. Most authors have continued to recognize *E. mcmichaeli* as a valid species (Turgeon et al. 1988, 1998).

Elliptio mcmichaeli is not known to occur in the Escambia River basin in Florida, but it has been reported from upstream tributaries in Alabama (Williams et al. 2008).

Elliptio mcmichaeli was named in honor of Donald F. McMichael, a student and colleague of William J. Clench and Ruth D. Turner, who participated in the 1950s eastern Gulf Coast rivers mollusk surveys (Clench and Turner 1956).

Synonymy

Elliptio mcmichaeli Clench and Turner 1956 (Clench and Turner 1956:170). Holotype, MCZ 191922, length 91 mm. Type locality: Choctawhatchee River, 8 miles west of Miller Crossroads, State Route 2, Holmes County, Florida, [Choctawhatchee River basin].

Elliptio monroensis (Lea 1843)
St. Johns Elephantear

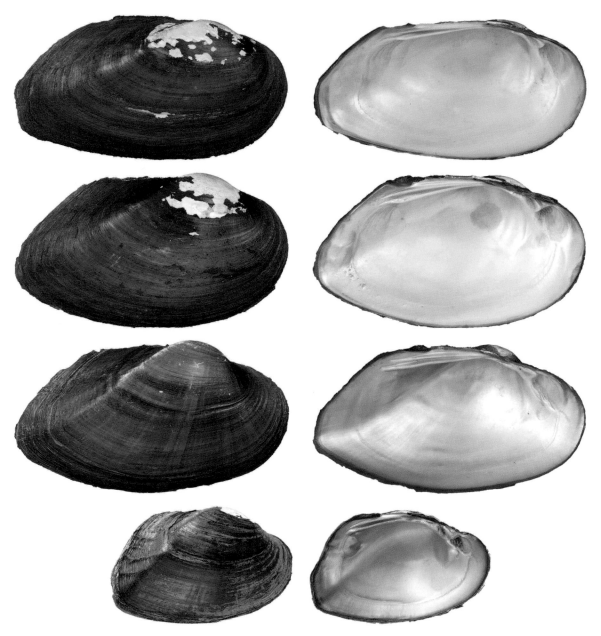

Elliptio monroensis – First image: length 96 mm, UF 2855. St. Johns River, Crows Bluff, Volusia County, Florida. Photograph by Zachary Randall. Second image: length 87 mm, UF 2855. St. Johns River, Crows Bluff, Volusia County, Florida. Photograph by Zachary Randall. Third image: length 61 mm, UF 2855. St. Johns River, Crows Bluff, Volusia County, Florida. Photograph by Zachary Randall. Fourth image: length 52 mm, UF 455456. St. Marys River, about 300 meters below mouth of Spanish Fork Creek, Nassau County, Florida, and Charlton County, Georgia, 23 June 2010. © Richard T. Bryant.

Description

Size: length to 128 mm.

Shell: thin to moderately thick; smooth; moderately compressed to moderately inflated, width 25%–45% of length; outline elliptical to subtriangular; anterior margin rounded; posterior margin rounded to obliquely truncate; dorsal margin straight to slightly convex; ventral margin straight to broadly rounded; posterior ridge sharp dorsally, rounded posterioventrally, occasionally weakly biangulate posterioventrally; posterior slope steep, concave, typically with thin corrugations from posterior ridge to posteriodorsal margin, may be absent in some individuals; umbo broad, inflated, typically elevated well above hinge line in riverine and lake populations, only slightly above in creek populations; umbo sculpture thin ridges; umbo cavity wide, shallow to moderately deep.

Teeth: pseudocardinal teeth moderately thick, triangular, 2 teeth in left valve, anterior tooth crest oriented anterioventrally, 1 tooth in right valve, occasionally with accessory denticle anteriorly; lateral teeth moderately long, straight to slightly curved, 2 in left valve, 1 in right valve; interdentum moderately short, narrow.

Nacre: bluish white, occasionally with pink or purple tint.

Periostracum: typically dull or clothlike, occasionally shiny; small individuals yellowish green to dark olive, with dark green rays of varying width and intensity, large individuals dark brown to black, usually obscuring rays.

Glochidium Description

The glochidium of *Elliptio monroensis* is unknown.

Similar Sympatric Species

Elliptio monroensis may superficially resemble some individuals of *Elliptio jayensis* and *Elliptio occulta* but differs in typically having a more inflated umbo that is elevated well above the hinge line, sharper posterior ridge, and corrugations on the posterior slope. In the St. Marys River, *E. monroensis* resembles *Elliptio crassidens*, but that species typically has a thicker shell, more rounded posterior ridge, and is more triangular in shape.

Distribution in Florida

Elliptio monroensis occurs in St. Marys and St. Johns River basins (Figure 10.31).

Ecology and Biology

Elliptio monroensis inhabits large creeks, rivers, and lakes in substrates of stable sand and sandy mud. In St. Johns basin it occurs in lakes along the main stem and lower reaches of some large tributaries.

Elliptio monroensis is a short-term brooder that is gravid in spring to summer. Glochidial hosts for *E. monroensis* are unknown. *Alosa* (shad and herring) have been confirmed or inferred as glochidial hosts in other species of corrugate *Elliptio* (e.g., *E. crassidens*).

Conservation Status

The conservation status of *Elliptio monroensis* was not evaluated in previous assessments since it was not recognized as a valid taxon until 1998 (Turgeon et al. 1998). The species was considered to be threatened by Williams et al. (in review).

Elliptio monroensis is extant in St. Marys and St. Johns River basins but is uncommon through much of its range.

Figure 10.31. Distribution of *Elliptio monroensis*.

Remarks

Elliptio monroensis shell morphology is highly variable within large river and lake habitats. Most notable is degree of sharpness of the posterior ridge and presence or absence of corrugations on the posterior ridge. *Elliptio monroensis* is the largest mussel in the St. Johns River basin. *Elliptio monroensis* is globally restricted to Florida streams, occurring in Georgia only where the border is formed by St. Marys River. Except for a disjunct population in the Black Creek drainage, Clay County, the species is not known from the lowermost portion of the basin.

Specimens of *Elliptio monroensis* from the lower reaches of Econlockhatchee River have been found in museum collections with byssal threads and occasional specimens of *Mytilopsis leucophaeata* attached to the posterior end of the shell. This area of St. Johns basin is known to have saline springs that may influence water quality.

In a review of the unionids of peninsular Florida, Johnson (1972) placed *Elliptio monroensis* in the synonymy of *Elliptio dariensis*, which was followed by Heard (1979a). It remained in synonymy of *E. dariensis* until Turgeon et al. (1998) recognized *E. monroensis* as valid.

Synonymy

Unio monroensis Lea 1843 (Lea 1843:[one page privately published]). Holotype by monotypy, USNM 85169, length 70 mm. Type locality: Lake Monroe, [Volusia and Seminole counties,] Florida, [St. Johns River basin].

Unio anthonyi Lea 1861 (Lea 1861:41). Type locality: published as Florida, restricted by Williams et al. (2011) to Lake Monroe, Seminole County, St. Johns River basin, Florida. The type locality was given as "Florida" but was subsequently restricted by Williams et al. (2011) based on a comparison of shell morphology and color of other corrugate *Elliptio* in Florida. The restriction of the type locality was deemed necessary to provide stability of nomenclature of corrugate *Elliptio* in Florida.

Unio websterii B.H. Wright 1888 (B.H. Wright 1888:113). Type locality: Lake Woodruff, Volusia County, Florida, [St. Johns River basin].

Unio hartwrightii B.H. Wright 1896 (B.H. Wright 1896:121). Type locality: Lake Beresford, [Volusia County,] Florida, [St. Johns River basin].

Unio polymorphus B.H. Wright 1899 (B.H. Wright 1899:42). Type locality: Spanish Creek, [tributary of St. Marys River, west of Folkston,] Okefenokee Swamp, Charlton County, Georgia, [St. Marys River basin].

Elliptio occulta (Lea 1843)
Hidden Spike

Elliptio occulta – First image: length 88 mm, UF 243959. St. Marys River, 0.5 miles east of St. George, upstream from Route 2 bridge, Nassau County, Florida, 30 June 1990. Photograph by Zachary Randall. Second image: length 81 mm, UF 243959. St. Marys River, 0.5 miles east of St. George, upstream from Route 2 bridge, Nassau County, Florida, 30 June 1990. Photograph by Zachary Randall. Third image: length 42 mm, UF 449433. St. Marys River slough, about 300 meters downstream of State Highway 2, Nassau County, Florida, 23 June 2010. © Richard T. Bryant. Fourth image: length 33 mm, UF 463465. Suwannee River at boat ramp in Suwannee River Music Campground Park, Suwannee and Hamilton counties, Florida, 6 June 2008. Photograph by Zachary Randall.

Description

Size: length to 88 mm.

Shell: moderately thin; smooth; compressed to moderately inflated, width 25%–45% of length; outline oval to subrhomboidal; anterior margin rounded; posterior margin rounded to bluntly pointed; dorsal margin convex; ventral margin convex; posterior ridge rounded, weakly biangulate posteriorly; posterior slope low, flat to slightly concave; umbo moderately broad, slightly compressed to inflated, even with hinge line or elevated slightly above; umbo sculpture thin ridges, with shallow ventral indentation; umbo cavity wide, shallow.

Teeth: pseudocardinal teeth moderately large, compressed, 2 teeth in left valve, occasionally joined at base, approximately equal in size or anterior tooth slightly larger, 1 tooth in right valve, with or without accessory denticle anteriorly; lateral teeth short to moderately long, thin to moderately thick, straight to slightly curved, 2 in left valve, 1 in right valve; interdentum moderately long, narrow.

Nacre: white to pinkish.

Periostracum: shiny, occasionally clothlike; yellowish green, olive to brown, typically with green rays varying in width and intensity, usually obscure in large individuals.

Glochidium Description

Outline depressed subelliptical; length 161–188 µm; height 157–196 µm; width 92–122 µm; ventral margin with internal and external irregularly arranged lanceolate micropoints (Figure 10.32).

Similar Sympatric Species

Elliptio occulta most closely resembles *Elliptio jayensis*, but *E. occulta* typically has thinner teeth, an umbo even with the hinge line or elevated slightly above, and a rounded posteriodorsal margin. The outline of *E. occulta* is typically more oval than that of *E. jayensis*.

Elliptio occulta may also resemble *Elliptio monroensis*, but *E. occulta* is typically more oval, has a more rounded posterior ridge, and lacks corrugations on the posterior slope.

Distribution in Florida

Elliptio occulta occurs in Steinhatchee and Suwannee River basins on the Gulf Coast and St. Marys and St. Johns River basins on the Atlantic Coast (Figure 10.33).

Ecology and Biology

Elliptio occulta occurs in creeks and rivers in sand and sandy mud substrates in slow to moderate current.

Elliptio occulta is a short-term brooder, gravid from spring to early summer. Glochidial hosts for *E. occulta* are unknown.

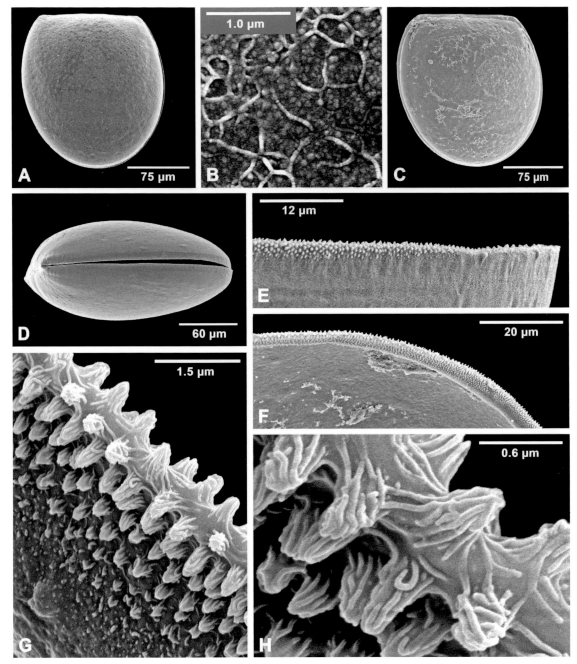

Figure 10.32. Scanning electron micrographs of glochidial valves of *Elliptio occulta* from St. Marys River, Nassau County, Florida, and Charlton County, Georgia, 22 May 2008—(A) external view; (B) surface sculpture; (C) internal view; (D) lateral view; (E) lateral view of ventral margin and micropoints; (F) internal view of micropoints along ventral margin; (G) internal view of micropoints; and (H) close-up of micropoints.

Conservation Status

Elliptio occulta was not recognized as a distinct taxon until 2011 (Williams et al. 2011) and therefore was not included in conservation status assessments of the 1990s. The species was considered currently stable by Williams et al. (in review). *Elliptio occulta* is extant throughout its range. It remains common at most localities.

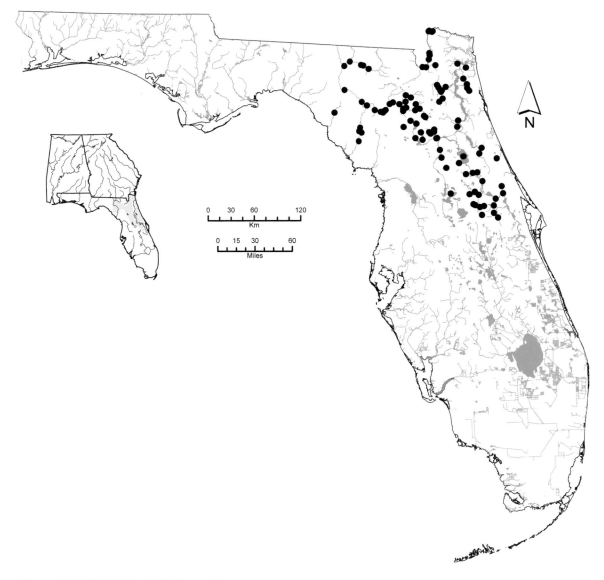

Figure 10.33. Distribution of *Elliptio occulta*.

Remarks

Elliptio occulta was not considered a valid taxon until recognized by Williams et al. (2011) and was previously treated as a synonym of *Elliptio icterina* (Johnson 1972). Many of the historical records of *E. icterina* were based on *E. occulta*.

Synonymy

Unio occultus Lea 1843 (Lea 1843:[one page privately published]). Lectotype (Johnson 1974), USNM 85247, length 53 mm. Type locality: published as Black Creek and Lake Monroe, Florida, restricted by Williams et al. (2011) to Lake Monroe, Volusia and Seminole counties, Florida, St. Johns River basin.

Unio fuscatus Lea 1843 (Lea 1843:[one page privately published]). Type locality: Black Creek, [Clay County,] Florida, [St. Johns River basin].

Unio fryanus B.H. Wright 1888 (B.H. Wright 1888:113). Type locality: Lake Ashby, Volusia County, Florida, [St. Johns River basin].

Unio burtchianus S.H. Wright 1897 (S.H. Wright 1897:137). Type locality: St. Marys River, Nassau County, Florida, [St. Marys River basin].

Unio lehmanii S.H. Wright 1897 (S.H. Wright 1897:138). Type locality: St. Marys River, [Nassau County,] Florida, [St. Marys River basin].

Unio dispalans B.H. Wright 1899 (B.H. Wright 1899:50). Type locality: Suwannee River, Florida, [Suwannee River basin].

Unio rafinesquei Vanatta 1915 (Vanatta 1915:559) [replacement name for *Unio fuscatus* Lea 1843].

Elliptio pullata (Lea 1856)
Gulf Spike

Elliptio pullata – First image: length 99 mm, UF 8351. Ponce Creek, Walton County, Florida, 1933. Photograph by Zachary Randall. Second image: length 93 mm, UF 64880. Little Patsaliga Creek, Crenshaw County, Alabama. © Richard T. Bryant. Third image: length 87 mm, UF 388512. Canal tributary to Smokehouse Lake (off Choctawhatchee River), 200 meters downstream and southwest of Smokehouse Landing, Walton County, Florida, 22 March 2000. © Richard T. Bryant. Fourth image: length 58 mm, UF 375380. A canal tributary to Smokehouse Lake, Choctawhatchee River drainage [basin], Walton County, Florida, 22 March 2000. © Richard T. Bryant.

Description

Size: length to 110 mm.

Shell: small individuals thin, large individuals moderately thick; smooth; moderately compressed to moderately inflated, width 20%–40% of length; outline elongate, elliptical, occasional large individuals arcuate; anterior margin typically rounded, occasionally truncate; posterior margin bluntly pointed to narrowly rounded; dorsal margin straight to slightly convex; ventral margin straight to slightly convex, occasional large individuals slightly concave; posterior ridge low, rounded, typically biangulate posterioventrally; posterior slope low, flat to slightly convex; umbo broad, slightly inflated, even with hinge line or elevated slightly above; umbo sculpture moderately thick ridges (Figure 10.34); umbo cavity wide, shallow.

Teeth: pseudocardinal teeth triangular, 2 teeth in left valve, anterior tooth often slightly compressed, crest oriented anterioventrally to ventrally, 1 thick tooth in right valve; lateral teeth long, straight to slightly curved, 2 in left valve, 1 in right valve; interdentum short to moderately long, narrow, occasionally moderately wide.

Nacre: purplish, occasionally white, with salmon tint.

Periostracum: small individuals shiny, large individuals dull to clothlike; small individuals greenish brown, occasionally with green rays of varying width and intensity, large individuals brown to black, typically obscuring rays.

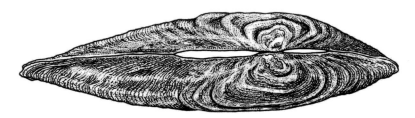

Figure 10.34. Umbo sculpture of *Elliptio pullata*, length 16 mm, UF 449117. Escambia River, about 0.3 miles above Chumuckla Springs boat ramp, Santa Rosa County, Florida, 24 September 2008. Illustration by Susan Trammell.

Glochidium Description

Outline depressed subelliptical; length 169–188 μm; height 183–210 μm; width 93–99 μm; ventral margin with internal and external irregularly arranged lanceolate micropoints (Figure 10.35).

Similar Sympatric Species

In Florida *Elliptio pullata* resembles several species of elongate, elliptical *Elliptio* inhabiting Gulf Coast rivers. *Elliptio pullata* most closely resembles *Elliptio fumata* but is more elongate in outline. *Elliptio arctata* typically has a thinner shell, smaller pseudocardinal teeth, and more arcuate ventral margin than *E. pullata*. *Elliptio purpurella* is more inflated and has darker purple nacre than *E. pullata*.

Elliptio pullata may resemble *Ptychobranchus jonesi* but is typically more compressed and has a more angular posterior margin. Also, *P. jonesi* has a triangulate posterior ridge with a shallow radial sulcus between each ridge.

Distribution in Florida

Elliptio pullata occurs in most basins from Escambia River east to Apalachicola River (Figure 10.36).

Ecology and Biology

Elliptio pullata occurs in a variety of aquatic habitats, including creeks, rivers, and floodplain lakes. It is common in streams with slow to moderate current and occurs in a variety of substrates ranging from fine to coarse sediments often with detritus. It is also found in gravel and rubble, and fissures of bedrock.

Elliptio pullata is a short-term brooder and has been reported gravid in June and July in the Apalachicola basin (Brim Box and Williams 2000). *Elliptio pullata* is typically a dioecious species, but a very small percentage may be hermaphroditic (Kotrla 1988; Heard 1979b). Glochidial hosts for *E. pullata* are unknown.

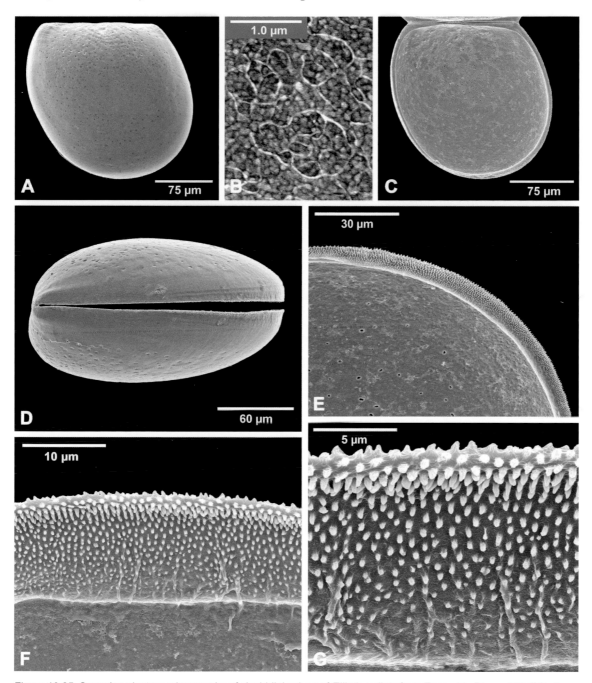

Figure 10.35. Scanning electron micrographs of glochidial valves of *Elliptio pullata* from Escambia River at Bluff Springs, Escambia County, Florida, 27 June 2010—(A) external view; (B) surface sculpture; (C) internal view; (D) lateral view; (E) internal view of ventral margin and micropoints; (F) internal view of micropoints along ventral margin; and (G) close-up of micropoints.

Conservation Status

Elliptio pullata was not recognized as a distinct species until 2008 (Williams et al. 2008) and was not included in conservation status assessments of the 1990s. The species was considered currently stable rangewide by Williams et al. (in review).

Elliptio pullata is extant throughout its Florida range. It is one of the most common mussels across its range in Florida.

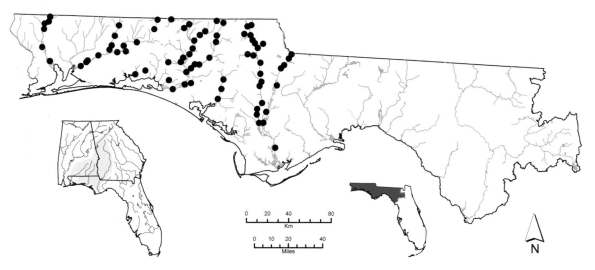

Figure 10.36. Distribution of *Elliptio pullata*.

Remarks

Elliptio pullata, like other *Elliptio* species, exhibits considerable variation in shell morphology throughout its range. This variability has resulted in its inconsistent taxonomic treatment in historical accounts, most notably *Elliptio strigosa* by Clench and Turner (1956) and *Elliptio icterina* by Johnson (1970). The use of *Elliptio strigosa* by Clench and Turner included all elongate *Elliptio* (e.g., *E. arctata*, *E. purpurella*) from the Escambia River basin east to Waccasassa River. *Elliptio icterina* is endemic to southern Atlantic Coast basins. Historical records of *E. icterina* from Escambia to Choctawhatchee Rivers are actually *E. pullata*. Also, some records of *E. icterina* from the Apalachicola basin are based on *E. pullata*.

Due to taxonomic confusion surrounding the species, the range of *Elliptio pullata* has not been consistently delineated (Williams et al. 2008, 2011). Additional analyses of the species and related forms of *Elliptio* are needed to ascertain its true taxonomic status and distribution. For example, *E. pullata* in the Choctawhatchee River basin is typically more inflated than in adjacent basins and may represent an undescribed species. As currently aligned, numerous synonyms have been assigned to *E. pullata* (Williams et al. 2011).

Synonymy

Unio pullatis Lea 1856 (Lea 1856:262) [spelling corrected to *Unio pullatus* Lea 1858c]. Lectotype (Johnson 1970), USNM 86020, length 84 mm. Type locality: creeks near Columbus, [Muscogee County,] Georgia, [Chattahoochee River drainage].

Unio aquilus Lea 1857 (Lea 1857a:172). Type locality: Flint River near Macon [County], Georgia, [Flint River drainage].

Unio extensus Lea 1857 (Lea 1857b:31). Type locality: Dry Creek, [tributary to Upatoi Creek,] near Columbus, [possibly Talbot County,] Georgia, [Chattahoochee River drainage].

Unio maconensis Lea 1857 (Lea 1857a:172). Type locality: Flint River near Macon [County], Georgia, [Flint River drainage].

Unio sublatus Lea 1857 (Lea 1857a:169). Type locality: Uchee Bar, [Chattahoochee River near mouth of Uchee Creek,] below Columbus, Georgia, [Chattahoochee River drainage].

Unio tetricus Lea 1857 (Lea 1857a:170). Type locality: Flint River near Albany, [Dougherty County,] Georgia, [Flint River drainage].

Unio roswellensis Lea 1858 (Lea 1858b:165). Type locality: Chattahoochee River at Roswell, Cobb County, Georgia, [Chattahoochee River drainage].

Unio verutus Lea 1859 (Lea 1859a:171). Type locality: Flat Rock [Flatrock] Creek near Columbus, [Muscogee County,] Georgia, [Chattahoochee River drainage].

Unio viridiradiatus Lea 1859 (Lea 1859b:154). Type locality: Big Uchee [Creek, Lee and Russell counties, Alabama,] near Columbus, [Muscogee County,] Georgia, [Chattahoochee River drainage].

Unio mercerii Lea 1862 (Lea 1862a:169). Type locality: Lee County, Georgia, [Flint River drainage].

Unio singularis B.H. Wright 1899 (B.H. Wright 1899:75). Type locality: Spring Creek, [tributary to the Flint River,] Decatur County, Georgia, [Flint River drainage].

Elliptio purpurella (Lea 1857)
Inflated Spike

Elliptio purpurella – Upper image: length 49 mm, UF 388492. Muckalee Creek, State Highway 195, 3.5 air miles northeast of Leesburg, Lee County, Georgia, 11 August 1992. © Richard T. Bryant. Lower image: length 46 mm, UF 373510. Wrights Creek at County Road 179, 3 air miles northeast of Westville, Holmes County, Florida, 29 July 1998. © Richard T. Bryant.

Description

Size: length to 90 mm.

Shell: moderately thin to moderately thick; smooth; moderately inflated, width 30%–40% of length; outline elongate, oval to elliptical, occasional large individuals slightly arcuate; anterior margin typically rounded, occasionally truncate; posterior margin obliquely truncate; dorsal margin straight to slightly convex; ventral margin straight to slightly convex, occasional large individuals slightly concave; posterior ridge rounded, typically biangulate posterioventrally; posterior slope moderately steep, flat; umbo broad, elevated slightly above hinge line; umbo sculpture unknown; umbo cavity wide, shallow.

Teeth: pseudocardinal teeth moderately thick, triangular, 2 divergent teeth in left valve, approximately equal in size or anterior tooth occasionally smaller, 1 tooth in right valve; lateral teeth short to moderately long, thin, straight to slightly curved, 2 in left valve, 1 in right valve; interdentum moderately long, narrow.

Nacre: purple, occasionally with salmon tint in umbo cavity.

Periostracum: dull to clothlike; small individuals olive brown to brown, often with dark green rays, large individuals dark brown to black, usually obscuring rays.

Glochidium Description

The glochidium of *Elliptio purpurella* is unknown.

Similar Sympatric Species

Elliptio purpurella most closely resembles *Elliptio arctata* but is more inflated and usually less arcuate along its ventral margin. The nacre is usually purple in *E. purpurella* and typically bluish white in *E. arctata*.

Elliptio purpurella also resembles *Elliptio fumata* and *Elliptio pullata* but is more inflated, has a thinner shell, and the nacre is more consistently purple.

Distribution in Florida

Elliptio purpurella occurs in Choctawhatchee, Apalachicola, and Ochlockonee River basins (Figure 10.37).

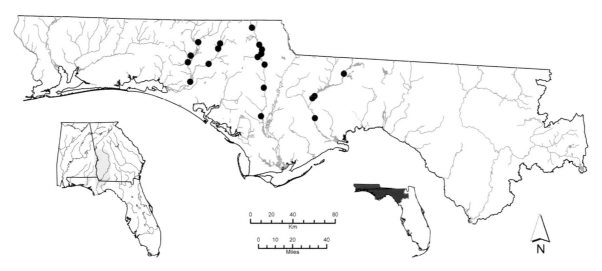

Figure 10.37. Distribution of *Elliptio purpurella*.

Ecology and Biology

Elliptio purpurella inhabits large creeks and rivers. It occurs in moderate current in sand, gravel, and rubble substrates.

Elliptio purpurella is presumably a short-term brooder, gravid in spring. An examination of specimens collected between May and September from the Apalachicola basin revealed no gravid females (Brim Box and Williams 2000). Glochidial hosts for *E. purpurella* are unknown.

Conservation Status

Conservation status of *Elliptio purpurella* had not been evaluated nationally or in Florida because it was not elevated from synonymy of other *Elliptio* species until 2000 (Brim Box and Williams 2000). It was assigned a rangewide status of threatened by Williams et al. (in review).

Elliptio purpurella is extant throughout its Florida range where it occurs in localized populations. The species is generally uncommon.

Remarks

Elliptio purpurella was recognized by Brim Box and Williams (2000) as a valid species based on shell morphology and electrophoresis data. Previous workers had treated *E. purpurella* as a synonym of *Elliptio arctata* (Simpson 1914), *Elliptio nigella* (Frierson

1927; Johnson 1968), and *Elliptio strigosa* (Clench and Turner 1956). Historical records of *Elliptio icterina* from Choctawhatchee to Apalachicola Rivers in part may represent *E. purpurella*.

Elliptio purpurella was considered to be endemic to the Apalachicola River basin by Brim Box and Williams (2000) and Williams et al. (2008).

Synonymy

Unio purpurellus Lea 1857 (Lea 1857a:171). Lectotype (Johnson 1968), USNM 85675, length 36 mm. Type locality: Flint River near Albany, [Dougherty County,] Georgia, [Flint River drainage].

Genus *Elliptoideus* Frierson 1927

Elliptoideus is monotypic and found only in the Apalachicola and Ochlockonee River basins. *Elliptoideus* is the only genus of Unionidae endemic to the Greater Floridan Region. It occurs in Chattahoochee River in Alabama and Georgia, Flint River in Georgia, and Apalachicola and Ochlockonee Rivers in Florida.

The distinctiveness of *Elliptoideus* has never been contested, but agreement to its generic placement has not been consistent. Conrad (1853) placed it and nine other plicate species in *Plectomerus*. *Elliptoideus sloatianus* utilizes all four gills as marsupia. Based primarily on this character, Frierson (1927) erected the subgenus *Elliptoideus*. Heard and Guckert (1970) elevated *Elliptoideus* to generic level. Phylogenetic analyses using molecular genetics have not consistently resolved the relationship of *Elliptoideus* to other genera (Serb et al. 2003; Campbell et al. 2005; Campbell and Lydeard 2012a).

Type species
Unio sloatianus Lea 1840 = *Elliptoideus sloatianus* (Lea 1840)

Diagnosis
Shell thick; sculptured with oblique ridges posteriorly; inflated; outline rhomboidal; umbo moderately inflated, elevated above hinge line; umbo sculpture 3–5 ridges, somewhat thickened where they cross posterior ridge, with slight indention ventrally; posterior ridge well developed; shell disk anterior to posterior ridge usually with corrugations; posterior slope with well-developed, irregular corrugations; periostracum dull to clothlike, brown to black; pseudocardinal teeth thick, rough, 2 in left valve, 1 in right valve; lateral teeth moderately thick, 2 in left valve, 1 in right valve; umbo cavity wide, deep; nacre purplish.

Incurrent aperture papillae arborescent; inner lamellae of inner gills usually connected to visceral mass only anteriorly; all 4 gills marsupial; marsupium not extended beyond original gill margin when gravid; glochidial outline elongate oval, without styliform hooks (Simpson 1914; Frierson 1927).

Elliptoideus sloatianus (Lea 1840)
Purple Bankclimber

Elliptoideus sloatianus – Upper image: length 159 mm, UF 243927. Ochlockonee River at U.S. Highway 90 bridge, 7 miles west of Tallahassee, Leon County, Florida, 29 July 1980. © *Richard T. Bryant.* Middle image: length 117 mm, UF 388494. Flint River, 2 air miles upstream of State Highway 345 boat ramp, Decatur County, Georgia, 24 September 1992. © *Richard T. Bryant.* Lower image: length 38 mm, UMMZ 234121. Ochlockonee River near Tallahassee, Leon County, Florida, May 1963. © Richard T. Bryant.

Description

Size: length to 215 mm.

Shell: thick; with variously developed low plications extending from posterior ridge to ventral margin, less pronounced to absent in large individuals; moderately inflated, width 25%–40% of length; outline rhomboidal; anterior margin rounded; posterior margin obliquely truncate, posterior edge of shell usually straight in large individuals, not undulating; dorsal margin straight to slightly convex; ventral margin straight, large individuals (≥140 mm) slightly concave; posterior ridge prominent in smaller individuals, typically obscured by plications in larger individuals, biangulate posterioventrally; posterior slope moderately steep, flat to slightly concave; posterior ridge and slope with oblique plications extending from posterior ridge to posteriodorsal margin; umbo broad, elevated slightly above hinge line; umbo sculpture 3–5 ridges, somewhat thickened where they cross posterior ridge, with slight indention ventrally; umbo cavity wide, shallow.

Teeth: pseudocardinal teeth large, rough, triangular, 2 divergent teeth in left valve, anterior tooth crest oriented ventrally, aligned along posterior margin of anterior adductor muscle scar, 1 tooth in right valve, occasionally with accessory denticle anteriorly; lateral teeth moderately long, thick, straight to slightly curved, 2 in left valve, 1 in right valve; interdentum moderately long, narrow to moderately wide.

Nacre: white to bluish white to purple.

Periostracum: small individuals shiny, large individuals dull or clothlike; small individuals yellowish to greenish yellow, with dark green rays of varying widths, large individuals dark olive to brownish black, obscuring rays.

Glochidium Description

Outline elongate oval; length 120–170 μm; height 100–150 μm; width 81 μm; ventral margin with internal irregularly arranged lanceolate micropoints (O'Brien and Williams 2002) (Figure 10.38).

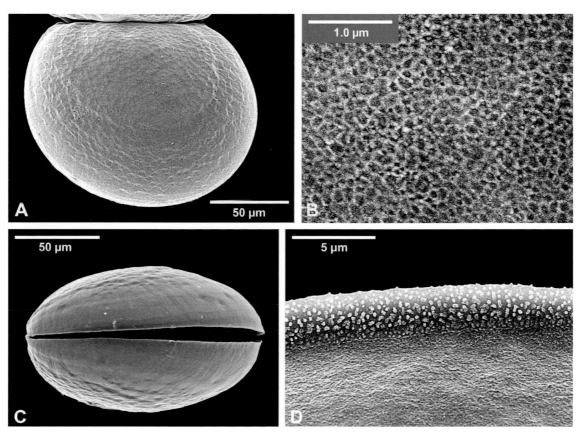

Figure 10.38. Scanning electron micrographs of glochidial valves of *Elliptoideus sloatianus* from Flint River, Cooks Landing, Macon County, Georgia, 21 March 2002—(A) external view; (B) surface sculpture; (C) lateral view; and (D) internal view of micropoints on ventral margin.

Similar Sympatric Species

Elliptoideus sloatianus resembles *Megalonaias nervosa* but typically has a more prominent posterior ridge that usually ends in a blunt point posterioventrally. The nacre is purple and the umbo cavity is shallower in *E. sloatianus* as opposed to white nacre and a deeper umbo cavity in *M. nervosa*. In large *E. sloatianus* the posterior edge of the shell is usually straight as opposed to that of *M. nervosa*, which is undulating. In very large individuals of *E. sloatianus* the ventral margin is often concave, while that of *M. nervosa* is straight or convex.

Elliptoideus sloatianus may superficially resemble some *Amblema neislerii*, but that species has plications parallel to the ventral margin.

Small *Elliptoideus sloatianus* may resemble large *Medionidus penicillatus* and *Medionidus simpsonianus* but differ in being more compressed, with greater shell height in relation to length, and having more pronounced sculpture, especially on the shell disk.

Distribution in Florida

Elliptoideus sloatianus occurs in Apalachicola and Ochlockonee Rivers and the lowermost reach of Chipola River (Figure 10.39).

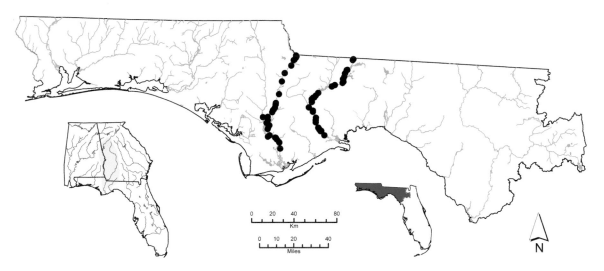

Figure 10.39. Distribution of *Elliptoideus sloatianus*.

Ecology and Biology

Elliptoideus sloatianus occurs in medium and large rivers in moderate to swift current in substrates composed of sand and gravel, often near rocky outcrops. It can be found in shallow areas along the bank but may occur in water several meters deep. Large individuals have been found in reservoirs, but there is no evidence to suggest that they reproduce in impounded waters.

Elliptoideus sloatianus appears to be a short-term brooder, gravid from late February through April, but the end of the brooding period was not determined (O'Brien and Williams 2002). *Elliptoideus sloatianus* were observed to abort rigid conglutinates comprised entirely of eggs in early February. When mature, the glochidia were discharged in loose, easily separated conglutinates that disintegrated upon release. The conglutinates were lanceolate, 10–15 mm long, 1.5 mm wide, and consisted of two layers of glochidia (O'Brien and Williams 2002). Examination of internal annuli in thin-sectioned *E. sloatianus* shells revealed individuals 80 mm and 184 mm in length were 3 and 14 or 15 years old, respectively (USFWS, Panama City, Florida, unpublished data).

In laboratory trials *Elliptoideus sloatianus* glochidia metamorphosed with limited success on two fishes that occur in Florida—Percidae: *Percina nigrofasciata* (Black-banded Darter) and Poeciliidae: *Gambusia holbrooki* (Eastern Mosquitofish) (O'Brien and Williams 2002).

Recent host fish trials (Fritts et al. 2012) exposed a total of 29 fish species in 7 families. Five fishes were found to be highly successful as potential hosts—Acipenseridae: *Acipenser brevirostrum* (Shortnose Sturgeon), *Acipenser fulvescens* (Lake Sturgeon), *Acipenser oxyrinchus desotoi* (Gulf Sturgeon), and *Acipenser oxyrinchus oxyrinchus* (Atlantic Sturgeon); and Percidae: *Percina nigrofasciata* (Blackbanded Darter). Transformation rates ranged from 79% to 89% for sturgeons and 34% ± 18% for *P. nigrofasciata* (Fritts et al. 2012). Of these five fishes, only two—*A. o. desotoi* and *P. nigrofasciata*—occur in Florida.

Elliptoideus sloatianus has been observed to be infested with mites (Unionicolidae). Of 73 individuals examined from two localities on Apalachicola and Ochlockonee Rivers, 2 were parasitized by *Unionicola tupara* (Vidrine 1996).

Conservation Status

Elliptoideus sloatianus was considered endangered throughout its range in the first conservation status assessments conducted (Athearn 1970; Stansbery 1971). In more recent status assessments, the species was considered threatened by Williams et al. (1993) and endangered by Williams et al. (in review). In Florida Williams and Butler (1994) considered it to be threatened. In 1998 *E. sloatianus* was listed as threatened under the authority of the federal ESA (USFWS 1998). Critical habitat was designated in 2007 and included a total of 1,493 km of streams in Florida and Georgia (USFWS 2007). In Florida it included portions of the main channels of the Apalachicola and Ochlockonee Rivers. The recovery plan was approved in 2003 (USFWS 2003).

Elliptoideus sloatianus is extant throughout its Florida range. It is uncommon at most localities and appears to be declining at most Ochlockonee River sites. Many populations are dominated by older cohorts with little evidence of recruitment.

Remarks

An *Elliptoideus sloatianus* shell reported from Escambia River in Florida was probably *Plectomerus dombeyanus* (Williams and Butler 1994; Williams et al. 2008).

Fossils of *Elliptoideus sloatianus*, radiocarbon dated to 17,500 BP, are known from a Pleistocene deposit on Suwannee River at Troy Spring, indicating a more widespread prehistoric occurrence in Florida. The fossil shell of *E. sloatianus* reported from Leisey Shell Pits near Tampa by Bogan and Portell (1995) has been reidentified as *Elliptio crassidens*.

Elliptoideus sloatianus shells are common in archaeological material from Apalachicola River in Florida (Williams and Fradkin 1999), suggesting that it was likely a source of food for Native Americans.

Synonymy

Unio sloatianus Lea 1840 (Lea 1840:287). Lectotype (Johnson 1974), AMNH 56104, length 107 mm. Type locality: published as Chattahoochee River, Georgia, restricted by Clench and Turner (1956) to Columbus, Georgia, [Chattahoochee River drainage]. Restriction of the type locality by Clench and Turner appears to be un-justified as they provided no evidence that the primary type came from Columbus.

Unio atromarginatus Lea 1840 (Lea 1840:288). Type locality: Chattahoochee River, Columbus, [Muscogee County,] Georgia, [Chattahoochee River drainage].

Unio aratus Conrad 1849, *non* Lea 1843 (Conrad 1849:302) [replacement name *Unio plectrophorus* Conrad 1849]. Type locality: Flint River, Georgia, [Flint River drainage].

Unio plectophorus Conrad 1850 (Conrad 1850:277) [spelling corrected from *Unio plectrophorus* Conrad 1849].

Genus *Fusconaia* Simpson 1900

There were 13 species assigned to *Fusconaia* by Turgeon et al. (1998). One additional species, *Quincuncina burkei*, was placed into *Fusconaia* (Lydeard et al. 2000). Five species have been subsequently moved to other genera—*Fusconaia barnesiana* into *Pleuronaia*; *Fusconaia succissa* into *Quadrula*; and *Fusconaia apalachicola*, *Fusconaia ebenus*, and *Fusconaia rotulata* into *Reginaia*. The genus occurs throughout the Hudson Bay, Great Lakes, and Mississippi River basins; in Gulf Coast basins from Texas to Florida; and southern Atlantic Coast basins from Virginia to Georgia.

Type Species
Unio trigonus Lea 1831 = *Fusconaia flava* (Rafinesque 1820)

Diagnosis
Shell moderately thin to moderately thick; typically smooth, one species with sculpture; outline quadrate to triangular; umbo elevated above hinge line; umbo sculpture thick, nodulous, irregular ridges (Figure 10.43); periostracum shiny, dull to clothlike in a few species; pseudocardinal and lateral teeth well developed, 2 in left valve, 1 in right valve; umbo cavity deep; nacre white to salmon.

Inner lamellae of inner gills only connected to visceral mass anteriorly; excurrent aperture smooth; mantle bridge separating excurrent and supra-anal apertures very short or absent; all 4 gills marsupial; glochidia held across entire gill; marsupium not extended beyond original gill margin when gravid; ova and embryos pink or purplish, visible through the marsupia walls, becoming white or tan as glochidia mature; glochidium outline depressed subelliptical, without styliform hooks (Simpson 1900a, 1914; Ortmann 1912; Baker 1928).

Fusconaia burkei (Walker 1922)
Tapered Pigtoe

Fusconaia burkei – Upper image: length 55 mm, UF 64972. Holmes Creek, Jackson County, Florida. © Richard T. Bryant. Middle image: length 45 mm, UF 64977. Little Choctawhatchee River, Dale County, Alabama, March–May 1916. © Richard T. Bryant. Lower image: length 27 mm, UF 449434. Bruce Creek on Walton Bridge Road, about 14 kilometers (air) south-southwest of Ponce de Leon, Walton County, Florida, 22 April 2011. © Richard T. Bryant.

Description

Size: length to 78 mm.

Shell: moderately thin to moderately thick; small individuals usually with short, raised chevrons dorsally, corrugations variously developed over entire shell, large individuals often with subtle sculpture; compressed to moderately inflated, width 30%–50% of length; outline oval to subtriangular; anterior margin rounded; posterior margin obliquely truncate to bluntly pointed; dorsal margin straight to slightly rounded; ventral margin straight to convex, large individuals occasionally slightly concave; posterior ridge moderately sharp dorsally, rounded posterioventrally, occasionally weakly biangulate posterioventrally; posterior slope flat to slightly concave, with variously developed corrugations extending from posterior ridge to posteriodorsal margin; umbo broad, elevated slightly above hinge line; umbo sculpture moderately thick ridges; umbo cavity wide, shallow to moderately deep.

Teeth: pseudocardinal teeth small, moderately thick, triangular, 2 divergent teeth in left valve, anterior tooth crest almost parallel to shell margin, 1 tooth in right valve; lateral teeth moderately long, thin, straight to slightly curved, 2 in left valve, 1 in right valve; interdentum short, narrow.

Nacre: white to bluish white, usually iridescent.

Periostracum: shiny to clothlike; small individuals yellowish brown to olive brown, occasionally with dark green rays, large individuals dark brown to black, usually obscuring rays.

Glochidium Description

Outline depressed subelliptical; mean length 167 µm; mean height 160 µm (Pilarczyk et al. 2005) (Figure 10.40).

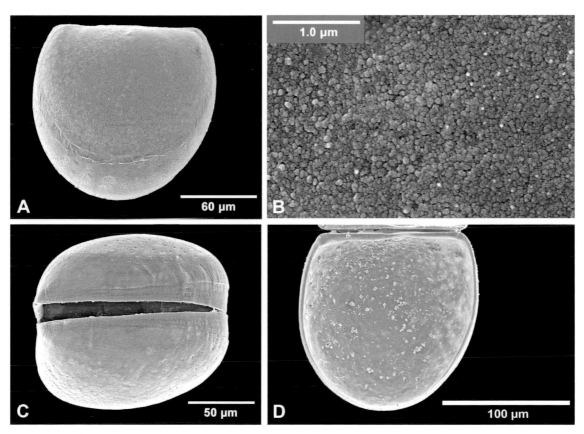

Figure 10.40. Scanning electron micrographs of glochidial valves of *Fusconaia burkei* from Bruce Creek, south-southwest of Ponce de Leon, Walton County, Florida, 22 April 2011—(A) external view; (B) surface sculpture; (C) lateral view; and (D) internal view.

Similar Sympatric Species

Fusconaia burkei resembles small (length up to 40 mm) *Elliptio crassidens* but differs in having a thinner shell and pseudocardinal teeth. Large *F. burkei* are more elongate than similar size *E. crassidens*.

Fusconaia burkei can be confused with small individuals of *Elliptio mcmichaeli* but can be distinguished by a more inflated shell, often with sculpture on the shell disk, and smaller pseudocardinal and lateral teeth.

Fusconaia burkei may resemble *Medionidus acutissimus* but differs in being more inflated and less elongate.

Distribution in Florida

Fusconaia burkei occurs only in Choctawhatchee River basin (Figure 10.41).

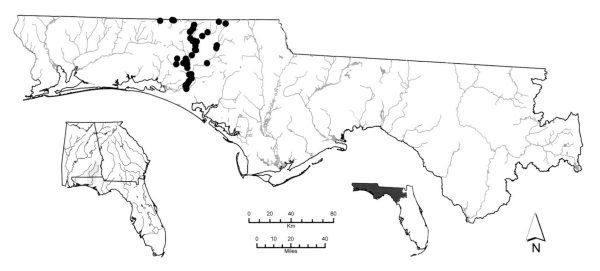

Figure 10.41. Distribution of *Fusconaia burkei*.

Ecology and Biology

Fusconaia burkei typically inhabits medium to large creeks and rivers in slow to moderate current in stable substrates ranging from silty sand to sandy gravel. It can be found in smaller creeks with stable groundwater flow and has occasionally been found in sloughs and floodplain lakes.

Fusconaia burkei is a short-term brooder and has been found gravid from mid-March through May, possibly into June, in Eightmile Creek, a Pea River tributary, Walton County, Florida. The conglutinates are cylindrical and tapered at both ends and pinkish in color due to undeveloped eggs. Fecundity in four females ranged from 3,880–10,395 per year, with an average of 6,058 glochidia per individual (Pilarczyk et al. 2005).

In laboratory trials *Fusconaia burkei* glochidia metamorphosed on only one fish species that occurs in Florida—Cyprinidae: *Cyprinella venusta* (Blacktail Shiner). Five additional fishes that occur in Florida were exposed, but no metamorphosed glochidia were recovered—Centrarchidae: *Lepomis macrochirus* (Bluegill); Cyprinidae: *Notropis texanus* (Weed Shiner) and *Pteronotropis hypselopterus* (Sailfin Shiner); Fundulidae: *Fundulus olivaceus* (Blackspotted Topminnow); and Ictaluridae: *Noturus leptacanthus* (Speckled Madtom) (White et al. 2008). *Fusconaia* species are known to be host specialists on members of Cyprinidae (shiners) (Williams et al. 2008).

Fusconaia burkei has been observed to be infested with mites (Unionicolidae). Of 45 individuals examined from four localities in Choctawhatchee River basin, 20 were parasitized by *Unionicola gowani* (Vidrine 1996).

Conservation Status

Fusconaia burkei (as *Quincuncina burkei*) was considered endangered throughout its range by Athearn (1970) and threatened by Williams et al. (1993, in review). In Florida it was assigned a status of threatened (Williams and Butler 1994). *Fusconaia burkei* was listed as threatened with critical habitat designated under the authority of the federal ESA in 2012 (USFWS 2012). Critical habitat included a total of 1,131 km of streams in Alabama and Florida. In Florida, critical habitat was designated in most of the main channel of the Choctawhatchee River and selected tributaries.

Fusconaia burkei is extant in some Choctawhatchee River reaches and its larger tributaries. The species is generally uncommon through most of its range.

Remarks

Fusconaia burkei was originally described and placed in a new genus, *Quincuncina*, in a paper published by Ortmann and Walker (1922). In the original description Ortmann pointed out the similarity of soft anatomy between *Quincuncina* and *Fusconaia*, stating that the former "resembles only one genus, *Fusconaia*" but was reluctant to place it in that genus because it had sculpturing on the shell. Frierson (1927) placed *Quincuncina* as a subgenus of *Quadrula*, but subsequent workers continued to recognize *Quincuncina* as a genus. A phylogenetic analysis based on genetic sequence data determined *Quincuncina burkei* to be sister to *Fusconaia escambia* (Lydeard et al. 2000; Serb et al. 2003). This species appears to be the only member of *Fusconaia* with shell sculpture.

At the request of H.H. Smith of the ALMNH, *Fusconaia burkei* was named for Joseph B. Burke, who discovered the new species (Ortmann and Walker 1922). Smith hired Burke in the early 1900s to collect mollusks from south Alabama and west Florida.

Fusconaia burkei is only known from the Choctawhatchee River basin. However, a single valve from an archaeological site at the mouth of Omusee Creek on the Chattahoochee River, Houston County, Alabama, resembles *F. burkei* (Figure 10.42). Additional material is needed to determine the taxonomic status of this shell.

Figure 10.42. *Fusconaia* sp. – Length 41 mm, UF 445299 (left valve only). Archaeological specimen from Omusee Creek at mouth on B.W. Andrews Reservoir [Lake] in Omusee Park (area 1 northwest slope), on Chattahoochee River Drive, Houston County, Alabama. © Richard T. Bryant.

Synonymy

Quincuncina burkei Walker 1922 (Walker *in* Ortmann and Walker 1922:3). Holotype, UMMZ 94495, length 51 mm. Type locality: Sikes Creek, a tributary of Choctawhatchee River, Barbour County, Alabama, [Choctawhatchee River basin].

Fusconaia escambia Clench and Turner 1956
Narrow Pigtoe

Fusconaia escambia – Upper image: length 40 mm, UF 4997 (paratype). Escambia River, 3 miles southeast of Century, Escambia County, Florida, 27 August 1954. © Richard T. Bryant. Lower image: length 34 mm, UMMZ 163303. Mill or Youngblood Pond near Youngblood, Pike County, Alabama, June 1917. © Richard T. Bryant.

Description

Size: length to 75 mm, in Florida to 73 mm.

Shell: moderately thick; smooth; moderately compressed, width 35%–55% of length; outline subtriangular to quadrate; anterior margin rounded; posterior margin broadly rounded to obliquely truncate; dorsal margin slightly convex; ventral margin straight to broadly rounded, occasional large individuals slightly concave; posterior ridge sharp dorsally, rounded posterioventrally; posterior slope moderately steep, flat to slightly concave, some individuals with a very subtle radial depression just anterior of posterior ridge; umbo narrow to broad, slightly inflated, elevated slightly above hinge line; umbo sculpture thick, nodulous ridges (Figure 10.43); umbo cavity moderately wide, deep.

Teeth: pseudocardinal teeth moderately thin to thick, finely striate, 2 divergent teeth in left valve, anterior tooth smaller, moderately compressed, crest oriented anterioventrally, aligned along anterior margin to center of anterior adductor muscle scar,

1 tooth in right valve; lateral teeth short, straight to slightly curved, 2 in left valve, 1 in right valve, occasionally 2; interdentum short, moderately wide.

Nacre: usually white, small individuals (up to 50 mm) pinkish.

Periostracum: shiny to clothlike; reddish brown to black, small individuals occasionally with faint rays.

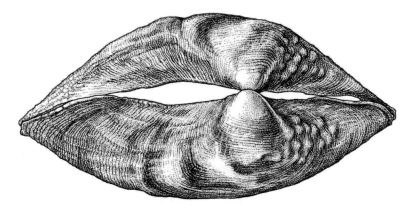

Figure 10.43. Umbo sculpture of *Fusconaia escambia*, length 14 mm, UF 449234. Escambia River, about 0.3 miles above Chumuckla Springs boat ramp, Santa Rosa County, Florida, 24 September 2008. Illustration by Susan Trammell.

Glochidium Description

The glochidium of *Fusconaia escambia* is unknown.

Similar Sympatric Species

Fusconaia escambia most closely resembles *Quadrula succissa* but can be distinguished by its sharper posterior ridge, less inflated umbo, wider umbo cavity, and typically white to pinkish nacre. Nacre of *Q. succissa* is typically purplish but is occasionally white.

Fusconaia escambia may resemble some individuals of *Pleurobema strodeanum* but has a deeper, more compressed umbo cavity. Most *P. strodeanum* are more elongate and pointed posteriorly as opposed to the subtriangular to quadrate shape of *F. escambia*. Nacre of *P. strodeanum* is always white.

Distribution in Florida

Fusconaia escambia occurs in Escambia and Yellow River basins (Figure 10.44).

Ecology and Biology

Fusconaia escambia occurs in large creeks to medium rivers with slow to moderate current. It inhabits substrates consisting of sand, silty sand, and gravel.

Fusconaia escambia is a short-term brooder, observed to be gravid in Escambia River in May and June, simultaneously brooding eggs and reddish immature glochidia. Glochidial hosts for *F. escambia* are unknown. Species of *Fusconaia* typically produce conglutinates and tend to be host specialists on members of Cyprinidae (shiners) (Williams et al. 2008).

Fusconaia escambia has been observed to be infested with mites (Unionicolidae). Of two individuals examined from Yellow River, one was parasitized by *Unionicola parkeri* (Vidrine 1996).

Conservation Status

Fusconaia escambia was assigned a threatened status throughout its range by Williams et al. (1993, in review). In Florida it was considered threatened by Williams and Butler (1994). *Fusconaia escambia* was listed as threatened with critical habitat designated under the authority of the federal ESA in 2012 (USFWS 2012). Critical habitat included a total of 1,112 km of streams in Alabama and Florida. In Florida, critical habitat was designated in most of the main channel of Escambia River and the main channel of Yellow River and selected tributaries.

Fusconaia escambia is extant in both the Escambia and Yellow Rivers, but its occurrence is sporadic. The species is uncommon at most localities.

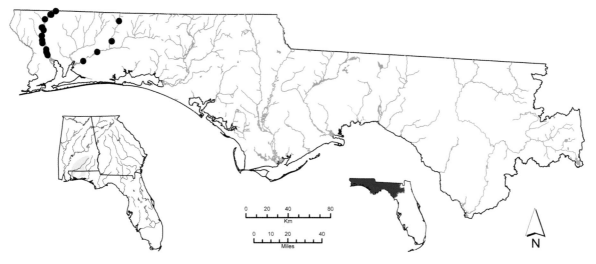

Figure 10.44. Distribution of *Fusconaia escambia*.

Remarks

Fusconaia escambia is unusual in being restricted to the Escambia and Yellow River basins. Typically, mussels endemic to eastern Gulf Coast rivers share similar distribution patterns in that they occur across all three major basins of the western Florida panhandle—Escambia, Yellow, and Choctawhatchee (e.g., *Hamiota australis*, *Obovaria choctawensis*, *Pleurobema strodeanum*, *Ptychobranchus jonesi*, and *Quadrula succissa*).

Synonymy

Fusconaia escambia Clench and Turner 1956 (Clench and Turner 1956:152). Holotype,
    MCZ 191470, length 46 mm. Type locality: Escambia River, 3 miles southeast of
    Century, Escambia County, Florida, [Escambia River basin].

Genus *Glebula* Conrad 1853

Glebula is monotypic (Turgeon et al. 1998). It occurs in Gulf Coast basins from Texas to the Florida panhandle. It is also found sporadically in Arkansas and eastern Oklahoma and historically in the Ohio River (Watters et al. 2009). *Glebula rotundata* is somewhat unusual in that it typically occurs in the extreme lower reaches of coastal rivers in areas that may occasionally experience intrusion of brackish water. *Glebula rotundata* is likely the most euryhaline species of mussel in Florida. In a controlled laboratory experiment, *G. rotundata* was able to survive at least 24 days in salinities up to 12 ppt (Johnson et al. in review).

Type Species
Unio rotundatus Lamarck 1819 = *Glebula rotundata* (Lamarck 1819)

Diagnosis
Shell moderately thick; smooth; outline elliptical; female shells swollen posterio-ventrally; posterior ridge low; periostracum clothlike to dull; umbo moderately to highly inflated; pseudocardinal teeth divided into small, radially arranged serrate ridges, 2 in left valve, 1 in right valve; lateral teeth thin to moderately thick, 2 in left valve, 1 in right valve.

Inner lamellae of inner gills completely connected to visceral mass; posterior half of outer gills marsupial; glochidia held in posterior portion of gill; marsupium extended beyond original gill margin when gravid; glochidial outline subelliptical, without styliform hooks (Conrad 1853; Simpson 1900a, 1914).

Glebula rotundata (Lamarck 1819)
Round Pearlshell

Glebula rotundata – Upper image: female, length 103 mm, UF 375985. Slough off Tensaw Lake, 1 air mile southwest of Hubbard Fish Camp and Landing, Baldwin County, Alabama, 18 September 1999. © Richard T. Bryant. Middle image: male, length 77 mm, UF 375985. Slough off Tensaw Lake, 1 air mile southwest of Hubbard Fish Camp and Landing, Baldwin County, Alabama, 18 September 1999. © Richard T. Bryant. Lower image: length 39 mm, UF 449288. Apalachicola River at river mile 46.8, Gulf County, Florida, 7 August 2006. © Richard T. Bryant.

Description

Size: length to 150 mm, in Florida to 96 mm.

Shell: small individuals thin, large individuals moderately thick; smooth; moderately to highly inflated, width 30%–60% of length, small individuals more compressed; outline oval to round; anterior margin rounded; posterior margin rounded to obliquely truncate; dorsal margin straight in small individuals, rounded in large individuals; ventral margin broadly rounded; sexual dimorphism subtle, females swollen posteriorly, slightly less elongate than males; posterior ridge rounded to weakly angular; posterior slope steep, slightly concave to slightly convex; umbo broad, inflated, even with hinge line in small individuals, elevated slightly above in large individuals; umbo sculpture thin, irregular ridges; umbo cavity wide, shallow to moderately deep.

Teeth: pseudocardinal teeth moderately thick, with radially arranged serrate ridges, 2 teeth in left valve, 1 tooth in right valve, small individuals with 2 thin, bladelike teeth in each valve; lateral teeth short, thick, straight to slightly curved, 2 in left valve, 1 in right valve; interdentum long, narrow.

Nacre: white, pinkish, or purplish.

Periostracum: typically clothlike; small individuals tan to brown, underlying purple nacre may give pinkish cast, occasionally with rays, large individuals usually dark brown to black.

Glochidium Description

Outline subelliptical; length 169–188 μm; height 183–198 μm; width 113–118 μm; ventral margin with internal irregularly perpendicular rows of lanceolate micropoints (Figure 10.45).

Similar Sympatric Species

Glebula rotundata can be distinguished from all Florida mussels with which it co-occurs by its oval to round outline, umbo even with hinge line or elevated only slightly above, brown periostracum without rays, narrow interdentum, shallow umbo cavity, and unique pseudocardinal teeth that have numerous radially arranged serrate ridges.

Distribution in Florida

Glebula rotundata occurs in Escambia, Choctawhatchee, Apalachicola, and Ochlockonee River basins (Figure 10.46).

Ecology and Biology

Glebula rotundata occurs in large creeks to rivers, sloughs, and floodplain lakes. It occurs in moderate to no current in substrates of silt, mud, and sand, often with detritus. It is tolerant of brackish water and co-occurs with euryhaline bivalves such as *Mytilopsis leucophaeata* and *Rangia cuneata*. *Glebula rotundata* has generally been considered to be a long-term brooder, overwintering their larvae and releasing them with increasing temperature. However, brooding period of *G. rotundata* appears to be atypical from those of other long-term brooders or at least exhibits regional differences. In Louisiana it was reported to be gravid from March to early October (Parker et al. 1984). In Florida, populations in Apalachicola River were gravid from June to August (Brim Box and Williams 2000) and in Escambia River from June to mid-August. In Texas Howells (2000) reported gravid females in June and July. In a study of a *G. rotundata* population in Louisiana, Parker et al. (1984) made biweekly observations during a 15-month period and found that females recharge soon after discharging glochidia and reported that each female may reproduce three times per year. Five females, ranging in length from 9.35–11.93 cm, contained an average of 531,000 glochidia in the Louisiana population (Parker et al. 1984). Such fecundity, multiplied by three reproductive periods per year, extrapolates to over 1.5 million glochidia per female annually.

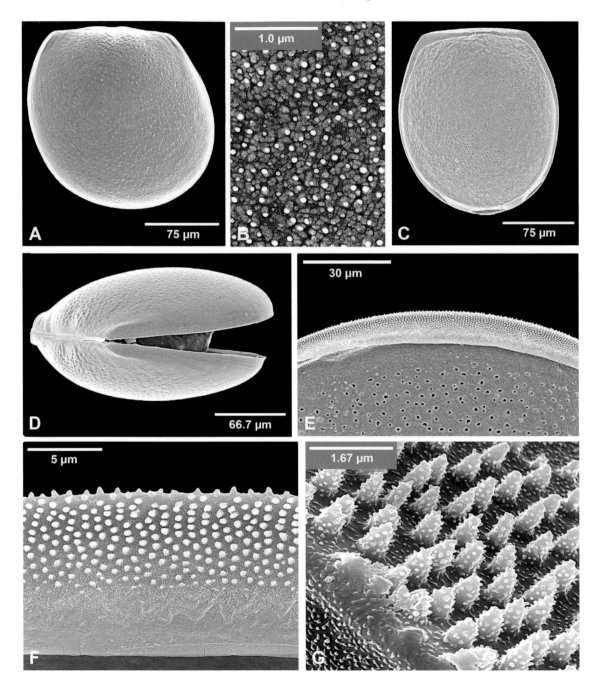

Figure 10.45. Scanning electron micrographs of glochidial valves of *Glebula rotundata* from Hitchcock Lake (Slough), Liberty County, Florida, 30 July 2008—(A) external view; (B) surface sculpture; (C) internal view; (D) lateral view; (E) internal view of ventral margin and micropoints; (F) internal view of micropoints on ventral margin; and (G) close-up of micropoints.

Glebula rotundata glochidia have been observed to metamorphose on seven fishes that occur in Florida. Natural infestations and transformations were observed on—Achiridae: *Trinectes maculatus* (Hogchoker); Cyprinidae: *Cyprinus carpio* (Common Carp); Engraulidae: *Anchoa mitchilli* (Bay Anchovy); Lepisosteidae: *Lepisosteus oculatus* (Spotted Gar); and Moronidae: *Morone chrysops* (White Bass) (Parker et al.

1984). Laboratory transformations were observed on two fishes that occur in Florida—Centrarchidae: *Lepomis cyanellus* (Green Sunfish) and *Lepomis macrochirus* (Bluegill). Four additional fishes that occur in Florida were exposed in laboratory trials, but no metamorphosed glochidia were recovered—Centrarchidae: *Lepomis humilis* (Orangespotted Sunfish), *Lepomis megalotis* (Longear Sunfish), and *Micropterus salmoides* (Largemouth Bass); and Ictaluridae: *Ictalurus punctatus* (Channel Catfish) (Parker et al. 1984).

Conservation Status

Glebula rotundata was determined by Williams et al. (1993, in review) to be currently stable throughout its range. In Florida it was reported as a species of special concern by Williams and Butler (1994).

Glebula rotundata is extant throughout its range in Florida and common at most localities. *Glebula rotundata* is generally distributed in Apalachicola and lower Chipola Rivers, but its occurrence in Escambia, Choctawhatchee, and Ochlockonee basins is sporadic.

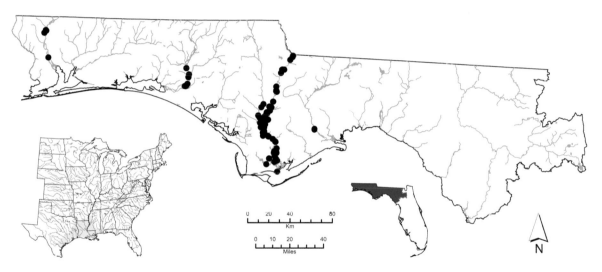

Figure 10.46. Distribution of *Glebula rotundata*.

Remarks

Preliminary analyses of mtDNA indicate that very little genetic differentiation exists within and among populations of *Glebula rotundata* from Texas to Florida (N.A. Johnson unpublished data).

The nacre in *Glebula rotundata* varies from white to shades of pink or purple. In a collection of approximately 100 shells from one site in Apalachicola River, half had white nacre and half were pink or purple.

Synonymy

Unio rotundatus Lamarck 1819 (Lamarck 1819:75). Syntype(s) lost, not figured. Type locality: unknown, restricted by Clench and Turner (1956) to Bayou Teche, St. Mary Parish, Louisiana. Restriction of the type locality by Clench and Turner appears to be unjustified as they provided no evidence that the primary type came from Bayou Teche.

Genus *Hamiota* Roe and Hartfield 2005

Hamiota was erected for four mussels restricted to eastern Gulf Coast basins from Mobile to Apalachicola. The species—*Hamiota altilis*, *Hamiota australis*, *Hamiota perovalis*, and *Hamiota subangulata*—were previously included in *Lampsilis* or *Villosa*. The unifying characteristic distinguishing them from all other unionids is the production of a superconglutinate (Haag et al. 1995) (Figure 10.47). Based on genetic analyses, it appears that the group forms a monophyletic clade distinct from *Lampsilis* and *Villosa* (Roe et al. 2001).

Type Species
Unio subangulatus Lea 1840 = *Hamiota subangulata* (Lea 1840)

Diagnosis
Shell thin to moderately thick; smooth; somewhat compressed to inflated; outline oval to elliptical; females often slightly more inflated posterioventrally than males; periostracum typically shiny, often with rays; pseudocardinal and lateral teeth moderately thick, compressed, 2 in left valve, 1 in right valve.

Females with mantle margin anterior to incurrent aperture, developed into flap or papillate fold, variously pigmented, rudimentary in males. Inner lamellae of inner gills completely connected to visceral mass or only connected anteriorly; outer gills marsupial; glochidia held in posterior portion of gill; marsupium broadest anteriorly, tapering posteriorly, extended beyond original gill margin when gravid, ventral margin darkly pigmented; glochidia discharged as superconglutinates; glochidium outline subelliptical to subspatulate, without styliform hooks (Haag et al. 1995; O'Brien and Brim Box 1999; Roe and Hartfield 2005).

Figure 10.47. *Hamiota subangulata* in an artificial stream discharging a superconglutinate tethered by a mucus strand. The individual was collected from the Flint River drainage, Georgia, May 1995. Photograph by Noel M. Burkhead.

Hamiota australis (Simpson 1900)
Southern Sandshell

Hamiota australis – First image: female, length 67 mm, UF 65312. Conecuh River near Searight, Crenshaw County, Alabama, June 1915. © Richard T. Bryant. Second image: male, length 59 mm, UF 65311. Little Choctawhatchee River near Drew›s Bridge, Houston County, Alabama, August 1916. © Richard T. Bryant. Third image: female, length 50 mm, UF 374675. Pea Creek on County Road 9, Barbour County, Alabama, 15 June 1998. © Richard T. Bryant. Fourth image: length 34 mm, UF 449435. Bruce Creek on Walton Bridge Road, about 14 kilometers (air) south-southwest of Ponce de Leon, Walton County, Florida, 22 April 2011. © Richard T. Bryant.

Description

Size: length to 83 mm.

Shell: moderately thin; smooth; moderately inflated, width 30%–40% of length; outline elliptical; anterior margin rounded; posterior margin narrowly rounded to bluntly pointed; dorsal margin straight to slightly rounded; ventral margin broadly rounded;

sexual dimorphism subtle, females slightly more inflated posterioventrally; posterior ridge rounded; posterior slope moderately steep, flat to slightly concave; umbo broad, elevated slightly above hinge line; umbo sculpture unknown; umbo cavity moderately wide, shallow to moderately deep.

Teeth: pseudocardinal teeth moderately thick, triangular, 2 slightly divergent teeth in left valve, anterior tooth compressed, crest almost parallel to shell margin, 1 tooth in right valve, occasionally with accessory denticle anteriorly; lateral teeth thin, straight, 2 in left valve, 1 in right valve; interdentum moderately long, narrow.

Nacre: white to bluish white, usually iridescent.

Periostracum: shiny; small individuals yellowish green to olive, with green rays of varying width and intensity, large individuals dark olive, brown to black, often obscuring rays.

Glochidium Description

Outline subspatulate; length 207–227 µm; height 260–294 µm; width 112–131 µm; ventral margin with internal and external irregularly arranged lanceolate micropoints (Blalock-Herod et al. 2002) (Figure 10.48).

Similar Sympatric Species

Hamiota australis resembles *Ptychobranchus jonesi* but typically has a more pointed posterior margin, more broadly rounded ventral margin, and more rounded posterior ridge. Also, the posterior ridge of *P. jonesi* is typically triangulate distally with shallow radial depressions between the ridges. Gravid *H. australis* have swollen but flat outer gills, unlike the folded outer gills of gravid *P. jonesi*.

Hamiota australis may resemble *Villosa villosa* but typically has a more pointed posterior margin. The periostracum of *V. villosa* is typically clothlike, unlike *H. australis*, which is typically shiny.

Hamiota australis superficially resemble male *Villosa lienosa* but are more elongate, have a shiny periostracum, and bluish-white nacre.

Distribution in Florida

Hamiota australis occurs in Yellow and Choctawhatchee River basins (Figure 10.49).

Ecology and Biology

Hamiota australis inhabits medium creeks to rivers in slow to moderate current in substrates of sand and sand mixed with fine gravel.

Hamiota australis is a long-term brooder, gravid from May to early August. A superconglutinate from a 62 mm long female was described by Blalock-Herod et al. (2002) as resembling a small fish and was composed of 36 pairs of oval conglutinates arranged in oblique rows. The superconglutinate was black dorsally, creamy white along the sides, and reddish brown ventrally, with a black eyespot anteriorly. It was attached to the female by two fused mucus tubes 45 mm long, but these were likely limited in length by its small aquarium. Based on the number of glochidia per conglutinate, fecundity was estimated at 61,200–122,400 (Blalock-Herod et al. 2002). Glochidial hosts for *H.*

australis are unknown; however, other *Hamiota* species utilize *Micropterus* spp. (Haag et al. 1995; O'Brien and Brim Box 1999).

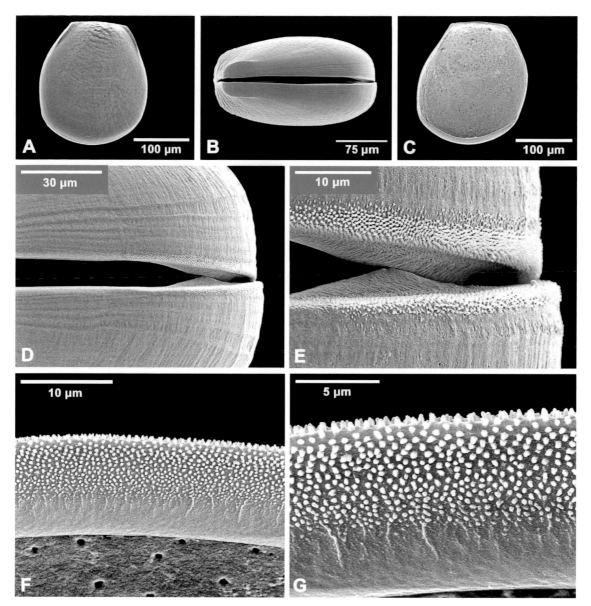

Figure 10.48. Scanning electron micrographs of glochidial valves of *Hamiota australis* from West Fork Choctawhatchee River, Dale County, Alabama, 3 June 1999—(A) external view; (B) lateral view; (C) internal view; (D) lateral view of ventral margin; (E) lateral view of micropoints along ventral margin; (F) internal view of micropoints on ventral margin; and (G) close-up of micropoints.

Conservation Status

Hamiota australis was included in the first lists of endangered mussels by Athearn (1970) and Stansbery (1971). It was reported as threatened throughout its range by Williams et al. (1993) and endangered by Williams et al. (in review). In Florida it was assigned a status of threatened by Williams and Butler (1994). *Hamiota australis* was listed as threatened with critical habitat designated under the authority of the federal ESA in 2012 (USFWS 2012). Critical habitat included a total of 2,222 km of streams in

Alabama and Florida. In Florida, critical habitat was designated in most of the main channel of Escambia River, as well as the main channel of Yellow and Choctawhatchee Rivers and selected tributaries.

Hamiota australis is extant in small, isolated populations throughout its Florida range and is generally uncommon.

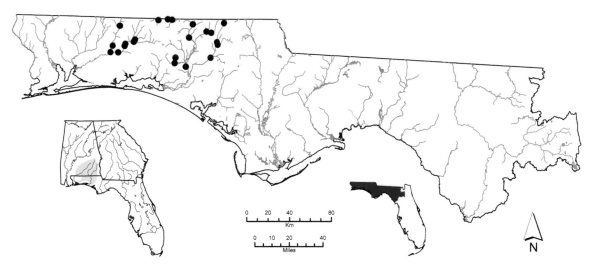

Figure 10.49. Distribution of *Hamiota australis*.

Remarks

Using molecular genetic analysis, Roe et al. (2001) discovered that *Hamiota australis* consists of two clades, one confined to Escambia and Yellow River basins and the other to Choctawhatchee River basin. Roe et al. (2001) concluded that only animals from the appropriate geographical source should be used during recovery activities that include captive propagation, translocation, or augmentation.

Hamiota australis is not known to occur in Florida reaches of Escambia River, but it has been reported from upstream tributaries in Alabama (Williams et al. 2008). It likely occurred in Florida reaches historically but became extirpated before being detected.

The older portions of some *Hamiota australis* shells typically have a darkened periostracum, while newer parts may be abruptly brighter with clearly visible rays. The cause of this phenomenon is unknown.

Synonymy

Lampsilis australis Simpson 1900 (Simpson 1900b:75). Holotype, USNM 150473, length 52 mm. Type locality: Little Patsaliga Creek, [Crenshaw County,] southeastern Alabama, [Escambia River basin].

Hamiota subangulata (Lea 1840)
Shinyrayed Pocketbook

Hamiota subangulata – Upper image: male, length 51 mm, UF 449436. Flint River at Highway 37 bridge, Baker and Mitchell counties, Georgia, 5 August 2009. © Richard T. Bryant. Middle image: length 48 mm, USNM 149648 (type of *Unio kirklandianus*). Ocklocknee [Ochlockonee] River, Leon County, Florida. Photograph by James D. Williams. Lower image: male, length 40 mm, UF 461893. Sugar Mill Spring, Florida Caverns State Park, about 200 yards upstream of paved road and where spring sinks, Jackson County, Florida, 10 June 2008. © Richard T. Bryant.

Description

Size: length to 85 mm.

Shell: moderately thin to moderately thick; smooth; moderately inflated, width 20%–45% of length; outline elliptical; anterior margin rounded; posterior margin narrowly rounded to bluntly pointed; dorsal margin straight to slightly convex; ventral margin broadly rounded; sexual dimorphism subtle, females slightly more inflated posterioventrally; posterior ridge sharp near umbo, rounded posterioventrally; posterior slope moderately steep, flat, slightly concave to slightly convex; umbo broad, moderately inflated, elevated slightly above hinge line; umbo sculpture unknown; umbo cavity moderately wide, shallow to moderately deep.

Teeth: pseudocardinal teeth moderately thick, 2 teeth in left valve, anterior tooth moderately compressed, larger than posterior tooth, crest almost parallel to shell margin, posterior tooth low, triangular, 1 tooth in right valve, occasionally with accessory denticle anteriorly; lateral teeth short, straight, 2 in left valve, 1 in right valve; interdentum long, narrow.

Nacre: white to bluish white, occasionally with salmon tint in umbo cavity.

Periostracum: shiny; yellowish, tan to dark brown, with dark green rays of varying width and intensity, occasionally almost rayless.

Glochidium Description

Outline subspatulate; length 184–211 μm; height 244–271 μm; width 93–112 μm; ventral margin with internal and external irregularly perpendicular rows of lanceolate micropoints (O'Brien and Brim Box 1999) (Figure 10.50).

Similar Sympatric Species

Hamiota subangulata may resemble *Lampsilis floridensis* but typically has well-developed green rays over most of the shell, while *L. floridensis* has rays (when present) confined to the posterior half of the shell. Also, the periostracum of *L. floridensis* is almost always yellow, whereas that of *H. subangulata* is typically dark yellowish brown. *Lampsilis floridensis* is usually more inflated and has a more rounded posterior margin than *H. subangulata*.

Individuals of *Hamiota subangulata* may resemble *Villosa vibex*, but that species has a thinner shell, more rounded posterior ridge, and thinner pseudocardinal and lateral teeth.

Distribution in Florida

Hamiota subangulata occurs in Econfina Creek and Apalachicola, Chipola, and Ochlockonee River basins (Figure 10.51).

Ecology and Biology

Hamiota subangulata inhabits medium creeks to rivers in slow to moderate current, typically in substrates of sandy clay, sand, and gravel.

Hamiota subangulata is a long-term brooder, with gravid females from December to the following August (O'Brien and Brim Box 1999). To attract host fishes, female *H. subangulata* produce superconglutinates, which are discharged from mid-May to mid-July (Figure 10.47). They range in length from 30 mm to 50 mm and are tethered to the parent by two fused mucus tubes up to 2.3 m in length. The mucus strand remains intact for several weeks, but the superconglutinate disintegrates about three days after being discharged. In the laboratory glochidia remained viable up to five days (O'Brien and Brim Box 1999).

In laboratory trials *Hamiota subangulata* glochidia metamorphosed on four fishes that occur in Florida—Centrarchidae: *Lepomis macrochirus* (Bluegill), *Micropterus punctulatus* (Spotted Bass), and *Micropterus salmoides* (Largemouth Bass); and Poeciliidae: *Gambusia holbrooki* (Eastern Mosquitofish). *Micropterus punctulatus* and *M. salmoides* were considered primary hosts based on the percentage of test fish allowing transformation, average number transformed per fish, and co-occurrence of the mussel and host fishes (O'Brien and Brim Box 1999).

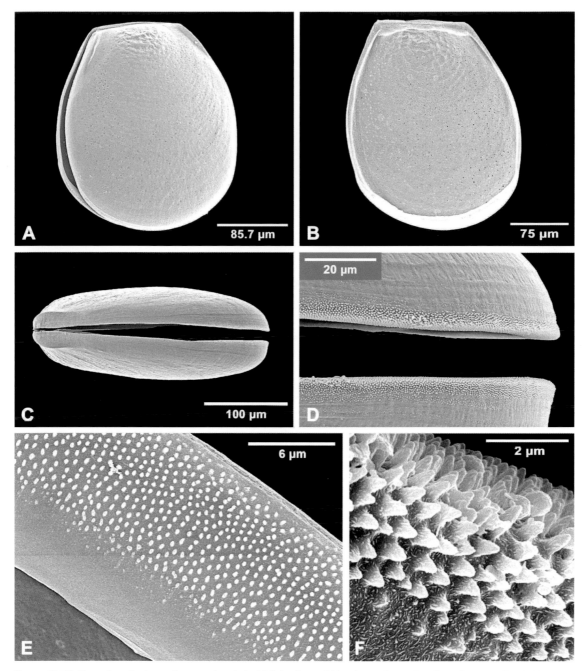

Figure 10.50. Scanning electron micrographs of glochidial valves of *Hamiota subangulata* from Cowarts Creek, Houston County, Alabama, 24 April 2007—(A) external view; (B) internal view; (C) lateral view; (D) lateral view of micropoints along ventral margin; (E) internal view of micropoints on ventral margin; and (F) close-up of micropoints.

Hamiota subangulata has been observed to be infested with mites (Unionicolidae). Of 14 individuals examined from one locality in Chipola River, 13 were parasitized by *Unionicola hoesei* (Vidrine 1996).

Conservation Status

Hamiota subangulata was reported as threatened throughout its range by Williams et al. (1993) and endangered by Williams et al. (in review). In Florida it was considered a species of special concern by Williams and Butler (1994). In 1998 *H. subangulata* was listed as endangered under the authority of the federal ESA (USFWS 1998). Critical habitat was designated in 2007 and included a total of 1557.4 km of streams in Alabama, Florida, and Georgia. In Florida, critical habitat included portions of the Chipola and Ochlockonee Rivers and selected tributaries. A recovery plan was approved in 2003 (USFWS 2003).

Hamiota subangulata is extant in small, isolated populations in the Chipola and Ochlockonee Rivers.

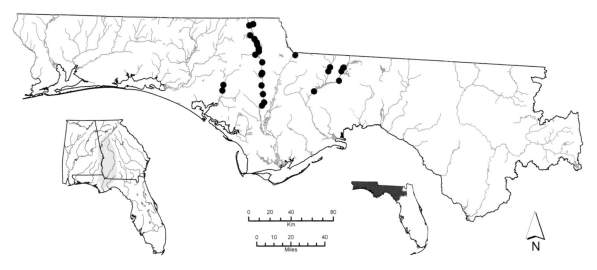

Figure 10.51. Distribution of *Hamiota subangulata*.

Remarks

Hamiota was erected for eastern Gulf Coast basin mussels characterized by their unique mode of glochidial release and host fish attraction. Glochidia are packaged as a superconglutinate that resembles a small fish and is released on the end of a long mucus strand that allows the lure to mimic a live fish and attract a predatory host.

Hamiota subangulata was erroneously reported to occur in Choctawhatchee River basin (Clench and Turner 1956; Burch 1975a).

Hamiota subangulata is known from archaeological samples along the Apalachicola River. It likely occurred throughout most of Apalachicola River prior to channel entrenchment following construction of Jim Woodruff Lock and Dam.

Synonymy

Unio subangulatus Lea 1840 (Lea 1840:287). Lectotype (Clench and Turner 1956), USNM 85081, length 48 mm (female). Type locality: Chattahoochee River, Columbus, [Muscogee County,] Georgia, [Chattahoochee River drainage].

Unio kirklandianus S.H. Wright 1897, *non* Lea 1834a (S.H. Wright 1897:136). Type locality: Ocklocknee [Ochlockonee] River, Leon County, Florida, [Ochlockonee River basin].

Genus *Lampsilis* Rafinesque 1820

Turgeon et al. (1998) recognized 28 *Lampsilis* species and 4 subspecies. Subsequently, four species—*Lampsilis altilis*, *Lampsilis australis*, *Lampsilis perovalis*, and *Lampsilis subangulata*—were moved to *Hamiota* when that genus was described (Roe and Hartfield 2005). *Lampsilis haddletoni* was moved to *Obovaria* by Williams et al. (2008). *Lampsilis floridensis* has been variously treated as a subspecies or synonym of *Lampsilis teres* until recognized as a valid species by Williams et al. (2008). *Lampsilis straminea claibornensis* was determined to be indistinguishable from *Lampsilis straminea straminea* and was placed in synonymy of that species by Williams et al. (2008). This leaves *Lampsilis* with a total of 24 recognized species and 3 subspecies.

Lampsilis is wide ranging, occurring from Hudson Bay, Great Lakes, and Mississippi basins east to Atlantic Coast basins from Nova Scotia to Georgia. On the Gulf Coast it occurs in basins from the Rio Grande east to Tampa Bay in peninsular Florida. There are three species of *Lampsilis* in Florida.

Type Species
Unio ovatus Say 1817 = *Lampsilis ovata* (Say 1817)

Diagnosis
Shell moderately thin to thick; smooth; outline oval to elliptical; sexual dimorphism distinct, females inflated posterioventrally; umbo sculpture undulating or weakly double-looped ridges; periostracum shiny, dull in a few species; pseudocardinal and lateral teeth thin to moderately thick, 2 in left valve, 1 in right valve.

Females usually with well-developed mantle flap just ventral to incurrent aperture (Figure 10.52), usually rudimentary in males. Inner lamellae of inner gills mostly or completely connected to visceral mass, may have small opening at posterior end; outer gills marsupial; glochidia held in posterior portion of gill; marsupium extended beyond original gill margin when gravid, ventral margin often pigmented; glochidium outline subspatulate, without styliform hooks (Simpson 1900a; Ortmann 1912; Kraemer 1970).

Figure 10.52. Female *Lampsilis ornata* mantle flap display—side view (left) and top view (right). Photographs by Paul L. Freeman and Paul D. Johnson.

Lampsilis floridensis (Lea 1852)
Florida Sandshell

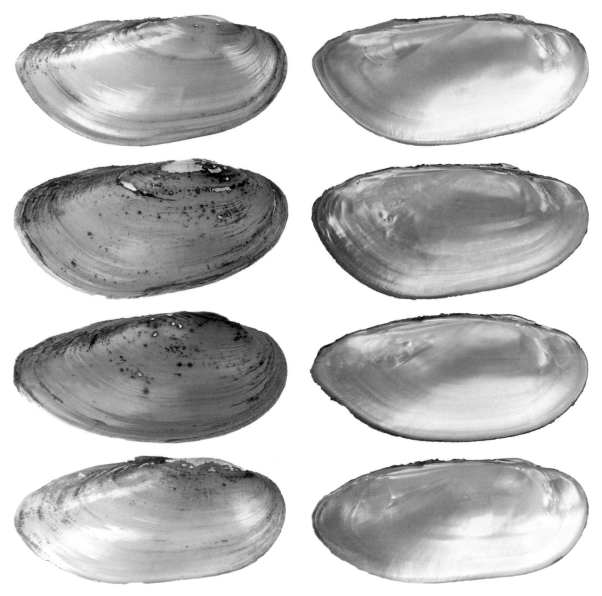

Lampsilis floridensis – First image: female, length 97 mm, UF 20437. Conecuh River, Gantt Reservoir [Lake], Covington County, Alabama, 1961. © Richard T. Bryant. Second image: female, length 80 mm, UF 4989. Mosquito Creek, 1 mile south of Chattahoochee, Gadsden County, Florida, October 1953. © Richard T. Bryant. Third image: male, length 86 mm, UF 4989. Mosquito Creek, 1 mile south of Chattahoochee, Gadsden County, Florida, October 1953. © Richard T. Bryant. Fourth image: male, length 71 mm, UF 65736. Escambia River near Flomaton, Escambia County, Alabama, May 1917. © Richard T. Bryant.

Description

Size: length to 131 mm, in Florida to 120 mm.

Shell: moderately thin; smooth; moderately compressed, width 20%–35% of length; outline elliptical; anterior margin rounded; posterior margin narrowly rounded to bluntly pointed in females, narrowly rounded in males; dorsal margin straight to slightly convex; ventral margin straight to convex; posterior ridge rounded; posterior slope flat to slightly convex; umbo broad, elevated slightly above hinge line; umbo sculpture moderately thick, undulating ridges, somewhat nodulous where they cross posterior ridge, with slight indentation just anterior to posterior ridge; umbo cavity wide, shallow.

Teeth: pseudocardinal teeth moderately small, triangular, occasionally compressed, 2 teeth in left valve, anterior tooth larger, crest oriented anterioventrally, aligned through center of anterior adductor muscle scar, 1 tooth in right valve, occasionally with accessory denticle anteriorly; lateral teeth moderately long, thin, straight to slightly curved, 2 in left valve, 1 in right valve; interdentum moderately long, narrow.

Nacre: white, usually highly iridescent.

Periostracum: shiny, occasionally dull; dark yellow, usually rayless or with few green rays on posterior slope in Escambia, Choctawhatchee, Apalachicola, and Ochlockonee River basins, light yellow, with green rays on posterior portion of shell in Suwannee, Withlacoochee (southern), and Hillsborough River basins.

Glochidium Description

Outline subspatulate; length 180–210 µm; height 213–262 µm; width 120 µm; ventral margin with internal irregularly perpendicular rows of spatulate micropoints (Figure 10.53). The unusual spatulate micropoints were also observed in *Lampsilis teres* by Hoggarth (1999).

Similar Sympatric Species

Lampsilis floridensis superficially resembles *Lampsilis straminea* but is more elongate and males are more pointed posteriorly.

Lampsilis floridensis may resemble some individuals of *Hamiota subangulata* but is more elongate and usually has fewer green rays.

Distribution in Florida

Lampsilis floridensis occurs in most major Gulf Coast basins from Escambia River to Hillsborough River (Figure 10.54).

Ecology and Biology

Lampsilis floridensis inhabits medium creeks, rivers, lakes, and floodplain sloughs in slow to moderate current in substrates of sand and sandy mud. It is occasionally found in reservoirs.

Lampsilis floridensis is a long-term brooder, gravid from late summer or autumn to the following spring or early summer. Nocturnal mantle flap displays have been observed in Apalachicola and New Rivers (Figure 10.55). The display consists of pulsating movements that are similar to those described for *Lampsilis teres* by Rypel (2008) in an Alabama reservoir.

In laboratory trials *Lampsilis floridensis* glochidia metamorphosed on three fishes that occur in Florida—Centrarchidae: *Micropterus salmoides* (Largemouth Bass); and Lepisosteidae: *Lepisosteus osseus* (Longnose Gar) and *Lepisosteus platyrhincus* (Florida Gar). These were reported by Keller and Ruessler (1997a) as hosts for *Lampsilis teres*, but the study specimens were actually *L. floridensis* collected from Suwannee River basin in Florida (Williams et al. 2008).

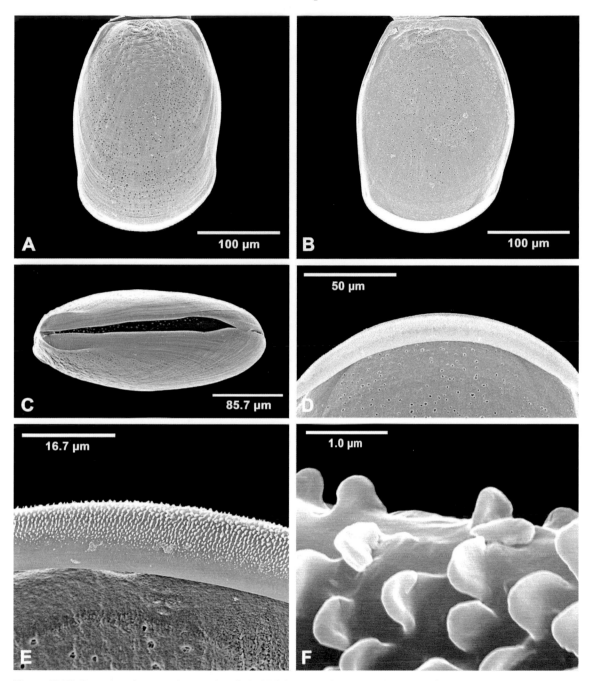

Figure 10.53. Scanning electron micrographs of glochidial valves of *Lampsilis floridensis* from boat basin off Suwannee River in the Original Suwannee River Campground, west of Fanning Springs, Dixie County, Florida, 23 January 2008— (A) external view; (B) internal view; (C) lateral view; (D) internal view of ventral margin; (E) internal view of micropoints on ventral margin; and (F) close-up of micropoints.

Lampsilis floridensis has been observed to be infested with mites (Unionicolidae). Of 188 individuals examined from 11 localities in north Florida rivers (Choctawhatchee east to Withlacoochee [southern]), 95 were parasitized by one or more of the following species: *Unionicola abnormipes*, *Unionicola aculeata*, *Unionicola hoesei*, *Unionicola sakantaka*, and *Unionicola vamana* (Vidrine 1996).

Conservation Status

Lampsilis floridensis was not previously evaluated as it was considered a synonym or subspecies of the wide-ranging *Lampsilis teres*. The conservation status of *L. floridensis* throughout its range was reported as currently stable (Williams et al. in review).

Lampsilis floridensis is extant throughout its Florida range. It remains relatively common, with the exception of Escambia River where it appears to be declining.

Figure 10.54. Distribution of *Lampsilis floridensis*.

Remarks

Lampsilis floridensis has been variously treated as a subspecies (Clench and Turner 1956) or a synonym of *Lampsilis teres* (Burch 1975a; Heard 1979a; Brim Box and Williams 2000). It was recognized as a distinct species by Williams et al. (2008) based on genetic analyses and subtle differences in shell morphology.

Figure 10.55. *Lampsilis floridensis* gravid female displaying—mantle flaps extended over charged gills (left) and exposed charged gills (right). Santa Fe River, Bradford County, Florida, 20 April 1996. Photographs by David S. Ruessler.

Synonymy

Unio floridensis Lea 1852 (Lea 1852b:274). Holotype by monotypy, ANSP 42081, length 76 mm. Type locality: published as Chácktaháchi [Choctawhatchee] River, west Florida, restricted by Clench and Turner (1956) to Choctawhatchee River, 1 mile west of Caryville, Holmes County, Florida, [Choctawhatchee River basin].

Lampsilis ornata (Conrad 1835)
Southern Pocketbook

Lampsilis ornata – Upper image: female, length 68 mm, UMMZ 163576. Escambia River near McDavid, Escambia County, Florida, May 1917. Photograph by Zachary Randall. Lower image: male, length 72 mm, UMMZ 163576. Escambia River near McDavid, Escambia County, Florida, May 1917. Photograph by Zachary Randall.

Description

Size: length to 122 mm, in Florida to 88 mm.

Shell: small individuals thin, large individuals moderately thick; smooth; inflated, width 45%–70% of length; outline oval to subtriangular; anterior margin rounded; posterior margin broadly rounded in females, narrowly rounded to bluntly pointed in males; dorsal margin slightly convex to rounded; ventral margin broadly rounded; valves gape anteriorly and posteriorly; sexual dimorphism subtle, females more inflated posterioventrally; posterior ridge sharp dorsally, rounded posterioventrally; posterior slope steep, flat to slightly concave; umbo broad, highly inflated, elevated well above hinge line; umbo sculpture unknown; umbo cavity wide, deep.

Teeth: pseudocardinal teeth moderately thick, 2 teeth in left valve, anterior tooth larger, moderately compressed, crest oriented anterioventrally, aligned through center of anterior adductor muscle scar, posterior tooth usually partially fused with anterior tooth, 1 tooth in right valve, with well-developed accessory denticle anteriorly; lateral teeth short, straight to slightly curved, 2 in left valve, 1 in right valve; interdentum moderately long, narrow.

Nacre: white, often with pink or salmon tint in umbo cavity, iridescent posteriorly.

Periostracum: shiny, occasionally dull; yellowish to brown, typically with dark green rays of varying width and intensity posteriorly.

Glochidium Description

Outline subspatulate; length 173–225 μm; height 223–287 μm; width 83–101 μm; dorsal margin 86–125 μm; ventral margin with internal irregularly arranged micropoints (Kennedy and Haag 2005) (Figure 10.56).

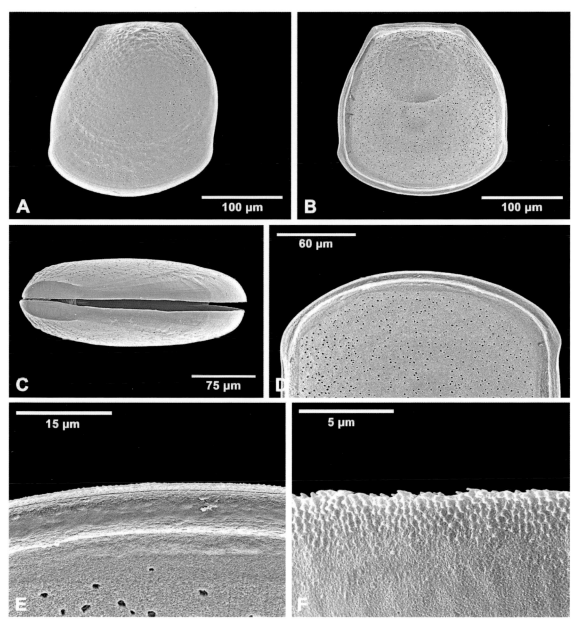

Figure 10.56. Scanning electron micrographs of glochidial valves of *Lampsilis ornata* from Black Warrior River [Locust Fork] near Warrior, Jefferson County, Alabama, 11 November 1998—(A) external view; (B) internal view; (C) lateral view; (D) internal view of ventral margin; (E) internal view of micropoints on ventral margin; and (F) close-up of micropoints.

Similar Sympatric Species

Lampsilis ornata can be distinguished from all mussels in the Escambia basin by its highly inflated umbo that is elevated well above the hinge line, sharp posterior ridge, and yellow periostracum with green rays.

Distribution in Florida

Lampsilis ornata occurs in Escambia River (Figure 10.57).

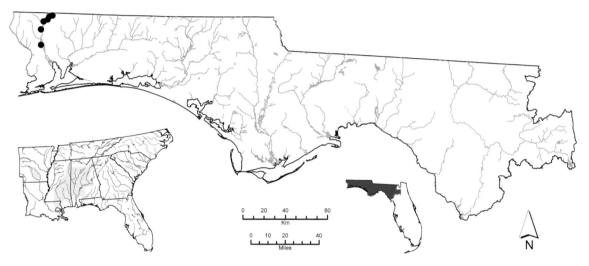

Figure 10.57. Distribution of *Lampsilis ornata*.

Ecology and Biology

Lampsilis ornata inhabits medium creeks to rivers in slow to moderate current in sandy mud to sand and gravel substrate but has been observed in fairly swift water in gravel in Escambia River.

Lampsilis ornata is a long-term brooder, gravid from August to June of the following year (Haag and Warren 2003). It is dioecious, but a hermaphroditic individual was reported from Little Tallahatchie River, Mississippi (Haag and Staton 2003). Sexual maturity was reported in some individuals from a Sipsey River, Alabama, population during their first year and in all individuals two or more years of age. In the Sipsey River population, 94 percent of sexually mature females were reported gravid, with only 1.5 percent of the eggs unfertilized. *Lampsilis ornata* fecundity was reported to average 281,776 glochidia per year but ranged from 48,625 to 739,600 per individual (Haag and Staton 2003). *Lampsilis ornata* is a host specialist that displays a large and elaborate mantle flap, presumably to attract large piscivorous hosts, as do other *Lampsilis* species having large mantle flaps. These hosts typically extract glochidia by striking bulging ovisacs extruded between the pulsating flaps (Barnhart et al. 2008).

In laboratory trials *Lampsilis ornata* glochidia metamorphosed on only one host fish that occurs in Florida—Centrarchidae: *Micropterus salmoides* (Largemouth Bass). Nine additional fishes that occur in Florida were exposed, but no metamorphosed glochidia were recovered—Centrarchidae: *Lepomis cyanellus* (Green Sunfish), *Lepomis macrochirus* (Bluegill), and *Lepomis megalotis* (Longear Sunfish); Cyprinidae: *Campostoma oligolepis* (Largescale Stoneroller), *Cyprinella venusta* (Blacktail Shiner), and *Notemigonus crysoleucas* (Golden Shiner); Esocidae: *Esox americanus* (Redfin

Pickerel); Ictaluridae: *Ictalurus punctatus* (Channel Catfish); and Percidae: *Percina nigrofasciata* (Blackbanded Darter) (Haag and Warren 2003).

Conservation Status

Lampsilis ornata was considered a species of special concern throughout its range by Williams et al. (1993) and currently stable by Williams et al. (in review). In Florida Williams and Butler (1994) assigned it a conservation status of threatened.

Lampsilis ornata is extant in isolated reaches of the Escambia River where it is uncommon.

Remarks

Lampsilis ornata was reported from Florida by Clench and Turner (1956) and Heard (1979a) as *Lampsilis excavatus*, a name now recognized as a synonym of *L. ornata*. In the early 1970s D.H. Stansbery began using *L. ornata* in various unpublished documents after pointing out that Conrad (1835a) had introduced the name *Unio ornata* for this species.

Two valves of *Lampsilis* (Figure 10.58) were recovered from an archaeological site on the east side of Lake Blackshear on the Flint River, Crisp County, Georgia. These valves are *Lampsilis*, but it is unclear if they represent an undescribed species, a disjunct population of *L. ornata*, or a downstream population of *L. binominata*. These specimens have much heavier teeth and hinge plates, and if their margins were extrapolated to their approximate size, they would be larger than any known specimen of *L. binominata*.

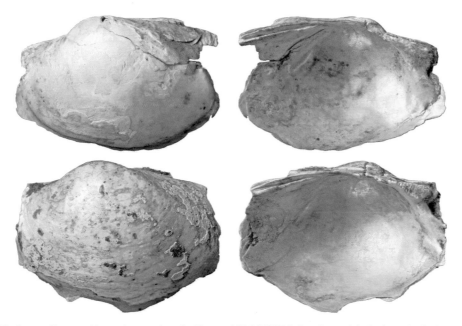

Figure 10.58. *Lampsilis* sp. – Upper image: length 45 mm, UF 445297 (left valve only). Archaeological specimen from Flint River, east side of Lake Blackshear (9cP108 site), Crisp County, Georgia, Fall 1973. © Richard T. Bryant. Lower image: length 41 mm, UF 445297 (left valve only). Archaeological specimen from Flint River, east side of Lake Blackshear (9cP108 site), Crisp County, Georgia, Fall 1973. © Richard T. Bryant.

Synonymy

Unio ovatus var. *ornatus* Conrad 1835 (Conrad 1835a:4). Type lost. Type locality: rivers of south Alabama.

Lampsilis straminea (Conrad 1834)
Southern Fatmucket

Lampsilis straminea – Upper image: female, length 66 mm, UF 388505. Kinchafoonee Creek at State Highway 49, 9 air miles northeast of Dawson, Terrell and Sumter counties, Georgia, 11 August 1992. © Richard T. Bryant. Middle image: female, length 58 mm, UF 375555. Chipola River, east bank at southern tip of cutoff island, 0.1 nautical mile above Apalachicola River, Gulf County, Florida, 11 August 1990. © Richard T. Bryant. Lower image: male, length 67 mm, UF 65338. Patsaliga River, Horten›s Lake, 10 miles north of Searight, Crenshaw County, Alabama, July 1915. © Richard T. Bryant.

Description

Size: length to 119 mm.

Shell: moderately thin to thick; smooth; moderately to highly inflated, width 35%–60% of length; outline oval to elliptical; anterior margin rounded; posterior margin broadly rounded to truncate in females, narrowly rounded in males; dorsal margin slightly convex to broadly rounded; ventral margin straight to broadly rounded; sexual dimorphism prominent in large individuals, females highly inflated posterioventrally;

posterior ridge rounded; posterior slope flat to slightly concave; umbo broad, moderately inflated, elevated slightly to well above hinge line; umbo sculpture weakly double-looped ridges; umbo cavity wide, moderately deep.

Teeth: pseudocardinal teeth triangular, 2 teeth in left valve, anterior tooth crest oriented anterioventrally, 1 tooth in right valve, occasionally with accessory denticle anteriorly; lateral teeth moderately long, thick, slightly curved, 2 in left valve, 1 in right valve; interdentum short, narrow.

Nacre: white, occasionally with orange tint.

Periostracum: shiny, occasionally dull; yellowish to brown, typically rayless, except Escambia River population may have rays posteriorly.

Glochidium Description

Outline subspatulate; length 190–209 μm; height 240–278 μm; width 103–115 μm; dorsal margin 94–113 μm; ventral margin with internal and external irregularly perpendicular rows of lanceolate micropoints (Kennedy and Haag 2005) (Figure 10.59).

Similar Sympatric Species

Lampsilis straminea may resemble *Lampsilis floridensis* but is less elongate. Male *L. floridensis* are typically more pointed posteriorly than *L. straminea*.

Distribution in Florida

Lampsilis straminea occurs in most major basins from Escambia River to Suwannee River (Figure 10.60).

Ecology and Biology

Lampsilis straminea inhabits medium to large creeks and rivers in sandy mud, sand, and gravel substrates in slow to moderate current. It is tolerant of some reservoir conditions as it has been found in Lake Talquin, an impoundment on the Ochlockonee River.

Lampsilis straminea is a long-term brooder that is gravid from April to late August (Brim Box and Williams 2000). *Lampsilis straminea* displays a large and elaborate mantle flap, likely to attract piscivorous hosts.

In laboratory trials *Lampsilis straminea* glochidia metamorphosed on six fishes that occur in Florida—Centrarchidae: *Lepomis macrochirus* (Bluegill) and *Micropterus salmoides* (Largemouth Bass); Cyprinidae: *Notemigonus crysoleucas* (Golden Shiner) and *Notropis texanus* (Weed Shiner); Ictaluridae: *Ictalurus punctatus* (Channel Catfish); and Poeciliidae: *Gambusia holbrooki* (Eastern Mosquitofish) (Keller and Ruessler 1997a; Williams et al. 2008). The highest rates of glochidial transformation were from *L. macrochirus* and *M. salmoides*.

Lampsilis straminea has been observed to be infested with mites (Unionicolidae). Of 206 individuals examined from nine localities in north Florida rivers (Escambia to Suwannee), 155 were parasitized by *Unionicola abnormipes* and/or *Unionicola hoesei* (Vidrine 1996).

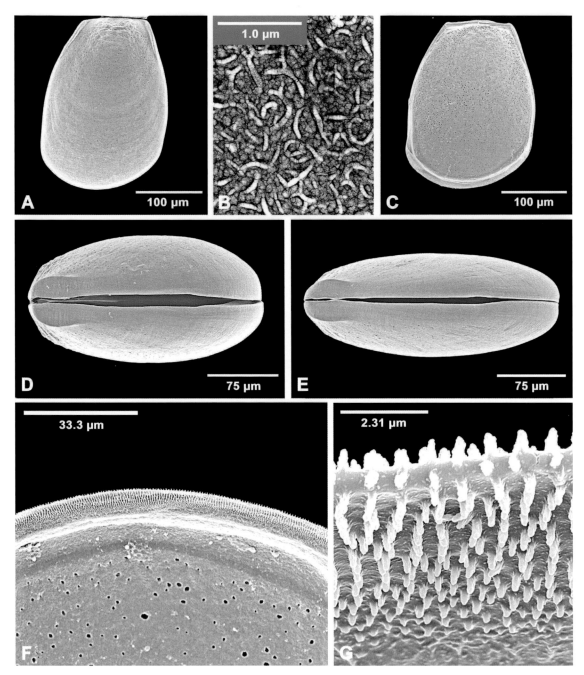

Figure 10.59. Scanning electron micrographs of glochidial valves of *Lampsilis straminea* from Escambia River at Molino, Escambia County, Florida, 18 April 2004—(A) external view; (B) surface sculpture; (C) internal view; (D) lateral view (inflated); (E) lateral view (compressed); (F) internal view of ventral margin and micropoints; and (G) close-up of micropoints.

Conservation Status

Lampsilis straminea was listed as currently stable throughout its range by Williams et al. (1993, in review).

Lampsilis straminea is extant throughout its Florida range and common in most basins.

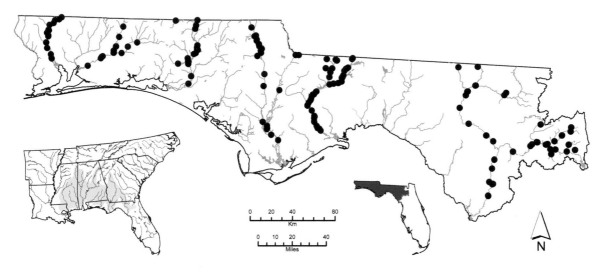

Figure 10.60. Distribution of *Lampsilis straminea*.

Remarks

Lampsilis straminea was reported from Florida by Clench and Turner (1956) and Heard (1979a) as *Lampsilis claibornensis*. Two subspecies, *Lampsilis straminea straminea* and *Lampsilis straminea claibornensis*, were recognized by Turgeon et al. (1998), but they are no longer considered valid (Williams et al. 2008).

Synonymy

Unio stramineus Conrad 1834 (Conrad 1834:339). Type lost: length about 64 mm. Type locality: small creeks of south Alabama.

Unio obtusus Lea 1840 (Lea 1840:287). Type locality: Chattahoochee River, Columbus, [Muscogee County,] Georgia, [Chattahoochee River drainage].
Unio contrarius Conrad 1849 (Conrad 1849:153). Type locality: published as Ogeechee River, Georgia, corrected and restricted by Johnson (1970) to Flint River, Georgia, [Flint River drainage].

Genus *Medionidus* Simpson 1900

Turgeon et al. (1998) recognized seven species of *Medionidus*, but one, *M. mcglameriae*, was placed in the synonymy of *Leptodea fragilis* by Williams et al. (2008). *Medionidus* occurs in Cumberland and Tennessee River drainages and eastern Gulf Coast basins from the Mobile to Suwannee.

Type Species
Unio conradicus Lea 1834 = *Medionidus conradicus* (Lea 1834)

Diagnosis
Shell thin to moderately thick; with sculpture on posterior slope consisting of fine wrinkles or corrugations; inflated; outline elongate; sexual dimorphism subtle, females somewhat inflated posterioventrally; periostracum shiny to dull; pseudocardinal teeth small, well developed, 2 in left valve, 1 in right valve; lateral teeth short, 2 in left valve, 1 in right valve; nacre typically bluish gray, occasionally with salmon or purplish tint.

Inner lamellae of inner gills completely or partially connected to visceral mass; mantle edge slightly thickened ventral to incurrent aperture, with finely crenulate, pigmented margin; outer gills marsupial; glochidia usually held in central portion of gill; marsupium extended beyond original gill margin when gravid; glochidium outline subspatulate to subelliptical, without styliform hooks (Simpson 1900a, 1914; Ortmann 1912; Johnson 1977).

Gravid females emerge from the substrate to display a modified mantle margin that is flickered against a contrasting background (Brim Box and Williams 2000; Haag and Warren 2001).

Medionidus acutissimus (Lea 1831)
Alabama Moccasinshell

Medionidus acutissimus. Upper image: length 41 mm, UF 89877. Little Choctawhatchee River, 5 miles south of Pinckard, Dale County, Alabama, April 1916. © Richard T. Bryant. Middle image: length 30 mm, UMMZ 163536. East Choctawhatchee River at Brevents Bridge, Dale County, Alabama, November 1915. © Richard T. Bryant. Lower image: length 26 mm, FMNH 89892. Choctawhatchee River, Walton County, Florida, October 1933. © Richard T. Bryant.

Description

Size: length to 55 mm, in Florida to 40 mm.

Shell: thin to moderately thick; smooth; moderately inflated, width 25%–45% of length; outline oval to elliptical; anterior margin rounded; posterior margin obliquely truncate to narrowly rounded; dorsal margin straight to slightly convex; ventral margin straight to convex, small individuals broadly rounded; posterior ridge moderately sharp dorsally, rounded posterioventrally; posterior slope moderately steep, with fine corrugations extending from posterior ridge to posteriodorsal margin, occasionally extending anterioventrally on shell disk; umbo broad, elevated slightly above hinge line; umbo sculpture thin, undulating ridges; umbo cavity wide, shallow.

Teeth: pseudocardinal teeth small, thin in small individuals, thick in large individuals, slightly compressed, triangular, 2 divergent teeth in left valve, occasionally joined at base, aligned almost end to end, anterior tooth crest almost parallel to shell margin, 1 tooth in right valve; lateral teeth short, moderately thin, typically straight, 2 in left valve, 1 in right valve; interdentum short to moderately long, narrow.

Nacre: white to bluish gray.

Periostracum: shiny to dull; small individuals yellowish to olive brown, with green rays of varying width and intensity, green chevrons occasionally present, large individuals dark brown to black, often obscuring rays.

Glochidium Description

Outline subspatulate; length 167–215 μm; height 187–280 μm; width 92–103 μm; dorsal margin 77–102 μm; ventral margin with internal irregularly arranged, blunt micropoints (Kennedy and Haag 2005) (Figure 10.61).

Similar Sympatric Species

Medionidus acutissimus somewhat resembles small *Fusconaia burkei* but differs in being more elongate.

Distribution in Florida

Medionidus acutissimus occurs in Choctawhatchee River basin (Figure 10.62).

Ecology and Biology

Medionidus acutissimus inhabits medium creeks to rivers, usually in moderate to swift current, in substrates of sand, gravel, and rubble.

Medionidus acutissimus is gravid from late February to mid-March (Haag and Warren 1997). Female *M. acutissimus* emerge from the substrate and display a modified mantle margin consisting of a small white patch that is flickered against a black background (Haag and Warren 2001).

In laboratory trials *Medionidus acutissimus* glochidia metamorphosed on 13 host fishes, including members of Fundulidae (1 species) and Percidae (12 species) (Haag and Warren 1997, 2003). Five of these fishes are known to occur in Florida—Fundulidae: *Fundulus olivaceus* (Blackspotted Topminnow); and Percidae: *Etheostoma stigmaeum* (Speckled Darter), *Etheostoma swaini* (Gulf Darter), *Percina nigrofasciata* (Blackbanded Darter), and *Percina vigil* (Saddleback Darter). This information suggests that *M. acutissimus* is a host specialist primarily on darters, as is typical for other *Medionidus* species (Zale and Neves 1982a).

Conservation Status

Medionidus acutissimus was considered threatened throughout its range by Williams et al. (1993) and endangered by Williams et al. (in review). Florida populations of *M. acutissimus* (in part as *Medionidus penicillatus*) were assigned a status of threatened by Williams and Butler (1994). In 1993 it was listed as threatened under the authority of the federal ESA (USFWS 1993). Critical habitat was designated for *M. acutissimus* in 2004, but none was identified in Florida streams (USFWS 2004). A recovery plan was approved in 2000 (USFWS 2000).

Medionidus acutissimus may be extirpated from Florida as it has not been collected since 1933. The localities where *Medionidus* was collected in the 1930s have been resampled, but none were found.

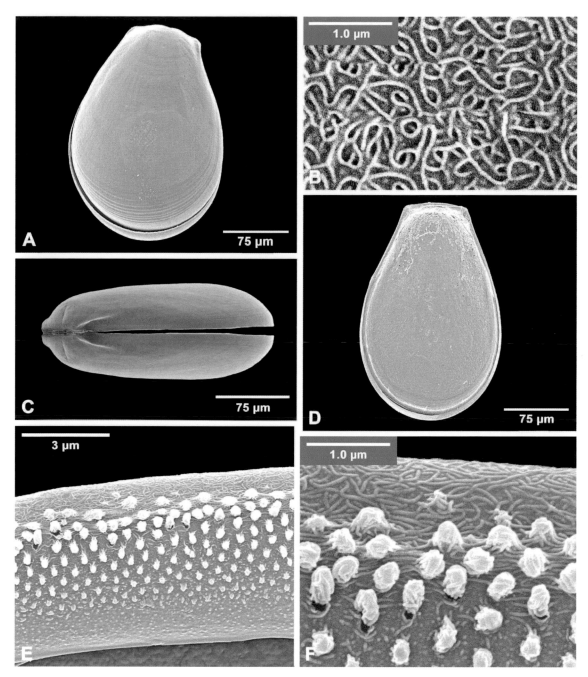

Figure 10.61. Scanning electron micrographs of glochidial valves of *Medionidus acutissimus* from Buttahatchee River near Caledonia, Monroe and Lowndes counties, Mississippi, 14 March 2007—(A) external view; (B) surface sculpture; (C) lateral view; (D) internal view; (E) internal view of micropoints on ventral margin; and (F) close-up of micropoints.

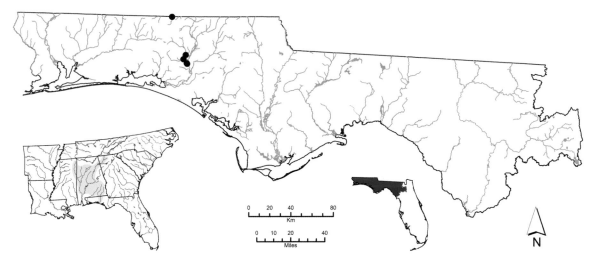

Figure 10.62. Distribution of *Medionidus acutissimus*.

Remarks

Medionidus acutissimus was long regarded as a Mobile basin endemic (Williams et al. 2008). Evaluation of *Medionidus* shells from Escambia, Yellow, and Choctawhatchee River basins during the past century calls into question the taxonomic status of these populations. Small numbers of museum specimens are inadequate for a thorough comparison with those from Mobile basin. Most authors (Johnson 1977; Butler 1990; Williams and Butler 1994) have included them as western populations of *Medionidus penicillatus*. Based on conchological characters, most specimens more closely resemble *M. acutissimus* than *M. penicillatus*. In addition, distribution patterns of other organisms suggest that species from Choctawhatchee, Yellow, and Escambia Rivers are likely more closely related to those of Mobile basin than those of Apalachicola basin (Williams et al. 2008).

Synonymy

Unio acutissimus Lea 1831 (Lea 1831:89). Type lost, length about 28 mm. Type locality: Alabama River, [Alabama, Alabama River drainage].

Medionidus penicillatus (Lea 1857)
Gulf Moccasinshell

Medionidus penicillatus – Upper image: length 42 mm, UF 411. Flint River at Hutchison's Ferry (State Highway 310) at Recovery, Decatur County, Georgia, 25 August 1954. © Richard T. Bryant. Middle image: length 36 mm, UF 388493. Kinchafoonee Creek at State Highway 45, 5.3 air miles southwest of Plains, Webster County, Georgia, 10 August 1992. © Richard T. Bryant. Lower image: length 29 mm, UMMZ 139210. Reedie [Reedy] Creek near Madrid, Houston County, Alabama, August 1916. © Richard T. Bryant.

Description

Size: length to 55 mm, in Florida to 52 mm.

Shell: moderately thin to moderately thick; disk anterior to posterior ridge smooth; moderately inflated, width 30%–45% of length; outline oval to elliptical; anterior margin rounded; posterior margin narrowly to broadly rounded, occasionally bluntly pointed; dorsal margin straight to slightly convex; ventral margin straight to convex, large individuals slightly concave just anterior to posterior ridge; posterior ridge moderately sharp dorsally, rounded posterioventrally; posterior slope moderately steep, typically with fine corrugations extending from posterior ridge to posteriodorsal margin, sometimes obscure in large individuals; umbo broad, moderately inflated, may be even with hinge line or elevated slightly to well above it; umbo sculpture 6–10 looped ridges, angular across posterior ridge; umbo cavity wide, shallow.

Teeth: pseudocardinal teeth moderately small, compressed, 2 teeth in left valve, occasionally joined at base, aligned almost end to end, anterior tooth crest almost parallel to shell margin, 1 tooth in right valve, moderately thick, triangular; lateral teeth moderately short, thin, typically straight to slightly curved, 2 in left valve, 1 in right valve; interdentum short, narrow.

Nacre: white to bluish gray, occasionally purplish.

Periostracum: shiny to dull; greenish yellow to dark brown, with green rays of varying width and intensity, especially prominent posteriorly, often obscure in large individuals.

Glochidium Description

Outline subspatulate; length 218–241 μm; height 280–310 μm; ventral margin with internal irregularly arranged micropoints (O'Brien and Williams 2002) (Figure 10.63).

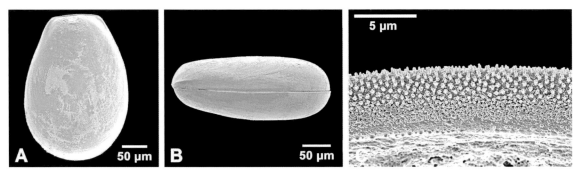

Figure 10.63. Scanning electron micrographs of glochidial valves of *Medionidus penicillatus* from Chickasawhatchee Creek, southwest of Dawson, Terrell County, Georgia, 21 March 1996—(A) external view; (B) lateral view; and (C) internal view of ventral margin and micropoints (modified from O'Brien and Williams 2002).

Similar Sympatric Species

Large *Medionidus penicillatus* may resemble small *Elliptoideus sloatianus* but has little or no sculpture anterior to the posterior ridge, rays of varying intensity, and thinner pseudocardinal and lateral teeth.

Medionidus penicillatus may resemble small individuals of several species in shape but differs in having sculpture on the posterior slope.

Distribution in Florida

Medionidus penicillatus occurs in Econfina Creek and Apalachicola River basins (Figure 10.64).

Ecology and Biology

Medionidus penicillatus inhabits small creeks to rivers, typically in slow to moderate current, in substrates of sandy mud, stable sand, and sand mixed with small gravel, as well as rocky shoals.

Medionidus penicillatus is a long-term brooder. A female was observed in March exposed on the substrate, displaying its mantle margin, suggestive of glochidial host attraction behavior (Brim Box and Williams 2000). Male *M. penicillatus* produce spermatozeugmata during late summer and autumn (C.A. O'Brien, personal communication).

In laboratory trials *Medionidus penicillatus* glochidia metamorphosed on four fishes that occur in Florida—Percidae: *Etheostoma edwini* (Brown Darter) and *Percina nigrofasciata* (Blackbanded Darter); and Poeciliidae: *Gambusia holbrooki* (Eastern Mosquitofish) and *Poecilia reticulata* (Guppy); however, the latter is not sympatric with *M. penicillatus*. Four additional fishes that occur in Florida were exposed, but no metamorphosed glochidia were recovered—Centrarchidae: *Lepomis macrochirus* (Bluegill) and *Micropterus salmoides* (Largemouth Bass); Cyprinidae: *Notropis texanus* (Weed Shiner); and Ictaluridae: *Ameiurus natalis* (Yellow Bullhead) (O'Brien and Williams 2002). This information and mantle display behavior suggest that *M. penicillatus* is a host specialist primarily on darters.

Conservation Status

Medionidus penicillatus was reported as endangered throughout its range by Athearn (1970) and Williams et al. (1993, in review). In Florida it was assigned a conservation status of threatened (Williams and Butler 1994). In 1998 *M. penicillatus* was listed as endangered under the authority of the federal ESA (USFWS 1998). Critical habitat was designated in 2007 and included a total of 1377.3 km in Alabama, Florida, and Georgia (USFWS 2007). In Florida, portions of Econfina Creek and the Chipola River and selected tributaries were designated as critical habitat. A recovery plan was approved in 2003 (USFWS 2003).

Medionidus penicillatus is extant in tributaries of upper Chipola River and Econfina Creek but is uncommon.

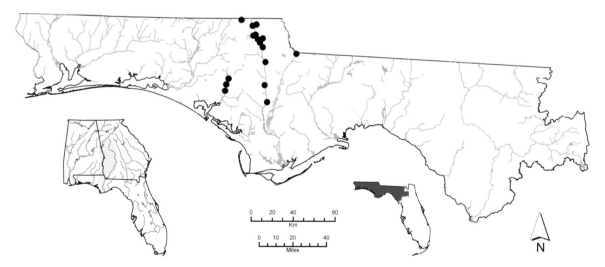

Figure 10.64. Distribution of *Medionidus penicillatus*.

Remarks

Medionidus penicillatus is known from archaeological samples from Apalachicola River suggesting that it likely occurred throughout the main stem prior to entrenchment and channel maintenance activities following construction of Jim Woodruff Lock and Dam.

Populations of *Medionidus* west of the Apalachicola basin were considered to be *M. penicillatus* by Clench and Turner (1956), Johnson (1977), Heard (1979a), and Butler

(1990). These populations have been reevaluated and are now considered to represent *Medionidus acutissimus* (Brim Box and Williams 2000; Williams et al. 2008).

Synonymy

Unio penicillatus Lea 1857 (Lea 1857a:171). Lectotype (Johnson 1974), USNM 84142, length 34 mm. Type locality: Flint River near Albany, [Dougherty County,] Georgia, [Flint River drainage].

Unio kingii B.H. Wright 1900 (B.H. Wright 1900:138). Type locality: a branch of the Flint River, Baker County, Georgia, [Flint River drainage].

Medionidus simpsonianus Walker 1905
Ochlockonee Moccasinshell

Medionidus simpsonianus – Upper image: length 44 mm, UF 469. Ochlockonee River, between Reno and Beachton, Grady County, Georgia, 4 September 1954. Photograph by Zachary Randall. Middle image: length 42 mm, UF 469. Ochlockonee River, between Reno and Beachton, Grady County, Georgia, 4 September 1954. Photograph by Zachary Randall. Lower image: length 41 mm, UF 229357. Ochlockonee River, 10 miles northeast of Tallahassee, Leon County, Florida. © Richard T. Bryant.

Description

Size: length to 56 mm.

Shell: thin to moderately thick; disk typically smooth anterior to posterior ridge; moderately inflated, width 30%–45% of length; outline oval to elliptical; anterior margin rounded; posterior margin narrowly rounded to bluntly pointed; dorsal margin straight to slightly convex; ventral margin straight to convex; posterior ridge moderately sharp dorsally, rounded posterioventrally; posterior slope moderately steep, typically with corrugations extending from posterior ridge to posteriodorsal margin; umbo broad, inflated, elevated slightly above hinge line; umbo sculpture unknown; umbo cavity wide, shallow.

Teeth: pseudocardinal teeth moderately small, compressed, 2 teeth in left valve, occasionally joined at base, aligned almost end to end, anterior tooth crest almost parallel to shell margin, 1 triangular tooth in right valve; lateral teeth short to moderately long, thin, straight to slightly curved, 2 in left valve, 1 in right valve; interdentum short, narrow.

Nacre: white to bluish gray.

Periostracum: shiny to dull, occasionally clothlike; small individuals greenish yellow, with green rays of varying width and intensity, occasionally dark green chevrons, large individuals brown to dark brown, often obscuring rays.

Glochidium Description

The glochidium of *Medionidus simpsonianus* is unknown.

Similar Sympatric Species

Large *Medionidus simpsonianus* may resemble small *Elliptoideus sloatianus* but has little or no sculpture anterior to the posterior ridge, rays of varying intensity, and thinner pseudocardinal and lateral teeth.

Distribution in Florida

Medionidus simpsonianus is known only from Ochlockonee River basin (Figure 10.65).

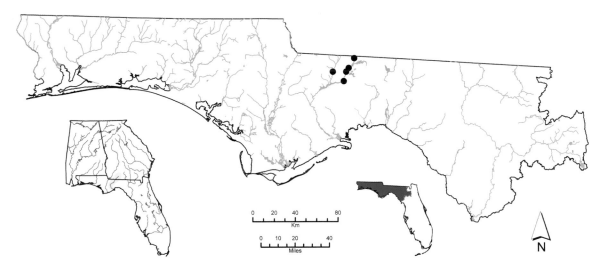

Figure 10.65. Distribution of *Medionidus simpsonianus*.

Ecology and Biology

Medionidus simpsonianus inhabits large creeks to rivers, typically in slow to moderate current, in substrates of stable sand and sand mixed with small gravel.

Medionidus simpsonianus, like other *Medionidus* in Florida, is presumably gravid from late January to March. Glochidial hosts for *M. simpsonianus* are unknown. Other species of *Medionidus* appear to be host specialists on darters (Haag and Warren 2001).

Conservation Status

Medionidus simpsonianus was assigned a conservation status of endangered throughout its range by Williams et al. (1993, in review). In Florida Williams and Butler (1994) considered it to be endangered. In 1998 it was listed as endangered under the authority of the federal ESA (USFWS 1998). In 2007 critical habitat was designated for *M. simpsonianus* and included a total of 177.3 km in the Ochlockonee River in Florida and Georgia (USFWS 2007). A recovery plan was approved in 2003 (USFWS 2003).

Medionidus simpsonianus is possibly extant in a short reach of the main stem of the Ochlockonee River upstream of Lake Talquin. The species has been very rare for decades and the last live individual was collected in 1993. Habitat in Ochlockonee River has been degraded considerably due primarily to increasing levels of sediment. *Medionidus simpsonianus* is on the brink of extinction, if not already extinct.

Remarks

Medionidus simpsonianus is endemic to the Ochlockonee River basin. All records are from the main stem of Ochlockonee River with the exception of a single collection from Little River. The record from downstream of Jackson Bluff Dam reported by Johnson (1977) is questionable as it was not included in the list of material examined.

The species was named in honor of Charles T. Simpson (1846–1932), a biologist and malacologist who worked at the Smithsonian Institution in Washington, DC, from 1889 to 1902.

Synonymy

Unio simpsonianus Walker 1905 (Walker 1905b:136). Lectotype (Clench and Turner 1956), UMMZ 98510, length 36 mm. Type locality: [Ochlockonee River,] Calvary, [Grady County,] Georgia, [Ochlockonee River basin].

Medionidus walkeri (B.H. Wright 1897)
Suwannee Moccasinshell

Medionidus walkeri – Upper image: length 36 mm, UF 4140. Santa Fe River at bridge north of Bland, Alachua County, Florida, 28 April 1934. © Richard T. Bryant. Lower image: length 33 mm, UF 4140. Santa Fe River at bridge north of Bland, Alachua County, Florida, 28 April 1934. © Richard T. Bryant.

Description

Size: length to 53 mm.

Shell: thin to moderately thick; disk smooth, occasionally with sculpture posteriorly; moderately inflated, width 35%–45% of length; outline oval; anterior margin rounded; posterior margin obliquely truncate to narrowly rounded; dorsal margin straight to convex; ventral margin straight to convex, occasional large individuals arcuate; posterior ridge moderately sharp dorsally, rounded posterioventrally; posterior slope moderately steep, with corrugations extending from posterior ridge to posteriodorsal margin, extending anterioventrally on shell disk in some individuals; umbo broad, moderately inflated, elevated slightly above hinge line; umbo sculpture 4–6 looped ridges, first 2–4 with slight indentation ventrally, angular across posterior ridge; umbo cavity wide, shallow.

Teeth: pseudocardinal teeth moderately small, compressed, 2 teeth in left valve, aligned almost end to end, crest almost parallel to shell margin, 1 triangular tooth in right valve; lateral teeth short, thin to moderately thick, straight, 2 in left valve, 1 in right valve; interdentum short, narrow.

Nacre: white to bluish gray, occasionally with salmon tint.

Periostracum: shiny to dull, often clothlike; small individuals greenish yellow to brown, with green rays of varying width and intensity, large individuals olive brown to brownish black, often obscuring rays.

Glochidium Description

The glochidium of *Medionidus walkeri* is unknown.

Similar Sympatric Species

Medionidus walkeri is easily distinguished from all other mussels in the Suwannee River basin by having an oval outline and sculpture on the posterior slope. In the Suwannee River basin, the only other mussel with sculpture, *Quadrula kleiniana*, is round to subtriangular in shape.

Distribution in Florida

Medionidus walkeri occurs in Suwannee and Hillsborough River basins (Figure 10.66).

Figure 10.66. Distribution of *Medionidus walkeri*.

Ecology and Biology

Medionidus walkeri inhabits small creeks to rivers in slow to moderate current in substrates of stable sand and gravel.

Medionidus walkeri has been observed to be gravid in late January. The period of gravidity probably extends into March. Glochidial hosts for *M. walkeri* are unknown. Other species of *Medionidus* appear to be host specialists using darters (Haag and Warren 2001).

Conservation Status

Medionidus walkeri was assigned a conservation status of threatened throughout its range by Williams et al. (1993) and endangered by Williams et al. (in review). In Florida Williams and Butler (1994) considered it to be endangered.

Medionidus walkeri is extremely rare and critically imperiled in the Suwannee River basin with fewer than 10 individuals found during the past 25 years. Its status in Hillsborough River is unknown.

Remarks

Medionidus walkeri was considered to be a synonym of *Medionidus penicillatus* (Clench and Turner 1956) before being recognized by Johnson (1977). *Medionidus walkeri* has generally been regarded as a Suwannee River basin endemic. However, there is a single record of *M. walkeri* from Hillsborough River, Morris Bridge, U.S. Highway 301, collected by T.H. Van Hyning in 1932. This disjunct population extends the range of *M. walkeri* southward into peninsular Florida.

Synonymy

Unio walkeri B.H. Wright 1897 (B.H. Wright 1897c:91). Lectotype (Simpson 1900b), USNM 150506, length 43 mm. Type locality: published as Suwannee River, Madison County, Florida, restricted by Johnson (1967a) to Suwannee River, Ellaville, Madison [Suwannee] County, Florida, [Suwannee River basin].

Genus *Megalonaias* Utterback 1915

There are currently two recognized species of *Megalonaias* in North America, *M. nervosa* and *M. nickliniana*. *Megalonaias nervosa* is the more widespread species, occurring in the Mississippi River basin and Gulf Coast basins from northeast Mexico east to the Ochlockonee River in Florida. Fossil records from the Pleistocene extend its prehistoric range eastward into the Suwannee basin and Tampa Bay area of peninsular Florida. *Megalonaias nickliniana* occurs in Mexico and Central America. *Megalonaias boykiniana* was formerly recognized from the Apalachicola and Ochlockonee River basins until it was placed in synonymy of *M. nervosa* by Mulvey et al. (1997) based on genetic data. Turgeon et al. (1998) recognized only *M. nervosa* from the United States. *Megalonaias nervosa* attains a length of 280 mm and is the largest mussel in North America.

Type Species
Unio heros Say 1829 = *Megalonaias nervosa* (Rafinesque 1820)

Diagnosis
Shell thick; sculptured with various plications and wrinkles; outline oval to trapezoidal; low posteriodorsal wing may be present; umbo sculpture thick, double-looped ridges, nodulous near posterior ridge; periostracum dull to clothlike; pseudocardinal teeth massive, 2 in left valve, 1 in right valve; lateral teeth long, straight, 2 in left valve, 1 in right valve; umbo cavity wide, deep.

Excurrent aperture almost smooth; mantle bridge separating excurrent and supra-anal apertures short; inner lamellae of inner gills usually completely connected to visceral mass; all 4 gills marsupial; glochidia held throughout gill; marsupium thickened, not extended beyond original gill margin when gravid; glochidial outline subelliptical, without styliform hooks (Utterback 1915).

Megalonaias nervosa (Rafinesque 1820)
Washboard

Megalonaias nervosa – Upper image: length 127 mm, UF 243972. Ochlockonee River at State Highway 20 bridge, 12 miles west of Tallahassee, Leon County, Florida, 12 July 1980. © Richard T. Bryant. Middle image: length 99 mm, UF 269752. Escambia County, Florida. © Richard T. Bryant. Lower image: length 29, UMMZ 197242. Ochlockonee River, 11 miles northwest of Tallahassee, Leon County, Florida. © Richard T. Bryant.

Description

Size: length to 280 mm, in Florida to 216 mm.

Shell: thick; with 3–5 large, oblique, parallel plications from near umbo to posterioventral margin, umbo with wrinkled sculpture that may extend onto disk, wrinkles occasionally present across entire disk; moderately compressed to moderately inflated, width 30%–50% of length; outline trapezoidal to oval; anterior margin rounded; posterior margin obliquely truncate to rounded, posterior edge of shell usually undulating; dorsal margin straight to slightly rounded, often obliquely angled as part of low dorsal wing; ventral margin straight to convex, occasional large individuals concave; posterior ridge indistinct, obscured by plications; posterior slope low, typically with arcuate corrugations extending from posterior ridge to posteriodorsal margin, often with smaller, irregular ridges; umbo broad, slightly to moderately inflated, elevated slightly above hinge line; umbo sculpture thick, double-looped ridges, nodulous near posterior ridge; umbo cavity moderately wide, deep.

Teeth: pseudocardinal teeth very large, triangular, 2 divergent teeth in left valve, 1 tooth in right valve, occasionally with accessory denticle anteriorly and/or posteriorly; lateral teeth moderately long, thick, straight to slightly curved, 2 in left valve, 1 in right valve; interdentum short to long, moderately wide.

Nacre: white to bluish white, large individuals occasionally with purple tint, iridescent.

Periostracum: dull to clothlike; dark olive brown to black.

Glochidium Description

Outline subelliptical; length 235–280 μm; height 310–380 μm; dorsal margin 135–179 μm; ventral margin with internal irregularly perpendicular rows of micropoints (Surber 1912, 1915; Utterback 1915; Hoggarth 1999; Kennedy and Haag 2005).

Similar Sympatric Species

Megalonaias nervosa most closely resembles *Amblema neislerii* and *Amblema plicata* but differs from those species in having wrinkled sculpture present on the umbo.

Megalonaias nervosa may also resemble *Elliptoideus sloatianus*, but the posterior ridge of *Elliptoideus* is typically more prominent and usually ends in a blunt point posterioventrally. The umbo cavity of *M. nervosa* is deeper and the nacre is white, usually bluish white to purplish in *E. sloatianus*. In large *M. nervosa* the posterior edge of the shell is usually undulating compared to the relatively straight edge in *Elliptoideus*.

Megalonaias nervosa also resembles *Plectomerus dombeyanus* but usually has a more rounded posterior margin and less pronounced posterior ridge. *Megalonaias nervosa* has a deeper umbo cavity and white nacre.

Distribution in Florida

Megalonaias nervosa occurs in Escambia, Apalachicola, and Ochlockonee River basins (Figure 10.67).

Ecology and Biology

Megalonaias nervosa inhabits large creeks to rivers in slow to moderate current in sand, sandy mud, gravel, and rubble substrates. In some portions of its range, it inhabits reservoirs but not in Florida.

Megalonaias nervosa has generally been considered a short-term brooder. In upper Mississippi River, *M. nervosa* was reported to reach sexual maturity by age eight (Woody and Holland-Bartels 1993). Most gonadal activity was reported to occur in July, followed by spawning in August. In a Tennessee River population, most gonadal activity occurred from mid-July to mid-September, followed by spawning during a two-week period in late September. Average brood size of 15 *M. nervosa* (100 to 150 mm in length) in the

Tennessee River population was 750,000 glochidia (Haggerty et al. 2005). *Megalonaias nervosa* from the Apalachicola River basin held in captivity were observed to broadcast glochidia embedded in a weblike mucus mass (C.A. O'Brien, personal communication), presumably to ensnare host fishes. This species is a host generalist. Glochidia of *Megalonaias nervosa* have been reported to infest both gills and fins of host fishes (Howard 1914b; Weiss and Layzer 1995). *Megalonaias nervosa* is the only species with hookless glochidia reported to attach to fins (Arey 1924).

Twenty-nine fish species belonging to 13 families, as well as the salamander *Necturus maculosus* (Mudpuppy), have been identified as *Megalonaias nervosa* hosts using observations of natural infestations and laboratory transformations (OSU Museum of Biological Diversity 2012).

Natural infestations of *Megalonaias nervosa* glochidia on potential hosts have been reported for 12 fishes that occur in Florida—Amiidae: *Amia calva* (Bowfin); Anguillidae: *Anguilla rostrata* (American Eel); Centrarchidae: *Lepomis gulosus* (Warmouth), *Lepomis macrochirus* (Bluegill), *Micropterus punctulatus* (Spotted Bass), and *Pomoxis annularis* (White Crappie); Clupeidae: *Alosa chrysochloris* (Skipjack Herring) and *Dorosoma cepedianum* (Gizzard Shad); Ictaluridae: *Noturus gyrinus* (Tadpole Madtom) and *Pylodictis olivaris* (Flathead Catfish); Lepisosteidae: *Lepisosteus osseus* (Longnose Gar); and Moronidae: *Morone chrysops* (White Bass). Natural transformation of glochidia of *M. nervosa* has been observed for one fish—Centrarchidae: *Lepomis macrochirus* (Bluegill) (OSU Museum of Biological Diversity 2012).

In laboratory trials *Megalonaias nervosa* glochidia metamorphosed on 14 fishes that occur in Florida—Centrarchidae: *Lepomis cyanellus* (Green Sunfish), *Lepomis macrochirus* (Bluegill), *Lepomis megalotis* (Longear Sunfish), *Micropterus salmoides* (Largemouth Bass), *Pomoxis annularis* (White Crappie), and *Pomoxis nigromaculatus* (Black Crappie); Cyprinidae: *Notemigonus crysoleucas* (Golden Shiner); Ictaluridae: *Ameiurus natalis* (Yellow Bullhead), *Ameiurus nebulosus* (Brown Bullhead), *Ictalurus punctatus* (Channel Catfish), and *Pylodictis olivaris* (Flathead Catfish); Lepisosteidae: *Lepisosteus osseus* (Longnose Gar); Moronidae: *Morone chrysops* (White Bass); and Percidae: *Perca flavescens* (Yellow Perch) (OSU Museum of Biological Diversity 2012).

Megalonaias nervosa has been observed to be infested with mites (Unionicolidae). Of 47 individuals examined from five localities in Escambia, Apalachicola, and Ochlockonee River basins, 45 were parasitized by *Unionicola tupara* (Vidrine 1996).

Conservation Status

Megalonaias nervosa was reported as currently stable throughout its range by Williams et al. (1993, in review). It is one of the most important commercially harvested species in the United States. There was apparently a small commercial harvest in Florida in the early 1900s in Chipola River, but none in recent decades.

Megalonaias nervosa is extant throughout its Florida range. It is common in Apalachicola, lower Chipola, and Ochlockonee Rivers but is uncommon in Escambia River.

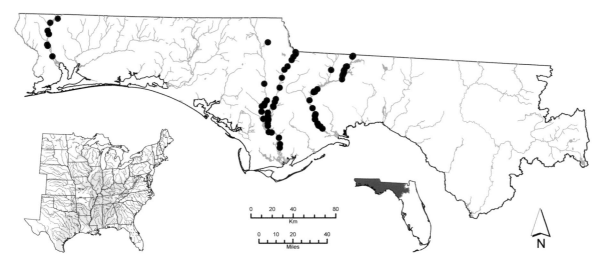

Figure 10.67. Distribution of *Megalonaias nervosa*.

Remarks

Prior to 1997 *Megalonaias boykiniana* was recognized as a valid species endemic to the Apalachicola and Ochlockonee River basins. A preliminary genetic analysis resulted in it being placed in the synonymy of *Megalonaias nervosa* (Mulvey et al. 1997). Unfortunately, they used a small number of individuals in their analysis and a gene (16s) that evolves relatively slowly in bivalve mollusks compared to other mitochondrial markers. In a rebuttal Berg and Berg (2000) demonstrated that the results could have been very different had a larger sample size and a more appropriate gene (e.g., CO1) been selected. The conclusions of Mulvey et al. (1997) and rejection of *M. boykiniana* by Turgeon et al. (1998) were followed herein, though it is clear that additional research is needed to resolve the taxonomic status of *Megalonaias* in eastern Gulf Coast basins.

Megalonaias nervosa is known from Pleistocene fossils from two localities in Gulf Coast basins of Florida. It was reported from an early Pleistocene site, Leisey Shell Pits, Hillsborough County, by Bogan and Portell (1995) and Suwannee River at Troy Spring, Lafayette County. Radiocarbon dated specimens from Troy Spring were determined to be late Pleistocene, about 17,000 BP.

Megalonaias nervosa was reported from a single Choctawhatchee River site by Shelton et al. (2006). The Choctawhatchee River has not been included in the range of *M. nervosa*. A concerted effort to verify this record was unsuccessful.

Synonymy

Unio (*Leptodea*) *nervosa* Rafinesque 1820 (Rafinesque 1820:296). Type lost. Type locality: "rapides de l'Ohio," [Ohio River drainage].

Unio boykinianus Lea 1840 (Lea 1840:288). Type locality: Chattahoochee River, Columbus, [Muscogee County,] Georgia, [Chattahoochee River drainage].

Genus *Obovaria* Rafinesque 1819

Obovaria was recognized as having six species by Turgeon et al. (1998). *Lampsilis haddletoni* was moved to *Obovaria* based on its circular shape, shallow umbo cavity, and triangular, divergent, striated pseudocardinal teeth (Williams et al. 2008). *Villosa choctawensis* was placed in *Obovaria* based on marsupial morphology of gravid females and mtDNA data (Williams et al. 2011). With these realignments the number of species in *Obovaria* increases to eight. *Obovaria* ranges from the Great Lakes and upper Mississippi River basins south to the Gulf Coast from eastern Texas east to the Choctawhatchee River basin in Alabama and Florida.

Type Species
Unio retusa Lamarck 1819 = *Obovaria retusa* (Lamarck 1819)

Diagnosis
Shell moderately thin to thick; smooth; inflated; outline round to oval or subtriangular; sexual dimorphism subtle, mature females somewhat inflated posterioventrally; posterior ridge weak or absent; periostracum shiny to clothlike, with or without rays; pseudocardinal and lateral teeth thick, 2 in left valve, 1 in right valve.

Inner lamellae of inner gills completely attached to visceral mass; mantle margin just ventral to incurrent aperture slightly thickened in females, finely crenulate and pigmented; outer gills marsupial; glochidia held in posterior portion of gill; marsupium extended beyond original gill margin when gravid; glochidium outline subelliptical, without styliform hooks (Rafinesque 1819, 1820; Simpson 1900a, 1914; Ortmann 1912).

Obovaria choctawensis (Athearn 1964)
Choctaw Bean

Obovaria choctawensis – Upper image: female, length 30 mm, UF 57076. Patsaliga Creek, 7.6 miles north-northwest of Dozier, Crenshaw County, Alabama, 24 April 1970. © Richard T. Bryant. Middle image: male, length 41 mm, UF 455457. Bruce Creek on Walton Bridge Road, about 14 kilometers (air) south-southwest of Ponce de Leon, Walton County, Florida, 22 April 2011. Photograph by Zachary Randall. Lower image: male, length 25 mm, UMMZ 197257. East Fork Choctawhatchee River, 8 miles west of Abbeville, Henry County, Alabama. © Richard T. Bryant.

Description
Size: length to 49 mm, in Florida to 43 mm.

Shell: thin to moderately thick; smooth; moderately inflated, width 35%–50% of length; outline oval; anterior margin rounded, occasionally slightly truncate; posterior margin narrowly to broadly rounded; dorsal margin slightly convex; ventral margin convex to broadly rounded; sexual dimorphism subtle, females slightly more inflated posterioventrally; posterior ridge rounded; posterior slope moderately steep, flat to slightly convex; umbo broad, moderately inflated, elevated slightly above hinge line; umbo sculpture 3–5 moderately thick, parallel ridges; umbo cavity wide, moderately shallow.

Teeth: pseudocardinal teeth moderately thick, 2 divergent teeth in left valve, anterior tooth moderately compressed, crest oriented anterioventrally, posterior tooth triangular, 1 tooth in right valve, occasionally with accessory denticle anteriorly; lateral teeth moderately long, straight to slightly curved, 2 in left valve, 1 in right valve; interdentum short, narrow to moderately wide.

Nacre: white to bluish white, iridescent posteriorly.

Periostracum: shiny to clothlike; small individuals yellowish green, with dark green rays, large individuals dark olive, brown, or black, often obscuring rays.

Glochidium Description

Outline subelliptical; length 175–200 μm; height 225–237 μm; width 88–101 μm; ventral margin with internal and external perpendicular rows of lanceolate micropoints (Williams et al. 2008) (Figure 10.68).

Similar Sympatric Species

Obovaria choctawensis resembles *Obovaria haddletoni* but differs in being more elongate and having more prominent rays.

Obovaria choctawensis resembles some individuals of *Villosa lienosa* but is less elongate, has thin green rays on the periostracum (though often only visible near the umbo), and pearly white nacre instead of purplish or salmon. Sexual dimorphism is much more pronounced in *V. lienosa* females.

Obovaria choctawensis resembles small *Pleurobema strodeanum* but differs in having a rounded, less defined posterior ridge, and thinner pseudocardinal teeth. Also, the periostracum of *O. choctawensis* is yellowish, with green rays, compared to the darker, rayless periostracum of *P. strodeanum*.

Obovaria choctawensis resembles some small individuals of *Toxolasma* but differs in having thin green rays on the periostracum and umbo sculpture consisting of 3–5 moderately thick, parallel ridges instead of single-looped ridges.

Distribution in Florida

Obovaria choctawensis occurs in Escambia, Yellow, and Choctawhatchee River basins (Figure 10.69).

Ecology and Biology

Obovaria choctawensis inhabits large creeks and rivers in slow to moderate current in silty sand to sandy clay substrates.

Obovaria choctawensis is presumably a long-term brooder, gravid from late summer or autumn to the following spring. Glochidial hosts for *O. choctawensis* are unknown. The species is probably a host specialist based on observations of other *Obovaria* (Haag and Warren 2003). A closely related species, *Obovaria arkansasensis*, apparently attracts fishes with a subtle mantle lure (C.M. Barnhart, personal communication).

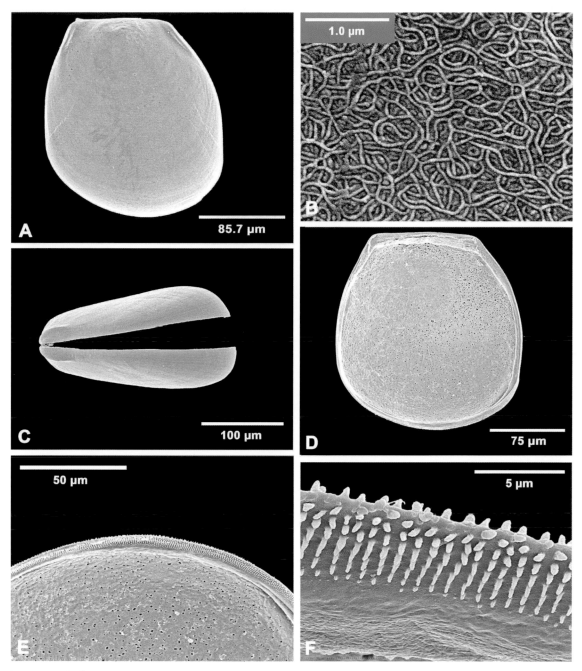

Figure 10.68. Scanning electron micrographs of glochidial valves of *Obovaria choctawensis* from Choctawhatchee River, Geneva County, Alabama, 18 May 1999—(A) external view; (B) surface sculpture; (C) lateral view; (D) internal view; (E) internal view of ventral margin and micropoints; and (F) close-up of micropoints.

Conservation Status

Obovaria choctawensis was assigned a conservation status of threatened throughout its range by Williams et al. (1993) and endangered by Williams et al. (in review). In Florida the species was also considered to be threatened by Thompson (1982) and Williams and Butler (1994). *Obovaria choctawensis* was listed as endangered with critical habitat designated under the authority of the federal ESA in 2012 (USFWS 2012).

Critical habitat included a total of 2,222 km of streams in Alabama and Florida. In Florida, critical habitat was designated in most of the main channel of the Escambia River, as well as the main channels of Yellow and Choctawhatchee Rivers and selected tributaries.

Obovaria choctawensis is extant throughout its Florida range but is uncommon.

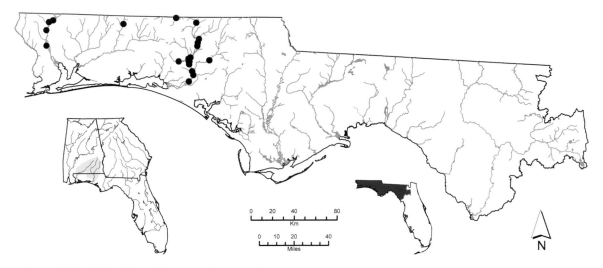

Figure 10.69. Distribution of *Obovaria choctawensis*.

Remarks

Prior to its description *Obovaria choctawensis* was often confused with *Pleurobema strodeanum*. The species is considered to be an *Obovaria* based on position, shape, and color of the marsupium (Williams et al. 2011).

The species name *choctawensis* was derived from Choctawhatchee River, in which the species was originally thought to be endemic (Athearn 1964). The Choctawhatchee River itself was named for a tribe of southeastern Native Americans, the Choctaws.

Synonymy

Villosa choctawensis Athearn 1964 (Athearn 1964:137). Holotype, CMNML 20096, length 37 mm. Type locality: Choctawhatchee River, 2 miles southwest of Caryville, about 1 mile downstream of U.S. Highway 90, Holmes County, Florida, [Choctawhatchee River basin,] 28 November 1958. © Richard T. Bryant.

Obovaria haddletoni (Athearn 1964)
Haddleton Lampmussel

Obovaria haddletoni – Upper image: length 31 mm, NCSM 55869 (paratype). Choctawhatchee River, West Fork, 7 miles southeast of Ozark, Dale County, Alabama, 23 September 1956. © Richard T. Bryant. Middle image: length 28 mm, UMMZ 246879 (right valve only). Choctawhatchee River at U.S. Highway 90, west of Caryville, Washington County, Florida, 25 May 1963. © Richard T. Bryant. Lower image: length 27 mm, UMMZ 246879 (left valve only). Choctawhatchee River at U.S. Highway 90, west of Caryville, Washington County, Florida, 25 May 1963. © Richard T. Bryant.

Description
Size: length to 31 mm, in Florida to 28 mm.

Shell: moderately thin; smooth; moderately inflated; outline round to oval; anterior margin rounded; posterior margin rounded; dorsal margin convex; ventral margin convex; posterior ridge broad, rounded; posterior slope low, flat; umbo moderately inflated, even with hinge line or elevated slightly above; umbo sculpture unknown; umbo cavity shallow.

Teeth: pseudocardinal teeth moderately thick, triangular, 2 divergent teeth in left valve, anterior tooth somewhat compressed, roughly parallel to shell margin, aligned through center of anterior adductor muscle scar, posterior tooth smaller, somewhat knobby, 1 tooth in right valve, with accessory denticle anteriorly and posteriorly; lateral teeth thin, slightly curved, 2 in left valve, 1 in right valve; interdentum moderately long, very narrow.

Nacre: bluish white to white.

Periostracum: shiny to dull; yellowish brown to brown, with thin dark green rays posteriorly.

Glochidium Description

The glochidium of *Obovaria haddletoni* is unknown.

Similar Sympatric Species

Obovaria haddletoni most closely resembles *Obovaria choctawensis* but differs in being more round and having less prominent rays.

Distribution in Florida

Obovaria haddletoni is known only from Choctawhatchee River basin (Figure 10.70).

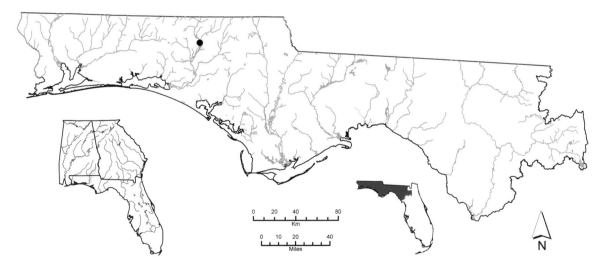

Figure 10.70. Distribution of *Obovaria haddletoni*.

Ecology and Biology

Information on the habitat of *Obovaria haddletoni* is extremely limited. The type specimens were collected in West Fork Choctawhatchee River, a small river with moderate current and sandy substrate with mixtures of gravel and rubble. The second locality, Choctawhatchee River at U.S. Highway 90, is a large river with moderate current and sandy substrate.

Obovaria haddletoni is probably a long-term brooder, gravid from late summer or autumn to the following summer. Glochidial hosts for *O. haddletoni* are unknown. Other members of the genus appear to be host specialists that attract host fishes with a small mantle lure (Haag and Warren 2003; C.M. Barnhart, personal communication).

Conservation Status

Obovaria haddletoni (as *Lampsilis haddletoni*) was considered endangered throughout its range by Athearn (1970), Stansbery (1971), and Williams et al. (1993, 2011). It was considered to be extinct by Williams et al. (in review). In Florida *Obovaria haddletoni* was assigned a status of special concern by Thompson (1982).

There are only four *Obovaria haddletoni* specimens known, including the two types. Two unpaired valves are known from Florida, both collected on the same date in 1963 in Choctawhatchee River at U.S. Highway 90. Some authors have considered the species extinct (Garner et al. 2004).

Remarks

Obovaria haddletoni was originally described as a species of *Lampsilis* with no justification for its generic placement (Athearn 1964). Based on its circular shape, shallow umbo cavity, and triangular, divergent pseudocardinal teeth, it was placed in *Obovaria* by Williams et al. (2008). Its shell morphology appears most similar to that of *Obovaria unicolor*, which occurs in Pearl and Pascagoula Rivers in Mississippi and Mobile basin in Alabama and Mississippi.

The type locality was originally published as "Choctawhatchee River, West Fork, 7 miles southwest of Ozark…" but was corrected by Butler (1990) to read "…southeast of Ozark."

Obovaria haddletoni was named in honor of Arthur Haddleton Clarke Jr., an esteemed colleague of Herbert D. Athearn, who has made significant contributions to our knowledge of unionid mollusks.

Synonymy

Lampsilis haddletoni Athearn 1964 (Athearn 1964:135). Holotype, CMNML 20095, length 30 mm. Type locality: Choctawhatchee River, West Fork, 7 miles southwest [southeast] of Ozark, Dale County, Alabama, [Choctawhatchee River basin,] 23 September 1956. © Richard T. Bryant.

Genus *Plectomerus* Conrad 1853

Plectomerus is a monotypic genus (Turgeon et al. 1998) and occurs in Gulf Coast rivers from the San Jacinto in eastern Texas to the Escambia in Florida. In the Mississippi River basin, it is found south of the Ohio River in west Tennessee and Kentucky and in Missouri, Arkansas, Louisiana, and Mississippi. Its presence in lower Tennessee River in Tennessee and Kentucky is likely the result of a recent expansion of its range following impoundment (Pharris et al. 1982).

Plectomerus was described by Conrad (1853), but no type species was designated. *Unio trapezoides* (= *Plectomerus dombeyanus*) was subsequently designated as the type species of *Plectomerus* by Ortmann and Walker (1922).

The phylogenetic relationships of *Plectomerus* have been investigated but remain unclear. Using immunoelectrophoresis, Davis and Fuller (1981) found *Plectomerus*, *Megalonaias*, and *Amblema* to be very closely related and suggested that they should be placed in a single genus. In an analysis using mtDNA, *Plectomerus* and *Elliptoideus* were recovered as sister taxa but weakly supported (Serb et al. 2003). In a subsequent genetic analysis that included most North American unionid genera, *Plectomerus* and *Elliptoideus* were recovered in widely separated clades (Campbell et al. 2005).

Type Species
Unio trapezoides Lea 1831 = *Unio dombeyana* Valenciennes (1827)

Diagnosis
Shell moderately thick to thick; sculptured with oblique plications; moderately inflated; outline rhomboidal to rectangular; umbo moderately inflated, elevated slightly above hinge line; posterior margin obliquely truncate, often pointed posterioventrally; posterior ridge well developed, with variously developed plications; periostracum dull, occasionally clothlike, dark olive brown to black; umbo cavity wide, moderately deep; nacre white, pinkish, or purplish; pseudocardinal teeth thick, 2 in left valve, 1 in right valve; lateral teeth moderately long, straight, 2 in left valve, 1 in right valve.

Incurrent aperture papillae simple; inner lamellae of inner gills usually connected to visceral mass anteriorly; typically all 4 gills marsupial; water tubes crowded and narrow in outer gill, less so in inner gill; ventral margin of outer gill not extended ventrally below original edge when gravid; glochidial outline subelliptical, without styliform hooks (Ortmann 1912; Simpson 1914; Frierson 1927).

Plectomerus dombeyanus (Valenciennes 1827)
Bankclimber

Plectomerus dombeyanus – Upper image: length 92 mm, UF 375934. Alabama River at mouth of Holly Creek, Montgomery Hill Landing, Baldwin County, Alabama, 20 October 1976. Photograph by Zachary Randall. Lower image: length 65 mm, UF 375934. Alabama River at mouth of Holly Creek, Montgomery Hill Landing, Baldwin County, Alabama, 20 October 1976. Photograph by Zachary Randall.

Description

Size: length to 150 mm, in Florida to 121 mm.

Shell: thick; with well-developed oblique plications, corrugations, and wrinkles on posterior 75% of shell, may be less prominent in large individuals; moderately inflated, width 25%–45% of length; outline rhomboidal to rectangular; anterior margin rounded; posterior margin obliquely truncate, often pointed posterioventrally; dorsal margin straight; ventral margin straight to slightly convex, occasionally concave just anterior to posterior ridge; posterior ridge usually high, moderately sharp; posterior slope moderately steep, slightly concave; posterior ridge and slope sculptured with moderately to well-developed plications, originating on crest of posterior ridge, extending to dorsal and posterior margins; umbo broad, slightly inflated, elevated slightly above hinge line; umbo sculpture irregular, nodulous ridges; umbo cavity wide, shallow to moderately deep.

Teeth: pseudocardinal teeth thick, 2 divergent teeth in left valve, crest of anterior tooth oriented ventrally to anterioventrally, may be separate dorsally, 1 tooth in right valve, thin accessory denticles may be present anteriorly and posteriorly; lateral teeth long, straight to slightly curved, 2 in left valve, 1 in right valve; interdentum moderately long, moderately wide; umbo cavity wide, moderately deep.

Nacre: small individuals white, becoming pinkish or purplish in large individuals.

Periostracum: small individuals shiny, large individuals dull to clothlike; small individuals greenish brown to brown, large individuals dark olive brown to black.

Glochidium Description

Outline subelliptical; length 223–231 μm; height 238–259 μm; ventral margin with internal irregularly perpendicular rows of lanceolate micropoints (Hoggarth 1999).

Similar Sympatric Species

Plectomerus dombeyanus resembles *Megalonaias nervosa* but usually has a more pronounced posterior ridge and obliquely truncate posterior margin that is pointed posterioventrally. *Plectomerus dombeyanus* typically has a shallower umbo cavity and pinkish or purplish nacre.

Distribution in Florida

Plectomerus dombeyanus occurs in Escambia River (Figure 10.71).

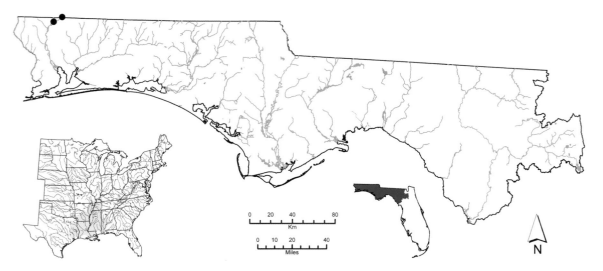

Figure 10.71. Distribution of *Plectomerus dombeyanus*.

Ecology and Biology

Plectomerus dombeyanus is typically found in medium to large rivers with slow to swift currents in sand and gravel substrates. It is also found in floodplain lakes. In Gulf Coast rivers it occurs in tidal freshwater reaches in areas that may periodically be inundated with low salinity water (Swingle and Bland 1974).

Plectomerus dombeyanus is presumably a short-term brooder and was reported to be gravid from May to September (Frierson 1904; Howells 2000). Mature glochidia were reported in specimens collected in late July (Hoggarth 1999). Glochidial hosts for *P. dombeyanus* are unknown.

Conservation Status

Plectomerus dombeyanus was considered currently stable throughout its range by Williams et al. (1993, in review). Based on lack of distribution and occurrence data, Williams and Butler (1994) listed its status as undetermined in Florida.

Plectomerus dombeyanus has not been found in Florida during the past 30 years and its current status is unknown.

Remarks

The first *Plectomerus dombeyanus* from Escambia River were collected near Century, Escambia County, and reported by Heard (1979a). These specimens were not vouchered in a museum collection but have been assumed to represent *P. dombeyanus* (Williams and Butler 1994). Another *P. dombeyanus* was found at a boat ramp on an oxbow lake of Escambia River just north of the state line in Alabama. This was reported to most likely be the result of an introduction (Williams et al. 2008). A reevaluation of the evidence suggests that the species was likely native to Escambia River. Similar distributional patterns of species shared between Mobile basin and Escambia River are known (e.g., *Anodonta hartfieldorum*, *Lampsilis ornata*, and several fishes).

Synonymy

Unio dombeyana Valenciennes 1827 (Valenciennes 1827:227). Type lost. Type locality: Peru. The type locality originally reported is erroneous (Williams et al. 2008).

Genus *Pleurobema* Rafinesque 1819

Taxonomy of species in *Pleurobema* has been the subject of change as additional information on genetics, distribution, and biology is published. In Turgeon et al. (1998), 32 species of *Pleurobema* were recognized as valid. Williams et al. (2008) made considerable changes in the number of recognized species of *Pleurobema* after extensive comparative study of primary type specimens of all Mobile basin and Gulf Coast species. Only one new species, *Pleurobema athearni*, has been described since publication of Turgeon et al. (1998). Currently, there are 25 species of *Pleurobema* recognized as valid.

Pleurobema is found in the Mississippi River and Great Lakes and on the Gulf Coast from Texas to Hillsborough River, Florida.

Type Species
Pleurobema mytiloides Rafinesque 1820 = *Pleurobema clava* (Lamarck 1819)

Diagnosis
Shell moderately thin to thick; smooth; outline subquadrate to oval or elongate oval; umbo often positioned near anterior end; posterior ridge low, rounded; periostracum shiny to dull, occasionally clothlike, typically without rays; umbo cavity shallow to moderately deep; hinge plate well developed; pseudocardinal teeth moderately thin to thick, 2 in left valve, 1 in right valve; lateral teeth moderately long, straight to slightly curved, 2 in left valve, 1 in right valve.

Inner lamellae of inner gills connected to visceral mass only anteriorly; outer gills marsupial, occasional individuals with all 4 marsupial; marsupium not extended beyond original gill margin when gravid; glochidium outline subelliptical, depressed subelliptical to subrotund, without styliform hooks (Rafinesque 1820; Simpson 1900a; Ortmann 1912).

Pleurobema pyriforme (Lea 1857)
Oval Pigtoe

Pleurobema pyriforme – Upper image: length 49 mm, UMMZ 57467. Hillsboro [Hillsborough] River at Morris Bridge, Hillsboro [Hillsborough] County, Florida, 30 May 1932. Photograph by Zachary Randall. Middle image: length 39 mm, UF 388504. Chipola River at mouth of Sink Creek, river mile 62.5, Jackson County, Florida, 14 August 1991. © Richard T. Bryant. Lower image: length 36 mm, UF 134940. Santa Fe River at Worthington Springs, Alachua County, Florida, 24 April 1949. © Richard T. Bryant.

Description

Size: length to 60 mm, in Florida to 49 mm.

Shell: moderately thin; smooth; moderately inflated, width 35%–50% of length; outline oval; anterior margin rounded; posterior margin narrowly rounded to obliquely truncate; dorsal margin slightly convex; ventral margin convex to broadly rounded; posterior ridge moderately sharp to rounded, occasionally biangulate posterioventrally; posterior slope moderately steep, flat to slightly concave; umbo narrow to moderately broad, moderately inflated, elevated slightly above hinge line; umbo sculpture 5–7 somewhat nodulous, undulating ridges (Figure 10.72); umbo cavity wide, shallow.

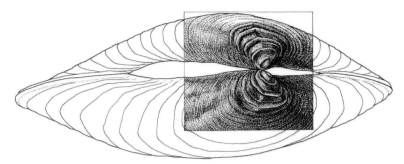

Figure 10.72. Umbo sculpture of *Pleurobema pyriforme*, length 29 mm, UF455458. Baker Creek on dirt road at first crossing above Bump Nose Road crossing, Jackson County, Florida, 19 July 2008. Illustration by Susan Trammell.

Teeth: pseudocardinal teeth moderately thick, low, triangular, 2 teeth in left valve, anterior tooth slightly smaller, compressed, crest almost parallel to shell margin, 1 tooth in right valve; lateral teeth short to moderately long, straight to slightly curved, 2 in left valve, 1 in right valve; interdentum short, narrow to moderately wide.

Nacre: white to bluish white, occasionally with pink tint in umbo cavity, iridescent posteriorly.

Periostracum: shiny to dull; small individuals yellowish brown, large individuals brown to black.

Glochidium Description

Outline depressed subelliptical; length 155–185 μm; height 152–180 μm; dorsal margin 120–145 μm; ventral margin with internal irregularly arranged micropoints (O'Brien and Williams 2002) (Figure 10.73).

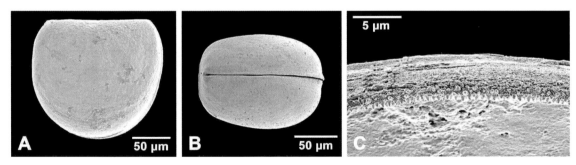

Figure 10.73. Scanning electron micrographs of glochidial valves of *Pleurobema pyriforme* from Kinchafoonee Creek, tributary of Flint River, Webster County, Georgia—(A) external view; (B) lateral view; and (C) internal view of ventral margin and micropoints (modified from O'Brien and Williams 2002).

Similar Sympatric Species

Pleurobema pyriforme can be distinguished from most other Florida mussels by having an oval outline, moderately thick shell and pseudocardinal teeth, yellowish-brown (large individuals may be brownish black) periostracum without rays, and umbo sculpture consisting of 5–7 somewhat nodulous, undulating ridges.

Small *Pleurobema pyriforme* may resemble small *Uniomerus* but have a more rounded posteriodorsal margin, thicker pseudocardinal teeth, and shinier periostracum.

Distribution in Florida

Pleurobema pyriforme occurs in Econfina Creek and Apalachicola, Ochlockonee, Suwannee, and Hillsborough River basins (Figure 10.74).

Figure 10.74. Distribution of *Pleurobema pyriforme*.

Ecology and Biology

Pleurobema pyriforme inhabits creeks and rivers in slow to moderate current in substrates of sandy mud, sand, and gravel. *Pleurobema pyriforme* is a short-term brooder that is gravid from March to July (Brim Box and Williams 2000; O'Brien and Williams 2002). Species of *Pleurobema* typically discharge conglutinates and tend to be host specialists (Haag 2012).

In laboratory trials *Pleurobema pyriforme* glochidia metamorphosed on two fishes that occur in Florida—Cyprinidae: *Pteronotropis hypselopterus* (Sailfin Shiner) and Poeciliidae: *Gambusia holbrooki* (Eastern Mosquitofish). Eight additional fishes that occur in Florida were exposed, but no metamorphosed glochidia were recovered—

Centrarchidae: *Lepomis macrochirus* (Bluegill) and *Micropterus salmoides* (Large-mouth Bass); Cyprinidae: *Notropis harperi* (Redeye Chub), *Notropis petersoni* (Coastal Shiner), *Notropis texanus* (Weed Shiner), and *Opsopoeodus emiliae* (Pugnose Minnow); Ictaluridae: *Ameiurus natalis* (Yellow Bullhead); and Percidae: *Percina nigrofasciata* (Blackbanded Darter) (O'Brien and Williams 2002).

Conservation Status

Pleurobema pyriforme was considered endangered throughout its range by Athearn (1970) and Williams et al. (1993, in review). In Florida it was assigned a conservation status of threatened (Williams and Butler 1994). In 1998 it was listed as endangered under the authority of the federal ESA (USFWS 1998). In 2007 critical habitat was designated for *P. pyriforme* and included a total of 1,637.7 km of streams in Alabama, Florida, and Georgia (USFWS 2007). In Florida, critical habitat included portions of Econfina Creek and Chipola, Ochlockonee, Santa Fe, and New Rivers. A recovery plan was approved in 2003 (USFWS 2003).

Pleurobema pyriforme is extant in isolated reaches of Econfina Creek, and Chipola and Santa Fe River drainages and is uncommon. It is becoming increasingly rare in Santa Fe River drainage. Extensive sampling of the Ochlockonee River basin during the past decade has failed to locate individuals of *P. pyriforme*.

Remarks

Pleurobema pyriforme is a variable species with many synonyms. During most of the 1900s, the distribution of *P. pyriforme* was reported as the Apalachicola River basin east to the Suwannee River basin (Clench and Turner 1956; Johnson 1970; Burch 1975a). Based on an examination of shells throughout its range, Williams and Butler (1994) recognized *Pleurobema reclusum* as valid and restricted to Ochlockonee and Suwannee River basins and *P. pyriforme* as endemic to Econfina Creek and Apalachicola River basins. Brim Box and Williams (2000) recognized only *P. pyriforme*, occurring from Econfina Creek to Suwannee River, based on genetic data presented by Kandl et al. (1997). Results of genetic analyses by Campbell et al. (2005, 2008) and Campbell and Lydeard (2012a) did not provide a clear resolution of the relationship of *P. pyriforme* with other *Pleurobema*. Additional research utilizing both mitochondrial and nuclear genes is needed to determine more precisely the levels of genetic divergence of *Pleurobema* in eastern Gulf Coast rivers as well as the Mobile and Mississippi basins.

The name *Unio striatus* proposed by Lea (1840) is senior to *Unio pyriformis* introduced by Lea (1857b) but was preoccupied by *Unio striatus* proposed by Rafinesque (1820), thus unavailable. Vanatta (1915) first recognized *Unio striatus* as being preoccupied and proposed the replacement name *Pleurobema simpsoni*.

Pleurobema pyriforme is known from archaeological sites along Apalachicola River (Percy 1976). It likely occurred through most of the middle and upper reaches of the river prior to entrenchment and channel maintenance activities following construction of Jim Woodruff Lock and Dam.

There is a single record of *Pleurobema pyriforme* from the Hillsborough River at Morris Bridge, U.S. Highway 301, collected by T.H. Van Hyning in 1932. This disjunct population extends the range of *P. pyriforme* southward into peninsular Florida.

Synonymy

Unio pyriformis Lea 1857 (Lea 1857b:31). Lectotype (Johnson 1974), USNM 84781, length 54 mm. Type locality: near Columbus, [Muscogee County,] Georgia, [Chattahoochee River drainage].

Unio striatus Lea 1840, *non* Rafinesque 1820 (Lea 1840:287). Type locality: Chattahoochee River, Columbus, [Muscogee County,] Georgia, [Chattahoochee River drainage].

Unio bulbosus Lea 1857 (Lea 1857a:172). Type locality: Flint River near Macon [County], Georgia, [Flint River drainage].

Unio modicus Lea 1857 (Lea 1857a:171). Type locality: Chattahoochee River near Columbus, [Muscogee County,] Georgia, [Chattahoochee River drainage].

Unio amabilis Lea 1865 (Lea 1865:89). Type locality: Butler, Taylor County, Georgia, [Flint River drainage].

Unio reclusus B.H. Wright 1898 (B.H. Wright 1898b:111). Type locality: Ocklocknee [Ochlockonee] River, Leon County, Florida, [Ochlockonee River basin].

Unio harperi B.H. Wright 1899 (B.H. Wright 1899:6). Type locality: published as Altamaha, Suwannee, and Flint Rivers. Johnson (1967a) incorrectly restricted the type locality to Spring Creek, a branch of the Flint River, Decatur County, Georgia, [Flint River drainage]. The Altamaha River locality is an error as *Pleurobema* has never been recorded from this river.

Pleurobema simpsoni Vanatta 1915 (Vanatta 1915:559) [replacement name for *Unio striatus* Lea 1840].

Pleurobema strodeanum (B.H. Wright 1898)
Fuzzy Pigtoe

Pleurobema strodeanum – Upper image: length 53 mm, UF 229566. Holmes Creek near Graceville, Jackson County, Florida. Photograph by Zachary Randall. Lower image: length 43 mm, UF 8362. Blue Creek, Walton County, Florida, July 1933. Photograph by Zachary Randall.

Description

Size: length to 75 mm, in Florida to 62 mm.

Shell: moderately thin to moderately thick; smooth; moderately inflated, width 30%–50% of length; outline oval to subtriangular; anterior margin rounded; posterior margin rounded to bluntly pointed; dorsal margin slightly convex; ventral margin convex to broadly rounded; posterior ridge moderately sharp dorsally, rounded posterioventrally, occasionally weakly biangulate posterioventrally; posterior slope steep dorsally, flat to slightly concave; umbo narrow to moderately broad, moderately inflated, elevated slightly above hinge line; umbo sculpture 3–4 thick, nodulous ridges; umbo cavity wide, shallow.

Teeth: pseudocardinal teeth triangular, 2 slightly divergent teeth in left valve, anterior tooth moderately compressed, crest oriented anterioventrally, aligned through center of anterior adductor muscle scar, 1 tooth in right valve, occasionally with accessory denticle posteriorly; lateral teeth moderately short to moderately long, straight, 2 in left valve, 1 in right valve; interdentum short, narrow.

Nacre: white, dull anteriorly, iridescent posteriorly.

Periostracum: typically clothlike, occasionally dull; small individuals yellowish brown to olive, large individuals dark olive to black.

Glochidium Description

Outline depressed subelliptical; mean length 176 µm; mean height 166 µm; mean dorsal margin 126 µm; ventral margin with internal and external irregularly arranged lanceolate micropoints (Pilarczyk et al. 2005) (Figure 10.75).

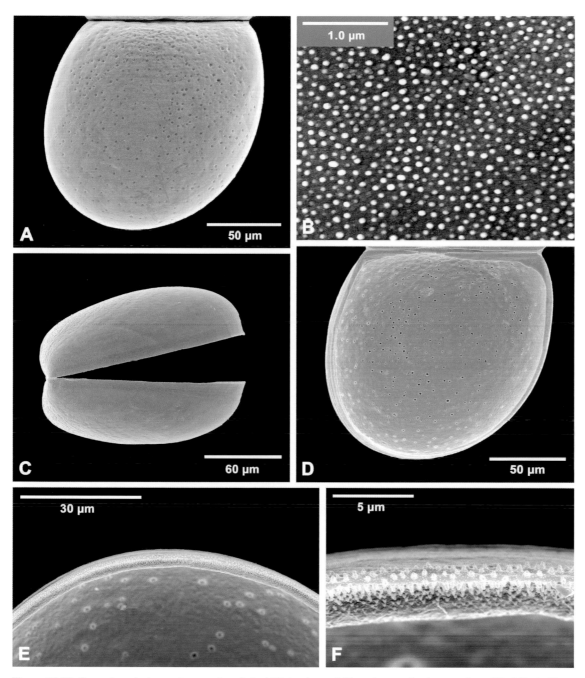

Figure 10.75. Scanning electron micrographs of glochidial valves of *Pleurobema strodeanum* from West Fork Choctawhatchee River near Blue Springs, Barbour County, Alabama, 20 May 2010—A) external view; (B) surface sculpture; (C) lateral view; (D) internal view; (E) internal view of ventral margin and micropoints; and (F) close-up of micropoints.

Similar Sympatric Species

Pleurobema strodeanum most closely resembles *Fusconaia escambia* but is usually more elongate and has a shallow umbo cavity. The posterior ridge of *P. strodeanum* is typically more rounded than that of most *F. escambia*. The nacre of *P. strodeanum* is white, but that of *F. escambia* is often salmon to pink.

Pleurobema strodeanum resembles some *Quadrula succissa* but is more elongate and has a shallower umbo cavity than that of *Q. succissa*, which is moderately deep. Shell nacre of *P. strodeanum* is white, but that of *Q. succissa* is typically purplish.

Small *Pleurobema strodeanum* may resemble *Obovaria choctawensis* but have a more prominent posterior ridge that terminates in a blunt point posterioventrally, giving it a more angular outline. The periostracum of *P. strodeanum* lacks rays, whereas that of *O. choctawensis* typically has green rays posteriorly.

Distribution in Florida

Pleurobema strodeanum occurs in Escambia, Yellow, and Choctawhatchee River basins (Figure 10.76).

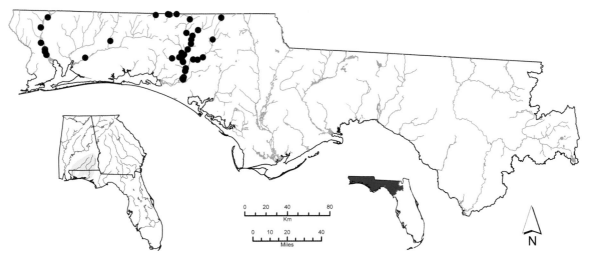

Figure 10.76. Distribution of *Pleurobema strodeanum*.

Ecology and Biology

Pleurobema strodeanum inhabits medium creeks to rivers in slow to moderate current in stable substrates of sand and silty sand.

Pleurobema strodeanum is a short-term brooder, gravid from mid-March through May, possibly June. Conglutinates are flattened, spindle-shaped, and pinkish orange, which fades to white as glochidia mature. Based on 11 individuals, fecundity ranged from 330 to 20,800, with an average of 15,173 glochidia (Pilarczyk et al. 2005).

In laboratory trials *Pleurobema strodeanum* glochidia metamorphosed on one fish that occurs in Florida—Cyprinidae: *Cyprinella venusta* (Blacktail Shiner). Nine additional fishes that occur in Florida were exposed, but no metamorphosed glochidia were recovered—Aphredoderidae: *Aphredoderus sayanus* (Pirate Perch); Centrarchidae: *Lepomis macrochirus* (Bluegill); Cyprinidae: *Lythrurus atrapiculus* (Blacktip Shiner), *Notropis texanus* (Weed Shiner), and *Pteronotropis hypselopterus* (Sailfin Shiner); Fundulidae: *Fundulus olivaceus* (Blackspotted Topminnow); Ictaluridae: *Noturus leptacanthus* (Speckled Madtom); and Percidae: *Etheostoma edwini* (Brown Darter) and *Percina nigrofasciata* (Blackbanded Darter) (White et al. 2008). Other species of *Pleurobema* tend to be host specialists (Haag 2012).

Conservation Status

Pleurobema strodeanum was assigned a conservation status of special concern throughout its range by Williams et al. (1993) and considered threatened by Williams et al. (in review). In Florida it was reported as threatened by Williams and Butler (1994). *Pleurobema strodeanum* was listed as threatened with critical habitat designated under the authority of the federal ESA in 2012 (USFWS 2012). Critical habitat included a total of 2,222 km of streams in Alabama and Florida. In Florida, critical habitat was designated in most of the main channel of the Escambia River, as well as the main channel of the Yellow and Choctawhatchee Rivers and selected tributaries.

Pleurobema strodeanum is extant throughout its Florida range and is uncommon. Its occurrence in Yellow River is sporadic where it is rare compared to other basins.

Remarks

In a phylogenetic analysis of *Pleurobema* in eastern Gulf Coast rivers, *P. strodeanum* was found to be sister to a clade containing *P. pyriforme* (Kandl et al. 2001). However, results of Campbell et al. (2005, 2008) and Campbell and Lydeard (2012a) indicate a close relationship with Mobile basin *Pleurobema*.

Pleurobema strodeanum was named for conchologist Dr. William S. Strode of Lewistown, Illinois.

Synonymy

Unio strodeanus B.H. Wright 1898 (B.H. Wright 1898a:5). Lectotype (Simpson 1900b), USNM 150498, length 38 mm. Type locality: Escambia River, [Escambia and Santa Rosa counties,] west Florida, [Escambia River basin].

Pleurobema patsaligensis Simpson 1900 (Simpson 1900b:82). Type locality: Little Patsaliga Creek, [Crenshaw County,] southeast Alabama, [Escambia River basin].

Genus *Ptychobranchus* Simpson 1900

Five species were recognized in *Ptychobranchus* by Turgeon et al. (1998). An additional species, *Ptychobranchus foremanianus*, from the eastern portion of the Mobile basin was elevated from synonymy by Williams et al. (2008), bringing the total recognized species in the genus to six. *Ptychobranchus jonesi* was described as a species of *Lampsilis* where it remained until a gravid female was examined, and the unique folded outer gill (Figure 10.77) was observed, confirming its placement in *Ptychobranchus* (Fuller and Bereza 1973). *Ptychobranchus* is found in Mississippi River and Great Lakes basins and eastern Gulf Coast rivers from Louisiana to Florida. Conglutinates produced by gravid females of *Ptychobranchus* are interesting mimics of insect larvae, fish eggs, and fish larvae (Hartfield and Hartfield 1996; Watters 1999). The same species are known to produce multiple types of conglutinates (Hartfield and Hartfield 1996; Haag and Warren 1997; Watters 1999).

Type Species
Unio phaseolus Hildreth 1828 = *Ptychobranchus fasciolaris* (Rafinesque 1820)

Diagnosis
Shell moderately thin to moderately thick; smooth; outline elliptical to triangular; female shell with oblique depression internally; male and female shells very similar externally, sexual dimorphism very subtle to nonexistent; periostracum shiny to dull; pseudocardinal and lateral teeth typically thick, 2 in left valve, 1 in right valve; umbo cavity very shallow.

Inner lamellae of inner gills completely connected to visceral mass or connected only anteriorly; mantle margin weakly modified ventral to incurrent aperture; mantle bridge separating supra-anal and excurrent apertures short; outer gills marsupial; entire marsupium developed into short parallel folds (Figure 10.77), extended beyond original gill margin when gravid; glochidia held in ventral portion of gill; conglutinates variable, in form of insect larvae, fish eggs, or fish larvae; glochidium outline subspatulate to subelliptical, without styliform hooks (Simpson 1900a, 1914; Ortmann 1912; Hartfield and Hartfield 1996; Watters 1999).

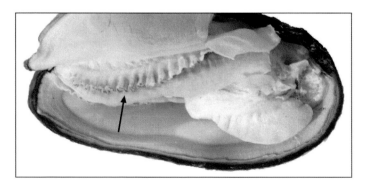

Figure 10.77. *Ptychobranchus foremanianus* gravid female. Note folded outer marsupial gill. Photograph by James D. Williams.

Ptychobranchus jonesi (van der Schalie 1934)
Southern Kidneyshell

Ptychobranchus jonesi – Upper image: length 61 mm, UF 65567. Conecuh River, Bozeman's Landing near Crenshaw County line, Covington County, Alabama, 1 July 1915. © Richard T. Bryant. Middle image: length 52 mm, ANSP 139374 (paratype). Pea River, Andrew's Fish Trap, Barbour County, Alabama, November 1915. Photograph by James D. Williams. Lower image: length 34 mm, UMMZ 234878. Yellow River at State Highway 55, 7 miles northwest of Lockhart, Covington County, Alabama, 28 October 1963. © Richard T. Bryant.

Description

Size: length to 72 mm, in Florida to 55 mm.

Shell: moderately thin; smooth; moderately inflated, width 30%–40% of length; outline elongate oval; anterior margin rounded; posterior margin narrowly rounded; dorsal margin straight to slightly convex; ventral margin straight to slightly convex; posterior ridge triangulate, prominent dorsally, with slight radial depression between ridges; posterior slope moderately steep, flat to slightly convex; umbo broad, elevated slightly above hinge line; umbo sculpture moderately thick ridges; umbo cavity wide, shallow.

Teeth: pseudocardinal teeth low, triangular, 2 divergent teeth in left valve, anterior tooth moderately compressed, crest almost parallel to shell margin, 1 tooth in right valve; lateral teeth moderately long, straight, 2 in left valve, 1 in right valve; interdentum moderately long, narrow.

Nacre: bluish white.

Periostracum: typically shiny, occasionally dull; small individuals greenish yellow to olive, occasionally with faint green rays, large individuals olive brown to black, obscuring rays.

Glochidium Description

Outline subspatulate; length 155–163 µm; height 191–212 µm; width 98–126 µm (Figure 10.78).

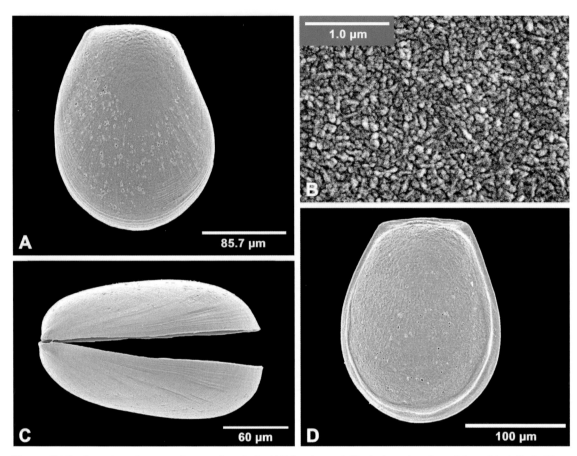

Figure 10.78. Scanning electron micrographs of glochidial valves of *Ptychobranchus jonesi* from West Fork Choctawhatchee River, Blue Springs State Park, Barbour County, Alabama, April 2007—(A) external view; (B) surface sculpture; (C) lateral view; and (D) internal view.

Similar Sympatric Species

Ptychobranchus jonesi most closely resembles *Hamiota australis* but is more inflated, has a more rounded posterior margin, and the ventral margin is typically straighter. Also, the posterior ridge of *P. jonesi* is typically triangulate distally with a shallow radial sulcus between the ridges. Gravid individuals of *P. jonesi* have swollen and folded outer gills, while the marsupia of *Hamiota* are swollen but not folded.

Ptychobranchus jonesi may resemble *Elliptio pullata* but is typically much more inflated, has a more rounded posterior margin, and posterior ridge that is typically triangulate distally with a shallow radial sulcus between the ridges.

Distribution in Florida

Ptychobranchus jonesi occurs in Choctawhatchee River basin (Figure 10.79).

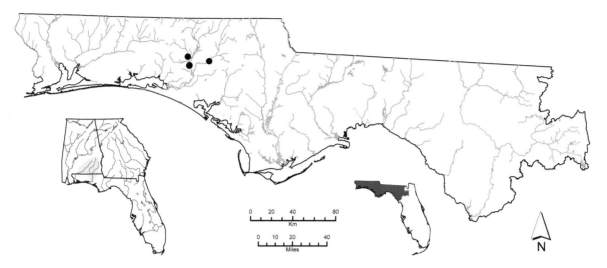

Figure 10.79. Distribution of *Ptychobranchus jonesi*.

Ecology and Biology

Ptychobranchus jonesi inhabits large creeks to rivers, typically in slow to moderate current in substrates of firm sand and fine gravel, often associated with bedrock.

A study determined that *Ptychobranchus jonesi* was often in geomorphically stable substrates associated with limestone outcroppings at depths ranging from 0.4–1.6 m. It comprised 0.27 percent (n = 13) of the mussels collected at 21 sites surveyed in the Choctawhatchee River basin. The stable substrate is likely an important constituent element of *P. jonesi* host fish habitat (Gangloff and Hartfield 2009).

Ptychobranchus jonesi is presumably a long-term brooder, gravid from autumn to the following spring or summer. The conglutinate of *P. jonesi* is approximately 3 mm in length, slightly bulbous on the distal end, and tapered on the proximal end (Figure 10.80). The distal half of the conglutinate is surrounded by a membrane that may have adhesive properties (P.D. Johnson and M.L. Buntin, personal communications). Glochidial hosts of *P. jonesi* are unknown, but hosts for other species of *Ptychobranchus* include sculpins (Cottidae) and darters (Percidae). While there are no sculpins in coastal streams in west Florida, there are more than 10 species of darters that have habitat requirements similar to those of *P. jonesi*.

Conservation Status

Ptychobranchus jonesi was first recognized as endangered by Athearn (1970). The species was assigned a conservation status of threatened throughout its range by Williams et al. (1993) and endangered by Williams et al. (in review). In Florida it was reported as special concern by Thompson (1982) and undetermined status by Williams and Butler (1994) based on the limited number of Florida records. *Ptychobranchus jonesi* was listed as endangered with critical habitat designated under the authority of the federal ESA in 2012 (USFWS 2012). Critical habitat included a total of 2,222 km of streams in Alabama and Florida. In Florida, critical habitat was designated in most of the main channel of

Escambia River, as well as the main channel of Choctawhatchee River and selected tributaries.

Figure 10.80. *Ptychobranchus jonesi* conglutinate, approximately 2.5 millimeters in length, released from captive female on 25 March 2007. Photograph by Michael Buntin.

Ptychobranchus jonesi is extant in the Choctawhatchee River basin. The only recent record of *P. jonesi* in Florida is from Holmes Creek where the species is extremely rare.

Excessive amounts of sand and silt that have accumulated in Pea and Choctawhatchee Rivers in Alabama were noted by Gangloff and Hartfield (2009) as a factor contributing to the decline of *Ptychobranchus jonesi*.

Remarks

Ptychobranchus jonesi was originally included in *Lampsilis*, where it remained for approximately four decades, until a gravid female was found to have folded marsupial gills, a characteristic of *Ptychobranchus* (Fuller and Bereza 1973; Williams et al. 2008).

Ptychobranchus jonesi was named for Walter Bryan Jones, Alabama state geologist and director of the ALMNH from 1927 to 1961.

Synonymy

Lampsilis jonesi van der Schalie 1934 (van der Schalie 1934:125). Holotype, UF 65558, length 46 mm (male). Type locality: Pea River [East Fork Choctawhatchee River], Pristons [Prestons] Mill, [near County Road 67 crossing, about 7 kilometers north of Midland City,] Dale County, Alabama, [Choctawhatchee River basin,] November 1915.

Genus *Pyganodon* Crosse and Fischer 1894

Turgeon et al. (1998) included five species in *Pyganodon*. The genus occurs in Hudson Bay, Great Lakes, and Mississippi River basins; Gulf Coast basins from Mexico to Florida; and Atlantic Coast basins from Newfoundland to Georgia. Only one species, *Pyganodon grandis*, is currently recognized as naturally occurring in Florida.

The report of *Pyganodon cataracta* from Apalachicola River in Florida was based on comparative shell morphology of specimens used in a genetic analysis from reservoirs in Georgia (Brim Box and Williams 2000). Based on genetic analyses, specimens from reservoirs in the Chattahoochee River in Alabama and Georgia were all *P. cataracta* (Williams et al. 2008). In recent DNA analyses of *Pyganodon* samples from the Apalachicola and Chipola Rivers in Florida, all were found to be *P. grandis*.

An introduced population of *Pyganodon cataracta* was discovered in the Suwannee River basin in 2002 in Warrior Creek, a tributary of Withlacoochee River (northern), Colquitt County, Georgia, and appears to be recruiting. *Pyganodon cataracta* has not been found in Florida but is included among mussels hypothetically occurring in the state.

Type Species
Anodonta globosa Lea 1841

Diagnosis
Shell thin to moderately thick; smooth; inflated; outline elliptical to oval; umbo inflated, elevated slightly to well above hinge line; umbo sculpture double-looped ridges; periostracum shiny to dull, occasionally clothlike; without pseudocardinal and lateral teeth.

Inner lamellae of inner gills connected to visceral mass only anteriorly; outer gills marsupial, with secondary water tubes when gravid; marsupium thickened, extended beyond original gill margin when gravid; glochidial outline subtriangular, with styliform hooks (Crosse and Fischer 1894; Frierson 1927; Hoeh 1990; Vidrine 1993).

Pyganodon grandis (Say 1829)
Giant Floater

Pyganodon grandis – Upper image: length 94 mm, UF 449290. Seminole Reservoir [Lake Seminole] near landing at Seminole Lodge Marina, about 1 air mile northeast of Sneads, Jackson County, Florida, 27 June 2006. © Richard T. Bryant. Middle image: length 84 mm, UF 388488. Harrison Creek, [tributary of Brothers River,] at first 180 degree turn, [10.8 miles north-northwest of Apalachicola,] north side of bend, Franklin County, Florida, 7 September 1991. © Richard T. Bryant. Lower image: length 42 mm, UF 449291. Chipola River, about 2.9 miles east of Wewahitchka, Gulf County, Florida, 15 September 2010. © Richard T. Bryant.

Description

Size: length to 254 mm, in Florida to 126 mm.

Shell: small individuals thin, large individuals moderately thick; smooth; moderately to highly inflated, width 30%–50% of length; outline oval to elliptical; anterior margin rounded; posterior margin rounded to bluntly pointed; dorsal margin straight to slightly convex; ventral margin convex to broadly rounded; posterior ridge sharp dorsally, rounded posteroventrally; posterior slope steep dorsally, flat to slightly concave; umbo broad, moderately to highly inflated, elevated slightly to well above hinge line; umbo sculpture double-looped, nodulous ridges; umbo cavity wide, moderately deep.

Teeth: pseudocardinal and lateral teeth absent, dorsal margin thickened posterior of umbo.

Nacre: white to bluish white, occasionally with salmon tint in umbo cavity.

Periostracum: typically shiny, occasionally dull; small individuals yellowish green to green, large individuals dark olive brown to black, occasionally with faint green rays.

Glochidium Description

Outline subtriangular; length 338–410 μm; height 343–420 μm; dorsal margin 215–294 μm (Ortmann 1912; Surber 1912; Utterback 1916a; Tucker 1928; Hoggarth 1999; Kennedy and Haag 2005).

Similar Sympatric Species

Pyganodon grandis resembles *Anodonta hartfieldorum*, *Anodonta heardi*, and *Anodonta suborbiculata* but has a more inflated umbo that is elevated well above the hinge line. The umbo sculpture of *P. grandis* is double-looped and nodulous ridges, while those of *A. hartfieldorum*, *A. heardi*, and *A. suborbiculata* are parallel.

Some small *Pyganodon grandis* may resemble *Anodontoides radiatus* but are more inflated, have higher umbos, lack lateral and pseudocardinal teeth, and are usually rayless.

Pyganodon grandis may resemble some *Utterbackia imbecillis* and *Utterbackia peggyae* but differs in being less elongate and having a more inflated umbo that is elevated well above the hinge line.

Distribution in Florida

Pyganodon grandis occurs in Escambia, Choctawhatchee, Apalachicola, and Ochlockonee River basins (Figure 10.81).

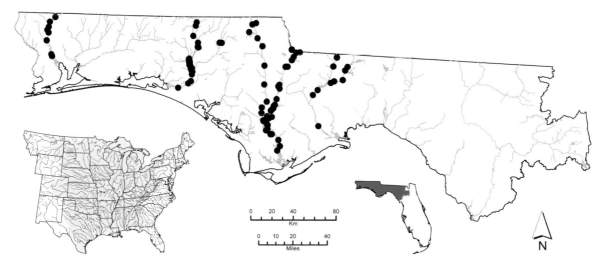

Figure 10.81. Distribution of *Pyganodon grandis*.

Ecology and Biology

Pyganodon grandis inhabits ponds, lakes, reservoirs, oxbows, sloughs, and slow-moving waters of creeks and rivers in mud or sandy mud substrates.

Pyganodon grandis is reported to be a long-term brooder, but brooding period may vary across its range. In Florida it appears to be gravid from late summer to the following spring. Fecundity was reported to be more than 400,000 glochidia by Haag (2012). It is usually dioecious, but hermaphroditic individuals have been observed and entire populations have been hypothesized (van der Schalie and Locke 1941; Jansen and Hanson 1991). *Pyganodon grandis* sperm are released as spermatozeugmata 40–50 μm in diameter (Lynn 1994).

Pyganodon grandis is a host generalist that broadcasts glochidia with larval threads embedded in a mucus web (Watters et al. 2009). A total of 43 fish species belonging to 15 families have been identified as *P. grandis* glochidial hosts with observations of natural infestations as well as laboratory transformations, though methods of some reports were unspecified (Williams et al. 2008; OSU Museum of Biological Diversity 2012).

Natural infestations of glochidia on potential hosts have been reported for 15 fishes that occur in Florida—Atherinopsidae: *Labidesthes sicculus* (Brook Silverside); Centrarchidae: *Lepomis cyanellus* (Green Sunfish), *Lepomis macrochirus* (Bluegill), *Lepomis megalotis* (Longear Sunfish), *Micropterus salmoides* (Largemouth Bass), *Pomoxis annularis* (White Crappie), and *Pomoxis nigromaculatus* (Black Crappie); Clupeidae: *Alosa chrysochloris* (Skipjack Herring) and *Dorosoma cepedianum* (Gizzard Shad); Cyprinidae: *Cyprinus carpio* (Common Carp) and *Notemigonus crysoleucas* (Golden Shiner); Fundulidae: *Fundulus chrysotus* (Golden Topminnow); Ictaluridae: *Ameiurus natalis* (Yellow Bullhead); Moronidae: *Morone chrysops* (White Bass); and Percidae: *Perca flavescens* (Yellow Perch) (OSU Museum of Biological Diversity 2012).

In laboratory trials *Pyganodon grandis* glochidia metamorphosed on 14 fishes that occur in Florida—Atherinopsidae: *Labidesthes sicculus* (Brook Silverside); Centrarchidae: *Lepomis cyanellus* (Green Sunfish), *Lepomis humilis* (Orangespotted Sunfish), *Lepomis macrochirus* (Bluegill), *Lepomis megalotis* (Longear Sunfish), *Micropterus salmoides* (Largemouth Bass), *Pomoxis annularis* (White Crappie), and *Pomoxis nigromaculatus* (Black Crappie); Cichlidae: *Herichthys cyanoguttatus* (Rio Grande Cichlid); Cyprinidae: *Luxilus chrysocephalus* (Striped Shiner) and *Notemigonus crysoleucas* (Golden Shiner); Fundulidae: *Fundulus chrysotus* (Golden Topminnow); Lepisosteidae: *Lepisosteus osseus* (Longnose Gar); and Percidae: *Perca flavescens* (Yellow Perch) (OSU Museum of Biological Diversity 2012).

Pyganodon grandis has been observed to be infested with mites (Unionicolidae). Of 42 individuals examined from four localities in the Choctawhatchee and Apalachicola River basins, 16 were parasitized by one or more of the following species: *Unionicola formosa*, *Unionicola mitchelli*, *Unionicola sakantaka*, *Unionicola serrata*, and *Unionicola wolcotti* (Vidrine 1996).

Conservation Status

Pyganodon grandis was assigned a conservation status of currently stable throughout its range by Williams et al. (1993, in review).

Pyganodon grandis is extant throughout its Florida range. The species is generally uncommon.

Remarks

Shell morphology of *Pyganodon grandis* is highly variable, which has led to the description of more than 50 nominal species and/or subspecies. At this time only one species of *Pyganodon*, *P. grandis*, is recognized in Florida waters. *Pyganodon grandis* is the most widely distributed mussel in North America (Williams et al. 2008). The species has been introduced outside of its known native range into Arizona (Bequaert and Miller 1973) and New Mexico (Lang and Mehlhop 1996).

Synonymy

Anodonta grandis Say 1829 (Say 1829:341). Neotype (Haas 1930), SMF 4300, length 170 mm, not figured. Type locality: Ohio River, [Ohio River drainage].

Genus *Quadrula* Rafinesque 1820

There were 18 species and 2 subspecies recognized in *Quadrula* by Turgeon et al. (1998). An additional species, *Quadrula nobilis*, was elevated from synonymy by Howells et al. (1996) but was not included in Turgeon et al. (1998), though it is now widely recognized as a valid species. *Quadrula kieneriana* was elevated from synonymy and the subspecies *Quadrula asperata archeri* was recognized as valid by Williams et al. (2008). Two additional species—*Fusconaia succissa* and *Quincuncina infucata*—were moved to *Quadrula* based on genetic analysis and shell morphology; a third species, *Quadrula kleiniana*, was removed from synonymy (Lydeard et al. 2000). The monotypic genus *Tritogonia* was moved into *Quadrula* based on genetic analysis (Serb et al. 2003; Campbell et al. 2005). Davis and Fuller (1981) previously noted that *Tritogonia* was immunologically closest to two groups within *Quadrula*. With the recognition of additional species and generic realignments, the total number of taxa recognized in *Quadrula* is currently 27, including 24 species and 3 subspecies.

Quadrula occurs in Hudson Bay, Great Lakes, and Mississippi River basins and Gulf Coast rivers from Texas to Florida.

Type Species

Obliquaria (*Quadrula*) *quadrula* Rafinesque 1820 = *Quadrula quadrula* (Rafinesque
    1820)

Diagnosis

Shell moderately thick; usually sculptured, may be smooth; outline rectangular, quadrate, triangular, or round; umbo elevated slightly to well above hinge line; posterior ridge usually well developed; periostracum shiny, dull, or clothlike; hinge plate well developed; pseudocardinal and lateral teeth thick, 2 in left valve, 1 in right valve; umbo cavity usually deep.

Supra-anal aperture usually long; mantle bridge separating supra-anal and excurrent apertures short, sometimes absent; excurrent aperture smooth; inner lamellae of inner gills connected to visceral mass only anteriorly; all 4 gills marsupial; glochidia held across entire gill; marsupium slightly thickened, not extended beyond original gill margin when gravid; glochidium outline subelliptical, without styliform hooks (Rafinesque 1820; Simpson 1900a, 1914; Ortmann 1912).

Quadrula infucata (Conrad 1834)
Sculptured Pigtoe

Quadrula infucata – Upper image: length 51 mm, UF 243963. Ochlockonee River at U.S. Highway 84, 9 miles east of Cairo, Thomas County, Georgia, 19 November 1988. © Richard T. Bryant. Lower image: length 38 mm, UF 388503. Muckalee Creek at State Highway 195, 3.5 air miles northeast of Leesburg, Lee County, Georgia, 11 August 1992. © Richard T. Bryant.

Description

Size: length to 68mm.

Shell: moderately thin to thick; with small nodulous chevrons mixed with thin corrugations, occasionally smooth; moderately compressed, width 35%–60% of length; outline rhomboidal to subtriangular; anterior margin rounded; posterior margin broadly rounded to obliquely truncate; dorsal margin convex; ventral margin straight to convex, occasional large individuals slightly concave just anterior to posterior ridge; posterior ridge low, rounded, often obscured by sculpture; posterior slope low, flat to slightly convex, occasionally with thin corrugations; umbo broad, elevated slightly above hinge line; umbo sculpture nodulous ridges; umbo cavity wide, moderately shallow.

Teeth: pseudocardinal teeth thick, 2 divergent teeth in left valve, anterior tooth larger, crest oriented anteroventrally along posterior margin of anterior adductor muscle scar, 1 tooth in right valve, occasionally with accessory denticle anteriorly; lateral teeth short, straight to slightly curved, 2 in left valve, 1 in right valve; interdentum short to moderately long, moderately wide.

Nacre: white to bluish white.

Periostracum: small individuals shiny, large individuals dull to clothlike; small individuals dark olive to brown, large individuals dark brown to black.

Glochidium Description

The glochidium of *Quadrula infucata* is unknown.

Similar Sympatric Species

Quadrula infucata has a distinctive rhomboidal to subtriangular outline, typically with sculpture on the shell disk, which distinguishes it from all sympatric mussels.

Distribution in Florida

Quadrula infucata occurs in Apalachicola and Ochlockonee River basins (Figure 10.82).

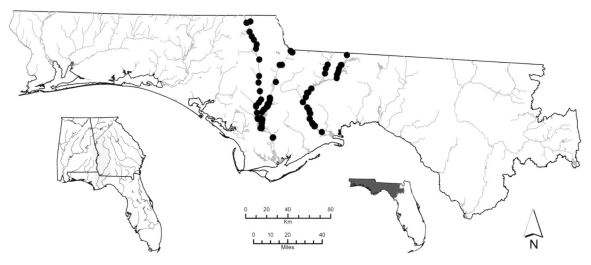

Figure 10.82. Distribution of *Quadrula infucata*.

Ecology and Biology

Quadrula infucata inhabits large creeks and rivers in pools and runs with moderate to swift current in stable substrates of sand, gravel, and rubble. *Quadrula infucata* is rarely found in impounded waters (Brim Box and Williams 2000).

Quadrula infucata is a short-term brooder, gravid in spring and summer (Brim Box and Williams 2000). Glochidial hosts for *Q. infucata* are unknown. Other species of *Quadrula* typically use members of Ictaluridae and Cyprinidae as hosts (Williams et al. 2008; Haag 2012).

Quadrula infucata has been observed to be infested with mites (Unionicolidae). Of 209 individuals examined from four localities in Apalachicola and Ochlockonee River basins, 174 were parasitized by one or more of the following species: *Unionicola hoesei*, *Unionicola sakantaka*, and *Unionicola vamana* (Vidrine 1996).

Conservation Status

Quadrula infucata was assigned a conservation status of special concern throughout its range by Williams et al. (1993) and the equivalent status of vulnerable by Williams et al. (in review).

Quadrula infucata is extant throughout its Florida range and is generally uncommon.

Remarks

Quadrula infucata was reported as present in "considerable quantity" in Apalachicola River at Chattahoochee, Florida, by Heard (1964). Following establishment of the nonindigenous *Corbicula fluminea*, Heard (1975a) reported that the population of *Q. infucata* was greatly reduced at this site. Reduction of the population of *Q. infucata* was likely due to a combination of competition with *C. fluminea* and destabilization of the river channel associated with channel incision following completion of Jim Woodruff Lock and Dam in 1957.

Native Americans harvested large numbers of *Quadrula infucata* from Apalachicola River. In a midden at a Weeden Island archaeological site, Jackson County, Florida, 86 percent of the shells excavated were *Q. infucata* (Percy 1976).

Taxonomic placement of *Quadrula infucata* has varied during the past century. Simpson (1900a, 1914) placed *infucata* in *Quadrula* where it remained until Ortmann and Walker (1922) included it in their new genus *Quincuncina*. Lydeard et al. (2000) and Serb et al. (2003) returned *infucata* to *Quadrula* based on genetic analyses.

Synonymy

Unio infucatus Conrad 1834 (Conrad 1834:45). Type lost. Type locality: published as Flint River, Georgia, restricted by Clench and Turner (1956) to Flint River, Albany, Dougherty County, Georgia, [Flint River drainage]. The restriction of the type locality by Clench and Turner is unjustified and invalid. They presented no evidence to support their designation other than to say that Albany was "probably" the place where the primary type was collected. Based on the route of Conrad's expedition through Georgia in 1833, he crossed the Flint River southwest of Knoxville, Georgia (Wheeler 1935).

Unio securiformis Conrad 1849 (Conrad 1849:152). Type locality: Flint River, Georgia, [Flint River drainage].

Quadrula kleiniana (Lea 1852)
Florida Mapleleaf

Quadrula kleiniana – Upper image: length 58 mm, UF 20341. Alapaha River, 0.3 miles west of Statenville, Echols County, Georgia, 2 July 1969. © Richard T. Bryant. Middle image: length 41 mm, UF 27826. Santa Fe River, 0.4 to 0.5 miles upstream of confluence of Olustee Creek, Alachua and Union counties, Florida, 22 July 1980. © Richard T. Bryant. Lower image: length 27 mm, UF 4006. Santa Fe River at Worthington Springs, Union County, Florida, 26 May 1932. © Richard T. Bryant.

Description
    Size: length to 65 mm.

Shell: moderately thick; typically with corrugations radiating from posterior ridge anteriorly to ventral margin, often interrupted to form pustules, some individuals with little to no sculpturing; moderately compressed to inflated, width 35%–57% of length; outline round to subtriangular; anterior margin rounded; posterior margin broadly rounded to obliquely truncate; dorsal margin convex; ventral margin straight to convex; posterior ridge moderately sharp dorsally, rounded posterioventrally, often obscured by corrugations; posterior slope low, flat to slightly concave, typically with fine corrugations extending from posterior ridge to posteriodorsal margin; umbo broad, moderately inflated, elevated slightly above hinge line; umbo sculpture 3–4 thick, nodulous ridges (Figure 10.83); umbo cavity wide, moderately deep.

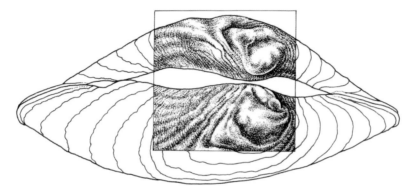

Figure 10.83. Umbo sculpture of *Quadrula kleiniana,* length 16 mm, UF 455459. Santa Fe River at Worthington Springs, Highway 121, Alachua and Union counties, Florida, 9 June 2011. Illustration by Susan Trammell.

Teeth: pseudocardinal teeth moderately thick, triangular, finely striate, 2 divergent teeth in left valve, anterior tooth larger, crest oriented anterioventrally along posterior margin of anterior adductor muscle scar, posterior tooth low, 1 low tooth in right valve, occasionally with accessory denticle anteriorly and/or posteriorly; lateral teeth moderately short, straight to slightly curved, 2 in left valve, 1 in right valve; interdentum short, moderately narrow to wide.

Nacre: white to purplish white.

Periostracum: small individuals shiny, large individuals dull to clothlike; brown to black.

Glochidium Description

Outline subelliptical; length 234–242 μm; height 275–287 μm; dorsal margin 102–105 μm; ventral margin with variable internal and external micropoints consisting of simple, irregularly arranged lanceolate points, individual crowns with 6–9 multiple points, and perpendicular rows of fused crowns (Hoggarth 1999). Hoggarth (1999) illustrated and described glochidia of *Quadrula kleiniana* (as *Quincuncina infucata*) from the Suwannee River, Florida.

Similar Sympatric Species

Quadrula kleiniana differs from all other mussels in Suwannee River basin in having sculpture on the shell disk, thick pseudocardinal teeth, and moderately deep umbo cavity.

Distribution in Florida

Quadrula kleiniana occurs in Suwannee River basin (Figure 10.84).

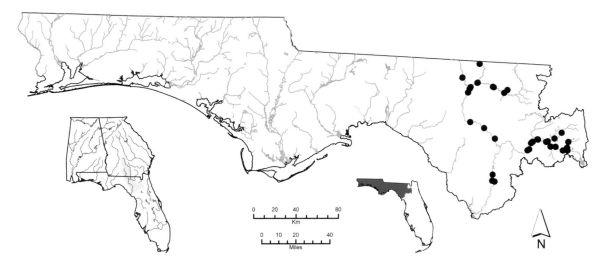

Figure 10.84. Distribution of *Quadrula kleiniana*.

Ecology and Biology

Quadrula kleiniana inhabits creeks and rivers in slow to moderate current in substrates of sand, sandy mud, and fine gravel.

Quadrula kleiniana is a short-term brooder, gravid in April and May. Glochidial hosts for *Q. kleiniana* are unknown. Other species of *Quadrula* typically use members of Ictaluridae and Cyprinidae as hosts (Williams et al. 2008; Haag 2012).

Conservation Status

Quadrula kleiniana was not considered a valid species until recently so was not included in any of the previous conservation assessments. It will be included in an upcoming evaluation, in which it will be considered vulnerable throughout its range (Williams et al. in review).

Quadrula kleiniana is extant throughout its Florida range and is generally uncommon. The species is reduced at some sites in the lower reaches of Suwannee River. Sample sizes in museum collections suggest that it is also less common now than it was historically.

Remarks

Quadrula kleiniana was included in *Quincuncina* until results of phylogenetic analyses placed it in *Quadrula* and distinguished it from *infucata* (Lydeard et al. 2000; Campbell and Lydeard 2012a). The species is endemic to Suwannee River making it the easternmost representative of *Quadrula*. It is most closely related to *Quadrula infucata*, which is allopatrically distributed to the west in Apalachicola and Ochlockonee River basins.

Synonymy

Unio kleinianus Lea 1852 (Lea 1852b:265). Lectotype (Johnson 1974), USNM 84041, length 42 mm. Type locality: Suwannee River, Florida, [Suwannee River basin].

Quadrula succissa (Lea 1852)
Purple Pigtoe

Quadrula succissa – Upper image: length 45 mm, UF 65274. Conecuh River, Searight, Crenshaw County, Alabama, October 1915. © Richard T. Bryant. Lower image: length 41 mm, UF 376881. Shoal River, U.S. Highway 90, about 3.5 miles east of junction of U.S. Highway 90 and State Highway 85 in Crestview, Okaloosa County, Florida, 21 August 1996. Photograph by Zachary Randall.

Description

Size: length to 65 mm, in Florida to 63 mm.

Shell: moderately thin to thick; smooth; usually moderately compressed, width 35%–70% of length, large individuals occasionally inflated; outline round to quadrate; anterior margin rounded; posterior margin rounded to obliquely truncate; dorsal margin slightly convex in small individuals, rounded in large individuals; ventral margin straight to convex; posterior ridge rounded; posterior slope moderately steep, flat to slightly convex; umbo broad, elevated slightly above hinge line; umbo sculpture 3–4 thick ridges, nodulous where they cross posterior ridge; umbo cavity wide, moderately deep.

Teeth: pseudocardinal teeth moderately thick, triangular, finely striate, 2 divergent teeth in left valve, anterior tooth larger, crest oriented anterioventrally along posterior margin of anterior adductor muscle scar, posterior tooth low, 1 low tooth in right valve, occasionally with accessory denticle anteriorly and/or posteriorly; lateral teeth moderately short, straight to slightly curved, 2 in left valve, 1 in right valve; interdentum short, moderately wide.

Nacre: purplish, occasionally white.

Periostracum: typically clothlike, occasionally dull; small individuals dark olive to brown, large individuals dark brown to black.

Glochidium Description

Outline subelliptical; length 192–230 μm; height 269–280 μm; width 126–142 μm; ventral margin with variable internal and external micropoints consisting of simple, irregularly arranged lanceolate points, individual crowns with 6–9 multiple points, and perpendicular rows of fused crowns (Figure 10.85).

Similar Sympatric Species

Quadrula succissa resembles *Fusconaia escambia* but typically has a more rounded posterior ridge, less steep posterior slope, rounded posterior margin, and purplish nacre. The anterior pseudocardinal tooth in the left valve of *Q. succissa* is oriented more ventrally, whereas that of *F. escambia* is oriented more anteriorly.

Quadrula succissa also resembles some *Pleurobema strodeanum* but is less elongate and typically round to quadrate in outline, whereas the latter is oval to subtriangular. The umbo cavity of *Q. succissa* is deep, while that of *P. strodeanum* is shallow. Nacre of *Q. succissa* is purplish, whereas that of *P. strodeanum* is white.

Quadrula succissa may superficially resemble *Reginaia rotulata* but is more compressed, has a less rounded posterior ridge, and clothlike periostracum.

Distribution in Florida

Quadrula succissa occurs in Escambia, Yellow, and Choctawhatchee River basins (Figure 10.86).

Ecology and Biology

Quadrula succissa occurs in medium creeks to rivers in slow to moderate current in sand or sandy mud substrates.

Quadrula succissa is a short-term brooder, gravid April to June. Glochidial hosts for *Q. succissa* are unknown. Other species of *Quadrula* typically use members of Ictaluridae and Cyprinidae as hosts (Williams et al. 2008; Haag 2012).

Quadrula succissa has been observed to be infested with mites (Unionicolidae). Of 43 individuals examined from two localities in Escambia and Choctawhatchee River basins, 37 were parasitized by *Unionicola sakantaka* (Vidrine 1996).

Conservation Status

Quadrula succissa was assigned a conservation status of special concern throughout its range by Williams et al. (1993). It was assigned the status of currently stable by Williams et al. (in review).

Quadrula succissa is extant throughout its Florida range and is moderately common.

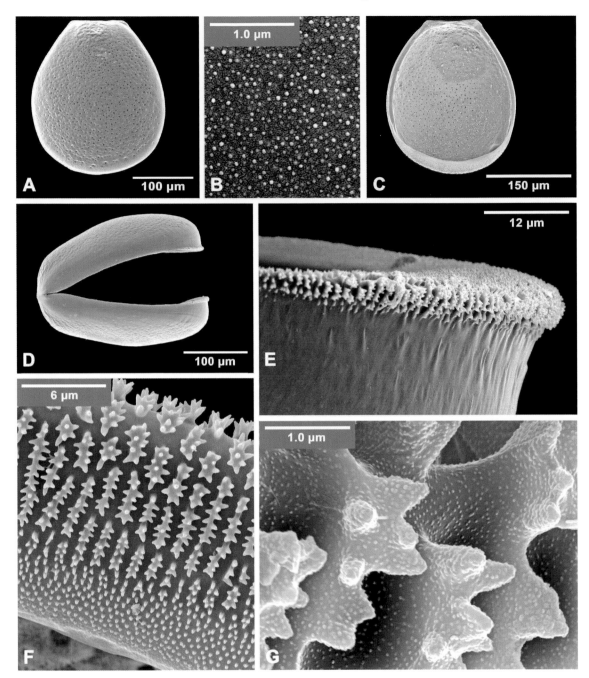

Figure 10.85. Scanning electron micrographs of glochidial valves of *Quadrula succissa* from Pea River at confluence of Choctawhatchee River, Geneva County, Alabama, 22 April 2011—(A) external view; (B) surface sculpture; (C) internal view; (D) lateral view; (E) lateral view of micropoints along ventral margin; (F) internal view of micropoints on ventral margin; and (G) close-up of micropoints.

Remarks

For most of the 1900s, *Quadrula succissa* was recognized as a species of *Fusconaia*. A comparative study of mtDNA found *succissa* to be most closely related to species of *Quadrula*, not *Fusconaia* (Lydeard et al. 2000). This study also found that *Q. succissa* was most closely related to *Quadrula infucata*. The presence of darkly pigmented gills in females of both species also supports this conclusion.

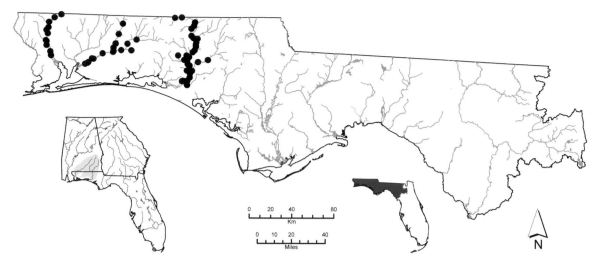

Figure 10.86. Distribution of *Quadrula succissa*.

Synonymy

Unio succissus Lea 1852 (Lea 1852b:275). Holotype by monotypy, USNM 84574, length 43 mm. Type locality: published as west Florida, restricted by Clench and Turner (1956) to Choctawhatchee River, Caryville, Holmes County, Florida, [Choctawhatchee River basin].

Unio cacao Lea 1859 (Lea 1859b:154). Type locality: Chacktahatchie [Choctawhatchee] River, west Florida, [Choctawhatchee River basin].

Quadrula wrightii Simpson 1914 (Simpson 1914:868). Type locality: Pine Barren Creek, Escambia County, Florida, [Escambia River basin].

Genus *Reginaia* Campbell and Lydeard 2012

Reginaia was recently described based on a phylogenetic analysis of the genera of the subfamily Ambleminae (Campbell and Lydeard 2012a). They removed three species from *Fusconaia* and assigned them to *Reginaia*. Their analysis established the distinctiveness of *Reginaia*; however, it was not clearly assignable to any recognized tribe within the subfamily Ambleminae (Campbell and Lydeard 2012a).

Reginaia is distributed in Mississippi River basin and Gulf Coast rivers from Louisiana to Florida.

Type Species
Unio ebenus Lea 1831 = *Reginaia ebenus* (Lea 1831)

Diagnosis
Shell thick; smooth; outline oval to round; umbo moderately to highly inflated, elevated slightly to well above hinge line; umbo sculpture a few weak ridges; posterior ridge rounded; periostracum shiny or dull, brown to black, without rays; pseudocardinal and lateral teeth well developed, 2 in left valve, 1 in right valve; umbo cavity deep; nacre white.

Inner lamellae of inner gills only connected to visceral mass anteriorly; excurrent aperture smooth or with crenulations or small papillae; mantle bridge separating excurrent and supra-anal apertures very short or absent; all 4 gills marsupial; glochidia held across entire gill; marsupium not extended beyond original gill margin when gravid; ova and embryos bright pink, becoming white as glochidia mature; glochidium depressed subelliptical, without styliform hooks (Williams et al. 2008; Campbell and Lydeard 2012a).

Reginaia apalachicola (Williams and Fradkin 1999)
Apalachicola Ebonyshell

Reginaia apalachicola – Upper image: length 34 mm, UF 358659 (left valve only; outside image reversed). Mouth of Omusee Creek [Chattahoochee River,] Omusee Park, Houston County, Alabama. © Richard T. Bryant. Lower image: length 40 mm, UF archaeological collection (left valve only; outside image reversed). Archaeological Site 8LI76, 500 meters east of Apalachicola River (T1N; R8W; SE ¼ Sec. 1) near river mile 88, about 5 miles north of Bristol, Liberty County, Florida. © Richard T. Bryant.

Description
   Size: length to 50 mm.
   Shell: moderately thick; smooth; compressed; outline round; anterior margin broadly rounded; posterior margin broadly rounded; ventral margin broadly rounded; dorsal margin rounded; posterior ridge rounded, indistinct; posterior slope flat to slightly convex; umbo broad, moderately inflated, elevated slightly above hinge line; umbo sculpture unknown; umbo cavity moderately wide, deep.
   Teeth: pseudocardinal teeth thick, low, triangular, 2 divergent teeth in left valve, 1 tooth in right valve; lateral teeth short to moderately long, straight to slightly curved, 2 in left valve, 1 in right valve; interdentum short, wide.

Nacre: unknown.

Periostracum: unknown.

Glochidium Description

The glochidium of *Reginaia apalachicola* is unknown.

Similar Sympatric Species

Reginaia apalachicola may resemble young *Glebula rotundata* but can be distinguished by the presence of a wide interdentum, deep umbo cavity, and pseudocardinal teeth that are thick and triangular, not radially arranged serrate ridges.

Distribution in Florida

Reginaia apalachicola is known only from Apalachicola River (Figure 10.87).

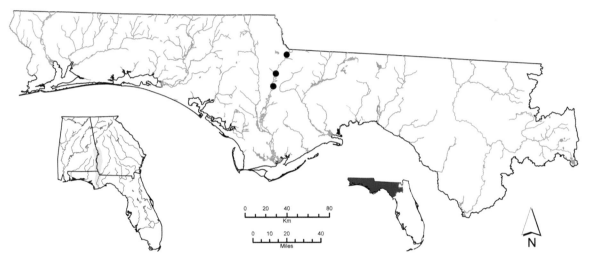

Figure 10.87. Distribution of *Reginaia apalachicola*.

Ecology and Biology

All archaeological excavations containing *Reginaia apalachicola* are adjacent to large rivers with moderate current in the Apalachicola basin. Currently, substrates in these rivers consist of a mixture of clay, silty sand, sand and gravel, and occasionally bedrock.

Reginaia apalachicola was likely a short-term brooder based on what is known about a related species, *Reginaia ebenus*. Glochidial hosts for *R. apalachicola* are unknown.

Conservation Status

Reginaia apalachicola was recently described and has not been previously evaluated. It was considered to be extinct by Williams and Fradkin (1999) and Williams et al. (in review).

Remarks

Reginaia apalachicola was originally placed in *Fusconaia*. Phylogenetic analysis of the Pleurobemini by Campbell and Lydeard (2012a, 2012b) recovered a clade containing *Fusconaia ebenus* and *Fusconaia rotulata* outside of the remainder of *Fusconaia*. Based on this, the new genus *Reginaia* was erected for *F. ebenus* and *F. rotulata*. Similarity of *apalachicola* shell characteristics to the other two *Reginaia* species led to its placement in the new genus. Based on shell morphology *R. apalachicola* appears to be most closely related to *Reginaia rotulata*, an Escambia River basin endemic (Williams et al. 2008).

Reginaia apalachicola was apparently endemic to large rivers in Apalachicola River basin. In Florida the species is known only from three archaeological sites along Apalachicola River in Gadsden, Jackson, and Liberty counties. Archaeological samples were from the Weeden Island culture period and ranged in age from A.D. 500–1000. There is no reason to believe that *R. apalachicola* was not extant when Europeans arrived in North America. Extinction of *R. apalachicola* may have been the result of habitat destruction that occurred in upper and middle reaches of Apalachicola basin beginning in late 1800s (Glenn 1911; Brim Box and Williams 2000). There is a single collection from outside of Florida, Chattahoochee River at the mouth of Omuscc Creek in Houston County, Alabama (Williams et al. 2008). *Reginaia apalachicola* has been misidentified as *Quincuncina infucata* (= *Quadrula infucata*) in the archaeological literature (Milanich 1974; Percy 1976).

The Florida archaeological sites where *Reginaia apalachicola* were found also contained 13 other unionids—*Amblema neislerii, Elliptio crassidens, Elliptio pullata, Elliptoideus sloatianus, Glebula rotundata, Lampsilis straminea, Lampsilis floridensis, Medionidus penicillatus, Megalonaias nervosa, Pleurobema pyriforme, Quadrula infucata, Villosa lienosa,* and *Villosa vibex.*

Reginaia apalachicola was named in reference to the species' restricted distribution to Apalachicola basin (Williams and Fradkin 1999).

Synonymy

Fusconaia apalachicola Williams and Fradkin 1999 (Williams and Fradkin 1999:54). Holotype, UF 05260690.1 (right valve only; archaeological collection), length 40 mm. Type locality: Archeological Site 8LI76, located 500 meters east of the Apalachicola River (T1N; R8W; SE 1/4 Sec. 1) near river mile 88 (USACE), about 5 miles north of Bristol, Liberty County, Florida, [Apalachicola River basin]. © Richard T. Bryant.

Reginaia rotulata (B.H. Wright 1899)
Round Ebonyshell

Reginaia rotulata – Upper image: length 61 mm, Doug N. Shelton personal collection, catalogue number 4997.1. Cone-cuh River, 1 mile upstream of State Highway 41, south of East Brewton, Escambia County, Alabama, 17 August 1996. © Richard T. Bryant. Lower image: length 50 mm, UF 358660. Escambia River, boat ramp 0.8 miles southeast of Bluff Springs, Escambia County, Florida, 29 June 1995. © Richard T. Bryant.

Description

Size: length to 70 mm, in Florida to 54 mm.

Shell: thick; smooth; moderately inflated, width 45%–60% of length; outline typically round, occasionally oval; anterior margin rounded; posterior margin rounded; dorsal margin rounded; ventral margin broadly rounded; posterior ridge rounded, indistinct; posterior slope convex; umbo broad, inflated, elevated slightly above hinge line; umbo sculpture unknown; umbo cavity moderately wide, deep.

Teeth: pseudocardinal teeth thick, low, triangular, 2 divergent teeth in left valve, anterior tooth smaller, 1 tooth in right valve; lateral teeth short, straight to slightly curved, 2 in left valve, 1 in right valve; interdentum short, wide.

Nacre: white, usually iridescent.

Periostracum: shiny to dull; dark olive brown to black.

Glochidium Description

The glochidium of *Reginaia rotulata* is unknown.

Similar Sympatric Species

Reginaia rotulata superficially resembles small *Glebula rotundata* but has a wide interdentum and deep umbo cavity. Also, the pseudocardinal teeth of *R. rotulata* are triangular, whereas those of *G. rotundata* are comprised of radially arranged serrate ridges.

Reginaia rotulata may resemble *Quadrula succissa* but is more inflated and has a more rounded posterior ridge.

Distribution in Florida

Reginaia rotulata occurs in Escambia River (Figure 10.88).

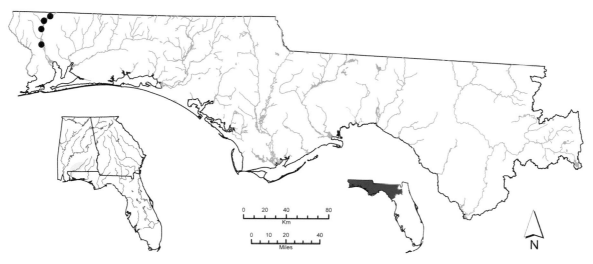

Figure 10.88. Distribution of *Reginaia rotulata*.

Ecology and Biology

Reginaia rotulata occurs in rivers in slow to moderate current in substrates of gravel, sand, and occasionally sandy mud.

Reginaia rotulata is presumably a short-term brooder, gravid in spring and summer. Glochidial hosts for *R. rotulata* are unknown. A related species, *Reginaia ebenus*, appears to be a host specialist on *Alosa* spp., migratory members of the Clupeidae (Surber 1913; Howard 1914b).

Conservation Status

In all previous assessments this species was included as *Fusconaia rotulata*. *Reginaia rotulata* was considered endangered throughout its range by Williams et al. (1993, in review). In Florida it was assigned a conservation status of endangered by Williams and Butler (1994). *Reginaia rotulata* was listed as endangered with critical habitat designated under the authority of the federal ESA in 2012 (USFWS 2012). Critical habitat included a total of 558 km of streams in Alabama and Florida. In Florida most of the main channel of the Escambia River was designated as critical habitat.

Reginaia rotulata is extant throughout its Florida range but is extremely rare.

Remarks

Reginaia rotulata was described in the genus *Unio* but was subsequently placed in *Obovaria* (Frierson 1927; Johnson 1967b; Turgeon et al. 1988, 1998). Williams and Butler (1994) placed this species in the genus *Fusconaia* based on shell morphology. Phylogenetic analysis of the Pleurobemini by Campbell and Lydeard (2012a, 2012b) recovered a clade containing *Fusconaia ebenus* and *Fusconaia rotulata* outside of the remainder of *Fusconaia*. Based on this analysis, the new genus *Reginaia* was erected for *F. ebenus* and *F. rotulata*.

Reginaia rotulata is the only mussel species known to be endemic to Escambia River basin.

Synonymy

Unio rotulatus B.H. Wright 1899 (B.H. Wright 1899:22). Holotype by monotypy, USNM 159969, length 49 mm. Type locality: Escambia River, Escambia County, Florida, [Escambia River basin].

Genus *Toxolasma* Rafinesque 1831

Toxolasma was recognized as having eight species by Turgeon et al. (1998). An additional undescribed species, *Toxolasma* sp. cf. *corvunculus*, which occurs in Escambia, Yellow, and Choctawhatchee River basins in Alabama and Florida, was recognized by Williams et al. (2008). Genetic analysis using mtDNA suggests there is another distinct lineage in Choctawhatchee River and upper portion of Chipola River (N.A. Johnson unpublished data).

Rafinesque (1831) introduced the genus name *Toxolasma* but did not designate a gender, and it was not clear from the species included in the genus. Valentine and Stansbery (1971) first recognized *Toxolasma* as the senior synonym of *Carunculina*, but they did not address the gender question. A recent review determined that the gender of *Toxolasma* is neuter. While three species—*Toxolasma corvunculus, Toxolasma cylindrellus*, and *Toxolasma pullus*—required no change from the current spelling, endings for four species were changed: *Toxolasma lividus* to *lividum*; *Toxolasma parvus* to *parvum*; *Toxolasma paulus* to *paulum*; and *Toxolasma texasensis* to *texasense* (Lee 2006).

Toxolasma are among the smallest of North American unionids, some rarely exceeding 40 mm in length. The genus is characterized by caruncles on the mantle margin, which move like "twiddling thumbs" that attract fish to eat the "wriggling red worms" (Morrison 1973). Host fishes probably include sunfishes of the Centrarchidae (Stern and Felder 1978; Watters et al. 2009).

Toxolasma occurs in Great Lakes and Mississippi River basins and Gulf Coast rivers from Texas to Florida. It also occurs in southern Atlantic Coast rivers from North Carolina to Florida.

Type Species
Unio lividus Rafinesque 1831 = *Toxolasma lividum* (Rafinesque 1831)

Diagnosis
Shell thin to moderately thick; smooth; somewhat compressed to inflated; outline oval to elliptical; females usually sexually dimorphic, often more truncate posteriorly than males; umbo sculpture thick, single-looped ridges; periostracum dull to clothlike, yellowish green, dark olive, brown to black, with or without green rays.

Mantle margin just ventral to incurrent aperture papillate; 1 caruncle per mantle, reduced or absent in males; inner lamellae of inner gills connected to visceral mass only anteriorly; outer gills marsupial; glochidia held in posterior portion of gill; marsupium extended beyond original gill margin when gravid; glochidium outline subelliptical, without styliform hooks (Ortmann 1912; Simpson 1914; Britton and Fuller 1980).

Toxolasma parvum (Barnes 1823)
Lilliput

Toxolasma parvum – Upper image: length 29 mm, UF 370407. Commercial sport and bait fish ponds near Elliotts Creek, about 2 miles southwest of Moundville, Hale County, Alabama, August 1997. © Richard T. Bryant. Lower image: length 19 mm, UF 449289. Apalachicola River at river mile 46.8, Gulf County, Florida, 7 August 2006. © Richard T. Bryant.

Description

Size: length to 40 mm.

Shell: moderately thin; smooth; moderately inflated; width 35%–45% of length; outline oval to elliptical; anterior margin rounded; posterior margin rounded, large females obliquely truncate; dorsal margin straight to slightly convex; ventral margin straight to slightly convex; sexual dimorphism subtle, with females slightly more inflated posteriorly than males; posterior ridge rounded; posterior slope moderately steep, slightly concave; umbo broad, moderately inflated; umbo sculpture moderately thick, single-looped ridges; umbo cavity shallow.

Teeth: pseudocardinal teeth small, compressed, 2 teeth in left valve, crests almost parallel to shell margin, 1 tooth in right valve, often with accessory denticle anteriorly; lateral teeth thin, straight to slightly curved, 2 in left valve, 1 in right valve; interdentum short, very narrow to narrow.

Nacre: white to bluish white, iridescent.

Periostracum: typically clothlike, occasionally dull; small individuals greenish yellow to dark brown, large individuals black, typically without rays.

Glochidium Description

Outline subelliptical; length 170–180 µm; height 200 µm (Surber 1915; Utterback 1916a; Ortmann 1919).

Similar Sympatric Species

Toxolasma parvum most closely resembles other *Toxolasma* species and is difficult to distinguish based on shell characters alone. The best character to distinguish live female *T. parvum* from all native Florida *Toxolasma* is the white marsupial gill, compared to the dark gray to black margin in both gravid and nongravid females of the other species. Pseudocardinal teeth of *T. parvum* are typically thinner than those of *Toxolasma paulum*. Many populations of *T. parvum* are not sexually dimorphic or if they are it is far more subtle than in *T. paulum*.

Toxolasma parvum resembles some small *Villosa lienosa* but has white nacre, while that of *V. lienosa* is usually purplish. Live female *T. parvum* have a well-developed caruncle, which is absent in *V. lienosa*. Caruncles are poorly developed in male *T. parvum* but may be visible as a small darkly pigmented bump on the mantle fold.

Toxolasma parvum may resemble small *Villosa villosa* but differs in having a well-developed caruncle in females, visible as a small darkly pigmented bump on the mantle fold in males. The umbo sculpture of *T. parvum* consists of single-looped ridges instead of being strongly double looped as those of *V. villosa*.

Distribution in Florida

Toxolasma parvum is nonindigenous in Florida and occurs in Apalachicola and Ochlockonee River basins (Figure 10.89).

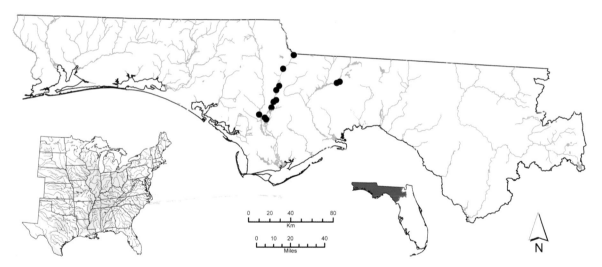

Figure 10.89. Distribution of *Toxolasma parvum*.

Ecology and Biology

Toxolasma parvum occurs in creeks, rivers, ponds, and lakes. It is also found in artificial habitats such as reservoirs, borrow pits, and fish culture ponds. It typically inhabits areas with little or no current in substrates of mud and sand. It is often found along the shore where it tracks water level changes.

Toxolasma parvum is a long-term brooder but with a highly variable brooding period (Williams et al. 2008). Histological examinations of gonadal tissues of some populations of *T. parvum* revealed hermaphroditism (van der Schalie 1970). Sexual dimorphism is subtle, often evident only in very large individuals (Ortmann 1919).

A combination of natural infestation observations and laboratory infestations have been used to identify *Toxolasma parvum* hosts. These investigations included five species of fishes belonging to the family Centrarchidae and one species belonging to Percidae (OSU Museum of Biological Diversity 2012).

A natural infestation of glochidia of *T. parvum* was reported for one fish that occurs in Florida—Centrarchidae: *Lepomis gulosus* (Warmouth) (OSU Museum of Biological Diversity 2012).

In laboratory trials *Toxolasma parvum* glochidia metamorphosed on five fishes that occur in Florida—Centrarchidae: *Lepomis cyanellus* (Green Sunfish), *Lepomis gulosus* (Warmouth), *Lepomis humilis* (Orangespotted Sunfish), *Lepomis macrochirus* (Bluegill), and *Pomoxis annularis* (White Crappie) (OSU Museum of Biological Diversity 2012).

Conservation Status

Toxolasma parvum was assigned a conservation status of currently stable throughout its range by Williams et al. (1993, in review). *Toxolasma parvum* is a nonindigenous species and has no assigned conservation status in Florida.

Toxolasma parvum is extant throughout its current range in Florida where it is relatively common.

Remarks

Toxolasma parvum is not native in Gulf Coast basins east of Mobile basin (Williams et al. 2008). The general expansion of its range and occurrence in Florida is most likely due to the movement of fishes with encysted glochidia. Records of its introduction via sport fishes have been verified in Alabama and Georgia (Williams et al. 2008). It is likely to continue to increase its distribution and abundance in Florida and elsewhere considering the ease of transfer via glochidia on its host fishes. One of the fish hosts for *T. parvum*, *Lepomis macrochirus* (Bluegill), is among the most widely distributed cultured sport fishes in the eastern United States.

Clench and Turner (1956) considered all Florida *Toxolasma* to be *T. paulum*. Conversely, Heard (1979a) considered all *Toxolasma* in Florida to be *T. parvum* (as *Carunculina parva*).

Synonymy

Unio parvus Barnes 1823 (Barnes 1823:174). Type lost. Type locality: Fox River, Wisconsin [Great Lakes or Mississippi basin].

Toxolasma paulum (Lea 1840)
Iridescent Lilliput

Toxolasma paulum – Upper image: female, length 34 mm, UF 455460. Boat basin off Suwannee River in the Original Suwannee River Campground, about 1 mile west of Fanning Springs, Dixie County, Florida, 23 January 2008. © Richard T. Bryant. Middle image: male, length 37 mm, UF 455460. Boat basin off Suwannee River in the Original Suwannee River Campground, about 1 mile west of Fanning Springs, Dixie County, Florida, 23 January 2008. © Richard T. Bryant. Lower image: length 24 mm, UF 455460. Boat basin off Suwannee River in the Original Suwannee River Campground, about 1 mile west of Fanning Springs, Dixie County, Florida, 23 January 2008. © Richard T. Bryant.

Description

Size: length to 49 mm, in Florida to 42 mm.

Shell: thin to moderately thick; smooth; moderately inflated, width 35%–50% of length; outline elliptical; anterior margin rounded; posterior margin obliquely truncate in females, narrowly to broadly rounded in males; dorsal margin slightly convex; ventral margin straight to convex; posterior ridge rounded; posterior slope moderately steep, flat to slightly convex; umbo broad, moderately inflated, elevated slightly above hinge line; umbo sculpture single-looped ridges; umbo cavity wide, shallow.

Teeth: pseudocardinal teeth moderately thick, triangular, 2 divergent teeth in left valve, occasionally joined at base, 1 tooth in right valve; lateral teeth short to moderately long, thin to moderately thick, usually straight, 2 in left valve, 1 in right valve; interdentum short, narrow to moderately wide.

Nacre: white to bluish white, occasionally purple, often with salmon tint in umbo cavity.

Periostracum: shiny to clothlike; small individuals greenish yellow to olive, often with green rays, large individuals dark olive, brown, or black, typically obscuring rays.

Glochidium Description

Outline subelliptical; length 166–188 µm; height 190–211 µm; width 108–128 µm; ventral margin with internal and external irregularly arranged lanceolate micropoints (Figure 10.90).

Similar Sympatric Species

Toxolasma paulum resembles *Toxolasma parvum*, but many populations of *T. parvum* are not sexually dimorphic, or if they are it is far more subtle than in *T. paulum*. Also, pseudocardinal teeth of *T. paulum* are typically thicker than those of *T. parvum*.

Toxolasma paulum resembles some small *Villosa lienosa* but are more inflated and have thicker pseudocardinal teeth. Live female *T. paulum* have caruncles on the mantle margin just ventral to the incurrent aperture, whereas *V. lienosa* only have a papillate mantle fold. Caruncles and papillate mantle folds are poorly developed in males of both species.

Toxolasma paulum may resemble small *Villosa amygdalum* and *Villosa villosa*, but its umbo sculpture consists of single-looped ridges instead of being strongly double looped. Also, the pseudocardinal teeth of *Toxolasma* are typically much thicker than those of *V. amygdalum* and *V. villosa*.

Distribution in Florida

Toxolasma paulum occurs in most major basins from Econfina Creek to St. Johns River and peninsular Florida to Lake Okeechobee (Figure 10.91).

Ecology and Biology

Toxolasma paulum typically occurs in creeks, rivers, and occasionally lakes with little or no current in sand and sandy mud substrates. It is often found along the shore where it tracks water level changes.

Toxolasma paulum is a long-term brooder, with gravid females reported from May to July (Brim Box and Williams 2000). Mature ova and spermatozoa have been reported to be present in the gonad throughout the year (Heard 1969). Glochidial hosts for *T. paulum* are unknown but presumably include sunfishes of the family Centrarchidae.

Toxolasma paulum has been observed to be infested with mites (Unionicolidae). Of 170 individuals examined from six localities in north Florida river basins (Choctawhatchee to Suwannee), 50 were parasitized by one or more of the following species: *Unionicola abnormipes*, *Unionicola formosa*, *Unionicola latipalpa*, *Unionicola sakantaka*, and *Unionicola vamana* (Vidrine 1996).

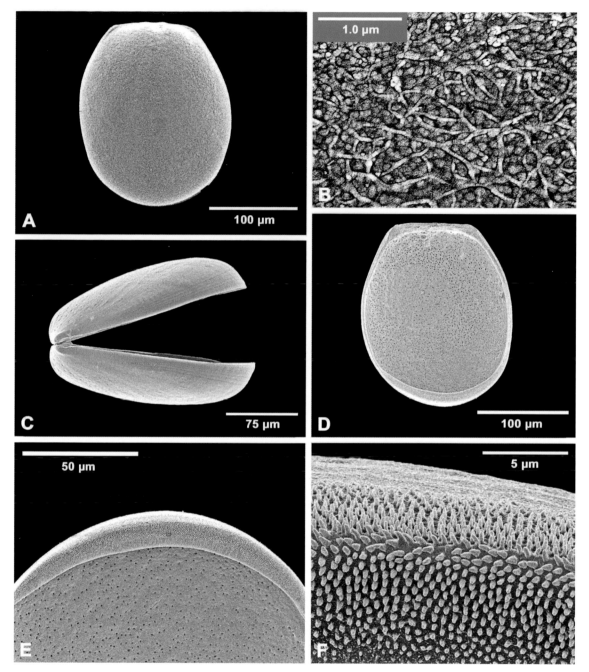

Figure 10.90. Scanning electron micrographs of glochidial valves of *Toxolasma paulum* from Flint River, south side of Newton, Baker and Mitchell counties, Georgia, 8 August 2007—(A) external view; (B) surface sculpture; (C) lateral view; (D) internal view; (E) internal view of ventral margin and micropoints; and (F) close-up of micropoints.

Conservation Status

Toxolasma paulum was assigned a conservation status of currently stable throughout its range by Williams et al. (1993, in review).

Toxolasma paulum is extant throughout most of its Florida range and is generally common. Its abundance has decreased from much of Apalachicola River downstream of Jim Woodruff Lock and Dam possibly due to rapidly fluctuating water levels associated with power generation.

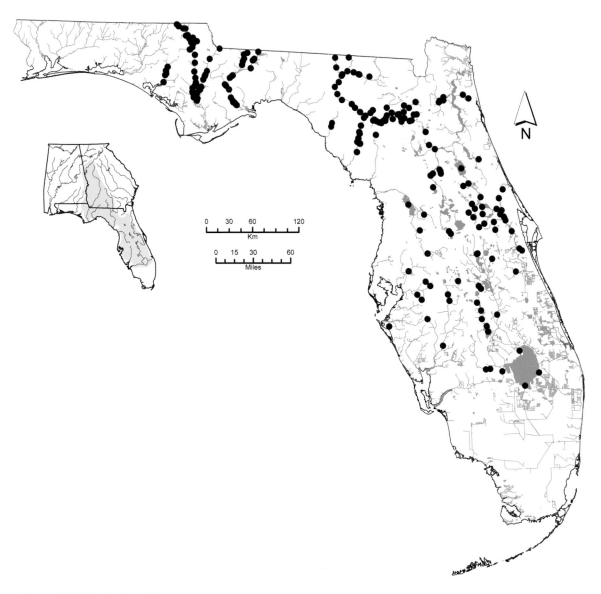

Figure 10.91. Distribution of *Toxolasma paulum*.

Remarks

The taxonomic status of *Toxolasma* species in the eastern Gulf Coast basins remains poorly understood. While *Toxolasma paulum* was recognized as a distinct species by some authors (Clench and Turner 1956), others have treated it as a synonym of *Toxolasma parvum* (Johnson 1972; Heard 1979a); however, *T. paulum* does appear to be a valid species. The range of *T. paulum* given by Clench and Turner (1956) included Yellow River to peninsular Florida. All *Toxolasma* in Florida were considered by Heard (1979a) to be *T. parvum* (as *Carunculina parva*), a species that has been introduced in

Apalachicola and Ochlockonee River basins. The Choctawhatchee River basin has not been considered to be part of the range of *T. paulum* (Brim Box and Williams 2000; Williams et al. 2008). However, recent genetic analysis has revealed a cryptic lineage in the upper Chipola River drainage and Choctawhatchee River basin exclusive of Pea River drainage. This putative cryptic species appears more closely related to *T. paulum* than *Toxolasma* sp. cf. *corvunculus*, which occurs in Escambia, Yellow, and Pea Rivers (N.A. Johnson unpublished data). Additional research is needed to determine the taxonomic status and distributional limits of this putative cryptic species.

Synonymy

Unio paulus Lea 1840 (Lea 1840:287). Holotype by monotypy, USNM 85274, length 24 mm. Type locality: Chattahoochee River, Columbus, [Muscogee County,] Georgia, [Chattahoochee River drainage].

Unio minor Lea 1843 (Lea 1843:[one page privately published]). Type locality: Lake George, [Putnam and Volusia counties,] Florida, [St. Johns River basin].

Unio cromwellii Lea 1865 (Lea 1865:89). Type locality: Kiokee Creek near Albany, Dougherty County, Georgia, [Flint River drainage].

Unio marginis Lea 1865 (Lea 1865:89). Type locality: Blue Springs, [Albany,] Dougherty County, Georgia, [Flint River drainage].

Unio corvinus Lea 1868 (Lea 1868a:144). Type locality: Flint River, Georgia, [Flint River drainage].

Unio corvunculus Lea 1868 (Lea 1868a:144). Type locality: Flint River, Georgia, [Flint River drainage].

Toxolasma sp. cf. *corvunculus*
Gulf Lilliput

Toxolasma sp. cf. *corvunculus* – Upper image: female, length 30 mm, UF 69099. Yellow River near Harmony and Andalusia, [County Route 70,] Covington County, Alabama. © Richard T. Bryant. Lower image: male, length 28 mm, UF 69101. Conecuh River, Bozeman's Landing near Crenshaw County line, Covington County, Alabama, July 1915. © Richard T. Bryant.

Description

Size: length to 41 mm, in Florida to 36 mm.

Shell: thin to moderately thick; smooth; moderately inflated, width 35%–50% of length, females more inflated posteriorly than males; outline oval to elliptical; anterior margin rounded; posterior margin broadly rounded in females, more narrowly rounded in males, large females obliquely truncate; dorsal margin slightly convex; ventral margin straight to convex; posterior ridge rounded; posterior slope moderately steep, flat to slightly convex; umbo broad, moderately inflated, elevated slightly above hinge line; umbo sculpture single-looped ridges (Figure 10.92); umbo cavity wide, shallow.

Teeth: pseudocardinal teeth thick, triangular, 2 teeth in left valve, occasionally joined at base, anterior tooth moderately compressed, 1 tooth in right valve; lateral teeth short to moderately long, straight to slightly curved, 2 in left valve, 1 in right valve; interdentum moderately long, narrow.

Nacre: bluish white to salmon, iridescent posteriorly.

Periostracum: shiny to clothlike; small individuals yellowish green to olive brown, occasionally 2–3 dark green rays on posterior slope, large individuals dark brown, typically without rays.

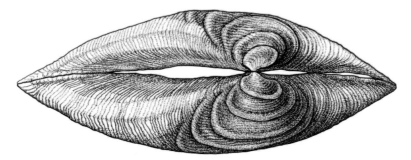

Figure 10.92. Umbo sculpture of *Toxolasma* sp. cf. *corvunculus*, length 14 mm, UF 449118. Escambia River, about 0.3 miles above Chumuckla Springs boat ramp, Santa Rosa County, Florida, 24 September 2008. Illustration by Susan Trammell.

Glochidium Description

The glochidium of *Toxolasma* sp. cf. *corvunculus* is unknown.

Similar Sympatric Species

Toxolasma sp. cf. *corvunculus* resembles *Obovaria choctawensis* but is more elongate. Live female *Toxolasma* sp. cf. *corvunculus* have caruncles, which are absent in *O. choctawensis*.

Toxolasma sp. cf. *corvunculus* may also resemble *Villosa lienosa* of similar length but are generally more inflated. Live female *Toxolasma* sp. cf. *corvunculus* also differ from *V. lienosa* by the presence of caruncles, which are absent in *V. lienosa*.

Toxolasma sp. cf. *corvunculus* may resemble small *Villosa villosa* but has umbo sculpture comprised of single-looped ridges instead of being strongly double looped. Also, the teeth of native *Toxolasma* in Florida are typically much thicker than those of *V. villosa*.

Distribution in Florida

Toxolasma sp. cf. *corvunculus* occurs in Escambia, Yellow, and Choctawhatchee River basins (Figure 10.93).

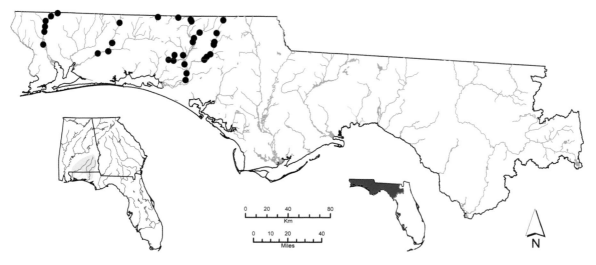

Figure 10.93. Distribution of *Toxolasma* sp. cf. *corvunculus*.

Ecology and Biology

Toxolasma sp. cf. *corvunculus* typically occurs in medium creeks to rivers in slow to moderate current in substrates of sand and sandy mud. It inhabits stream margins and occasionally backwater habitats.

Toxolasma sp. cf. *corvunculus* is presumably a long-term brooder, gravid from summer to the following spring or summer. Glochidial hosts for *Toxolasma* sp. cf. *corvunculus* are unknown but presumably include sunfishes of the family Centrarchidae.

Conservation Status

Toxolasma sp. cf. *corvunculus* has not been evaluated during previous conservation status assessments due to its uncertain taxonomic status.

Toxolasma sp. cf. *corvunculus* is extant in Escambia, Yellow, and Choctawhatchee River basins. It is generally common in Escambia basin but is less frequently encountered in Yellow and Choctawhatchee basins.

Remarks

This undescribed species of *Toxolasma* in Gulf Coast basins of west Florida and south Alabama was first recognized as a putative species by Williams et al. (2008) and appears most closely related to *Toxolasma corvunculus*. Prior to 2008 the species was reported as *Toxolasma paulum* by Clench and Turner (1956) and *Toxolasma parvum* by Heard (1979a).

Synonymy

This species is undescribed.

Genus *Uniomerus* Conrad 1853

Three species were recognized in *Uniomerus* by Turgeon et al. (1998). An additional species, *Uniomerus columbensis*, was elevated from synonymy based primarily on zoogeographic patterns and preliminary genetic data (Williams et al. 2008). A more comprehensive examination of mitochondrial markers in *Uniomerus* from southern Atlantic and eastern Gulf Coast basins further supports the decision to recognize *U. columbensis* (N.A. Johnson unpublished data). *Uniomerus* occurs in the Great Lakes, Mississippi and Ohio basins, and Gulf Coast basins from Texas to Florida. *Uniomerus* also occurs in Atlantic Coast basins from Virginia to Florida.

The variable shell morphology and weak genetic structuring in *Uniomerus* have created considerable confusion concerning the delineation of distribution boundaries and taxonomy in the genus. *Uniomerus* in Florida have been variously recognized under several names—*U. obesus*, *U. tetralasmus*, *U. declivis*, *U. carolinianus*, and *U. columbensis* (Clench and Turner 1956; Johnson 1972; Heard 1979a; Davis 1983; Brim Box and Williams 2000; Williams et al. 2008). Recent preliminary genetic data suggests that three species of *Uniomerus* occur in Florida—*U. carolinianus* from Ochlockonee River to St. Johns River and peninsular Florida; *U. columbensis* in Choctawhatchee River, Econfina Creek, and Apalachicola River; and *U. tetralasmus* in Escambia and Yellow Rivers (N.A. Johnson unpublished data). More detailed analyses are needed to resolve the taxonomic status and distribution of species of *Uniomerus* in Florida and elsewhere.

Species of *Uniomerus* are among the most drought-resistant mussels in North America. The ability of *Uniomerus* to withstand periods of time without water has been observed on numerous occasions (Bosc 1803; Simpson 1892b; Frierson 1903a).

> In Florida Simpson (1892b) reported "I found thousands of the *Blandingianus* form [= *Uniomerus carolinianus*] in a little drain in the piney woods near Braidenton [sic], which was always dry except during the rainy season, and its banks of slightly damp sand were full of dormant specimens. A number of these survived after lying out in the sun for months."

Type Species
Unio excultus Conrad 1838 = *Uniomerus tetralasmus* (Say 1831)

Diagnosis
Shell thin to thick; smooth; outline quadrate to elliptical or subrhomboidal; dorsal and ventral margins almost parallel; umbo even with hinge line or elevated slightly above; umbo sculpture single-looped ridges; periostracum typically clothlike, yellow to brown or black, typically without rays; hinge plate well developed, thin; pseudocardinal teeth compressed, 2 in left valve, 1 in right valve; lateral teeth straight to slightly curved, 2 in left valve, 1 in right valve; nacre bluish white, gray, or pale purple.

Incurrent aperture with arborescent papillae; mantle bridge separating excurrent and supra-anal apertures long; inner lamellae of inner gills connected to visceral mass only anteriorly; outer gills marsupial; glochidia held across entire gill; glochidium outline subelliptical, without styliform hooks (Conrad 1853; Ortmann 1912; Simpson 1914; Britton and Fuller 1980).

Uniomerus carolinianus (Bosc 1801)
Eastern Pondhorn

Uniomerus carolinianus – Upper image: length 58 mm, USNM 85715. St. Johns River, [near St. Augustine,] Florida. Photograph by James D. Williams. Middle image: length 51 mm, USNM 85374. Sink of Noonan's [Newnans] Lake, [Alachua County,] Florida. Photograph by James D. Williams. Lower image: length 21 mm, UF 449437. Santa Fe River at Worthington Springs, Highway 121 bridge, Alachua and Union counties, Florida, 9 June 2011. © Richard T. Bryant.

Description

Size: length to 125 mm.

Shell: moderately thin to thick; smooth; moderately inflated, width 30%–45% of length; outline quadrate to subrhomboidal; anterior margin rounded; posterior margin obliquely truncate to rounded; dorsal margin straight to convex; ventral margin straight to convex, occasional large individuals concave just anterior to posterior ridge; posterior ridge rounded; posterior slope moderately steep, flat to slightly concave; umbo broad, slightly inflated, elevated slightly above hinge line; umbo sculpture single-looped ridges (Figure 10.94); umbo cavity wide, shallow.

Teeth: pseudocardinal teeth moderately thick, compressed, 2 teeth in left valve, crests almost parallel to shell margin, 1 triangular tooth in right valve; lateral teeth short, moderately thin, straight to slightly curved, 2 in left valve, 1 in right valve; interdentum moderately long, narrow.

Nacre: bluish white, gray, or pale purple.

Periostracum: typically clothlike, occasionally dull; yellowish green, olive brown to black, small individuals occasionally with faint green rays.

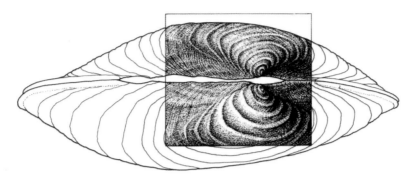

Figure 10.94. Umbo sculpture of *Uniomerus carolinianus*, length 23 mm, UF 441122. Canal off Withlacoochee River [southern], at the county boat ramp off Highway 39, Citrus County, Florida, 6 February 2008. Illustration by Susan Trammell.

Glochidium Description

The glochidium of *Uniomerus carolinianus* is unknown.

Similar Sympatric Species

Uniomerus carolinianus most closely resembles several species of *Elliptio* but differs in having a clothlike periostracum. Papillae along the incurrent aperture of live *U. carolinianus* are typically arborescent, while those of *Elliptio* are usually simple or bifid. *Uniomerus* umbo sculpture consists of moderately thick, single-looped ridges (Figure 10.94), differentiating it from Florida unionid genera except *Pleurobema* and *Toxolasma*. These two genera can be easily distinguished from *Uniomerus* based on their shell shape and size.

Distribution in Florida

Uniomerus carolinianus occurs from Ochlockonee River to St. Marys River and throughout peninsular Florida (Figure 10.95).

Ecology and Biology

Uniomerus carolinianus occurs in a wide variety of aquatic habitats including marshes, swamps, creeks, spring runs, rivers, ponds, lakes, and tidally influenced lower reaches of rivers. It also occurs in artificial habitats such as roadside ditches, canals, borrow pits, and reservoirs. It inhabits standing backwaters or flowing waters with slow to moderate current, in mud, sandy mud, and sand substrates.

Uniomerus carolinianus appears to be a short-term brooder in Florida, gravid in February and March. Glochidial hosts for *U. carolinianus* are unknown.

Uniomerus carolinianus has been observed to be infested with mites (Unionicolidae). Of 25 individuals examined from two localities in Suwannee and Withlacoochee (southern) Rivers, 15 were parasitized by *Unionicola causeyae* and/or *Unionicola lasallei* (Vidrine 1996).

Conservation Status

Uniomerus carolinianus was considered currently stable throughout its range by Williams et al. (1993, in review).

Uniomerus carolinianus is extant throughout its range in Florida and is generally common.

Figure 10.95. Distribution of *Uniomerus carolinianus*.

Remarks

Heard (1979a) reported two species of *Uniomerus* to occur in Florida, *U. carolinianus* from throughout the state and *U. declivis* from Apalachicola River. Preliminary analysis of mtDNA data suggests that all *Uniomerus* from Ochlockonee River to St. Marys River and throughout peninsular Florida arc *U. carolinianus* (N.A. Johnson unpublished data).

Uniomerus carolinianus, like other species of *Uniomerus*, are known to be extremely tolerant of drought and can survive long periods of desiccation of its habitat. *Uniomerus carolinianus* (as *Uniomerus obesus*) is the only native mussel in Florida reported from subterranean waters. They were found live in Rose Creek Swallet, Columbia County, in water up to 12 m deep and 60 m inside a cave (Franz et al. 1994).

Synonymy

Unio carolinianus Bosc 1801 (Bosc 1801:142). Type lost. Type locality: "en Caroline," the Carolinas.

Unio blandingianus Lea 1834 (Lea 1834a:101). Type locality: St. Johns River, Florida, [St. Johns River basin].

Unio paludicolus Gould 1845 (Gould 1845:53). Type locality: Everglades of Florida, [Everglades basin].

Unio jewettii Lea 1867 (Lea 1867b:81). Type locality: published as Florida, restricted by Lea (1868b) to Sink of Noonan's [Newnans] Lake, [Alachua County,] Florida, [St. Johns River basin].

Unio rivicolus Conrad 1868 (Conrad 1868:280). Type locality: brook near Tampa, [Hillsborough County,] Florida, [Hillsborough River basin].

Uniomerus columbensis (Lea 1857)
Apalachicola Pondhorn

Uniomerus columbensis – Upper image: length 68 mm, UF 399. Mosquito Creek at dam, Chattahoochee, Gadsden County, Florida, 12 September 1954. Photograph by Zachary Randall. Lower image: length 41 mm, UF 366678. Mercers Mill Pond, unnamed tributary to Mill Creek, County Road 12, about 7 air miles south of Oakfield, Worth County, Georgia, 10 June 1992. © Richard T. Bryant.

Description

Size: length to 125 mm.

Shell: moderately thin to thick; smooth; moderately inflated, width 25%–40% of length; outline quadrate to subrhomboidal; anterior margin rounded; posterior margin obliquely truncate to rounded; dorsal margin straight to convex; ventral margin straight to convex, occasional large individuals concave just anterior to posterior ridge; posterior ridge rounded; posterior slope moderately steep, flat to slightly concave; umbo broad, slightly inflated, elevated slightly above hinge line; umbo sculpture single-looped ridges; umbo cavity wide, shallow.

Teeth: pseudocardinal teeth moderately thick, compressed, 2 teeth in left valve, crests almost parallel to shell margin, 1 triangular tooth in right valve; lateral teeth short, moderately thin, straight to slightly curved, 2 in left valve, 1 in right valve; interdentum moderately long, narrow.

Nacre: bluish white, gray, or pale purple.

Periostracum: typically clothlike, occasionally dull; yellowish green to olive brown or black, small individuals occasionally with faint green rays.

Glochidium Description

The glochidium of *Uniomerus columbensis* is unknown.

Similar Sympatric Species

Uniomerus columbensis resembles several species of *Elliptio* but differs in having a clothlike periostracum. Incurrent aperture papillae of live *U. columbensis* are arborescent, while those of *Elliptio* are typically simple or bifid. *Uniomerus* umbo sculpture consists of moderately thick, single-looped ridges (Figure 10.94), differentiating it from Florida unionid genera except *Pleurobema* and *Toxolasma*. These two genera can be easily distinguished from *Uniomerus* based on their shell shape and size.

Distribution in Florida

Uniomerus columbensis occurs in Choctawhatchee River, Econfina Creek, and Apalachicola River basins (Figure 10.96).

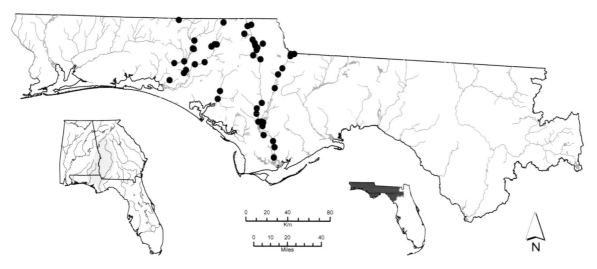

Figure 10.96. Distribution of *Uniomerus columbensis*.

Ecology and Biology

Uniomerus columbensis inhabits swamps, creeks, rivers, sloughs, ponds, and lakes. It is found in substrates of mud, sandy mud, and sand. It also occurs in pockets and crevices of limestone bedrock. *Uniomerus columbensis*, like other species of *Uniomerus*, are known to be extremely tolerant of drought and can survive long periods of desiccation of its habitat.

Uniomerus columbensis is presumably a short-term brooder, with gravid females collected in May (Brim Box and Williams 2000). Glochidial hosts for *U. columbensis* are unknown.

Uniomerus columbensis has been observed to be infested with mites (Unionicolidae). Of 11 individuals examined from two localities in Choctawhatchee and Apalachicola River basins, 7 were parasitized by *Unionicola alleni* and/or *Unionicola lasallei* (Vidrine 1996).

Conservation Status

The conservation status of *Uniomerus columbensis* was not evaluated in early assessments since it was only recently recognized as valid. The species was considered to be currently stable rangewide by Williams et al. (in review).

Uniomerus columbensis is extant throughout its Florida range but is generally uncommon.

Remarks

Uniomerus columbensis was elevated from the synonymy of *Uniomerus carolinianus* by Williams et al. (2008). *Uniomerus* in the eastern Gulf Coast basins have been variously recognized as *U. carolinianus* (Heard 1979a; Brim Box and Williams 2000) and/or *U. declivis* (Davis 1983). Johnson (1972) recognized only *Uniomerus tetralasmus* from Florida, considering the genus to be monotypic. *Uniomerus tetralasmus* was reported to occur from Escambia, Yellow, and Choctawhatchee River basins, and *U. columbensis* to be endemic to Apalachicola River basin by Williams et al. (2008).

Preliminary genetic data suggest that the *Uniomerus* population in Choctawhatchee River is distinct from *U. tetralasmus* and possibly distinct from *U. columbensis* (N.A. Johnson unpublished data).

Synonymy

Unio columbensis Lea 1857 (Lea 1857b:31). Lectotype (Johnson 1974), USNM 85360, length 88 mm. Type locality: creeks near Columbus, [Muscogee County,] Georgia, [Chattahoochee River drainage].

Uniomerus tetralasmus (Say 1831)
Pondhorn

Uniomerus tetralasmus – Upper image: length 102 mm, AUM 1489. Calebee Creek, Macon County, Alabama. © Richard T. Bryant. Lower image: length 39 mm, UF 374093. Yellow River, below a spring tributary, north of U.S. Highway 84, Santa Rosa County, Florida, 19 August 1996. © Richard T. Bryant.

Description

Size: length to 135 mm, in Florida to 89 mm.

Shell: moderately thin; smooth; moderately inflated, width 30%–45% of length; outline elliptical to subrhomboidal; anterior margin rounded; posterior margin obliquely truncate to narrowly rounded or bluntly pointed; dorsal margin straight to convex; ventral margin straight to convex, occasional large individuals slightly concave; posterior ridge low, rounded; posterior slope moderately steep, flat to slightly concave; umbo broad, slightly inflated, elevated slightly above hinge line; umbo sculpture single-looped ridges; umbo cavity wide, shallow.

Teeth: pseudocardinal teeth moderately thick, compressed, 2 teeth in left valve, crests almost parallel to shell margin, 1 tooth in right valve; lateral teeth short, moderately thin, straight to slightly curved, 2 in left valve, 1 in right valve; interdentum long, narrow.

Nacre: bluish white, gray, or pale purple.

Periostracum: typically clothlike, occasionally dull; yellowish brown to olive or black, small individuals occasionally with faint green rays.

Glochidium Description

Outline subelliptical; length 150–175 μm; height 200–225 μm (Williams et al. 2008).

Similar Sympatric Species

Uniomerus tetralasmus resembles several species of *Elliptio* but differs in having a clothlike periostracum. Incurrent aperture papillae of *U. tetralasmus* are usually arborescent, while those of *Elliptio* are typically simple or bifid. *Uniomerus* umbo sculpture consists of moderately thick, single-looped ridges (Figure 10.94), differentiating it from Florida unionid genera except *Pleurobema* and *Toxolasma*. These two genera can be easily distinguished from *Uniomerus* based on their shell shape and size.

Distribution in Florida

Uniomerus tetralasmus occurs in Escambia and Yellow River basins (Figure 10.97).

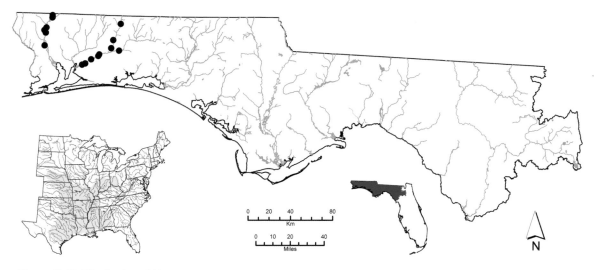

Figure 10.97. Distribution of *Uniomerus tetralasmus*.

Ecology and Biology

Uniomerus tetralasmus is found in swamps, creeks, rivers, ponds, and floodplain lakes in pools or moderate current in mud, sandy mud, and sand substrates. *Uniomerus tetralasmus*, like other species of *Uniomerus*, are known to be extremely tolerant of drought and can survive long periods of desiccation of its habitat.

Uniomerus tetralasmus is presumably a short-term brooder. Conglutinates were described as sole-shaped with transparent areas arranged crosswise at regular intervals (Utterback 1915).

A host fish investigation for glochidia of *Uniomerus tetralasmus* included only observations of natural infestations. *Uniomerus tetralasmus* glochidia encysted on one fish that occurs in Florida—Cyprinidae: *Notemigonus crysoleucas* (Golden Shiner) (Stern and Felder 1978).

Conservation Status

Uniomerus tetralasmus was assigned a conservation status of currently stable throughout its range by Williams et al. (1993, in review).

Uniomerus tetralasmus is extant throughout its Florida range but is generally uncommon.

Remarks

Uniomerus tetralasmus, a widespread species in Mississippi and Gulf Coast basins, was reported to be the only species of *Uniomerus* in the state (Johnson 1972). Heard (1979a) reported two species of *Uniomerus* from Florida—*U. carolinianus* found throughout the state and *U. declivis* from Apalachicola River. However, *Uniomerus tetralasmus* was reported to occur from Escambia, Yellow, and Choctawhatchee River basins and *U. columbensis* to be endemic to Apalachicola River basin by Williams et al. (2008).

Synonymy

Unio tetralasmus Say 1831 (Say 1831:[no pagination]). Type lost. Type locality: Bayou St. John, New Orleans, [Orleans Parish, Louisiana, Lake Pontchartrain basin].

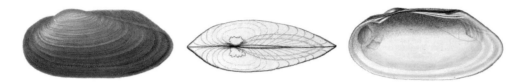

Genus *Utterbackia* Baker 1927

Turgeon et al. (1998) recognized three species in *Utterbackia*. *Utterbackia imbecillis* is the most wide-ranging North American mussel species, occurring in Mississippi and Great Lakes basins, Gulf Coast basins from Mexico to Florida, and Atlantic Coast basins from Delaware to Florida. It has apparently been introduced in some Atlantic Coast rivers and peninsular Florida. *Utterbackia* is also known from early Pleistocene fossil deposits in the Leisey Shell Pits in the Tampa Bay area (Bogan and Portell 1995).

Type Species
Anodonta imbecillis Say 1829 = *Utterbackia imbecillis* (Say 1829)

Diagnosis
Shell thin; smooth; inflated; outline oval to elliptical; umbo even with hinge line; umbo sculpture undulating, irregular ridges; periostracum shiny, occasionally dull, green to brown, with variable green rays; pseudocardinal and lateral teeth absent; nacre bluish white.

Excurrent aperture smooth; mantle bridge separating excurrent and supra-anal apertures long; inner lamellae of inner gills connected to visceral mass only anteriorly; outer gills marsupial, with secondary water tubes when gravid; glochidia held across entire gill; marsupium thickened, not extended beyond original gill margin when gravid, marsupia creamy white when holding embryos, becoming rusty brown when brooding glochidia mature; glochidial outline subtriangular, with styliform hooks (Baker 1927, 1928; Britton and Fuller 1980).

Utterbackia imbecillis (Say 1829)
Paper Pondshell

Utterbackia imbecillis – Upper image: length 84 mm, UF 47241. Apalachicola River below Jim Woodruff [Lock and] Dam at Chattahoochee, Gadsden County, Florida, 18 May 1974. © Richard T. Bryant. Middle image: length 58 mm, UF 449292. Flint Creek on U.S. Highway 301, about 4.4 miles northeast of junction of U.S. Highway 301 and State Highway 582, Hillsborough County, Florida, 20 May 2004. © Richard T. Bryant. Lower image: length 26 mm, UF 449292. Flint Creek on U.S. Highway 301, about 4.4 miles northeast of junction of U.S. Highway 301 and State Highway 582, Hillsborough County, Florida, 20 May 2004. © Richard T. Bryant.

Description

Size: length to 120 mm, in Florida to 102 mm.

Shell: thin; smooth; small individuals compressed, large individuals inflated, width 25%–40% of length; outline oval to elliptical; anterior margin rounded; posterior margin narrowly rounded to bluntly pointed; dorsal margin straight, low dorsal wing typically present in small to medium individuals, usually absent in large individuals; ventral margin straight to broadly rounded; posterior ridge low, rounded; posterior slope moderately steep, flat to slightly concave; umbo broad, moderately inflated, typically even with hinge line, occasionally slightly above in large individuals; umbo sculpture undulating, irregular ridges (Figure 10.98); umbo cavity wide, shallow.

Teeth: pseudocardinal and lateral teeth absent.

Nacre: white to bluish white, iridescent.

Periostracum: shiny, occasionally dull to clothlike; small individuals yellowish tan or green to greenish brown, may be weakly rayed, large individuals dark olive brown, often obscuring rays.

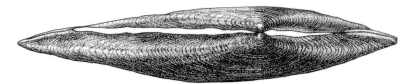

Figure 10.98. Umbo sculpture of *Utterbackia imbecillis*, length 22 mm, UF 455461. Peace River on County Route 640 bridge, about 5.5 air miles south-southeast of Bartow, Polk County, Florida, 24 May 2006. Illustration by Susan Trammell.

Glochidium Description

Outline subtriangular; length 248–313 μm; height 249–310 μm; width 148 μm; with styliform hooks (Ortmann 1912; Surber 1912; Hoggarth 1999; Schwartz and Dimock 2001) (Figure 10.99).

Similar Sympatric Species

Utterbackia imbecillis closely resembles *Utterbackia peggyae* and *Utterbackia peninsularis* but typically has a more elongate shell and is usually more pointed and narrow posteriorly. *Utterbackia imbecillis* typically has less prominent rays than *U. peggyae* and *U. peninsularis*. The mantle margin of live *U. imbecillis* is typically unpigmented, whereas those of *U. peggyae* and *U. peninsularis* have small dark spots, giving them a peppered appearance.

Utterbackia imbecillis may resemble some *Pyganodon grandis* but differs in being more elongate and having a less inflated umbo that is even with the hinge line.

Small *Utterbackia imbecillis* may resemble *Anodonta couperiana*, *Anodonta hartfieldorum*, *Anodonta heardi*, and *Anodonta suborbiculata* but lack the fine green rays on the umbo and upper portion of the shell disk, which are typically present in those species.

Distribution in Florida

Utterbackia imbecillis was historically distributed from Escambia to Apalachicola River basins. Its occurrence in the remainder of the state is the result of introductions and subsequent invasions (Figure 10.100).

Ecology and Biology

Utterbackia imbecillis inhabits a wide variety of aquatic habitats including wetlands, marshes, swamps, creeks, rivers, ponds, and lakes. It also occurs in artificial habitats such as roadside ditches, canals, borrow pits, and reservoirs. It inhabits standing or slow-flowing waters in mud, sandy mud, and sand substrates.

Utterbackia imbecillis is generally a long-term brooder and may be gravid any month of the year (Utterback 1915, 1916b; Ortmann 1919; Baker 1928; Heard 1975b; Howells 2000). Individuals are known to produce multiple broods per year (Allen 1924). Several populations have been reported to have simultaneously hermaphroditic individuals (Sterki 1898b; van der Schalie 1966, 1970; Heard 1975b; Kotrla 1988). However, one Florida panhandle population of *U. imbecillis* was found to have females in addition to hermaphrodites (Heard 1975b; Kat 1983b; Kotrla 1988). Individuals may reach sexual maturity as early as their second year (Allen 1924).

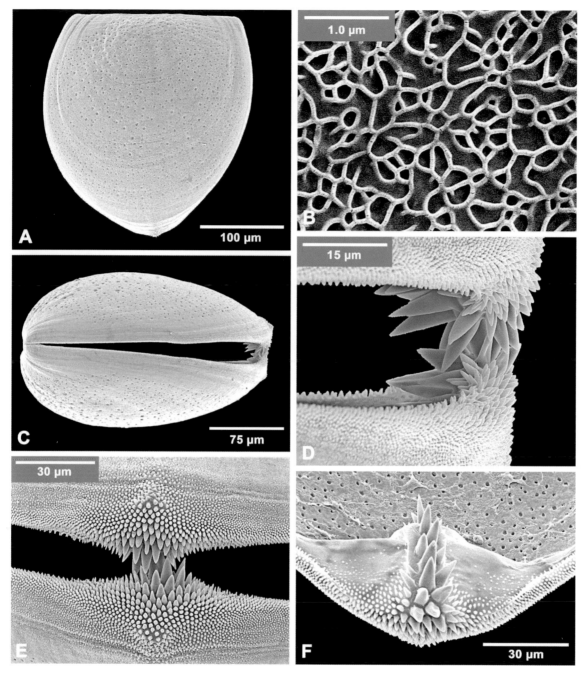

Figure 10.99. Scanning electron micrographs of glochidial valves of *Utterbackia imbecillis* from Peace River, south-southeast of Bartow, Polk County, Florida, 24 May 2006—(A) external view; (B) surface sculpture; (C) lateral view; (D) lateral view of styliform hooks; (E) ventral view of styliform hooks; and (F) close-up of styliform hook on ventral margin.

Female *Utterbackia imbecillis* release strands of mucus with embedded glochidia, which become entangled to form netlike structures. Fishes are parasitized, on their fins, as they swim through the netlike structures (Watters et al. 2009). It was reported by

Howard (1914a) to undergo direct development from glochidium to juvenile stage within the marsupium of the parent, thus bypassing the parasitic stage (making it a facultative parasite). These observations have been subsequently confirmed (Heard 1975b; Watters and O'Dee 1998; Dickinson and Sietman 2008).

Host fish investigations for glochidia of *Utterbackia imbecillis* include observations of natural infestations and laboratory transformations. These included 17 native fish species belonging to six families, 26 nonindigenous aquarium fishes, one salamander, and three frogs (OSU Museum of Biological Diversity 2012). Of the 17 native potential host fishes, 10 are known to occur in Florida.

Natural infestations of *Utterbackia imbecillis* glochidia have been reported on three potential host fishes that occur in Florida—Centrarchidae: *Lepomis gulosus* (Warmouth), *Lepomis macrochirus* (Bluegill), and *Lepomis marginatus* (Dollar Sunfish) (OSU Museum of Biological Diversity 2012).

Successful laboratory transformations of *Utterbackia imbecillis* glochidia have occurred on eight fishes that occur in Florida—Centrarchidae: *Lepomis cyanellus* (Green Sunfish), *Lepomis macrochirus* (Bluegill), *Lepomis megalotis* (Longear Sunfish), *Micropterus salmoides* (Largemouth Bass), and *Pomoxis nigromaculatus* (Black Crappie); Cyprinidae: *Notemigonus crysoleucas* (Golden Shiner); Ictaluridae: *Ictalurus punctatus* (Channel Catfish); and Percidae: *Perca flavescens* (Yellow Perch) (OSU Museum of Biological Diversity 2012).

In addition, 17 fish species belonging to eight families, collected 13 April 1988 from a borrow pit on Shoal River, Highway 285, Walton County, Florida, were examined and 10 had encysted glochidia of *Utterbackia imbecillis*—Atherinopsidae: *Labidesthes sicculus* (Brook Silverside); Centrarchidae: *Lepomis gulosus* (Warmouth) and *Lepomis punctatus* (Spotted Sunfish); Cyprinidae: *Notemigonus crysoleucas* (Golden Shiner); Elassomatidae: *Elassoma zonatum* (Banded Pygmy Sunfish); Esocidae: *Esox americanus* (Redfin Pickerel); Fundulidae: *Fundulus escambiae* (Russetfin Topminnow) and *Fundulus olivaceus* (Blackspotted Topminnow); Percidae: *Etheostoma fusiforme* (Swamp Darter); and Poeciliidae: *Gambusia holbrooki* (Eastern Mosquitofish).

Utterbackia imbecillis has been observed to be infested with mites (Unionicolidae). Of 18 individuals examined from three localities in Apalachicola and Ochlockonee River basins, 15 were parasitized by *Unionicola abnormipes* and/or *Unionicola foili* (Vidrine 1996).

Conservation Status

Utterbackia imbecillis was assigned a conservation status of currently stable throughout its range by Williams et al. (1993, in review).

Utterbackia imbecillis is extant throughout its historical range in Florida and has dispersed to drainages beyond. The species may continue to invade new watersheds.

Remarks

Early 1900s museum records suggest that *Utterbackia imbecillis* was distributed in Florida only from Escambia to Apalachicola Rivers. *Utterbackia imbecillis* was reported to "possess exceptional means of dispersal" by Ortmann (1925) and has extended its geographic range considerably throughout peninsular Florida since the mid-1960s. Its dispersal has presumably been through the introduction of sport and bait fishes with encysted glochidia. The network of man-made canals, many of which cross drainage divides, has allowed *U. imbecillis* to disperse widely in peninsular Florida.

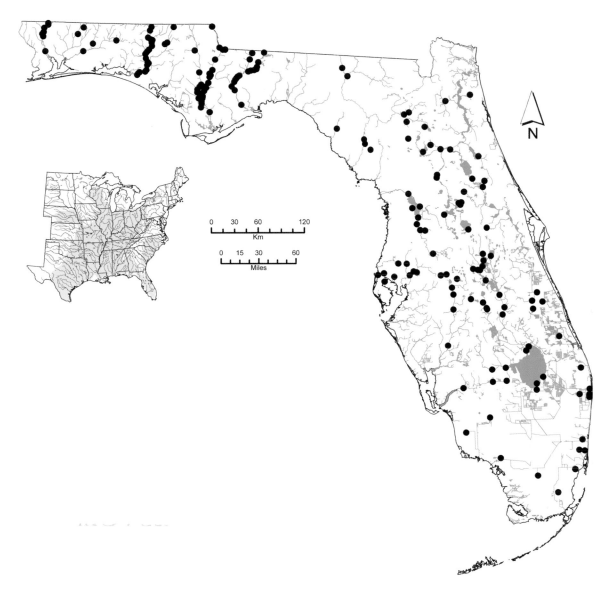

Figure 10.100. Distribution of *Utterbackia imbecillis*.

Synonymy

Anodonta imbecillis Say 1829 (Say 1829:355). Neotype (Haas 1930), SMF 4301, length 58 mm, not figured. Type locality: published as Wabash River, Indiana, restricted by Clench and Turner (1956) to Wabash River, New Harmony, Indiana, [Ohio River drainage].

Utterbackia peggyae (Johnson 1965)
Florida Floater

Utterbackia peggyae – Upper image: length 57 mm, UF 449439. Yellow River at Little Gin Hole Landing boat ramp, Okaloosa County, Florida, 22 September 2008. © Richard T. Bryant. Lower image: length 36 mm, UF 449438. Lake Talquin, about 4 kilometers south-southeast of Midway, Gadsden County, Florida, 24 August 2010. © Richard T. Bryant.

Description

Size: length to 89 mm, in Florida to 87 mm.

Shell: thin; smooth; moderately inflated, width 25%–45% of length; outline oval to elliptical, higher posteriorly; anterior margin rounded; posterior margin rounded to obliquely truncate; dorsal margin straight, low dorsal wing typically present in small to medium individuals, usually absent in large individuals; ventral margin broadly rounded; posterior ridge low, rounded, occasionally weakly biangulate posteriorly; posterior slope flat to slightly concave; umbo broad, slightly inflated, even with hinge line; umbo sculpture undulating, irregular ridges; umbo cavity wide, shallow.

Teeth: pseudocardinal and lateral teeth absent.

Nacre: white to bluish white, iridescent.

Periostracum: shiny; yellowish green to greenish brown, with green rays of varying width and intensity.

Glochidium Description

Outline subtriangular; length 276–312 μm; height 300–325 μm; width 149 μm; with styliform hooks (Williams et al. 2008) (Figure 10.101).

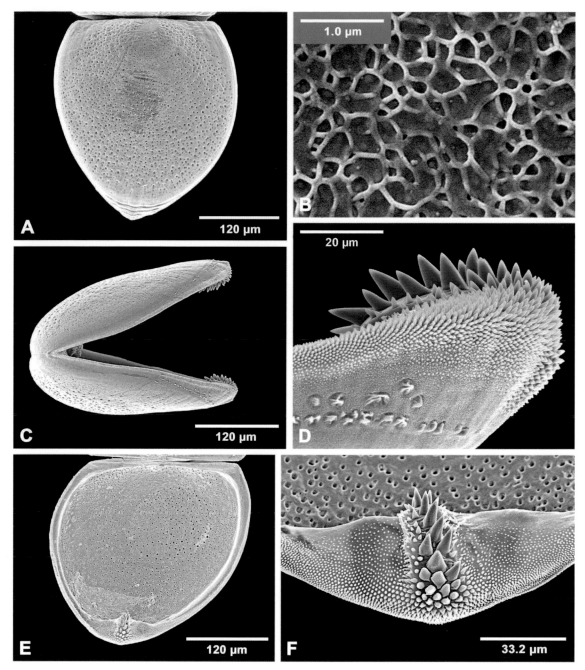

Figure 10.101. Scanning electron micrographs of glochidial valves of *Utterbackia peggyae* from Choctawhatchee River, just above State Highway 20 bridge, Washington County, Florida, 5 September 2007—(A) external view; (B) surface sculpture; (C) lateral view; (D) lateral view of styliform hook on ventral margin; (E) internal view of styliform hook on ventral margin; and (F) close-up of styliform hook.

Similar Sympatric Species

Utterbackia peggyae resembles *Utterbackia imbecillis* but is less elongate and higher posteriorly. *Utterbackia peggyae* typically has more prominent green rays than that of *U. imbecillis*. The mantle margin of live *U. peggyae* usually has small dark spots that give it a peppered appearance, which are absent in *U. imbecillis*.

Utterbackia peggyae may resemble some *Pyganodon grandis* but differs in being more elongate and having a less inflated umbo that is even with the hinge line instead of elevated above.

Small *Utterbackia peggyae* may resemble *Anodonta hartfieldorum*, *Anodonta heardi*, and *Anodonta suborbiculata* but lack the fine green rays on the umbo and upper portion of the shell disk, which are typically present in those species.

Distribution in Florida

Utterbackia peggyae occurs in most major basins from Escambia River to St. Marks River (Figure 10.102).

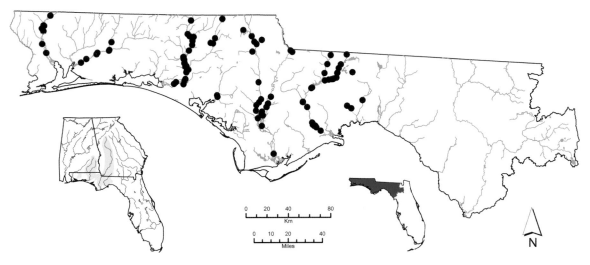

Figure 10.102. Distribution of *Utterbackia peggyae*.

Ecology and Biology

Utterbackia peggyae inhabits swamps, creeks, rivers, sloughs, ponds, and lakes. It also occurs in artificial habitats such as borrow pits and reservoirs. It inhabits standing or slow-flowing waters in mud, sandy mud, and sand substrates.

Utterbackia peggyae is presumably a long-term brooder, gravid from August to the following summer. However, one Florida population produced two consecutive broods in a year (Heard 1975b). Mature ova and spermatozoa have been reported to be present in the gonad throughout the year (Heard 1969). Glochidial hosts for *U. peggyae* are unknown. A closely related species, *Utterbackia imbecillis*, is a host generalist.

Utterbackia peggyae has been observed to be infested with mites (Unionicolidae). Of 100 individuals examined from six localities in north Florida rivers (Yellow to Ochlockonee), 28 were parasitized by one or more of the following species: *Unionicola abnormipes*, *Unionicola dobsoni*, *Unionicola foili*, *Unionicola hoesei*, *Unionicola mitchelli*, *Unionicola serrata*, and *Unionicola wolcotti* (Vidrine 1996).

Conservation Status

Utterbackia peggyae was assigned a conservation status of currently stable throughout its range by Williams et al. (1993, in review).

Utterbackia peggyae is extant throughout its Florida range and is relatively common.

Remarks

In the original description *Utterbackia peggyae* was reported to occur in coastal streams from Choctawhatchee to Hillsborough River basins (Johnson 1965, 1972). Populations of *U. peggyae* from Suwannee to Hillsborough River basins were subsequently recognized as a new species, *Utterbackia peninsularis* (Bogan and Hoeh 1995).

Utterbackia peggyae and *Utterbackia peninsularis* are superficially very similar and cannot be differentiated using shell morphology and color characters alone. However, the two species are allopatrically distributed and can be distinguished using molecular genetics and soft anatomy (Kat 1983c; Bogan and Hoeh 1995).

Utterbackia peggyae was described by Richard I. Johnson, who named the species after his first wife.

Synonymy

Anodonta peggyae Johnson 1965 (Johnson 1965:1). Holotype, MCZ 251040, length 71 mm. Type locality: southeast shore of Lake Talquin (formed by a dam on the Ochlockonee River), Leon County public fishing ground, Leon County, Florida, [Ochlockonee River basin].

Utterbackia peninsularis Bogan and Hoeh 1995
Peninsular Floater

Utterbackia peninsularis – Upper image: length 46 mm, UF 455462. Withlacoochee River [southern] on State Highway 48, about 4 miles southeast of Floral City, Sumter and Citrus counties, Florida, 31 January 2007. © Richard T. Bryant. Lower image: length 38 mm, UF 455462. Withlacoochee River [southern] on State Highway 48, about 4 miles southeast of Floral City, Sumter and Citrus counties, Florida, 31 January 2007. © Richard T. Bryant.

Description

Size: length to 79 mm.

Shell: thin; smooth; moderately inflated, width 30%–40% of length; outline oval to elliptical, higher posteriorly; anterior margin rounded; posterior margin rounded to obliquely truncate; dorsal margin straight, low dorsal wing typically present in small to medium individuals, usually absent in large individuals; ventral margin broadly rounded; posterior ridge low, rounded, occasionally weakly biangulate posteriorly; posterior slope flat to slightly concave; umbo broad, slightly inflated, even with hinge line; umbo sculpture undulating, irregular ridges; umbo cavity wide, shallow.

Teeth: pseudocardinal and lateral teeth absent.

Nacre: white to bluish white, iridescent.

Periostracum: shiny; yellowish green to greenish brown, with green rays of varying width and intensity.

Glochidium Description

The glochidium of *Utterbackia peninsularis* is unknown.

Similar Sympatric Species

Utterbackia peninsularis resembles *Utterbackia imbecillis* but is less elongate and higher posteriorly. *Utterbackia peninsularis* typically has more prominent green rays than *U. imbecillis*. The mantle margin of live *U. peninsularis* usually has small dark spots that give it a peppered appearance, which are absent in *U. imbecillis*.

Small *Utterbackia peninsularis* may resemble *Anodonta couperiana* but lack the fine green rays on the umbo and upper portion of the shell disk, which are typically present in those species.

Distribution in Florida

Utterbackia peninsularis occurs in Suwannee, Withlacoochee (southern), and Hillsborough River basins (Figure 10.103).

Figure 10.103. Distribution of *Utterbackia peninsularis*.

Ecology and Biology

Utterbackia peninsularis inhabits swamps, creeks, rivers, sloughs, ponds, and lakes. It also occurs in artificial habitats such as borrow pits and reservoirs. It inhabits standing or slow-flowing waters in mud, sandy mud, and sand substrates.

Utterbackia peninsularis is generally considered to be a long-term brooder, gravid from August to the following summer. Glochidial hosts for *U. peninsularis* are unknown. A closely related species, *Utterbackia imbecillis*, is a host generalist.

Utterbackia peninsularis has been observed to be infested with mites (Unionicolidae). Of seven individuals examined from Withlacoochee River (southern) basin, six were parasitized by *Unionicola foili* and/or *Unionicola poundsi* (Vidrine 1996).

Conservation Status

The recent description of *Utterbackia peninsularis* precluded its inclusion in previous conservation status reviews. *Utterbackia peninsularis* was evaluated by Williams et al. (in review) and its status was considered to be currently stable.

Utterbackia peninsularis is extant throughout its Florida range and is common but may be declining in some areas.

Remarks

Prior to its description *Utterbackia peninsularis* was considered to be a peninsular population of *Utterbackia peggyae*. *Utterbackia peninsularis* and *U. peggyae* are superficially very similar and cannot be differentiated using shell morphology and color characters alone. However, the two species are allopatrically distributed and can be distinguished using molecular genetics and soft anatomy (Kat 1983c; Bogan and Hoeh 1995).

Synonymy

Utterbackia peninsularis Bogan and Hoeh 1995 (Bogan and Hoeh 1995:275). Holotype, UMMZ 253583, length 56 mm. Type locality: canal off of Suwannee River, at Dilger's Campground, Dixie County, Florida, [Suwannee River basin,] 17 October 1988.

Genus *Villosa* Frierson 1927

Turgeon et al. (1998) recognized 18 taxa in *Villosa*. Two subspecies, *Villosa vanuxemensis vanuxemensis* and *Villosa vanuxemensis umbrans*, were elevated to full species by Williams et al. (2008), and *Villosa choctawensis* was reassigned to *Obovaria* (Williams et al. 2011). *Nephronaias gundlachi* in Cuba was reported to be a *Villosa* (Johnson 1981) based only on shell morphology; however, this move has not been accepted by other workers. A fossil *Villosa* sp. was reported from the early Pleistocene Leisey Shell Pits near Tampa, Florida (Bogan and Portell 1995). There are currently 17 species recognized in the genus *Villosa* (Williams et al. in review).

Recent molecular phylogenetic analysis indicates that *Villosa*, as used today, is not monophyletic (Buhay et al. 2003; Kuehnl 2009). In addition there is the potential for wide-ranging species in the genus (e.g., *Villosa lienosa* and *Villosa vibex*) to represent multiple species. Future studies will likely result in taxonomic changes at the genus and species level.

Villosa mantle folds, located just ventral to their incurrent apertures, are modifications believed to function as host-attracting lures. They are variously pigmented and have papillate margins often resembling prey items of predatory fishes.

Villosa occurs in Mississippi and Great Lakes basins, Gulf Coast basins from Texas to Florida, and southern Atlantic Coast basins from northern Virginia to Florida.

Type Species

Unio villosus B.H. Wright 1898 = *Villosa villosa* (B.H. Wright 1898)

Diagnosis

Shell thin to moderately thick; smooth; moderately inflated; outline elliptical to oval; sexually dimorphic, females swollen posteriorly, posterior margin often becoming truncate; umbo elevated slightly above hinge line; umbo sculpture variable, thin to moderately thick, undulating ridges, with shallow or deep indentation ventrally; posterior ridge indistinct or evenly rounded; periostracum shiny to dull, occasionally clothlike.

Mantle margin with well-developed papillate fold just ventral to incurrent aperture, fold rudimentary in males; excurrent aperture crenulate; mantle bridge separating excurrent and supra-anal apertures variable; inner lamellae of inner gills completely connected to visceral mass or connected only anteriorly; outer gills marsupial; glochidia held in posterior portion of gill; marsupium extended beyond original gill margin when gravid; glochidium outline subspatulate to subelliptical, without styliform hooks (Frierson 1927; Britton and Fuller 1980; Brim Box and Williams 2000).

Villosa amygdalum (Lea 1843)
Florida Rainbow

Villosa amygdalum – Upper image: female, length 36 mm, UF 455463. Flint Creek, on U.S. Highway 301, about 4.4 miles northeast of junction of U.S. Highway 301 and State Highway 582, Hillsborough County, Florida, 20 May 2004. © Richard T. Bryant. Middle image: male, length 36 mm, UF 455463. Flint Creek, on U.S. Highway 301, about 4.4 miles northeast of junction of U.S. Highway 301 and State Highway 582, Hillsborough County, Florida, 20 May 2004. © Richard T. Bryant. Lower image: length 33 mm, UF 455463. Flint Creek, on U.S. Highway 301, about 4.4 miles northeast of junction of U.S. Highway 301 and State Highway 582, Hillsborough County, Florida, 20 May 2004. © Richard T. Bryant.

Description

Size: length to 74 mm.

Shell: thin to moderately thick; smooth; slightly inflated, width 30%–45% of length, females slightly more inflated posterioventrally; outline oval to elliptical; anterior margin rounded; posterior margin broadly to narrowly rounded, slightly more broadly rounded in females; dorsal margin straight to slightly convex; ventral margin broadly rounded; posterior ridge rounded; posterior slope moderately steep, flat to slightly concave; umbo moderately broad, inflated, elevated slightly above hinge line; umbo sculpture 6–8 thin, double-looped ridges, with deep indentation ventrally (Figure 10.104); umbo cavity wide, shallow.

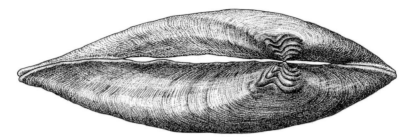

Figure 10.104. Umbo sculpture of *Villosa amygdalum*, length 34 mm, UF 441123. Caloosahatchee River at Ortona Lock boat ramp below the spillway, Glades County, Florida, 4 June 2008. Illustration by Susan Trammell.

Teeth: pseudocardinal teeth small, compressed, bladelike, 2 teeth in left valve, occasionally joined at base, crests almost parallel to shell margin, anterior tooth larger, 1 thin tooth in right valve, occasionally with accessory denticle anteriorly; lateral teeth moderately long, thin, straight to slightly curved, 2 in left valve, 1 in right valve; interdentum moderately long, narrow.

Nacre: white to bluish white, usually iridescent posteriorly.

Periostracum: shiny to clothlike; small individuals yellowish green to olive or brown, large individuals dark olive to black, typically with green rays of varying width and intensity.

Glochidium Description

Outline subspatulate; length 189–230 μm; height 230–263 μm; width 111–132 μm; ventral margin with internal and external irregularly perpendicular rows of lanceolate micropoints (Figure 10.105).

Similar Sympatric Species

Villosa amygdalum resembles *Villosa vibex* but is less elongate and has a more clothlike periostracum with thinner dark green rays. *Villosa amygdalum* is usually slightly more inflated and females exhibit more pronounced sexual dimorphism.

Small *Villosa amygdalum* superficially resemble *Toxolasma paulum*, but the teeth are typically much thinner. Small *V. amygdalum* can also be distinguished by the double-looped sculpture as opposed to the single-looped ridges of *T. paulum*.

Villosa amygdalum may superficially resemble *Anodonta couperiana* but has well-developed dark green rays of various widths. *Anodonta couperiana* lacks lateral and pseudocardinal teeth, which are always present in *V. amygdalum*.

Distribution in Florida

Villosa amygdalum occurs in Waccasassa and St. Johns River basins and throughout peninsular Florida (Figure 10.106).

Ecology and Biology

Villosa amygdalum inhabits marshes, creeks, spring runs, rivers, and lakes. It is also found in artificial habitats including canals, roadside ditches, and reservoirs. It is found in still water or slow to moderate current in substrates of sandy mud, sand, and sand mixed with detritus. Its occurrence in marsh habitat is usually near a stream or canal.

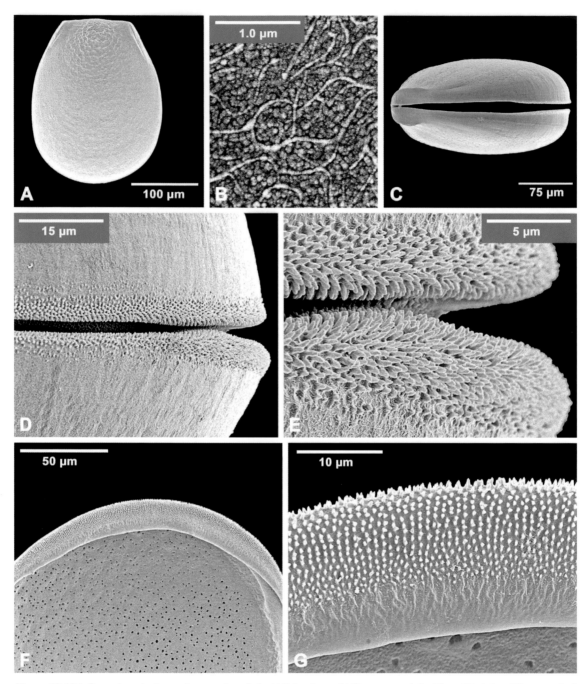

Figure 10.105. Scanning electron micrographs of glochidial valves of *Villosa amygdalum* from St. Johns River, west of Deland, Volusia County, Florida, 15 May 2008 (A, B, F, G); Upper Myakka Lake in Myakka State Park, Sarasota County, Florida, 4 June 2008 (C); Silver Lake on Withlacoochee River (southern), Hernando County, Florida, 21 February 2008 (D, E)—(A) external view; (B) surface sculpture; (C) lateral view; (D) lateral view of micropoints along ventral margin; (E) close-up of micropoints on ventral margin; (F) internal view of ventral margin and micropoints; and (G) close-up of micropoints.

Villosa amygdalum is a long-term brooder. Glochidial hosts for *V. amygdalum* are unknown. Species of *Villosa* tend to be host specialists (Haag 2012). The mantle margins of *V. amygdalum* are characterized by variously colored "brown and green papillae that together resemble a caterpillar that has fallen into the water and is lying on the bottom" (Morrison 1973).

Villosa amygdalum has been observed to be infested with mites (Unionicolidae). Of 75 individuals examined from eight localities in peninsular Florida, 65 were parasitized by one or more of the following species: *Najadicola ingens*, *Unionicola abnormipes*, *Unionicola aculeata*, and *Unionicola poundsi* (Vidrine 1996).

Conservation Status

Villosa amygdalum was assigned a conservation status of currently stable throughout its range by Williams et al. (1993, in review).

Villosa amygdalum is extant throughout its range and is generally common.

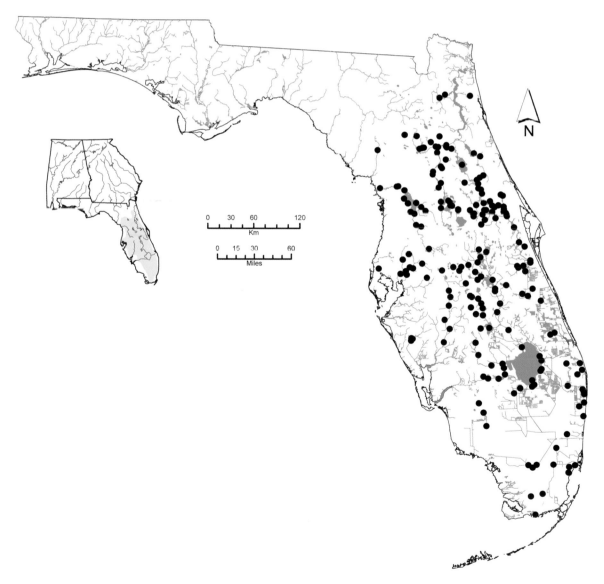

Figure 10.106. Distribution of *Villosa amygdalum*.

Remarks

Villosa amygdalum is the only mussel whose distribution is restricted entirely to the state of Florida. It displays varied shell morphology, periostracum texture, color, and raying pattern, which may be explained by its diverse habitats.

Synonymy

Unio amygdalum Lea 1843 (Lea 1843:[one page privately published]). Lectotype (Johnson 1974), USNM 86127, length 32 mm. Type locality: Lake George, [Putnam and Volusia counties,] Florida, [St. Johns River basin].

Unio trosculus Lea 1843 (Lea 1843:[one page privately published]) [corrected spelling to *Unio trossulus* Lea 1846]. Type locality: Lake Monroe, [Volusia and Seminole counties,] Florida, [St. Johns River basin].

Unio papyraceus Gould 1845 (Gould 1845:53). Type locality: Everglades of Florida, [Everglades basin].

Unio lepidus Gould 1856 (Gould 1856:15). Type locality: from a creek near Lake Monroe, [Volusia and Seminole counties,] Florida, [St. Johns River basin].

Unio vesicularis Lea 1872 (Lea 1872:156). Type locality: Lake Okeechobee, Florida, [Everglades basin].

Unio singleyanus Marsh 1891 (Marsh 1891:29). Type locality: a small creek near Pilatka [Palatka], [Putnam County,] Florida, [St. Johns River basin].

Lampsilis wrightiana Frierson 1927 (Frierson 1927:81). Type locality: Volusia County, Florida, [St. Johns River basin].

Villosa lienosa (Conrad 1834)
Little Spectaclecase

Villosa lienosa – Upper image: female, length 43 mm, UF 365272. Mosquito Creek at State Highway 97, 20 air miles southwest of Bainbridge, Decatur County, Georgia, 27 May 1992. © Richard T. Bryant. Middle image: female, length 34 mm, UF 449440. Bruce Creek on Walton Bridge Road, about 14 kilometers (air) south-southwest of Ponce de Leon, Walton County, Florida, 22 April 2011. © Richard T. Bryant. Lower image: male, length 39 mm, UF 365272. Mosquito Creek at State Highway 97, 20 air miles southwest of Bainbridge, Decatur County, Georgia, 27 May 1992. © Richard T. Bryant.

Description

Size: length to 84 mm, in Florida to 66 mm.

Shell: moderately thin to moderately thick; smooth; moderately inflated, width 35%–50% of length, females slightly more inflated posterioventrally; outline oval; anterior margin rounded; posterior margin broadly rounded to obliquely truncate in females, narrowly rounded to bluntly pointed in males; dorsal margin slightly convex; ventral margin straight to convex, occasionally concave in large females; posterior ridge low, rounded; posterior slope moderately steep, flat to slightly concave; umbo broad, moderately inflated, elevated slightly above hinge line; umbo sculpture few thin ridges, with shallow indentation along ventral margin; umbo cavity wide, shallow.

Teeth: pseudocardinal teeth small, occasionally moderately large, compressed, 2 teeth in left valve, anterior tooth larger, crest usually parallel to margin, 1 tooth in right valve, occasionally with accessory denticle anteriorly; lateral teeth moderately short, straight to slightly curved, 2 in left valve, 1 in right valve; interdentum long, narrow.

Nacre: purplish, populations from Escambia River to Apalachicola River occasionally with salmon tint, typically white to bluish white in Ochlockonee and Suwannee River populations.

Periostracum: shiny to clothlike; small individuals olive to brown, often with faint green rays, large individuals dark brown to black, often obscuring rays.

Glochidium Description

Outline subspatulate; length 181–220 μm; height 218–288 μm, width 135 μm; dorsal margin 96–111 μm; ventral margin with internal and external perpendicular rows of lanceolate micropoints (Ortmann 1912, 1916; Utterback 1916b; Kennedy and Haag 2005) (Figure 10.107).

Similar Sympatric Species

Villosa lienosa resembles *Villosa vibex* but lacks prominent dark green rays, has a thicker shell, thicker pseudocardinal and lateral teeth, and usually a less broadly rounded posterior margin. In live individuals pigmentation along the ventral mantle margin differs between the two species. In *V. lienosa* dark pigmentation ends abruptly just anterior of the incurrent aperture, whereas in *V. vibex* it extends along the entire margin (though tapered anteriorly).

Villosa lienosa may also resemble *Villosa villosa* but is less elongate and has thicker pseudocardinal teeth. In live individuals pigmentation along the ventral mantle margin also differs between the two species. In *V. lienosa* dark pigmentation ends abruptly just anterior of the incurrent aperture, whereas in *V. villosa* it extends along the entire margin (though tapered anteriorly).

Small *Villosa lienosa* may resemble some species of *Toxolasma* but are typically more compressed than *Toxolasma* of the same size. Umbo sculpture of *V. lienosa* consists of a few thin ridges, with shallow indentation along ventral margin, instead of moderately thick, single-looped ridges. Live female *Toxolasma* have well-developed caruncles on the posterioventral mantle margin, which are absent in *Villosa*.

Male *Villosa lienosa* resemble *Hamiota australis* but are less elongate, have a less shiny periostracum, and the nacre is purple or salmon instead of bluish white.

Villosa lienosa may resemble *Obovaria choctawensis* but is less rounded, usually has purple or salmon nacre instead of white, and has brown to black periostracum instead of yellowish to greenish yellow.

Distribution in Florida

Villosa lienosa occurs in most major basins from Escambia River to Suwannee River (Figure 10.108).

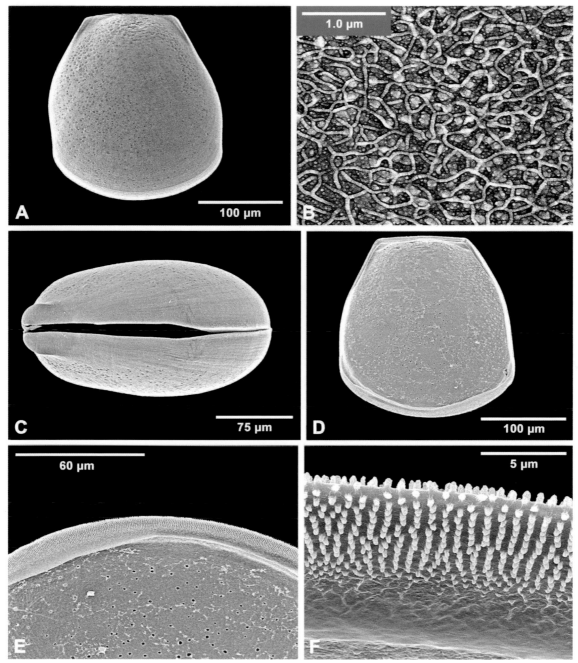

Figure 10.107. Scanning electron micrographs of glochidial valves of *Villosa lienosa* from Holmes Creek at Vernon, Washington County, Florida, 18 June 2009—(A) external view; (B) surface sculpture; (C) lateral view; (D) internal view; (E) internal view of ventral margin and micropoints; and (F) close-up of micropoints.

Ecology and Biology

Villosa lienosa inhabits creeks and rivers in slow to moderate current in substrates of stable sand, sandy mud, and fine gravel.

Villosa lienosa is a long-term brooder, gravid from late summer or autumn to the following summer (Ortmann 1912, 1924a).

In laboratory trials *Villosa lienosa* glochidia metamorphosed on four fishes that occur in Florida—Centrarchidae: *Lepomis macrochirus* (Bluegill) and *Micropterus salmoides* (Largemouth Bass); and Ictaluridae: *Ameiurus nebulosus* (Brown Bullhead) and *Ictalurus punctatus* (Channel Catfish) (Keller and Ruessler 1997a).

Villosa lienosa has been observed to be infested with mites (Unionicolidae). Of 169 individuals examined from five localities in north Florida river basins (Yellow to Ochlockonee), 77 were parasitized by one or more of the following species: *Unionicola gailae*, *Unionicola hoesei*, *Unionicola sakantaka*, *Unionicola serrata*, and *Unionicola tupara* (Vidrine 1996).

Conservation Status

Villosa lienosa was assigned a conservation status of currently stable throughout its range by Williams et al. (1993, in review).

Villosa lienosa is extant throughout its Florida range and is common.

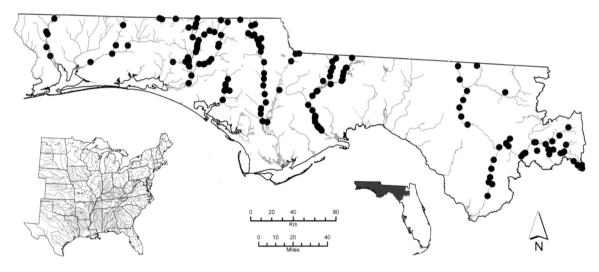

Figure 10.108. Distribution of *Villosa lienosa*.

Remarks

Villosa lienosa is the most wide-ranging species of the genus. Preliminary genetic analysis suggests that there may be cryptic species currently recognized as *V. lienosa* in eastern Gulf Coast basins (N.A. Johnson unpublished data).

The type locality of *Villosa lienosa* was published as "small streams in south Alabama." Williams et al. (2008) pointed out that a specimen labeled "Type" and with a notation of "Type *U. lienosus* Con. Big Pr. [Prairie] Creek, Marengo Co., Alabama" in the shell had been cataloged into the FLMNH as UF 174330 (length 59 mm; 2.3 inches) (Figure 10.109). The handwritten notation in the shell is consistent with other material labeled by Conrad. It is clear from the description that there were syntypes (e.g., Conrad gave a range of nacre colors). The length of the specimen given by Conrad (1834) in the original description was "2.8 inches"; however, it is unclear if he was referring to the figured specimen or to a maximum length of a syntype. Clench and Turner (1956) referred to ANSP 9747 as the holotype. However, there are multiple specimens in that lot so their reference to a holotype does not serve as an inadvertent lectotype designation. Johnson and Baker (1973) reported ANSP 9747 as syntypes without selecting a lectotype.

Figure 10.109. A possible syntype of *Villosa lienosa*, UF 174330, length 59 mm (male), Big Pr. [Prairie] Creek, Marengo County, Alabama. Photograph by Zachary Randall.

Synonymy

Unio lienosus Conrad 1834 (Conrad 1834:339). Syntypes, ANSP 9747, length 71 mm. Type locality: small streams in south Alabama.

Unio subellipsis Lea 1856 (Lea 1856:262). Type locality: creeks near Columbus, [Muscogee County,] Georgia, [Chattahoochee River drainage].

Unio concestator Lea 1857 (Lea 1857b:31). Type locality: creeks near Columbus, [Muscogee County,] Georgia, [Chattahoochee River drainage].

Unio fallax Lea 1857 (Lea 1857b:32). Type locality: streams near Columbus, [Muscogee County,] Georgia, [Chattahoochee River drainage].

Unio intercedens Lea 1857 (Lea 1857b:32). Type locality: streams near Columbus, [Muscogee County,] Georgia, [Chattahoochee River drainage].

Unio obfuscus Lea 1857 (Lea 1857a:172). Type locality: Flint River near Macon [County], Georgia, [Flint River drainage].

Unio sudus Lea 1857 (Lea 1857a:170). Type locality: Dry Creek near Columbus, [Muscogee County,] Georgia, [Chattahoochee River drainage].

Unio prattii Lea 1858 (Lea 1858b:166). Type locality: Chattahoochee River at Roswell, Cobb County, Georgia, [Chattahoochee River drainage].

Unio dispar Lea 1860 (Lea 1860:305). Type locality: [Chattahoochee River,] Columbus, [Muscogee County,] Georgia, [Chattahoochee River drainage].

Unio linguaeformis Lea 1860 (Lea 1860:305). Type locality: Columbus, [Muscogee County,] Georgia, [Chattahoochee River drainage].

Unio unicostatus B.H. Wright 1899 (B.H. Wright 1899:69). Type locality: Spring Creek, [tributary to Flint River,] Decatur County, Georgia, [Flint River drainage].

Villosa vibex (Conrad 1834)
Southern Rainbow

Villosa vibex – Upper image: female, length 61 mm, UF 388496. Spring Creek at Route 391, 1.3 miles west of State Highway 310, 12 air miles northwest of Bainbridge, Decatur County, Georgia, 29 May 1992. © Richard T. Bryant. Middle image: female, length 42 mm, UF 365560. Chickasawhatchee Creek at Route 121, 8.8 air miles northwest of Newton, Baker County, Georgia, 28 May 1992. © Richard T. Bryant. Lower image: male, length 32 mm, UF 365560. Chickasawhatchee Creek at Route 121, 8.8 air miles northwest of Newton, Baker County, Georgia, 28 May 1992. © Richard T. Bryant.

Description

Size: length to 103 mm, in Florida to 88 mm.

Shell: thin, males often thicker; smooth; slightly inflated, width 25%–40% of length, females slightly more inflated posterioventrally; outline oval; anterior margin rounded; posterior margin rounded, more broadly rounded posterioventrally in females; dorsal margin straight to slightly convex; ventral margin straight to convex; posterior ridge rounded; posterior slope low, flat to slightly concave; umbo broad, slightly inflated, elevated slightly above hinge line; umbo sculpture thin ridges, with shallow indentation along ventral margin (Figure 10.110); umbo cavity wide, moderately shallow.

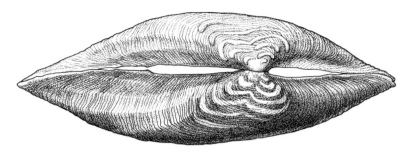

Figure 10.110. Umbo sculpture of *Villosa vibex*, length 22 mm, UF 449233. Escambia River, about 0.3 miles above Chumuckla Springs boat ramp, Santa Rosa County, Florida, 24 September 2008. Illustration by Susan Trammell.

Teeth: pseudocardinal teeth small, bladelike, 2 teeth in left valve, crests almost parallel to margin, anterior tooth considerably larger, 1 tooth in right valve, occasionally with accessory denticle anteriorly; lateral teeth moderately short, thin, straight to slightly curved, 2 in left valve, 1 in right valve; interdentum long, narrow.

Nacre: bluish white, iridescent.

Periostracum: typically shiny, occasionally dull; small individuals greenish yellow to olive, with green rays of varying width and intensity, usually wider and darker posteriorly, occasionally rayless in Escambia, Yellow, and Choctawhatchee River populations, large individuals dark olive to almost black, occasionally obscuring rays.

Glochidium Description

Outline subspatulate; length 203–265 μm; height 217–329 μm; width 109–137 μm; dorsal margin 105–133 μm; ventral margin with internal and external irregularly perpendicular rows of lanceolate micropoints (Ortmann 1924a; Hoggarth 1999; Kennedy and Haag 2005) (Figure 10.111).

Similar Sympatric Species

Villosa vibex resembles *Villosa lienosa* but typically has more prominent green rays, a thinner shell, and thinner, more bladelike pseudocardinal teeth. The posterior margin of the *V. vibex* shell is usually more broadly rounded than that of *V. lienosa*. The two differ in coloration of the ventral mantle margin, which is darkly pigmented along its entire length in *V. vibex* (though tapered anteriorly), but the dark pigment ends rather abruptly just anterior of the incurrent aperture in *V. lienosa*.

Villosa vibex may resemble *Hamiota subangulata*, but that species has a thicker shell, more prominent posterior ridge, and thicker pseudocardinal and lateral teeth.

Villosa vibex resembles *Villosa villosa* but typically has more prominent green rays, a thinner shell, and thinner, more bladelike pseudocardinal teeth. The posterior margin of the *V. vibex* shell is usually more broadly rounded than that of *V. Villosa*.

Villosa vibex resembles *Villosa amygdalum* but is more elongate and has a shinier periostracum with wider dark green rays. *Villosa vibex* is also usually slightly less inflated and females exhibit less pronounced sexual dimorphism.

Villosa vibex may superficially resemble *Anodonta couperiana* but has well-developed dark green rays of various widths. *Anodonta couperiana* lacks lateral and pseudocardinal teeth, which are always present in *V. vibex*.

Villosa vibex superficially resembles *Anodontoides radiatus* but has well-developed pseudocardinal and lateral teeth.

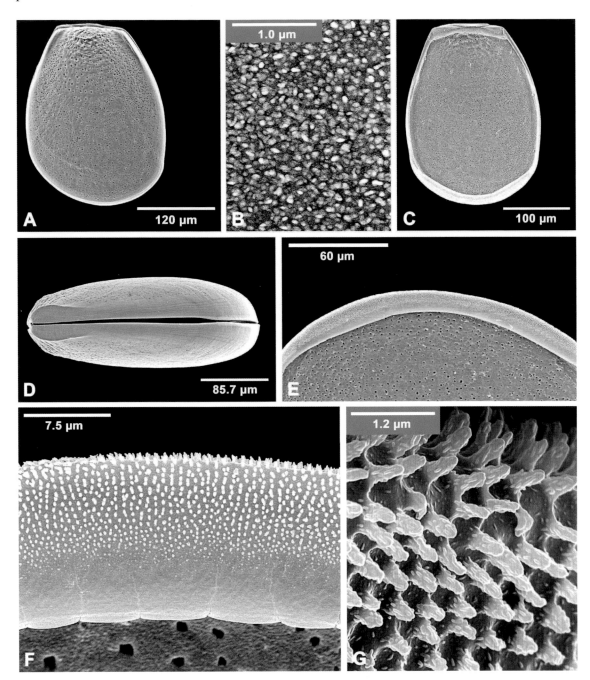

Figure 10.111. Scanning electron micrographs of glochidial valves of *Villosa vibex* from Hammock Branch, tributary of Dunn's Creek, northwest of Pomona Park, Putnam County, Florida, 11 January 2001 (A); Holmes Creek at Vernon, Washington County, Florida, 18 June 2009 (B); Escambia River at Molino, Escambia County, Florida, 18 April 2004 (C, D, E, F, G)—(A) external view; (B) surface sculpture; (C) internal view; (D) lateral view; (E) internal view of ventral margin; (F) internal view of micropoints on ventral margin; and (G) close-up of micropoints.

Distribution in Florida

Villosa vibex occurs in Gulf Coast basins from Escambia River to Alafia River and on the Atlantic Coast it occurs in St. Marys and St. Johns basins (Figure 10.112).

Figure 10.112. Distribution of *Villosa vibex*.

Ecology and Biology

Villosa vibex occurs in creeks and rivers in slow to moderate current in sandy mud, sand, and gravel substrates.

Villosa vibex is a long-term brooder, gravid from late summer or autumn to the following summer (Haag et al. 1999). In west Florida gravid females have been observed with immature glochidia in August and mature glochidia in late September. *Villosa vibex* from Swift Creek in the lower Choctawhatchee River basin were found to be gravid with mature glochidia in October 2012. *Villosa vibex* females display a dark papillate mantle fold with darker mottling, typically with an "eye spot," which is a host-attracting structure (Figure 10.113). The female usually moves to an upright position in the substrate to display its mantle fold. The display occurs primarily during the day (Haag et al. 1999).

Figure 10.113. *Villosa vibex* females with papillate mantle fold displayed—Econfina Creek, Bay County, Florida, April 2009 (left) and Flint River drainage, Georgia, May 1995 (right). Photographs by Sandra Pursifull and Noel M. Burkhead.

A host fish investigation for glochidia of *Villosa vibex* included only laboratory transformations. Of the four host fishes, only three occur in Florida—Centrarchidae: *Lepomis cyanellus* (Green Sunfish), *Micropterus punctulatus* (Spotted Bass), and *Micropterus salmoides* (Largemouth Bass). Eleven additional fishes that occur in Florida were exposed, but no metamorphosed glochidia were recovered—Centrarchidae: *Ambloplites ariommus* (Shadow Bass), *Lepomis auritus* (Redbreast Sunfish), *Lepomis cyanellus* (Green Sunfish), *Lepomis gulosus* (Warmouth), *Lepomis macrochirus* (Bluegill), *Lepomis marginatus* (Dollar Sunfish), and *Lepomis megalotis* (Longear Sunfish); Esocidae: *Esox niger* (Chain Pickerel); Fundulidae: *Fundulus olivaceus* (Blackspotted Topminnow); Ictaluridae: *Noturus leptacanthus* (Speckled Madtom); and Percidae: *Percina nigrofasciata* (Blackbanded Darter) (Haag et al. 1999).

Villosa vibex has been observed to be infested with mites (Unionicolidae). Of 225 individuals examined from five localities in north Florida river basins (Yellow to Apalachicola), 140 were parasitized by one or more of the following species: *Unionicola abnormipes*, *Unionicola hoesei*, *Unionicola serrata*, and *Unionicola vamana* (Vidrine 1996).

Conservation Status

Villosa vibex was assigned a conservation status of currently stable throughout its range by Williams et al. (1993, in review).

Villosa vibex is extant throughout its Florida range. It is generally common except in the St. Marys and St. Johns River basins where it is rare.

Remarks

Villosa vibex is unusual in that it occurs in St. Johns River and Tampa Bay basins but is absent from the remainder of peninsular Florida. Records for *Villosa villosa* in St. Marys River basin and possibly some localities in St. Johns River basin by Johnson (1972) are actually based on *V. vibex*.

Villosa vibex is one of the more common mussels in Florida panhandle streams. This may be due in part to its ability to tolerate unstable substrates associated with watershed disturbance.

Synonymy

Unio vibex Conrad 1834 (Conrad 1834:31). Lectotype (Johnson and Baker 1973), ANSP 56488a, length 59 mm (male); allotype, ANSP 56488, length 54 (female). Type locality: Black Warrior River, south of Blount's [Blount] Springs, [Blount County,] Alabama, [Tombigbee River drainage].

Unio exiguus Lea 1840 (Lea 1840:287). Type locality: Chattahoochee River, Columbus, [Muscogee County,] Georgia, [Chattahoochee River drainage].

Unio pellucidus Lea 1845 (Lea 1845:163). Type locality: Chattahoochee River, Georgia, [Chattahoochee River drainage].

Unio nigrinus Lea 1852 (Lea 1852b:252). Type locality: west Florida.

Unio averellii B.H. Wright 1888 (B.H. Wright 1888:115). Type locality: Lake Ashby, Volusia County, Florida, [St. Johns River basin].

Villosa villosa (B.H. Wright 1898)
Downy Rainbow

Villosa villosa – Upper image: female, length 60 mm, UF 455464. Seminole Reservoir [Lake Seminole] near landing at Seminole Lodge Marina, about 1 air mile northeast of Sneads, Jackson County, Florida, 27 June 2006. © Richard T. Bryant. Lower image: male, length 61 mm, UF 455464. Seminole Reservoir [Lake Seminole] near landing at Seminole Lodge Marina, about 1 air mile northeast of Sneads, Jackson County, Florida, 27 June 2006. © Richard T. Bryant.

Description

Size: length to 90 mm, in Florida to 85 mm.

Shell: thin to moderately thick; smooth; inflated, width 35%–45% of length; outline elliptical; anterior margin rounded; posterior margin broadly rounded to obliquely truncate in females, narrowly rounded to bluntly pointed in males; dorsal margin slightly convex; ventral margin straight to convex; posterior ridge low, rounded; posterior slope moderately steep, flat to concave; umbo broad, moderately inflated, elevated slightly to well above hinge line; umbo sculpture thin, double-looped ridges, with deep indentation ventrally (Figure 10.114); umbo cavity wide, shallow.

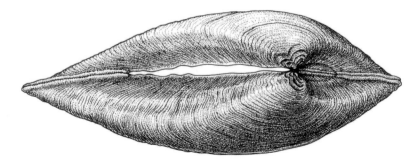

Figure 10.114. Umbo sculpture of *Villosa villosa,* female, length 35 mm, UF 449115. Boat basin off Suwannee River in the Original Suwannee River Campground, about 1 mile west of Fanning Springs, Dixie County, Florida, 23 January 2008. Illustration by Susan Trammell.

Teeth: pseudocardinal teeth small, moderately thick, compressed, 2 teeth in left valve, anterior tooth typically larger, crest almost parallel to shell margin, 1 tooth in right valve, occasionally with accessory denticle anteriorly; lateral teeth moderately long, thin, straight to slightly curved, 2 in left valve, 1 in right valve; interdentum moderately long, narrow.

Nacre: bluish white, highly iridescent posteriorly.

Periostracum: typically clothlike, occasionally dull; green, dark olive to brown, or black, with green rays of varying width and intensity, may be obscured by dark color.

Glochidium Description

Outline subelliptical; length 200–250 μm; height 233–308 μm; width 114–136 μm; ventral margin with internal and external irregularly perpendicular rows of lanceolate micropoints (Hoggarth 1999; Williams et al. 2008) (Figure 10.115).

Similar Sympatric Species

Villosa villosa resembles *Villosa lienosa* but is more elongate and has a thinner shell and pseudocardinal teeth. Dark pigment extends the entire length of the ventral mantle margin of *V. villosa* but ends abruptly just anterior of the papillate mantle fold in *V. lienosa*.

Villosa villosa resembles *Villosa vibex* but is more elongate, narrower posteriorly, more inflated, and typically has a slightly thicker shell. Green rays of *V. villosa* are less visible due to the overall darker periostracum than those of *V. vibex*, which are typically prominent.

Villosa villosa resembles some *Hamiota australis* but has a duller, clothlike periostracum and typically thinner pseudocardinal teeth.

Small *Villosa villosa* resemble *Toxolasma* but have umbo sculpture in the form of deeply indented ridges, which gives the appearance of strongly double-looped sculpture instead of single-looped ridges. Also, the teeth of *Toxolasma* in Florida are typically much thicker than those of *V. villosa*.

Distribution in Florida

Villosa villosa occurs in most major basins from Choctawhatchee River to Suwannee River (Figure 10.116).

Ecology and Biology

Villosa villosa typically inhabits creeks, rivers, and sloughs, in slow to moderate current, but may occur in some reservoirs. It is found in substrates of sandy mud, sand, and mixtures of sand and detritus. *Villosa villosa* from Mosquito Creek reservoir near Chattahoochee, Florida, are considerably larger than their counterparts from streams, potentially attributable to differences in productivity in these waters.

Villosa villosa is a long-term brooder, gravid from late summer or autumn to the following summer (Kotrla 1988). Females have a darkly mottled, rusty brown to black papillate mantle fold, which is likely used to lure glochidial hosts. Papillae may approach 1 cm in length during display.

In laboratory trials *Villosa villosa* glochidia metamorphosed on two fishes that occur in Florida—Centrarchidae: *Lepomis macrochirus* (Bluegill) and *Micropterus salmoides* (Largemouth Bass) (Keller and Ruessler 1997a).

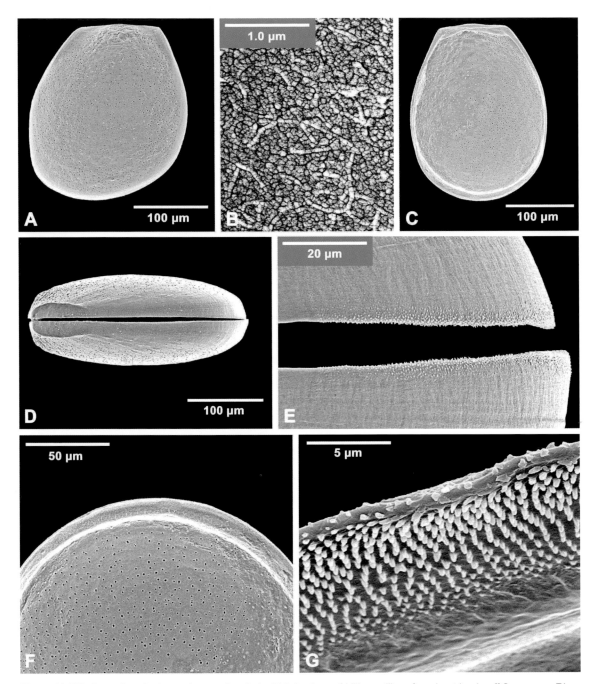

Figure 10.115. Scanning electron micrographs of glochidial valves of *Villosa villosa* from boat basin off Suwannee River in the Original Suwannee River Campground, west of Fanning Springs, Dixie County, Florida, 23 January 2008—(A) external view; (B) surface sculpture; (C) internal view; (D) lateral view; (E) lateral view of micropoints along ventral margin; (F) internal view of ventral margin; and (G) close-up of micropoints.

Villosa villosa has been observed to be infested with mites (Unionicolidae). Of 93 individuals examined from six localities in Apalachicola, Ochlockonee, and Suwannee

River basins, 50 were parasitized by one or more of the following species: *Najadicola ingens*, *Unionicola abnormipes*, *Unionicola hoesei*, *Unionicola lasallei*, *Unionicola poundsi*, and *Unionicola sakantaka* (Vidrine 1996).

Conservation Status

Villosa villosa was assigned a conservation status of special concern throughout its range by Williams et al. (1993) and the equivalent status of vulnerable by Williams et al. (in review).

Villosa villosa is extant throughout its range but is generally uncommon. This species appears to be rare in Choctawhatchee River basin.

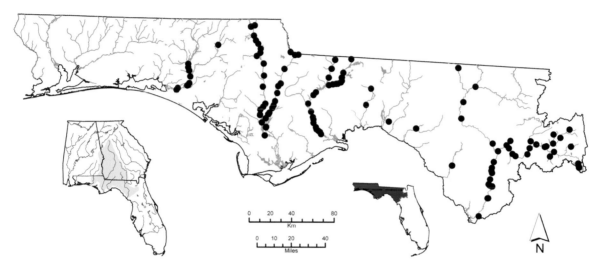

Figure 10.116. Distribution of *Villosa villosa*.

Remarks

Villosa villosa exhibits considerable variation in abundance across its range. It is more frequently encountered in Suwannee, Ochlockonee, and Apalachicola basins compared to Choctawhatchee basin. It is more common in the lower reaches of rivers.

Villosa villosa was first reported from Escambia River, Florida, by Simpson (1900a). It was reported by Williams et al. (2008) from a single locality in a tributary of Conecuh River in Alabama. This specimen has been reidentified as *Toxolasma* sp. cf. *corvunculus*. There are no known records of *V. villosa* from Escambia River basin in Alabama and Florida.

Small *Villosa villosa* are occasionally misidentified as *Toxolasma*. Some of the Chipola River mussels reported by van der Schalie (1940) as *Carunculina vesicularis* are based on *V. villosa*.

Villosa villosa was formerly considered to be distributed east and south of Suwannee River by Johnson (1972) and Heard (1979a). The *Villosa* population in St. Marys River basin is now considered to be *V. vibex*, while *V. villosa* records on the Gulf Coast south of Waccasassa River and probably some records in St. Johns River basin are actually based on *V. amygdalum*.

It was suggested by Williams et al. (2008) that the Choctawhatchee River population of *Villosa villosa* may represent an undescribed species. This population has a much shinier, less clothlike periostracum, with more prominent green rays than those from elsewhere in its range (Butler 1990). Genetic data corroborate the suggestion that the Choctawhatchee River population may represent a distinct taxon (N.A. Johnson unpublished data). There appears to be a faunal break between Choctawhatchee and Apalachicola River basins, which has previously been noted for mussels as well as other aquatic groups (Swift et al. 1986; Butler 1990).

Synonymy

Unio villosus B.H. Wright 1898 (B.H. Wright 1898c:32). Lectotype (Johnson 1967a), USNM 150503, length 49 mm (female). Type locality: Suwannee River, Luraville, Suwannee County, Florida, [Suwannee River basin].

Chapter 11

Species of Hypothetical Occurrence

There are five mussels that have never been collected in Florida but are known to occur in upstream reaches of rivers that flow into the state. The five species are *Margaritifera marrianae* of the family Margaritiferidae, and *Elliptio nigella*, *Lampsilis binominata*, *Lasmigona subviridis*, and *Pyganodon cataracta* of the family Unionidae. Although never documented from Florida, it is possible that one or more of them may have occurred in the state in the past or will potentially be found in the future. The primary basis for this assumption is the proximity of occurrences of these species to the state and presence of appropriate habitat in Florida streams.

Margaritifera marrianae Johnson 1983
Alabama Pearlshell

 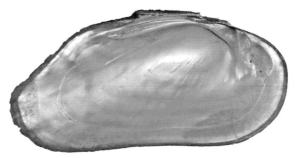

Margaritifera marrianae – Length 83 mm, MCZ 28491. Hunter's Creek, 5 miles southwest of Evergreen, Conecuh County, Alabama, February 1912. Photograph by James D. Williams.

Description

Size: length to 96 mm.

Shell: usually moderately thick; with plications, varying in size, typically more pronounced posterioventrally, well-developed plications originating on the posterior ridge radiate ventrally across posterior half of disk, anterior portion of disk typically without sculpture; moderately inflated, width 25%–35% of length; outline elliptical to trapezoidal; anterior margin rounded; posterior margin obliquely truncate to bluntly pointed; dorsal margin slightly convex; ventral margin straight to slightly concave; posterior ridge low, rounded; posterior slope flat to slightly concave, with corrugations; umbo moderately broad, not inflated, elevated very slightly above hinge line; umbo sculpture unknown; umbo cavity moderately wide, shallow.

Teeth: pseudocardinal teeth moderately thick, blunt, triangular, 2 divergent teeth in left valve, anterior tooth smaller, crest of anterior tooth oriented anterioventrally, 1 tooth in right valve; lateral teeth moderately thick, straight to slightly curved, 2 in left valve, 1 in right valve; interdentum long, narrow.

Nacre: white or bluish white.

Periostracum: dull to clothlike; small individuals olive brown to brown, large individuals dark brown to black.

General Distribution

Margaritifera marrianae occurs in tributaries of Conecuh River drainage in Conecuh and Crenshaw counties, Alabama, and tributaries of Alabama River in Monroe County, Alabama. An additional population is known from Buttahatchee River in Marion County, Alabama (Williams et al. 2008).

Habitat

Margaritifera marrianae inhabits creeks in slow to moderate current in sandy mud, sand, and gravel substrates. These streams typically have substantial groundwater input and are lined with stable riparian vegetation.

Reasons for Inclusion

Small-stream habitat in Escambia River basin in Florida is similar to that found in south Alabama, though it is limited. Streams in both states also support similar fish faunas including species of *Noturus*, a potential fish host for *Margaritifera marrianae* (P.D. Johnson, personal communication).

Elliptio nigella (Lea 1852)
Winged Spike

Elliptio nigella – Length 67 mm, UF 449604. Flint River about 200 meters downstream of Normans Ferry Landing boat ramp at end of County Road 55, Baker County, Georgia, 11 July 2011. Photograph by James D. Williams.

Description

Size: length to 61 mm.

Shell: moderately thin; smooth; moderately inflated, especially along posterior ridge, width 25%–40% of length; outline elliptical to subrhomboidal, greatest height midway between umbo and posterior end; anterior margin rounded; posterior margin broadly rounded to obliquely truncate; dorsal margin convex, with slight dorsal wing posteriorly; ventral margin straight to slightly convex; posterior ridge well developed, usually biangulate posterioventrally; posterior slope usually flat, slightly concave; umbo low, broad, elevated slightly above hinge line; umbo sculpture thick, single-looped, nodulous ridges; umbo cavity wide, shallow.

Teeth: pseudocardinal teeth small, compressed, 2 teeth in left valve, oriented almost parallel to shell margin, 1 tooth in right valve; lateral teeth moderately thick, straight to slightly curved, 2 in left valve, 1 in right valve; interdentum moderately long, narrow.

Nacre: white, bluish white to pale orange.

Periostracum: small individuals shiny, large individuals dull to clothlike; small individuals olive green with dark rays, large individuals olive brown or black, usually obscuring rays.

General Distribution

Elliptio nigella occurs in Apalachicola River basin, Alabama and Georgia.

Habitat

Elliptio nigella inhabits large rivers in moderate and swift currents in coarse sand substrates often associated with bedrock.

Reasons for Inclusion

Simpson (1914) reported *Elliptio nigella* from Chattahoochee River south into Florida, but there are no known specimens documenting its occurrence in the state (Brim Box and Williams 2000; Williams et al. 2008). *Elliptio nigella* historical records are known from within 20 river miles of the Florida border.

Elliptio nigella went undetected for half a century until recently when a population was found in Flint River, Baker County, Georgia. Habitat in Apalachicola River below Jim Woodruff Lock and Dam is similar to rocky outcrops with pockets of sand where it occurs in the Flint River, Georgia.

Lampsilis binominata Simpson 1900
Lined Pocketbook

Lampsilis binominata – Length 37 mm, male, USNM 84883. [Chattahoochee River,] Columbus, Muscogee County, Georgia. © Richard T. Bryant.

Description

Size: length to 60 mm.

Shell: thin; smooth; moderately inflated, width 40%–55% of length; outline oval, occasionally round; anterior margin rounded; posterior margin broadly rounded in females, narrowly rounded in males; dorsal margin convex; ventral margin convex; valves gape anteriorly and posteriorly; females more swollen posteriorly than males; posterior ridge well defined near umbo, broadly rounded posterioventrally, occasionally slightly biangulate; posterior slope flat to slightly concave, often with few fine wrinkles; umbo broad, inflated, elevated well above hinge line; umbo sculpture concentric ridges, with slight indentation on ventral margin; umbo cavity wide, moderately deep.

Teeth: pseudocardinal teeth compressed, 2 teeth in left valve, roughly parallel to dorsal margin, anterior tooth considerably larger, 1 tooth in right valve, usually with well-developed lamellar accessory denticle anteriorly along shell margin; lateral teeth thin to moderately thick, straight to slightly curved, 2 in left valve, 1 in right valve; interdentum moderately long, narrow.

Nacre: white to bluish white.

Periostracum: typically shiny; yellowish, with thin, irregularly spaced, green rays.

General Distribution

Lampsilis binominata occurs in Chattahoochee River in Alabama and Georgia, and Flint River and its large headwater tributary, Line Creek, in Georgia.

Habitat

Based on historical museum records, it appears *Lampsilis binominata* occurred in large creeks and rivers with moderate to swift current in substrates of sand and gravel (Williams et al. 2008).

Reasons for Inclusion

Although collection sites in Alabama and Georgia are more than 100 air miles from Florida, habitat in Apalachicola River below Jim Woodruff Lock and Dam are similar to sites where *Lampsilis binominata* has been found. Further, archaeological specimens that resemble the species are known from Flint River in Georgia.

Lasmigona subviridis (Conrad 1835)
Green Floater

Lasmigona subviridis – Length 48 mm, UMMZ 23324. Chattahoochee River, Columbus, Muscogee County, Georgia.
© Richard T. Bryant.

Description

Size: length to 63 mm.

Shell: thin; smooth; moderately compressed, width 35%–45% of length; outline elliptical; anterior margin rounded; posterior margin rounded, may be slightly truncate;

dorsal margin straight to slightly convex; ventral margin straight to slightly convex; posterior ridge rounded, may be weakly biangulate posterioventrally; posterior slope moderately steep, flat to slightly concave; umbo broad, elevated slightly above hinge line; umbo sculpture thick, double-looped, nodulous ridges; umbo cavity moderately wide, shallow.

Teeth: pseudocardinal teeth low, rudimentary, compressed, typically 2 teeth in left valve, 1 tooth in right valve, interdental projection in left valve poorly developed, confluent with pseudocardinal tooth; lateral teeth thin, straight, 2 teeth in left valve, 1 in right valve; interdentum moderately long, very narrow, interdentum of right valve may be divided by depression that accommodates interdental projection of opposing valve.

Nacre: white to bluish white.

Periostracum: shiny to dull; green, yellowish olive to brown, usually with green rays.

General Distribution

Lasmigona subviridis occurs along the Atlantic Coast in basins from the St. Lawrence River south to Cape Fear River in North Carolina. There is also an isolated population in Savannah River basin (Fuller 1971) and archaeological material has been recovered from a midden in the Altamaha River basin in Georgia (A.E. Bogan, personal communication). It also occurs in interior drainages in North Carolina, Virginia, and West Virginia (Clarke 1985; Pinder et al. 2003), and east Tennessee and west North Carolina (Parmalee and Bogan 1998). In eastern Gulf Coast rivers, it was found only in Apalachicola basin in Alabama and Georgia (Brim Box and Williams 2000).

Habitat

Lasmigona subviridis inhabits creeks and rivers in slow current in mud, sand, and gravel substrates.

Reasons for Inclusion

Habitats of *Lasmigona subviridis*, side channels of rivers and large creeks in slow currents, exist in Apalachicola River. The nearest localities for *L. subviridis* in Alabama and Georgia are from Chattahoochee and Flint Rivers, and are more than 100 air miles north of Florida.

Pyganodon cataracta (Say 1817)
Eastern Floater

Pyganodon cataracta – Length 105 mm, UF 359783. Lake Marion, Santee National Wildlife Refuge, Clarendon County, South Carolina, 7 October 2002. Photograph by James D. Williams.

Description

Size: length to 170 mm.

Shell: thin to moderately thick; smooth; inflated, width usually 30%–45% of length; outline oval to elliptical; anterior margin rounded; posterior margin rounded; dorsal margin straight to slightly convex; ventral margin straight to slightly convex; posterior ridge rounded; posterior slope moderately steep dorsally, flat to slightly concave; umbo broad, inflated, elevated above hinge line; umbo sculpture moderately thick, double-looped, nodulous ridges; umbo cavity moderately wide, deep.

Teeth: pseudocardinal and lateral teeth absent; hinge often thickened posterior to umbo.

Nacre: white to bluish white, occasionally with pinkish tint in umbo cavity.

Periostracum: small individuals shiny, large individuals dull to clothlike; small individuals yellowish to greenish brown, occasionally with green rays, large individuals yellowish brown to dark brown.

General Distribution

Pyganodon cataracta is native to the Atlantic Coast basins of North America, ranging from the St. Lawrence River, Canada, south to the Altamaha River, Georgia (Johnson 1970). In Gulf Coast basins, *Pyganodon cataracta* was reported in Apalachicola River in Alabama, Florida, and Georgia (Brim Box and Williams 2000) and a reservoir on Tallapoosa River in Alabama, which appears to be a recent introduction (Williams et al. 2008).

Habitat

Pyganodon cataracta inhabits creeks, rivers, ponds, lakes, and reservoirs. It is usually found in backwater areas with little or no current but can also be found along stream banks in sheltered areas in moderate current in mud, sandy mud, sand, and occasionally gravel substrates.

Reasons for Inclusion

Pyganodon cataracta was reported from Apalachicola River in Florida by Brim Box and Williams (2000) based on shell morphology of specimens used in a genetic analysis from reservoirs in Georgia. There are no reliable differences in shell morphology that distinguish *P. cataracta* from *Pyganodon grandis*. Samples from Apalachicola and Chipola Rivers exhibited shell morphology that graded imperceptibly between the extremes of the two species (Brim Box and Williams 2000). The Apalachicola basin *P. cataracta* populations may have resulted from introduction of glochidia-infested host fishes (e.g., *Morone saxatilis*) from an Atlantic Coast reservoir.

An introduced population of *Pyganodon cataracta* was discovered in the Suwannee River basin in 2002 in Warrior Creek, a tributary of Withlacoochee River (northern), Colquitt County, Georgia, and appears to be recruiting.

Chapter 12

Additional Bivalves in Inland Waters

Florida inland waters are inhabited by 16 species of bivalve mollusks that are not members of the families Margaritiferidae and Unionidae (Table 12.1). These additional bivalves belong to the Cyrenidae (formerly Corbiculidae), Dreissenidae, Mactridae, and Sphaeriidae, all families in the order Veneroida (Carter et al. 2011). Some members of these families may superficially resemble unionids, but they differ substantially in morphology, life history, and ecology. Perhaps the most profound differences are those associated with life history. Almost all unionids are dioecious, with a parasitic glochidium larval stage that requires a host, typically fish, to transform into a juvenile. Due to this dependence, most unionids have distributions that in part reflect the ranges of their host fishes (Haag and Warren 1998). By comparison, most other bivalves produce veligers that do not require a fish host and develop directly from larvae to adults. Bivalves producing veligers are often more abundant and have dispersal mechanisms that contribute to more cosmopolitan distributions than most unionids (McMahon and Bogan 2001).

Most bivalves (excluding Margaritiferidae and Unionidae) in Florida are capable of producing multiple generations per year and under certain environmental conditions can become so abundant that they are numerically dominant components of the benthic fauna. *Corbicula fluminea* often outnumbers other invertebrates in the sand sediments present in many Florida streams. *Corbicula fluminea* can also have substantial impacts on the ecological function of inland aquatic ecosystems in Florida. Their ability to attain high densities is generally viewed as disruptive and detrimental to other biota, especially benthic species. Huge accumulations of *Corbicula* shells may mimic gravel substrates that transform soft, penetrable bottom habitats, typically inhabited by infaunal benthic communities, into hard bottoms suitable only for epifaunal species. These transformations have occurred in such diverse locations as Little Wekiva River, Apalachicola River, and Lake Okeechobee. These bivalves have also been shown to substantially influence phytoplankton populations and nutrient cycling (McMahon and Bogan 2001), thus influencing the base of the food chain. These bivalves are nevertheless an important food resource in freshwaters and are consumed by fishes, turtles, muskrats, otters, waterfowl, and other organisms.

Three bivalves—*Mytilopsis leucophaeata*, *Polymesoda caroliniana*, and *Rangia cuneata*—are known in Florida from brackish water and freshwater habitats in tidally influenced streams. The transition zone between freshwater and brackish water in coastal rivers varies depending on tidal cycle, stream gradient, and discharge. During prolonged droughts brackish waters may extend farther inland in coastal streams than during periods of normal flow. In coastal areas where groundwater withdrawal exceeds recharge, salt water intrusion may penetrate low-lying streams, independent of stream gradient and discharge, and allow establishment of brackish water species. If sea levels continue to rise in response to climate change, populations of these bivalves may extend their distributions inland.

Family Cyrenidae – Marshclams

The family Cyrenidae (formerly Corbiculidae) is represented by three species within waters of the United States (Turgeon et al. 1998; Mikkelsen and Bieler 2008). Cyrenidae differ anatomically from Unionidae in having true siphons as opposed to in-current and excurrent apertures (McMahon and Bogan 2001). Two species, *Polymesoda caroliniana* and *Polymesoda maritima*, are predominantly found in estuarine and marine environments. However, *P. caroliniana* also occurs in brackish water and some freshwater habitats. The remaining species, *Corbicula fluminea*, is nonindigenous and occurs in freshwater habitats but may occasionally be found in brackish water.

Corbicula fluminea (Müller 1774)
Asian Clam

Description

Length to 50 mm; shell moderately thick, round or oval to triangular; numerous evenly spaced, thin, concentric ridges; umbo centrally positioned, moderately inflated, extending above shell margin; periostracum yellowish to light brown or dark olive, becoming dark brown to black in large individuals; hinge plate with three cardinal teeth in each valve; lateral teeth straight to slightly curved, serrated, 2 on each side of cardinal teeth in right valve, 1 on each side of cardinal teeth in left valve; pallial line without pallial sinus; nacre white to purple (Figure 12.1).

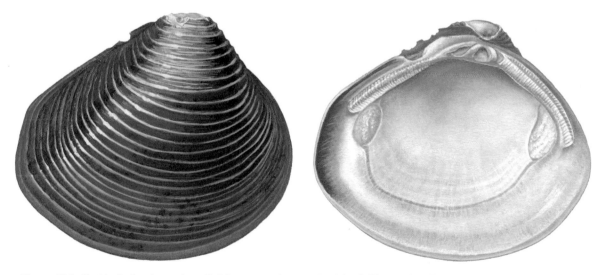

Figure 12.1. *Corbicula fluminea* – Length 30 mm, specimen not retained. Choctawhatchee River, 1.5 air miles south of Interstate Highway 10 bridge, Holmes and Washington counties, Florida, June 1998. Illustration by Susan Trammell.

Range

Corbicula fluminea is widely distributed in the United States except in the northernmost states where its distribution is limited by low water temperatures (Mackie 2007). In Florida it is distributed throughout the state.

Remarks

Corbicula fluminea occurs in a wide variety of freshwater habitats including springs, spring runs, creeks, rivers, ponds, lakes, canals, and reservoirs. The species represents the most abundant and ubiquitously distributed bivalve in inland waters. Its life history and morphological characteristics allow adaption to a variety of habitat conditions, including unstable sand bottoms. These characteristics contribute to its status as the most invasive of all North American freshwater bivalves (McMahon 1999; McMahon and Bogan 2001). In Florida *C. fluminea* has been documented in densities exceeding 1,000 per square meter in a variety of habitats (Warren and Vogel 1991; Warren et al. 2000). In tidal estuaries of southwest Florida, *C. fluminea* occurred at salinities of 7 ppt but was most common in areas less than 2 ppt (Montagna et al. 2008).

The date and location of the introduction of *Corbicula fluminea* into North America is not known with certainty. It may have been introduced into the northwestern United States as early as the 1840s (Fox 1970), but it most likely arrived in the Columbia River region in the 1920s or 1930s (Counts 1986; Warren 1997; McMahon 1999). It was first collected in the United States with certainty in 1938 from Columbia River (Sinclair and Isom 1963; Sinclair 1971) and spread rapidly across most of the United States. By 1982 it was present in 35 of the continental states. *Corbicula fluminea* was first observed in Florida panhandle streams in the early 1960s (Schneider 1967). By the mid-1970s, it had spread throughout Florida and was present in most major basins (Heard 1964, 1966, 1979a; Clench 1970; Bass and Hitt 1974).

There is uncertainty regarding *Corbicula* taxonomy and whether multiple species occur in the United States. Several investigations using morphology, karyology, electrophoresis, and genetic analysis have been conducted in an effort to resolve this problem (Hillis and Patton 1982; Britton and Morton 1986; Siripattrawan et al. 2000; Mackie 2007). Research by Lee et al. (2005) documented a more complex problem in *Corbicula* with the discovery of clonal lineages as well as populations of triploids. The taxonomic status of the species in the United States is far from settled. Therefore, only *Corbicula fluminea* is recognized herein as occurring in Florida.

Polymesoda caroliniana (Bosc 1801)
Carolina Marshclam

Description

Length to 90 mm; shell thick, oval to subtriangular; umbo centrally positioned, narrow, inflated, extending well above shell margin; posterior ridge absent; periostracum thin, brown, rough, with fine concentric ridges; external hinge ligament long; hinge plate with 3 cardinal teeth and 1 posterior lateral tooth in each valve; pallial line with deep, narrow pallial sinus; nacre white to purple (Figure 12.2).

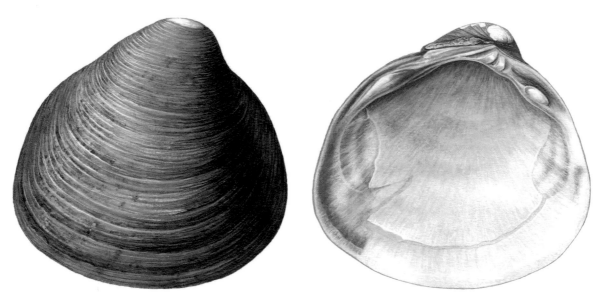

Figure 12.2. *Polymesoda caroliniana* – Length 43 mm, UF 181786. Mouth of Suwannee River, Dixie County, Florida, 1987. Illustration by Susan Trammell.

Range

Polymesoda caroliniana is distributed along the Atlantic Coast from Virginia to northern Florida and on the Gulf Coast from Texas to peninsular Florida (Abbott 1974). Museum records indicate that it occurs as far south as Naples and the Everglades on the Florida Gulf Coast.

Remarks

In Florida *Polymesoda caroliniana* occurs in estuaries and coastal stream reaches both within and upstream of tidal influence. It inhabits warm, shallow waters in mud and sandy mud without macrophytes or coarse organic debris (van der Schalie 1933; Andrews and Cook 1951). Distribution of *P. caroliniana* may be limited by infaunal competition with *Rangia cuneata* where they co-occur (Swingle and Bland 1974). In tidal estuaries of southwest Florida, *P. caroliniana* occurred at salinities between 1–20 ppt and was most abundant at salinities of about 5 ppt (Montagna et al. 2008).

Family Dreissenidae – Falsemussels

The family Dreissenidae is represented by four species in marine, estuarine, and freshwater environments of the United States (Turgeon et al. 1998). Two species are native, *Mytilopsis leucophaeata* and *Mytilopsis sallei*, and two species are introduced, *Dreissena polymorpha* and *Dreissena bugensis*. Both native species occur in Florida estuaries, but only *M. leucophaeata* is known to inhabit some freshwater habitats. Dreissenids are small bivalves that are sessile and use byssal threads to attach themselves to a wide range of natural and man-made substrates. They are considered fouling species due to their tendency to form large colonies on seawalls, water control structures, docks, pipes, and boats; dreissenids are rarely observed living as solitary individuals.

Introduced dreissenid species are known to have a major impact on native mussel populations by byssally attaching to their shells and competing for food and space (Strayer 1999a). When present in large numbers they have caused local extirpation of mussel populations (Martel et al. 2001; Haag 2012). Even more insidious is the fact that native mussel shells seem to be a preferred settling site for *Dreissena polymorpha* veligers (Ricciardi et al. 1996). After attachment of *D. polymorpha*, decline in mussel populations is almost immediate and once fouled, mussels can be extirpated in four to seven years (Martel et al. 2001).

Dreissena polymorpha and *Dreissena bugensis* were introduced into North America from Europe in the 1980s. *Dreissena polymorpha* was introduced into Lakes St. Claire and Erie in 1988 and rapidly spread across most of the eastern United States (Mackie and Schloesser 1996). *Dreissena bugensis* was found in the Erie Barge Canal and eastern Lake Erie in 1991 and had spread as far west as Lake Mead, Nevada, by 2007 (Mills et al. 1996; Stokstad 2007). Both species are believed to have been introduced into North America as veligers via ballast water pumped from ocean-going vessels originating in Ukraine (Mackie and Schloesser 1996).

Dreissena polymorpha and *Dreissena bugensis* have entered Florida but have not been found in open waters. Live *D. polymorpha* held in aquaria by individuals were destroyed by FWC. Dead *D. bugensis* have been found in the intake pipes of boats transported to Florida from Lake Mead, Nevada. Environmental conditions in some Florida freshwater habitats, including such diverse water bodies as Juniper Springs, Wekiva River, and Lake Okeechobee, are suitable for colonization by *D. polymorpha* (Hayward and Estevez 1997). Veliger transport in bilge water and live wells of boats is a likely pathway for introduction.

Mytilopsis leucophaeata (Conrad 1831)
Dark Falsemussel

Description

Length to 25 mm; shell byssate, moderately thin, elongate, narrow to bluntly pointed anteriorly, broadly rounded posteriorly; umbo positioned anteriorly, narrowly rounded to bluntly pointed, directed anterioventrally; ventral margin nearly straight, slightly convex posteriorly in large specimens; byssal notch slightly developed; dorsal margin broadly convex from umbo to posterior margin; posterior margin broadly rounded; periostracum cream-colored to dark brown, sometimes with zigzag color pattern similar to *Dreissena polymorpha*, especially in juveniles, also with thin concentric lines; apophysis small, rounded, occasionally pointed, closely associated with septum; pallial and extrapallial regions porcellaneous; nacre white with grayish areas posteriodorsally in some specimens (Figure 12.3).

Figure 12.3. *Mytilopsis leucophaeata* – Length 22 mm, UF 388524. St. Johns River, Florida, June 2002. Illustration by Susan Trammell.

Range

Mytilopsis leucophaeata is distributed along the Atlantic Coast from New York to Florida and on the Gulf Coast from Mexico to Florida (Abbott 1974).

Remarks

In Florida *Mytilopsis leucophaeata* occurs in freshwater reaches of streams, estuaries, brackish coastal lakes, and coastal stream reaches, both within and above tidal influence (Marelli and Gray 1983).

Upstream distribution limits of *Mytilopsis leucophaeata* may be controlled by the physiological tolerances of its larvae (Marelli and Gray 1983). It has probably spread to some freshwater localities in Florida via boats. Dreissenid mussels are extremely prolific with high fecundities and rapid growth and maturity rates. It has been observed in Florida freshwaters in such diverse locations as middle Suwannee River, Clewiston Lock on Lake Okeechobee, and Econlockhatchee River in the middle St. Johns River basin.

Mytilopsis leucophaeata has been confused with introduced species of *Dreissena* but can be distinguished by the presence of an apophysis, an internal shell character that is absent in *Dreissena* (Figure 12.4).

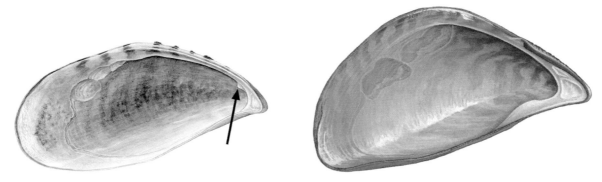

Figure 12.4. Interior of left valves of *Mytilopsis leucophaeata* (left) and *Dreissena polymorpha* (right). The arrow indicates the location of the apophysis in *M. leucophaeata*. Illustrations by Susan Trammell.

Family Mactridae – Surfclams

Nine genera and 10 species of the family Mactridae occur in the United States (Turgeon et al. 1998). They are primarily inhabitants of estuarine and marine environments, but *Rangia cuneata* is occasionally found in freshwater.

Rangia cuneata (Sowerby 1831)
Atlantic Rangia

Description
Length to 85 mm; shell thick, triangular to oval; umbo broad, inflated, extending well above shell margin, curved anteriorly; posterior ridge often well developed; posterior slope usually steep; periostracum thin, yellowish brown to grayish brown; hinge plate moderately wide; anterior and posterior lateral teeth well developed; hinge ligament fully internal, positioned in chondrophore; pallial line with moderately deep pallial sinus; nacre white to bluish white (Andrews 1971, 1981; Abbott 1974) (Figure 12.5).

Figure 12.5. *Rangia cuneata* – Length 50 mm, UF 360321. Choctawhatchee River at junction with Sister River, Walton County, Florida, March 2000. Illustration by Susan Trammell.

Range
Rangia cuneata occurs on the Atlantic Coast from New York to Palm Beach County, Florida, and on the Gulf Coast from Mexico to Lee County, Florida (Woodburn 1962; Gunter and Hall 1965; Abbott 1974).

Remarks
In Florida *Rangia cuneata* occurs in tidally influenced reaches of larger streams but has been found in the St. Johns River upstream of Lake Harney, Seminole and Volusia counties. The species inhabits soft mud and sand sediments in low salinity (0–18 ppt), shallow, brackish water, and freshwater habitats. Unlike *Polymesoda caroliniana*, *R. cuneata* may also be found among roots in vegetated habitats. In tidal estuaries of

southwest Florida, *R. cuneata* occurred at salinities of 1–16 ppt and was most abundant at 4–7 ppt (Montagna et al. 2008).

It was not collected from the Atlantic Coast north of Georgia prior to the 1950s. Its occurrence in the vicinity of Chesapeake Bay has been attributed to accidental introduction during oyster transplants (Pfitzenmeyer and Drobeck 1964).

Family Sphaeriidae – Fingernailclams, Peaclams, and Pillclams

The family Sphaeriidae includes small (length usually less than 25 mm), thin-shelled bivalves that are widely distributed across North America and occur in a variety of natural and man-made habitats, including wetlands, creeks, rivers, ponds, lakes, roadside ditches, canals, and reservoirs. In the United States and Canada, the family is represented by four genera (*Eupera, Musculium, Pisidium*, and *Sphaerium*) and 39 species (Mackie 2007). All genera and 12 species have been reported from Florida (Table 12.1). Species descriptions presented herein are focused upon external shell characters useful for differentiation of taxa. Detailed sphaeriid accounts, including species descriptions, life history, and ecological information, are presented by Herrington (1962) and Mackie (2007).

A comprehensive morphologically based treatment of Sphaeriidae was presented by Mackie (2007). However, the work was completed several years before funding was procured for its publication and does not include some of the more recent works addressing taxonomic and phylogenetic relationships. In 2003 most taxa of the Sphaeriidae were included in a molecular genetic analysis using mitochondrial and nuclear genes (Lee and O'Foighil 2003). Results of the molecular analysis did not recover the currently recognized genera *Musculium* and *Sphaerium* as monophyletic clades. The conflict between the morphological and phylogenetic studies has resulted in taxonomic uncertainty (Lee 2004). Generic names herein follow the traditional taxonomy using shell morphology and soft anatomy (Mackie 2007), but it should be pointed out that these are likely to change in the future based on phylogenetic studies.

The Sphaeriidae are characterized by small size, yellowish to brown periostracum, two cardinal teeth in the left valve (only one in *Eupera cubensis*), one cardinal tooth in the right valve, and nonserrated lateral teeth located anterior and posterior of the cardinal teeth. The ligament holding the two valves together is short and not well developed. The pallial line and muscle scars are indistinct.

Three of the four sphaeriid genera can usually be distinguished by the position of the umbo (Figure 12.6). The umbo is anterior in *Eupera*, posterior in *Pisidium*, and central in both *Sphaerium* and *Musculium*. Distinguishing *Musculium* and *Sphaerium* can be difficult, but *Musculium* has a very narrow hinge plate beneath the umbo and in *Sphaerium* it is two or more times wider than the thickened end of the cardinal tooth on the right valve. Taxonomic keys useful for identifying species of Sphaeriidae are included in Herrington (1962), Burch (1972), Heard (1979a), and Mackie (2007).

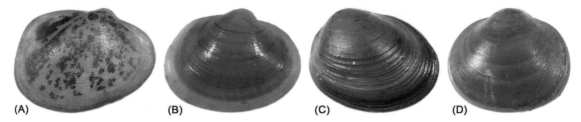

Figure 12.6. External shell morphology—(A) *Eupera cubensis*, length 4.1 mm, UF 247589. Apalachicola River at Race Shoals, Gadsden and Jackson counties, Florida, 26 July 1976. (B) *Musculium securis*, length 13.4 mm, UF 3749. Fish-eating Creek, 3 miles east of Palmdale, Glades County, Florida, 10 April 1939. (C) *Pisidium dubium*, length 6.2 mm, UF 247847. Choctawhatchee River, 1.5 miles west of Pittman, Holmes County, Florida, 27 May 1970. (D) *Sphaerium occidentale*, length 6.8 mm, UF 3758. Chipola River, Bellamy Bridge,12 miles above Marianna, Jackson County, Florida, 9 June 1930. Photographs by Gary L. Warren.

Eupera cubensis (Prime 1865)
Mottled Fingernailclam

Description

Length to 8 mm; shell oval; umbo distinctly anterior, low but prominent, inclined anteriorly; dorsal margin convex, joining anterior margin with little or no angle; ventral margin rounded; anterior margin narrowly rounded; posterior margin nearly truncate, joining dorsal margin at ~90° angle; height/length ratio ~0.77; width/length ratio ~0.56; periostracum usually mottled with dark pigment (Mackie 2007) (Figure 12.6-A).

Range

Eupera cubensis occurs from North Carolina to Texas and north to Kansas (Burch 1972; Mackie 2007). A nonindigenous population occurs in the Illinois River drainage in Illinois (Sneen et al. 2009). In Florida it is reported to occur in the Apalachicola, Ochlockonee, Waccasassa, St. Johns, and Hillsborough River basins, as well as Lake Okeechobee and its tributaries. It is also found in canals in Palm Beach, Broward, and Dade counties (Heard 1979a).

Remarks

Eupera cubensis occurs in creeks, rivers, and occasionally impoundments in mud and sand substrates. Using byssal threads it will attach to roots, submerged macrophytes, wood debris, and other hard surfaces but also occurs unanchored.

Musculium lacustre (Müller 1774)
Lake Fingernailclam

Description

Length to 10 mm; shell triangular, rhomboidal, or suborbicular; umbo subcentral, swollen, pointed, usually calyculate, directed medially and anteriorly; dorsal margin curved; ventral margin flattened or curved; anterior end narrow, rounded, joining dorsal margin at distinct angle, joining ventral margin at rounded angle; posterior end truncate or square, joining dorsal margin at sharp or slightly rounded angle, joining ventral margin with rounded angle; height/length ratio range 0.83–0.90; width/length ratio range 0.60–0.69; striae moderately fine to very fine (Mackie 2007).

Range

Musculium lacustre has the most extensive range of all Nearctic *Musculium*, occurring from Newfoundland and Hudson Bay west to the Northwest Territories and California to Florida (Mackie 2007). In Florida it has been reported from the Escambia, Apalachicola, St. Marks, Suwannee, and St. Johns River basins (Heard 1979a).

Remarks

Musculium lacustre occurs in marshes, swamps, creeks, rivers, ponds, lakes, and roadside ditches. It thrives in mud bottoms often among macrophytes (Herrington 1962; Heard 1979a; Mackie 2007). It has been collected at depths of 8 m (Herrington 1962).

Musculium partumeium (Say 1822)
Swamp Fingernailclam

Description

Length to 8 mm; shell subquadrate to oval, ends almost truncate; umbo subcentral to slightly anterior, extending slightly to moderately above shell margin; dorsal margin almost straight or only slightly curved; ventral margin curved; anterior end rounded, joining dorsal margin at angle; posterior end truncate to slightly rounded, joining dorsal margin at rounded but nearly 90° angle; height/length ratio ~0.85; width/length ratio ~0.60; striae fine, regularly spaced (Mackie 2007).

Range

Musculium partumeium is widespread in the United States, occurring in all states east of the Rocky Mountains and in Nevada and California. In Canada it occurs east of the Rocky Mountains. In Florida it has been found in Ochlockonee, Aucilla, Suwannee, Withlacoochee (southern), Peace, Hillsborough, and St. Johns Rivers, and canals in Dade County (Heard 1979a).

Remarks

Musculium partumeium inhabits shallow areas in swamps, creeks, rivers, ponds, lakes, and canals in mud substrates and often in association with organic debris (Herrington 1962; Mackie 2007). It may be associated with ephemeral habitats (Heard 1979a) and is relatively tolerant of anoxic conditions (Mackie 2007).

Musculium securis (Prime 1852)
Pond Fingernailclam

Description

Length to 12 mm; shell rhomboidal to suborbicular, inflated, higher on posterior end than anterior end; umbo subcentral to anterior, extending slightly above shell margin;

dorsal margin moderately long, convex; ventral margin curved; anterior end rounded, joining dorsal margin with or without angle; posterior end truncate, joining dorsal and ventral margins with rounded angles; height/length ratio ~0.88; width/length ratio ~0.64; striae fine, almost not distinguishable, irregularly to regularly spaced (Mackie 2007) (Figure 12.6-B).

Range

Musculium securis occurs in the eastern and central United States north to northern Canada (Mackie 2007). In Florida it has been found in Choctawhatchee, Ochlockonee, Suwannee, Peace, and St. Johns Rivers, Lake Okeechobee and its tributaries, and canals in Palm Beach, Broward, and Dade counties (Heard 1979a).

Remarks

Musculium securis is common in ephemeral waters, creeks, rivers, ponds, and canals (Heard 1979a; Mackie 2007). It is often found in macrophyte beds, substrates ranging from mud to fine sand, and mats of organic debris in anoxic conditions (Herrington 1962; Mackie 2007).

Musculium transversum (Say 1829)
Long Fingernailclam

Description

Length to 18 mm, most adults 12 mm; shell oval; umbo subcentral to anterior, extending slightly above shell margin; dorsal margin long, slightly convex; ventral margin long, curved; anterior end rounded to slightly convex, joining dorsal margin with angle; posterior end truncate, joining dorsal margin with right angle; height/length ratio ~0.72; width/length ratio ~0.47; striae fine, may be irregularly spaced (Mackie 2007).

Range

Musculium transversum occurs in the United States and Canada east of the Rocky Mountains but is not known from Virginia, Massachusetts, New Hampshire, and Vermont (Mackie 2007). In Florida it has been found in Escambia, Apalachicola, Suwannee, Peace, Caloosahatchee, and St. Johns Rivers, Lake Okeechobee and its tributaries, and canals in Palm Beach County (Heard 1979a).

Remarks

Musculium transversum occurs in creeks and rivers with slow to swift current and in sloughs, ponds, lakes, and canals. It is found in substrates of mud, sand, and gravel (Herrington 1962; Mackie 2007).

Pisidium adamsi Stimpson 1851
Adam Peaclam

Description

Length to 8 mm; shell oval to suboval, inflated; umbo somewhat posterior, extending well above shell margin; dorsal margin long, gently convex; ventral margin gently curved, joining anterior and posterior ends without angles; anterior end gently but steeply curved; posterior end broadly rounded; height/length ratio ~0.84; width/length ratio ~0.62; striae distinct, regularly spaced (Mackie 2007).

Range

Pisidium adamsi occurs sporadically east of the Rocky Mountains from Florida north to Nova Scotia (Mackie 2007). In Florida it has been reported from Suwannee, St. Johns, and Withlacoochee (southern) Rivers (Heard 1979a).

Remarks

Pisidium adamsi occurs in creeks, ponds, and lakes in mud substrates (Herrington 1962; Mackie 2007).

Pisidium casertanum (Poli 1791)
Ubiquitous Peaclam

Description

Length to 6 mm; shell usually oval to triangular; umbo posterior, broad, depressed, projecting posteriorly, extending slightly above shell margin; dorsal margin long, gently to strongly convex; ventral margin long, curved; anterior and posterior ends rounded, joining dorsal and ventral margins without angles; height/length ratio ~0.86; width/length ratio ~0.62; periostracum usually shiny, striae fine, faint, irregularly spaced (Mackie 2007).

Range

Pisidium casertanum is the most common and widely distributed *Pisidium*, occurring throughout North, Central, and South America (Herrington 1962; Mackie 2007). In Florida it has been found in Escambia, Apalachicola, Ochlockonee, Aucilla, Suwannee, Waccasassa, Withlacoochee (southern), Hillsborough, Myakka, Peace, and St. Johns Rivers, as well as Lake Okeechobee and its tributaries (Heard 1979a).

Remarks

Pisidium casertanum occurs in springs, creeks, lakes, and roadside ditches in mud, clay, and gravel substrates, and occasionally among macrophytes (Heard 1979a; Mackie 2007). It is an opportunistic strategist that can withstand seasonal desiccation and has been found at depths of 40 m in the Great Lakes (Herrington 1962; Mackie 2007).

Pisidium compressum Prime 1852
Ridgebeak Peaclam

Description

Length to 5 mm; shell triangular, moderately compressed; umbo posterior, narrow, usually with prominent apex, extending well above shell margin, typically with ridge on each side; dorsal margin short, round; ventral margin approximately three times as long as dorsal margin, curved, joining posterior end without angle; anterior end straight to slightly convex, joining ventral margin with angle; posterior end broadly rounded; height/length ratio ~0.9; width/length ratio ~0.6; striae fine to coarse, regularly to irregularly spaced (Mackie 2007).

Range

Pisidium compressum occurs in Georgia and Florida (Herrington 1962; Heard 1979a; Mackie 2007). In Florida it has been found in Choctawhatchee, Apalachicola, Ochlockonee, St. Marks, Suwannee, Withlacoochee (southern), and St. Johns Rivers (Heard 1979a).

Remarks

Pisidium compressum occurs in creeks, rivers, and lakes, and rarely in ponds. It has been collected from sand substrates in association with living and decaying vegetation and at depths up to 20 m (Herrington 1962; Mackie 2007).

Pisidium dubium (Say 1817)
Greater Eastern Peaclam

Description

Length to 10 mm; shell elongate, laterally compressed; umbo posterior, broad, smooth, extending slightly above shell margin; dorsal margin strongly convex posteriorly, gently convex anteriorly; ventral margin gently curved; anterior end straight to slightly convex; posterior end rounded to truncate, joining dorsal margin with rounded angle; height/length ratio ~0.8; width/length ratio ~0.5; striae coarse, inconspicuous dorsally, very faint on umbo, irregularly spaced ventrally (Mackie 2007) (Figure 12.6-C).

Range

Pisidium dubium occurs in the Great Lakes region, Atlantic Coast states, and Gulf Coast states from Texas to Florida. In Florida it is found in Escambia, Choctawhatchee, Apalachicola, Ochlockonee, St. Marks, Aucilla, Suwannee, and St. Johns Rivers (Heard 1979a).

Remarks

Pisidium dubium is found in springs, creeks, and small rivers in substrates of silty mud to fine sand at depths up to 10 m (Herrington 1962; Heard 1979a; Mackie 2007).

Pisidium punctiferum (Guppy 1867)
Striate Peaclam

Description

Length to 3.5 mm; shell round to somewhat suboval, moderately inflated; umbo positioned centrally, broad, low, extending slightly above shell margin; dorsal margin convex, joining posterior end with distinct angle and anterior end with rounded angle; anterior end short, broadly rounded, joining ventral margin with rounded angle; posterior end rounded; height/length ratio ~0.85; width/length ratio ~0.64; striae fine, irregularly spaced, absent on umbo (Mackie 2007).

Range

Pisidium punctiferum is native to Mexico and Central and South America but has been introduced in Arizona, New Mexico, Texas, and Florida (Mackie 2007). In Florida it is found in St. Johns, Withlacoochee (southern), and Peace Rivers (Heard 1979a).

Remarks

Pisidium punctiferum inhabits springs and streams with macrophytes, in slow currents and muddy substrates (Herrington 1962; Burch 1975b; Heard 1979a; Mackie 2007).

An erroneous record of *Pisidium nitidum* in Clay County is based on a misidentification of *Pisidium punctiferum* (Lee 2011, 2012).

Sphaerium occidentale Prime 1853
Herrington Fingernailclam

Description

Length to 8 mm; shell oval, moderately inflated; umbo central, broad, rounded, extending slightly above shell margin; dorsal margin long, convex, joining anterior and posterior ends with slight angle; ventral margin curved, joining anterior and posterior ends without angles; anterior and posterior ends curved; height/length ratio ~0.83; width/length ratio ~0.54; periostracum dull, striae fine, becoming faint near umbo (Mackie 2007) (Figure 12.6-D).

Range

Sphaerium occidentale occurs in Canada and the northwestern United States, east to the Atlantic Coast and south to Florida (Mackie 2007). In Florida it occurs in Chipola River (Heard 1979a) and Clay, Duval, and St. Johns counties (Lee 1979, 2011, 2012).

Remarks

Sphaerium occidentale occurs primarily in ephemeral habitats, such as pools and roadside ditches, but has also been found in marshes and creeks (Herrington 1962; Heard 1979a; Mackie 2007).

Sphaerium striatinum (Lamarck 1818)
Striated Fingernailclam

Description

Length to 15 mm; shell rectangular; umbo central to anterior, low, extending slightly above shell margin; dorsal margin slightly to strongly convex; ventral margin broadly rounded, joining anterior and posterior ends with rounded angles; height/length ratio ~0.79; width/length ratio range 0.56–0.68; periostracum dull, striae variable in coarseness, irregularly spaced, often prominent over entire shell (Mackie 2007).

Range

Sphaerium striatinum is the most common and widespread species of *Sphaerium* in North America (Mackie 2007). In Florida it occurs in St. Marks, Aucilla, and Suwannee Rivers (Heard 1979a).

Remarks

Sphaerium striatinum occurs in creeks, rivers, and lakes but is rare in ponds and swamps with stagnant water (Herrington 1962; Heard 1979a; Mackie 2007). *Sphaerium striatinum* inhabits substrates of mud, sand, gravel, and rubble (Mackie 2007).

Key to the Additional Bivalves in Inland Waters

Using a key to identify bivalves can be challenging. Identifications of three of the families—Dreissenidae, Mactridae, and Cyrenidae—included in the key are relatively easy to distinguish based on their distinctive shell characteristics. However, the fourth family, Sphaeriidae, can be very difficult to identify. Additional references may be needed to distinguish some species of this family (e.g., Heard 1979a, Mackie 2007).

The following key to the additional bivalves in Florida's inland waters is in the standard dichotomous format. Each numbered couplet provides a contrasting set of characteristics which leads to the identification of a particular taxa or to an additional couplet. The number in parentheses indicates the couplet that led to it. This allows the user to work backwards through the key to check a questionable identification.

1   Umbo terminal or subterminal; shell byssate .... family Dreissenidae, 2

             Umbo central or subcentral; byssal threads absent on living specimens 5

2(1)   Internal shell with apophysis anteriorly near septum.......genus *Mytilopsis*, 3

             Apophysis absent from anterior region of shell .... genus *Dreissena*, 4

3(2)   Shell elongated, height/width ratio <1.25; apophysis rounded and closely associated with septum... *Mytilopsis leucophaeata*

             Height/width ratio >1.30; apophysis pointed, often hook-shaped (may be found in inland waters) .... *Mytilopsis sallei*

4(2)   Venter of each valve with a sharply acute angle, resulting in the concave appearance of the ventral aspect of the shell (not presently known from open waters of Florida) .... *Dreissena polymorpha*

Venter of each valve with rounded obtuse angle, resulting in the convex appearance of the ventral aspect of the shell (not presently known from open waters of Florida) ....*Dreissena bugensis*

5(1)   Hinge internal, chondrophore between cardinal and posterior lateral teeth; shell thick, heavy......................................family Mactridae, *Rangia cuneata*

Hinge external, without chondrophore; shell variable, not thick and heavy.6

6(5)   Each valve with three distinct cardinal teeth on well-developed hinge plate . ...family Cyrenidae, 7

Neither valve contain three distinct cardinal teeth .... family Sphaeriidae, 9

7(6)   External shell with distinct, widely separated (>1 mm), concentric ridges; lateral teeth of internal shell serrated...............................*Corbicula fluminea*

Concentric ridges absent from external shell; lateral teeth not serrated......... .... genus *Polymesoda*, 8

8(7)   Shell oval shaped; periostracum dark with wavy, scaly ridges; beaks usually eroded in adults ....*Polymesoda caroliniana*

Shell wedge shaped, suboval; periostracum translucent; scalelike ridges absent; beaks not eroded in adults (may be found in inland waters) ....*Polymesoda maritima*

9(6)   Each valve with one cardinal tooth; shell mottled with dark color pattern..... ....*Eupera cubensis*

At least one valve (usually right) with two cardinal teeth; shell not mottled.. 10

10(9)   Umbos anterior, or if subcentral, located on anterior side of center...........11

Umbos posterior, or if subcentral, located on posterior side of center .... genus *Pisidium*, 17

11(10) Valves calyculate... genus *Musculium*, in part, 13

Valves not calyculate, although an annulus or rest mark may be present ...12

12(11) Hinge plate between umbos narrow, less than or equal to the width of the thickened end of the single cardinal tooth in the right valve .... genus *Musculium*, in part, 13

Hinge plate between umbos thick, greater than two times the width of the thickened end of the single cardinal tooth in the right valve ....
.... genus *Sphaerium*, 16

13(11,12) Shell long in outline, height/length ratio ≤0.75 .... *Musculium transversum*

Height/length ratio ≥0.80 14

14(13) Length of adult shell <8 mm; outline of shell not symmetrical, ventral margin slopes steeply upward .... *Musculium securis*

Length of adult shell >8 mm; shell outline more symmetrical, ventral margin slopes only slightly upward 15

15(14) Umbos pointed; posterior end forming obtuse angle with dorsal margin ......
.... *Musculium lacustre*

Umbos broad, but not pointed; posterior end nearly at right angle to dorsal margin, shell with square outline .... *Musculium partumeium*

16(12) Shell sculptured with unevenly spaced, coarse striae (≤8 striae per mm in middle of shell) .... *Sphaerium striatinum*

Shell sculptured with fine striae (≥12 striae per mm in middle of shell) ....
.... *Sphaerium occidentale*

17(10) Length of adult shell ≥6.0 mm 18

Length of adult shell <6.0 mm 19

18(17) Striae coarse, ≤10 per mm .... *Pisidium dubium*

Striae fine, ≥15 per mm .... *Pisidium adamsi*

19(17) Shell with straight or curved ridges on umbos .... *Pisidium compressum*

Shell without ridges on umbos 20

20(19) Cardinal teeth central or subcentral .... *Pisidium punctiferum*

Cardinal teeth located near anterior cusps .... *Pisidium casertanum*

Glossary

accessory denticle – a small, compressed or triangular prominence on the hinge plate anterior or posterior to pseudocardinal teeth.

adductor muscle – a large muscle attached internally to shell valves, located dorsally, one each near the anterior and posterior ends of the hinge plate, which serve to draw the two valves together.

allopatric – distributional relationship in which the ranges of two species do not overlap. Compare with **sympatric.**

Ambleminae, amblemine – a subfamily of the Unionidae as recognized by Ortmann (1911); typically characterized by well-developed hinge dentition, no mantle modification anterior to the incurrent aperture, marsupia formed by all four or only the outer two gills, glochidia without styliform hooks; type genus *Amblema*.

anadromous – pertaining to the movement of animals from a salt water environment to freshwater for the purpose of spawning.

angular – having a corner where two edges meet; not rounded.

annulus (annuli, pl.) – a ring; a line in the shell or periostracum, often visible as a dark line in the periostracum, and frequently presumed to be annual in occurrence; sometimes referred to as a growth ring or rest.

Anodontinae, anodontine – a subfamily of the Unionidae as recognized by Ortmann (1911); typically characterized by reduced or absent hinge dentition, no mantle modification anterior to the incurrent aperture, marsupia restricted to outer gills, and glochidia with styliform hooks; type genus *Anodonta*.

anteriodorsal – of or relating to the upper portion of the anterior end.

anterior – of or relating to the front or forward portion, typically the end nearer the umbo in Unionoida.

anterior shell margin – refers herein to the portion of shell lying forward of the umbo extending downward to the ventral margin.

anterioventral – of or relating to the lower portion of the anterior end.

anthropogenic – of, relating to, or involving the impact of humans on nature.

aperture – an opening between opposing mantle membranes (within each valve of a member of Unionoida) through which water is drawn or expelled from the mantle cavity; sometimes erroneously referred to as a siphon.

apex – a narrowed or pointed end.

apophysis – a small, triangular, downward protruding shell structure on the anterior septum of *Mytilopsis* valves, which serves as a muscle attachment.

aragonite – one of two crystalline forms of calcium carbonate.

arborescent – divided distally into multiple branches; treelike in form.

arcuate – curved or bent; in the shape of a bow.

augmentation – an action to increase the size of an existing population.

basin – a stream that empties directly into the marine environment (e.g., Escambia River basin and Econfina Creek basin). Compare with **drainage**.

biangulate – possessing two angles.

bifid – divided distally into two approximately equal parts.

bifurcate – to divide into two branches or parts.

binomial – a biological name consisting of two terms (genus and species).

bioturbation – the disturbance of sediments by organismal activity.

Bivalvia – a class of the phylum Mollusca, in which the primary diagnostic characteristic is two opposing valves; formerly Pelecypoda.

blackwater – water rich in humic acids and with low nutrient concentrations.

brackish – having salinity content intermediate between fresh and sea water.

broadcast – to release or scatter in all directions.

brood – to maintain eggs, embryos, and/or glochidia in the marsupia.

brooding period – the interval during which eggs, embryos, and glochidia are held in the marsupia of unionids; also called gravid period.

byssal thread, byssus – a tough, proteinaceous thread used to anchor a bivalve to a hard substrate, secreted by the byssal gland in the base of the foot; often occur in bundles (e.g., *Dreissena polymorpha*) but are single in Unionoida; not homologous to larval threads of glochidia.

calcareous – consisting of or containing calcium carbonate.

calyculate – growth variation of the umbo, typical of *Musculium* born at or near the end of summer, yielding a caplike appearance.

captive propagation – production of individuals in an artificial environment.

cardinal teeth – hinge teeth positioned between two sets of lateral teeth, which serve to stabilize opposing valves; found in some marine and freshwater families, including Cyrenidae and Sphaeriidae.

caruncle – a fleshy protuberance found on the posterioventral mantle margins of female *Toxolasma*; used to attract glochidial hosts.

channelization – the process of straightening and removing obstructions in the course of a stream.

chevron – a marking shaped like a wide-angled "V," usually inverted.

chitin, chitinous – hard, amorphous polysaccharide that comprises the covering of some invertebrates; the material forming the ligament of bivalve mollusks.

chondrophore – a deep pit or depression on the hinge plate ventral to the umbo.

cilium (cilia, pl.) – a small, hairlike process extending from cells, such as those on gills, labial palps, foot, and mantle surfaces, which transports water or small particles by repeated rhythmic movements.

clade – a natural, monophyletic group of two or more taxa believed to have evolved from a common ancestor.

clinal variation – gradual changes in characteristics (e.g., morphological or molecular) within a species over a geographic gradient (e.g., change in shell characteristics from headwaters to lower reaches of a river).

clothlike – having a surface texture of fabric.

commensal – a relation between two kinds of organisms in which one benefits from the other without harming or benefiting it.

compressed – laterally flattened.

concave – curved or rounded inward.

concentric – curving about a common center.

conchological – of or relating to shell characteristics.

conchology – a branch of zoology involving the study of mollusk shells.

conglutinate – a group of glochidia bound together in a gelatinous or mucous mass, which may mimic food items of glochidial hosts.

conspecific – belonging to the same species.

convex – curved or rounded outward.

corrugation – one of a series of small, roughly parallel ridges and intervening grooves on the surface of a shell.

cotype – a term formerly used for either syntype or paratype, not recognized by ICZN. See **syntype**.

crenulate – having a margin with fine notches or scallops.

cryptic species – two or more species that appear so similar as to be confused as a single species.

cultured pearl – a pearl produced by deliberate insertion of foreign material into a bivalve.

currently stable – a term used to describe the conservation status of a species which is not declining through much of its range and is not in imminent danger of extinction.

demibranch – a gill; one of the paired, membranous organs situated between the mantle and visceral mass of a bivalve.

dendritic – branching in form; treelike.

depressed – dorsoventrally flattened.

diaphragm – in unionids, the separation between the incurrent and excurrent apertures formed by a connection between the posteriodorsal surface of the gills and the mantle.

diaphragmatic septa – in margaritiferids, the posterior portions of the mantle which expand to meet the posterior end of the gill plate to form the diaphragm.

digestive gland – one of three glands external to the stomach, connected to the stomach by ducts which end in blind sacs (i.e., digestive diverticula); the location of extracellular and intercellular breakdown and absorption of food.

dioecious – having separate sexes; not hermaphroditic.

discharge (verb) – to expel; liberation of glochidia and/or conglutinates from marsupia in Unionoida.

discharge (noun) – the rate of stream flow at a given instant in terms of volume per unit of time.

disjunct – separated or isolated.

disk – the space between the umbo and the margin anterior of the posterior ridge of a bivalve shell.

distal – away from the point of attachment or origin; in reference to shell characters, it means away from the umbo. Compare with **proximal**.

distended – enlarged from internal pressure.

DNA – deoxyribonucleic acid, the molecule that encodes genetic information.

dorsal – of or relating to the back or top; represented by the hinge line in freshwater bivalves.

dorsal muscles – small muscles located in the umbo region, often in the umbo cavity, which serve to attach the mussel mantle to the shell.

dorsal shell margin – refers herein to the shell periphery along the back from the umbo extending posteriorly to slightly beyond the hinge.

dorsal wing – a thin, flat, dorsal projection of some bivalve shells.

double-looped – in the form of dual arcs, connected centrally.

drainage – a stream and its tributaries; refers herein to large tributaries within basins (e.g., Santa Fe River drainage within Suwannee River basin). Compare with **basin**.

drawdown – a reduction in water level or discharge due to manipulation of flow by dams or other control structures.

ecophenotypic variation – the nonhereditary divergence in shell morphology resulting from varying environmental conditions.

edentulous – without hinge teeth.

electrophoresis – the process of separating proteins suspended in a gelatinous medium using an electric current, with differential protein movement based on their size and charge (e.g., for screening of allozyme variation).

elliptical – elongate oval.

elongate – long in proportion to height.

embayment – permanently flooded indentation in a shoreline, generally at the mouth of a tributary.

embryo – an organism in the early stages of development; in Unionoida, the period from fertilization to the point where the glochidium is developed.

encyst – to become enclosed in a cyst; the process of a glochidium attaching to and becoming surrounded by tissue of the host.

endangered – in danger of extinction through all or a significant portion of its range; an official designation under the federal Endangered Species Act of 1973.

endemic – having a native range restricted to a particular area, region, or basin.

ephemeral – lasting for a short period of time; transient; temporary.

epifaunal – pertaining to fauna living on the surface of substrates.

euryhaline – able to live in waters of a wide range of salinity.

excurrent aperture – the posterior opening between opposing mantle membranes (within each valve of a member of Unionoida) through which water is expelled from the mantle cavity, located between the incurrent and supra-anal apertures; sometimes referred to as anal or exhalent aperture; erroneously referred to as a siphon.

excyst – the process of a juvenile unionid escaping from its cyst on a glochidial host.

extant – not extinct or regionally extirpated.

extinct – taxa which have been eliminated from their entire ranges, with no living representatives.

extirpated – eliminated from a particular area, but populations are extant elsewhere.

extrapallial – portion of the shell interior lying outside the pallial line.

facultative parasite – an organism with the capacity to undergo direct development or pass through a parasitic larval stage on a host.

Fall Line – the boundary between the Coastal Plain and upland physiographic provinces of the eastern United States.

family – a group of related taxa ranking above genus and below order.

first magnitude spring – a spring with an annual average discharge of 100 cubic feet per second or more.

flange – the curved, inward-projecting, anterioventral and posterioventral lateral margins of some glochidia.

foot – the large, muscular organ ventrally located in bivalves, used for locomotion and anchorage in the substrate.

fluvial – riverine; of or pertaining to flowing waters; lotic.

gamete – a mature germ cell (ovum or spermatozoan).

gape (verb) – to open or part.

gape (noun) – the space between two valves where they do not meet when closed.

gene – a hereditary unit that occupies a specific location (locus) on a chromosome.

genetic analysis – comparison of hereditary material among individuals and populations in an effort to determine evolutionary relationships.

genetic stock – a group of organisms that share a gene pool.

genus – a group of taxa with common characteristics, ranking above species and below family.

gill – an organ used to obtain oxygen from water, sort food particles, and incubate glochidia; in bivalves, they occur as paired, membranous organs situated between the visceral mass and mantle on each side of the body.

gill septa – the vertical membranes that divide water tubes within the gill, situated between the inner and outer gill lamellae.

glochidial host – an organism on which transformation of a glochidium to the juvenile stage is possible, typically a fish.

glochidial infection – see **glochidial infestation**.

glochidial infestation – the presence of glochidia on, or encysted in, tissues of a host; may be used interchangeably with **glochidial infection**.

glochidium (**glochidia**, pl.) – the bivalve larval stage of Unionidae, Margaritiferidae, and Hyriidae.

glochidium height – the distance between dorsal and ventral margins of a glochidium shell, measured approximately centrally, perpendicular to the dorsal margin.

glochidium length – the distance between anterior and posterior margins of a glochidium shell, measured parallel to the dorsal margin.

gonad – the organ in which gametes are formed.

gravid – containing unborn young; with regard to Unionidae, brooding eggs, embryos, or glochidia in marsupia (charged gills).

growth rest – see **annulus**.

haplotype – a unique pattern of nucleotide variation at a specific DNA locus.

headwaters – streams comprising the source of a creek or river.

hemolymph – the fluid in tissues, circulatory vessels, and hemocoels of mollusks; blood.

hermaphrodite – an individual capable of producing both eggs and sperm.

hinge ligament – the elongate, elastic structure that unites the dorsal margins of opposing valves of a bivalve mollusk, which acts to open valves when adductor muscles are relaxed.

hinge line – a hypothetical line passing through the hinge area from the anterior end of the pseudocardinal teeth to the posterior end of the lateral teeth, or the homologous area in shells with teeth reduced or absent.

hinge plate – the dorsal portion of the interior of a Unionoida shell, including pseudocardinal and lateral teeth, and interdentum if present.

historical – relating to the period of recorded history.

Holarctic – a biogeographical region encompassing the northern hemisphere.

holotype – the single specimen designated by the author in the original description as the primary type of a species or subspecies.

host generalist – refers to a parasite or parasitic stage that is capable of utilizing a number of potential hosts.

hybridization – the mixing of genetic material between two different species to produce offspring.

hypostracum – a layer of shell deposited only beneath muscle attachments.

hypoxia – a deficiency in the amount of oxygen reaching tissues of an organism.

impoundment – a body of water created by damming a stream; reservoir.

incurrent aperture – the posterior opening between opposing mantle membranes (within each valve of a member of Unionoida) through which water is drawn into the mantle cavity, located ventral to the excurrent aperture; sometimes referred to as inhalent aperture; erroneously referred to as a siphon.

infaunal – pertaining to fauna living within substrates.

inflated – swollen; expanded.

inflation – a measure of shell width in relation to height.

interdental projection – a small protuberance on the interdentum of some species (e.g., anodontines).

interdentum – the area between the posterior end of the pseudocardinal teeth and the anterior end of the lateral teeth.

interlamellar gill connection – see **gill septa**.

interrupted – not continuous; broken.

interstitial – of or relating to small spaces among substrate particles.

juvenile – a sexually immature individual.

karst – a limestone region with sinkholes, underground streams, and caverns formed through dissolution.

knob – a rounded protuberance; with regard to shell sculpture, usually refers to a large protuberance, as opposed to smaller tubercles or pustules.

labial palps – small, paired organs located near the anterior end of the gills which serve to separate food from nonfood particles and transfer them to the food groove.

laboratory infestation – the subjection of potential hosts to glochidia for determination of glochidial hosts or propagation of juveniles; may be used interchangeably with artificial infestation.

lamella (lamellae, pl.) – a membranous structure, such as one of the two gill layers.

lamellar – thin; membranous.

Lampsilinae, lampsiline – a subfamily of the Unionidae recognized by some authors (or used as a tribe, Lampsilini, by others); typically characterized by well-developed hinge dentition, a mantle modification anterior to the incurrent aperture, marsupia restricted to the outer gills and extended below ventral gill margin when gravid, and glochidia without styliform hooks; type genus *Lampsilis*.

lanceolate – narrow; tapering to a point at one or both ends.

larval thread – an elongate strand produced between the valves of some glochidia; not homologous to byssal threads.

larviparous – bearing and bringing forth young that are in a larval stage.

lateral – of or relating to the side, as opposed to medial.

lateral teeth – in most Unionoida, the linear, raised, interlocking processes positioned along the hinge line of a valve that function to stabilize opposing valves, located posterior to the pseudocardinal teeth and interdentum.

lectotype – a primary type designated from the syntypic series subsequent to the original description.

lentic – relating to still waters (e.g., lakes and reservoirs).

lime rock – a natural, partially consolidated form of limestone, consisting primarily of calcium carbonate but containing some silica.

locus (loci, pl.) – the site that a gene or molecular sequence of interest occupies on a chromosome.

long-term brooder – a species of Unionoida that broods glochidia for much of the year, typically from late summer or autumn to the following spring or summer; bradytictic.

lotic – relating to flowing waters (e.g., rivers and creeks).

lunule – the falcate or semicircular impression just anterior to the umbo, one-half being on each valve.

malacology – a branch of zoology which deals with the study of mollusks.

mantle – the membranous organ enveloping the body and lining the inside of the shell of a mollusk, which secretes the shell.

mantle attachment scars – in margaritiferids, a series of small pits near the center of the inside of the valve that are sites of mantle attachment to the shell.

mantle bridge – the connection between opposing mantle membranes (within each valve), located posteriodorsally and separating excurrent and supra-anal apertures; variable in length and found in most members of the Unionoida.

mantle flap – the specialized mantle modification of some Lampsilinae, such as *Lampsilis* and some *Hamiota*, located just ventral to the incurrent aperture, in the form of a variable expansion of tissue; serve as lures that rhythmically pulsate and mimic prey items of potential glochidial hosts.

mantle fold – the specialized mantle modification of the mantle margin of some Lampsilinae, principally *Medionidus*, *Ligumia*, and *Villosa*, that are generally not as broad or well developed as mantle flaps; serve as lures that may rhythmically pulsate and mimic prey items of potential glochidial hosts.

mantle margin – the outer edge of the mantle which actively secretes the shell and periostracum.

marginal glochidial host – see **secondary glochidial host**.

marsupium – the portion of bivalve gill in which embryos and glochidia are brooded.

medial – of or relating to the middle, as opposed to lateral.

micropoint – a small projection on the ventral margin of some glochidia, variable in numbers and arrangements depending on species.

midden – a pile of debris left by predators or humans often comprised of mollusk shells; refuse heap.

mitochondrial DNA – a circular DNA duplex found in the mitochondria of cells.

mitotype – a mitochondrial DNA sequence that differs from other sequences by one or more nucleotide positions.

molecular genetics – the study of the structure, function, and dynamics of genes at the molecular level.

monophyletic – a group consisting of a common ancestor and all descendent taxa.

monotypic – taxonomic groups above species represented by a single taxon.

mottling – colored spots or blotches.

muscle scar – the site of attachment for one of various muscles on the inside of a shell valve.

mushroom body – the cells comprising the mantle of encysted glochidia, enlarged and projected into the mantle cavity, involved in digestion of larval adductor muscle and tissues of the host trapped between the glochidial valves.

nacre, **nacreous layer** – the inner layer of Unionoida shells, composed primarily of calcium carbonate (aragonite) deposited in an organic matrix.

natural infestation – attachment of glochidia to hosts without human assistance; often used to suggest suitable hosts but does not necessarily indicate that successful transformation will occur.

Nearctic – the biogeographic region that includes Greenland and North America north of tropical Mexico.

neotype – a primary type designated subsequent to its original description if no syntypes are available.

nodulous – with small, irregularly rounded masses on the shell surface.

nominal – a taxon which is denoted by an available name.

nonindigenous – a species not native to an area; introduced.

nucleotide – the basic molecular unit of which DNA (and RNA) is composed.

obligate parasite – an organism with the capacity to exist only under one particularly characteristic mode of life. Compare with **facultative parasite**.

oblique – slanting; neither perpendicular nor parallel.

obscure – faint; lacking prominence; indistinct.

orbicular – round, circular; having the form of an orbit.

oval – oblong and curvilinear; broadly elliptical.

ovoviviparity – the condition of retaining eggs that hatch within the maternal body, with embryos obtaining no nourishment from outside the egg.

ovum (**ova**, pl.) – a female gamete; egg.

padded – slightly thickened or inflated.

pallial line – a thin, indented row of small muscle scars resulting from attachment of the mantle to the shell, approximately concentric to the ventral margin of a valve, between the anterior and posterior adductor muscle scars.

pallial muscles – a narrow row of small muscles approximately concentric to the ventral margin of the mantle which serve to attach the mantle to its respective valve and retract the mantle from the edge of the shell.

pallial sinus – an indentation in the posterior portion of the pallial line marking the space into which the siphons are withdrawn upon closure of the valves; not present in unionids or margaritiferids.

papilla (**papillae**, pl.) – a small projection, typically fingerlike in form; in Unionoida, present along margins of the incurrent aperture and sometimes the excurrent aperture, also found on modified mantle margins anterior to incurrent aperture in some species.

papillate – bearing papillae.

parasite – an organism that lives on or in another species, from which it obtains nutriment, usually to the detriment of the host.

paratype – a secondary type of the original type series, other than the holotype (primary type), designated by the author in the original description.

pearl – a mineralized deposit within soft tissues of mollusks in the form of a concretion of concentric layers of nacre, which may be attached to the shell or unattached; usually an abnormal growth around a nucleus of extraneous material.

pearl nucleus – the extraneous material around which concentric layers of nacre are laid to form a pearl.

pedal feeding – ingestion of food particles accumulated with a ciliated foot; utilized by juvenile mussels.

pedal protractor muscle – a small muscle used to extend the foot, located just posterioventral to the anterior adductor muscle.

pedal retractor muscle – one of two small muscles used to withdraw the foot into the shell, located near the anterior and posterior adductor muscles.

periostracum – the thin, uncalcified, proteinaceous outer layer of bivalve shells produced by the mantle margin which serves to protect underlying layers from erosion and dissolution.

phylogenetic – relating to the evolutionary relationship of organisms.

phylogeny – the evolutionary history of a group of organisms.

phylogeographic – prehistorical processes responsible for contemporary geographic distributions of populations, in light of the patterns associated with a gene genealogy.

plication – one of a series of parallel ridges; used herein to refer to relatively large ridges, as opposed to smaller corrugations.

polyphyletic – composed of unrelated organisms descended from more than one ancestor.

population fragmentation – division of a large connected interbreeding population into smaller isolated units; often associated with destruction or impairment of habitat in intervening areas.

posteriodorsal – of or relating to the upper part of the posterior end.

posterior – at or near the back or rear.

posterior ridge – the crest on the exterior of a bivalve shell extending from the umbo to the posterioventral margin, variously developed among species ranging from angular to rounded to almost nonexistent.

posterior shell margin – refers herein to the portion of shell from the dorsal margin extending downward to the ventral margin.

posterior slope – the area of the external surface of a shell valve between the posterior ridge and the posteriodorsal margin.

posterioventral – of or relating to the lower part of the posterior end.

prehistoric – relating to the period preceding recorded history.

primary host – a species on which glochidia are capable of transforming to the juvenile stage with optimal success.

primary type – the single individual (shell or soft parts) on which a species description is based; holotype, lectotype, or neotype.

prismatic layer – a layer in freshwater mussel shells comprised of minute calcium carbonate prisms oriented perpendicular to the shell surface.

proteinaceous – comprised primarily or entirely of proteins.

proximal – toward the point of attachment or origin; in reference to shell characters, it means towards to the umbo. Compare with **distal**.

pseudocardinal teeth – in most Unionoida, the raised, interlocking processes that serve to stabilize opposing valves, located anterior to the lateral teeth and interdentum.

pseudofeces – in suspension- or deposit-feeding bivalves, nonfood particles separated from food, bound in mucus in the mantle cavity, and usually excreted through the incurrent aperture or along the shell margin by rapid valve closure.

pustule – sculpture in the form of a small prominence on the surface of a shell, may be round or oblong.

pyriform – broad on one end and narrow on the other; pear-shaped.

quadrate – having four sides, approximately square or rectangular.

radial – proceeding out from a central point; from the umbo in bivalves.

range – the geographic area in which an organism occurs.

ray – a radial line of pigment on the periostracum, originating on the umbo, usually some shade of green, may be interrupted or continuous, and of varying width.

reach – a portion of a river.

reintroduction – the release of individuals into an area where they formerly occurred.

relative abundance – the number of individuals of a particular taxon present in a given area compared to numbers of other species.

reniform – kidney-shaped.

reservoir – an artificial lake created by impoundment of a creek or river.

restriction of type locality – reduction of multiple published type localities to a single location as a function of designating a primary type subsequent to the original description (lectotype or neotype), or narrowing a broad type locality based on discovery of more specific information regarding the origin of the primary type.

rhomboidal – having an outline in which opposite sides and angles are equal, but adjacent sides and angles are unequal.

riparian – pertaining to or situated on the bank of a stream or other water body.

riverine – found in or characteristic of rivers.

rotund – round or nearly round in outline.

rounded – having an evenly curved contour.

rudimentary – underdeveloped; poorly developed.

scarp – a steep slope of some extent along the margin of a plateau, terrace, or bench.

sculpture – impressed or raised structures on the surface of a shell.

secondary glochidial host – a species on which glochidia are capable of transforming to the juvenile stage but with low success relative to species identified as primary hosts; sometimes referred to as marginal host.

secondary water tubes – small tubes formed by accessory partitions in the gravid marsupial portion of anodontine gills, present just before and persisting until just after the gravid period, that allow for respiration during the period of gravidity.

sediment – geological material deposited by water, wind, or glaciers.

septum – a dividing wall or membrane.

serrated – notched or toothed on the edge.

sexually dimorphic – males and females of the same species differing in shell morphology (e.g., *Lampsilis* and *Villosa*).

shear stress – force exerted by flowing water on a stream bed or bank.

shell height – the distance between dorsal and ventral margins, measured near the midpoint of the hinge ligament, perpendicular to the shell length.

shell length – the distance between the anterior and posterior margins, measured parallel to the hinge ligament.

shell morphology – form and structure of a shell.

shell width – the distance between the outer surfaces of the paired valves.

short-term brooder – a species of Unionoida which broods glochidia from late winter or spring to summer; tachytictic.

simple papillae – papillae that are not divided or branched distally.

sinuous – having a serpentine or wavy form; broadly undulating.

siphon – a tubular organ formed by the fusion of mantle margins that aids in the intake or discharge of water; apertures of Unionoida are often erroneously referred to as siphons.

spatulate – oblong with an attenuated base opposite a rounded end.

spawning – discharge of gametes in aquatic organisms.

special concern – a conservation term used to describe species which are currently not imperiled but are thought to be declining in parts of their ranges and requiring diligent monitoring of their status; equivalent to vulnerable status.

species – a group of interbreeding natural populations that are reproductively isolated from all other such groups.

spermatozeugmata – unencapsulated sperm aggregates embedded in a semipermeable membrane.

spermatozoan (**spermatozoa**, pl.) – a male gamete, usually motile.

stochastic – a process or system that is connected with or involves random variables.

striae, **striate** – fine, raised lines.

styliform hook – an acute projection located on the ventral terminus or lateral edges of valves of some glochidia to aid host attachment.

sub- – a prefix meaning almost or nearly.

subspecies – a morphologically distinguishable and geographically isolated group of organisms capable of successfully interbreeding with other subspecies of the same species.

substrate – the surface on which an organism lives; material at the bottom of a body of water.

sulcus – a groove, furrow, or channel; in bivalves typically oriented from umbo to the ventral or posterior margin.

superconglutinate – the packet of glochidia containing the entire annual reproductive output of an individual, which mimics food of glochidial hosts, tethered to the excurrent aperture of the producing female by a hollow mucus tube; known to occur only in *Hamiota*.

supra-anal aperture – the posteriodorsal opening just dorsal to the excurrent aperture.

suprabranchial chamber – the narrow, elongate cavity dorsal to the connection of the gills; sometimes referred to as the epibranchial cavity.

sympatric – distributional relationship in which two or more species occur in the same area. Compare with **allopatric**.

synonym – one of two or more names for the same organism. **Junior synonym** – a published name for an organism that is not the oldest available. **Senior synonym** – the oldest available among two or more names for an organism.

synonymy – a list of synonyms for a given taxon.

syntype – one of a series of specimens used for an original species description when no holotype is designated; all have equal rank unless a primary type (lectotype) is chosen from among them, at which time the remainder become secondary types (paralectotypes).

taxon (taxa, pl.) – the name applied to any group of organisms in a formal system of nomenclature.

taxonomy – the science of finding, describing, classifying, and naming organisms, including the studying of the relationships between taxa and the principles of such a classification.

terrace – a relatively flat or gently inclined surface bounded by steeper slopes.

threatened – likely to become endangered in all or a significant portion of its range during the foreseeable future; an official designation under the federal Endangered Species Act of 1973.

transformation – the process of changing structure; refers herein to the transition of a glochidium to a free-living juvenile.

translocation – the movement of wild-caught animals from one location to another.

trapezoidal – an outline with four distinct sides, with only two sides parallel.

triangulate – possessing three angles.

tributary – a stream feeding a larger stream or lake.

trinomial – a biological name consisting of three terms (genus, species, and subspecies).

truncate – ending abruptly, more or less squarely.

tubercle – sculpture in the form of a small prominence on the surface of a shell, may be round or oblong.

type – a specimen upon which a taxon is described.

type locality – the geographical location of the collection of the name-bearing type of a species or subspecies.

type species – the species designated to characterize a genus or subgenus.

umbo – the raised, rounded, often inflated portion of a bivalve shell, located on the dorsal margin near the anterior end of the hinge ligament; the oldest part of a bivalve shell.

umbo cavity – the depression located inside each valve beneath the umbo.

umbo sculpture – the diminutive, natural, raised markings found on umbos of shells, generally in the form of small ridges or corrugations.

Unioninae – a subfamily of the Unionidae recognized by some authors; typically characterized by well-developed hinge dentition, no mantle modification anterior to the incurrent aperture, marsupia formed by all four or just the outer gills, marsupia are not distended when gravid and have a sharp ventral edge, glochidia with styliform hooks; type genus *Unio*.

Unionoida – the order comprised of six freshwater bivalve families: Etheriidae, Hyriidae, Iridinidae, Margaritiferidae, Mycetopodidae, and Unionidae; all with an obligate parasitic larval stage; the order contains about 180 genera and approximately 800 species that occur on six continents.

valve – one of the opposing shells of a bivalve mollusk.

veliger – a pelagic larval mollusk with a ciliated swimming membrane.

ventral – relating to or situated near the bottom or underside.

ventral shell margin – refers herein to the lower shell periphery between the anterior and posterior margins, opposite the umbo and dorsal margin.

visceral mass – the collective assemblage of internal organs, including the digestive system, kidney, and reproductive tissues, positioned inside the paired mantles.

watershed – the land area contained within a basin or drainage; catchment area.

water tube – one of a number of canals in the intralamellar space of a mussel gill, extending from the dorsal to the ventral margin.

wrinkle – a small furrow or ridge on a shell surface; refers herein to irregularly arranged sculpture.

Literature Cited

Abell, R.A., D.M. Olson, E. Dinerstein, P.T. Hurley, J.T. Diggs, W. Eichbaum, S. Walters, W. Wettengel, T. Allnutt, C.J. Loucks, and P. Hedao. 2000. Freshwater Ecoregions of North America: A Conservation Assessment. Island Press, Washington, DC. 319 pages.

Abbott, R.T. 1974. American Seashells. The Marine Mollusca of the Atlantic and Pacific Coasts of North America. Second edition. Van Nostrand Reinhold Company, New York. 663 pages.

Academy of Natural Sciences of Philadelphia. 1953. Escambia River, Florida: vicinity of the Chemstrand Corporation Plant, Fall 1952 – Spring 1953: stream survey report for the E.I. Du Pont de Nemours & Company. Academy of Natural Sciences of Philadelphia. 165 pages.

Adovasio, J.M., and A. Hemmings. 2008. The first snowbirds: the archaeology of inundated late Pleistocene landscapes in the northeastern Gulf of Mexico. Accessed 14 January 2013. http://oceanexplorer.noaa.gov/explorations/08negmexico/welcome.html.

Agassiz, L. 1852. Ueber die Gattungen unter den nordamerikanischen Najaden. Archiv für Naturgeschichte 18(1):41–52.

Akiyama, Y., and T. Iwakuma. 2007. Survival of glochidial larvae of the freshwater pearl mussel, *Margaritifera laevis* (Bivalvia: Unionoida), at different temperatures: a comparison between two populations with and without recruitment. Zoological Science 24:890–893.

Albertson, P.N., and L.J. Torak. 2002. Simulated effects of ground-water pumpage on stream-aquifer flow in the vicinity of federally protected species of freshwater mussels in the Lower Apalachicola-Chattahoochee-Flint River Basin (Subarea 4), southeastern Alabama, northeastern Florida, and southwestern, Georgia. U.S. Geological Survey Water Investigations Report 02-4016.

Aldrich, T.H. 1911. Notes on some Pliocene fossils from Georgia with descriptions of new species. The Nautilus 24(11):131–132; 24(12):138–140.

Aldridge, D.C., T.M. Fayle, and N. Jackson. 2007. Freshwater mussel abundance predicts biodiversity in lowland UK rivers. Aquatic Conservation: Marine and Freshwater Ecosystems 17:554–564.

Aldridge, D.C., and A.L. McIvor. 2003. Gill evacuation and release of glochidia by *Unio pictorum* and *Unio tumidus* (Bivalvia: Unionidae) under thermal and hypoxic stress. Journal of Molluscan Studies 69:55–59.

Alexander, M.K. 2007. Images of America: Muscatine's Pearl Button Industry. Arcadia Publishing, Charleston, South Carolina. 127 pages.

Allen, D.C., and C.C. Vaughn. 2009. Burrowing behavior of freshwater mussels in experimentally manipulated communities. Journal of the North American Benthological Society 28:93–100.

Allen, E. 1924. The existence of a short reproductive cycle in *Anodonta imbecillis*. Biological Bulletin 46:88–94.

Alt, D., and H.K. Brooks. 1965. Age of Florida marine terraces. Journal of Geology 73:406–411.

Amyot, J.-P., and J.A. Downing. 1991. Endo- and epibenthic distribution of the unionid mollusk *Elliptio complanata*. Journal of the North American Benthological Society 10:280–285.

Amyot, J.-P., and J.A. Downing. 1998. Locomotion in *Elliptio complanata* (Mollusca: Unionidae): a reproductive function? Freshwater Biology 39:351–358.

Anderson, R.V., and R.E. Ingham. 1978. Character variation in mollusk populations of *Lampsilis* Rafinesque, 1820. Transactions of the Illinois State Academy of Science 71:403–411.

Andrews, J.D. 1971. Sea Shells of the Texas Coast. University of Texas Press, Austin. 298 pages.

Andrews, J.D. 1981. Texas Shells: a Field Guide. University of Texas Press, Austin. 175 pages.

Andrews, J.D., and C. Cook. 1951. Range and habitat of the clam *Polymesoda caroliniana* (Bosc) in Virginia (family Cycladidae). Ecology 32(4):758–760.

Andrusov, N.I. 1897. Fossil and living Dreissenidae of Eurasia. Trudy Sanktpeterburgskogo Obshchestva Estestvoispytateley. Otdel Geologii i Mineralogii [Proceedings of the Saint Petersburg Society of Naturalists, Department of Geology and Mineralogy] 25:285–286.

Anthony, J.L., and J.A. Downing. 2001. Exploitation trajectory of a declining fauna: a century of freshwater mussel fisheries in North America. Canadian Journal of Fisheries and Aquatic Sciences 58(10):2071–2090.

Applin, P.L. 1951. A preliminary report on buried pre-Mesozoic rocks in Florida and adjacent states. U.S. Geological Survey Circular Number 91. 28 pages.

Arey, L.B. 1924. Glochidial cuticulae, teeth and the mechanics of attachment. Journal of Morphology and Physiology 39(2):332.

Arey, L.B. 1932a. A microscopical study of glochidial immunity. Journal of Morphology 53(2):367–379.

Arey, L.B. 1932b. The formation and structure of the glochidial cyst. Biological Bulletin 62(2):212–221.

Arey, L.B. 1932c. The nutrition of glochidia during metamorphosis. Journal of Morphology 53(1):201–221.

Arter, H.E. 1989. Effect of eutrophication on species composition and growth of freshwater mussels (Mollusca, Unionidae) in Lake Hallwil (Argau, Switzerland). Aquatic Sciences 51:87–99.

Arthur, J.D., P. Bond, E. Lane, F.R. Rupert, and T.M. Scott. 1994. Florida's global wanderings through the geological eras. Chapter 3, pages 11–35. In: E. Lane (editor). Florida's Geological History and Geological Resources. Florida Geological Survey, Special Publication Number 35. 46 pages.

Arthur, J.R., P. Bond, G.H. Means, F. Rupert, and T.M. Scott. 2007. State of Florida geological highway map and geoscience resource guide. Florida Geological Survey Map Series 150.

Athearn, H.D. 1964. Three new unionids from Alabama and Florida and a note on *Lampsilis jonesi*. The Nautilus 77(4):134–139.

Athearn, H.D. 1967. Changes and reductions in our freshwater molluscan populations. American Malacological Union, Inc. Annual Reports 1967:44–45.

Athearn, H.D. 1970. Discussion of Dr. Heard's paper. Malacologia 10(1):28–31.

Athearn, H.D. 1998. Additional records and notes on the unionid fauna of the Gulf drainage of Alabama, Florida, and Georgia. Occasional Papers on Mollusks, Museum of Comparative Zoology, Harvard University 5(76):465–466.

Atkinson, C.L., M.R. First, A.P. Covich, S.P. Opsahl, and S.W. Golladay. 2011. Suspended material availability and filtration-biodeposition processes performed by a native and invasive bivalve species in streams. Hydrobiologia 667(1):191–204.

Atkinson, C.L., S.P. Opsahl, A.P. Covich, S.W. Golladay, and L.M. Conner. 2010. Stable isotopic signatures, tissue stoichiometry, and nutrient cycling (C and N) of native and invasive freshwater bivalves. Journal of the North American Benthological Society 29(2):496–505.

Auffenberg, K., I.R. Quitmyer, J.D. Williams, and D.S. Jones. 2006. Non-marine Mollusca. Chapter 9, pages 247–261. In: S.D. Webb (editor). First Floridians and Last Mastodons: the Page-Ladson Site in the Aucilla River. Springer, The Netherlands.

Augspurger, T., A.E. Keller, M.C. Black, W.G. Cope, and F.J. Dwyer. 2003. Water quality guidance for protection of freshwater mussels (Unionidae) from ammonia exposure. Environmental Toxicology and Chemistry 22(11):2569–2575.

Aumen, N.G. 1995. The history of human impacts, lake management, and limnological research on Lake Okeechobee, Florida (USA). Pages 1–16 In: N.A. Aumen and R.G. Wetzel (editors). Ecological Studies on the Littoral and Pelagic Systems of Lake Okeechobee Florida (USA). Archive für Hydrobiologie Special Issues: Advancements in Limnology 45, Stuttgart, Germany.

Bailey, R.C. 1989. Habitat selection by a freshwater mussel: an experimental test. Malacologia 31:205–210.

Bailey, R.C., and R.H. Green. 1988. Within-basin variation in the shell morphology and growth rate of a freshwater mussel. Canadian Journal of Zoology 66:1704–1408.

Baker, F.C. 1898. The Mollusca of the Chicago area, Part I: the Pelecypoda. Bulletin of the Chicago Academy of Science 3(1):1–130, 27 plates.

Baker, F.C. 1927. On the division of the Sphaeriidae into two subfamilies: and the description of a new genus of Unionidae, with descriptions of new varieties. The American Midland Naturalist 10(7):220–223.

Baker, F.C. 1928. The fresh water Mollusca of Wisconsin. Part II. Pelecypoda. Bulletin of the University of Wisconsin, Serial Number 1527. Bulletin of the Wisconsin Geological and Natural History Survey 70(2):1–495.

Balfour, D.L., and L.A. Smock. 1995. Distribution, age structure, and movements of the freshwater mussel *Elliptio complanata* (Mollusca: Unionidae) in a headwater stream. Journal of Freshwater Ecology 10:254–268.

Ball, G.H. 1922. Variation in fresh-water mussels. Ecology 3(2):93–121.

Bärlocker, F., and H. Brendelberger. 2004. Clearance of aquatic hyphomycete spores by a benthic suspension feeder. Limnology and Oceanography 49:2292–2296.

Barnes, D.W. 1823. On the genera *Unio* and *Alasmodonta*; with introductory remarks. American Journal of Science and Arts 6(1):107–127; 6(2):258–280, 13 plates.

Barnett, R.S. 1975. Basement structure of Florida and its tectonic implications. Transactions of the Gulf Coast Association of Geological Societies 25:122–42.

Barnhart, M.C., W.R Haag, and W.N. Roston. 2008. Adaptations to host infection and larval parasitism in Unionoida. Journal of the North American Benthological Society 27(2):370–394.

Barnhart, M.C., and A. Roberts. 1997. Reproduction and fish hosts of unionids from the Ozark uplift. Pages 16–20. In: K.S. Cummings, A.C. Buchanan, C.A. Mayer, and T.J. Naimo (editors). Conservation and Management of Freshwater Mussels II: Initiatives for the Future. Proceedings of a UMRCC Symposium, 16–18 October 1995, St. Louis, Missouri. Upper Mississippi River Conservation Committee, Rock Island, Illinois.

Bass, D.G., Jr. 1983. Study III. Rivers of Florida and their fishes. Completion Report of Investigations Project. Dingell-Johnson Project F-36, North Florida Streams Project. Florida Game and Fresh Water Fish Commission, Tallahassee. 397 pages.

Bass, D.G., Jr., and V.G. Hitt. 1974. Ecological distribution of the introduced Asiatic clam, *Corbicula manilensis*, in Florida. Report to the Florida Game and Fresh Water Fish Commission, Tallahassee. 32 pages.

Bauer, G. 1987. Reproductive strategy of the fresh-water pearl mussel *Margaritifera margaritifera*. Journal of Animal Ecology 56:691–704.

Bauer, G. 1994. The adaptive value of offspring size among freshwater mussels (Bivalvia; Unionoidea). The Journal of Animal Ecology 63(4):933–944.

Bauer, G. 1998. Allocation policy of female freshwater pearl mussels. Oecologia (Berlin) 117:90–94.

Bauer, G. 2001. Factors affecting naiad occurrence and abundance. Pages 155–162. In: G. Bauer and K. Wächtler (editors). Ecology and Evolution of the Freshwater Mussels Unionoida. Ecological Studies, Volume 145. Springer-Verlag, Berlin.

Bauer, G., and C. Vogel. 1987. The parasitic stage of the freshwater pearl mussel (*Margaritifera margaritifera* L.) I. Host response to glochidiosis. Archiv für Hydrobiologie 76:393–402.

Beck, W.M., Jr. 1954. Studies in stream pollution biology. I. A simplified ecological classification of organisms. Quarterly Journal of the Florida Academy of Sciences 17(4):211–227.

Beck, W.M., Jr. 1965. The streams of Florida. Bulletin of the Florida State Museum of Biological Sciences 10(3):91–126.

Beckett, D.C., B.W. Green, S.A. Thomas, and A.C. Miller. 1996. Epizoic invertebrate communities on upper Mississippi River unionid bivalves. The American Midland Naturalist 135:102–114.

Bednarek, A.T. 2001. Undamming rivers: a review of the ecological impacts of dam removal. Environmental Management 27:803–814.

Beedham, G.E. 1971. The extrapallial cavity in *Anodonta cygnea* L. inhabited by an insect larva. Journal of Conchology 26:380–385.

Belanger, T.V., C.G. Annis, Jr., and D.D. VanEpps. 1990. Growth rates of the Asiatic clam, *Corbicula fluminea*, in the upper and middle St. Johns River, Florida. The Nautilus 104(1):4–9.

Benz, G.W., and S. Curren. 1997. Results of an ongoing survey of metazoan symbionts of freshwater mussels (Unionidae) from Kentucky Lake, Tennessee. Pages 39–66. In: A.F. Scott, S.W. Hamilton, E.W. Chester, and D.S. White (editors). Proceedings of the Seventh Symposium on the Natural History of the Lower Tennessee and Cumberland River Valleys. The Center for Field Biology, Austin Peay State University, Clarksville, Tennessee.

Bequaert, J.C., and W.B. Miller. 1973. The Mollusks of the Arid Southwest, with an Arizona Check List. The University of Arizona Press, Tucson. 271 pages.

Berg, D.J., and P.H. Berg. 2000. Conservation genetics of freshwater mussels: comments on Mulvey et al. Conservation Biology 14:1920–1923.

Berndt, M.P., E.T. Oaksford, G.L. Mahon, and W. Schmidt. 1998. Groundwater. Pages 38–63. In: E.A. Fernald and E.D. Purdum (editors). Water Resources Atlas of Florida. Institute of Science and Public Affairs, Florida State University. 312 pages.

Berner, L. 1950. The mayflies of Florida. University of Florida Studies in Biological Science Series 4(4):1–267.

Bernhardt, E.S., M.A. Palmer, J.D. Allan, G. Alexander, K. Barnas, S. Brooks, J. Carr, S. Clayton, C. Dahm, J. Follst-Shah, D. Galat, S. Gloss, P. Goodwin, D. Hart, B. Hassett, R. Jenkinson, S. Katz, G.M. Kondolf, P.S. Lake, R. Lave, J.L. Meyer, T.K. O'Donnell, L. Pagano, B. Powell, and E. Sudduth. 2005. Synthesizing U.S. river restoration efforts. Science 308:636–637.

Bieler R., J.G. Carter, and E.V. Coan. 2010. Classification of bivalve families. Pages 113–133. In: P. Bouchet and J.-P. Rocroi. Nomenclator of bivalve families. Malacologia 52(2):1–184.

Blalock, H.N., and J.J. Herod. 1999. A comparative study of stream habitat and substrate utilized by *Corbicula fluminea* in the New River, Florida. Florida Scientist 62(2):145–151.

Blalock-Herod, H.N. 2000. Community ecology of three freshwater mussel species (Bivalvia: Unionidae) from the New River, Suwannee drainage, Florida. Master's thesis, University of Florida, Gainesville. 72 pages.

Blalock-Herod, H.N., J.J. Herod, and J.D. Williams. 2002. Evaluation of conservation status, distribution, and reproductive characteristics of an endemic Gulf Coast freshwater mussel, *Lampsilis australis* (Bivalvia: Unionidae). Biodiversity and Conservation 11(10):1877–1887.

Blalock-Herod, H.N., J.J. Herod, J.D. Williams, B.N. Wilson, and S.W. McGregor. 2005. A historical and current perspective of the freshwater mussel fauna (Bivalvia: Unionidae) of the Choctawhatchee River drainage in Alabama and Florida. Alabama Museum of Natural History Bulletin 24:1–26.

Blažek, R., and M. Gelnar. 2006. Temporal and spatial distribution of glochidial larval stages of European unionid mussels (Mollusca: Unionidae) on host fishes. Folia Parasitologica 53:98–106.

Blystad, C.N. 1923. Significance of larval mantle of freshwater mussels during parasitism, with notes on a new mantle condition exhibited by *Lampsilis luteola*. Bulletin of the U.S. Bureau of Fisheries 39(1923–1924):203–219. [Issued separately as U.S. Bureau of Fisheries Document 950.]

Bogan, A.E. 2008. Global diversity of freshwater mussels (Mollusca, Bivalvia) in freshwater. Hydrobiologia 595:139–147.

Bogan, A.E., and C.M. Bogan. 2002. The development and evolution of Isaac Lea's publications on the Unionoida. Pages 363–375. In: M. Falkner, K. Groh, and M.C.D. Speight (editors). Collectanea Malacologica: Festschrift für Gerhard Falkner. ConchBooks and Verlag der Friedrich-Held-Gesellschaft, Hackenheim and Munich, Germany.

Bogan, A.E., and W.R. Hoeh. 1995. *Utterbackia peninsularis*, a newly recognized freshwater mussel (Bivalvia: Unionidae: Anodontinae) from Peninsular Florida, USA. Walkerana (for 1993–1994) 7(17–18):275–287.

Bogan, A.E., and R.W. Portell. 1995. Freshwater bivalves (Bivalvia: Unionidae) from the Early Pleistocene Leisey Shell Pits, Hillsborough County, Florida. Leisey Shell Pits Volume. Bulletin of the Florida State Museum, Biological Sciences 37(Part I) (6):171–182.

Bogan, A.E., and K.J. Roe. 2008. Freshwater bivalve (Unioniformes) diversity, systematics, and evolution: status and future directions. Journal of the North American Benthological Society 27(2):349–369.

Bolden, S.R., and K.M. Brown. 2002. Role of stream, habitat, and density in predicting translocation success in the threatened Louisiana pearlshell, *Margaritifera hembeli* (Conrad). Journal of the North American Benthological Society 21:89–96.

Bosc, L.A.G. 1801–1804. Histoire naturelle des coquilles, contenant leur description, les moeurs des animaux qui les habitent et leurs usages. Avec figures dessinées d'après nature. Volumes 1–5. Deterville, Paris, France.

Boyer S., A. Howe, N. Juergens, and M. Hove. 2011. A DNA-barcoding approach to identifying juvenile freshwater mussels (Bivalvia: Unionidae) recovered from naturally infested fishes. Journal of the North American Benthological Society 30:182–194.

Boyko, C.B., and W.E. Sage, III. 1996. Catalog of recent type specimens in the Department of Invertebrates, American Museum of Natural History. II. Mollusca Part I (Classes Aplacophora, Polyplacophora, Gastropoda [Subclass Opisthobranchia], Bivalvia and Scaphopoda). American Museum Novitates, Number 3170. 50 pages.

Brainwood, M., S. Burgin, and M. Byrne. 2008. The impact of small and large impoundments on freshwater mussel distribution in the Hawkesbury-Nepean River, Southeastern Australia. River Research and Applications 24:1325–1342.

Brenner, M., M.W. Binford, and E.S. Deevey. 1990. Lakes. Chapter 11, pages 364–391. In: R.L. Myers and J.J. Ewel (editors). Ecosystems of Florida. University of Central Florida Press, Orlando. 765 pages.

Breton, S., D.T. Steware, S. Shepardson, R.J. Trdan, A.E. Bogan, E.G. Chapman, A.J. Ruminas, H. Piontkivska, and W.R. Hoeh. 2011. Novel protein genes in animal mtDNA: a new sex determination system in freshwater mussels (Bivalvia: Unionoida)? Molecular Biology and Evolution 28(5):1645–1659.

Brim Box, J., R.B. Dorazio, and W.D. Liddell. 2002. Relationships between streambed substrate characteristics and freshwater mussels (Bivalvia: Unionidae) in Coastal Plain streams. Journal of the North American Benthological Society 21:253–260.

Brim Box, J., and J. Mossa. 1999. Sediment, land use, and freshwater mussels: prospects and problems. Journal of the North American Benthological Society 18(1):99–117.

Brim Box, J., and J.D. Williams. 2000. Unionid mollusks of the Apalachicola Basin in Alabama, Florida, and Georgia. Alabama Museum of Natural History Bulletin 21:1–143.

Bringolf, R.B., W.G. Cope, M.C. Barnhart, S. Mosher, P.R. Lazaro, and D. Shea. 2007a. Acute and chronic toxicity of pesticide formulations (atrazine, chlorpyrifos, and permethrin) to glochidia and juveniles of *Lampsilis siliquoidea*. Environmental Toxicology and Chemistry 26:2101–2107.

Bringolf, R.B., W.G. Cope, S. Mosher, M.C. Barnhart, and D. Shea. 2007b. Acute and chronic toxicity of glyphosate compounds to glochidia and juveniles of *Lampsilis siliquoidea* (Unionidae). Environmental Toxicology and Chemistry 26:2094–2100.

Bringolf, R.B., A.K. Fritts, M.C. Barnhart, and W.G. Cope. 2011. Is glochidia viability accurately determined by NaCl? (Abstract). Freshwater Mollusk Conservation Society Seventh Biennial Symposium, April 2011, Louisville, Kentucky.

Bringolf, R.B., R.M. Heltsley, T.J. Newton, C.B. Eads, S.J. Fraley, D. Shea, and W.G. Cope. 2010. Environmental occurrence and reproductive effects of the pharmaceutical fluoxetine in native freshwater mussels. Environmental Toxicology and Chemistry 29:1–8.

Britton, J.C., and S.L.H. Fuller. 1980. The freshwater bivalve Mollusca (Unionidae, Sphaeriidae, Corbiculidae) of the Savannah River Plant, South Carolina. Savannah River Plant, U.S. Department of Energy, SRO-NERP-3. 37 pages.

Britton, J.C., and B. Morton. 1986. Polymorphism in *Corbicula fluminea* (Bivalvia: Corbiculoidea) from North America. Malacological Review 19:1–43.

Brönmark, C., and B. Malmqvist 1982. Resource partitioning between unionid mussels in a Swedish lake outlet. Holarctic Ecology 5:389–395.

Brooks, H.K. 1974. Lake Okeechobee. Pages 256–286 In: P.J. Gleason (editor). Environments of South Florida: Present and Past. Memoir 2. Miami Geological Society, Miami, Florida.

Brooks, H.K. 1981. Physiographic divisions of Florida (map). Institute of Food and Agricultural Sciences, University of Florida Gainesville, Florida.

Brown, C.J.D., C. Clark, and B. Gleissner. 1938. The size of certain naiades from western Lake Erie in relation to shoal exposure. The American Midland Naturalist 19(3):682–701.

Brown, K.M., W. Daniela, and G. Taylor. 2011. Do unionid mussel life histories differ between headwater streams and large rivers? Data from Louisiana rivers. (Abstract). Freshwater Mollusk Conservation Society Seventh Biennial Symposium, April 2011, Louisville, Kentucky.

Brown, R.C. 1994. Florida's First People. Pineapple Press, Inc., Sarasota, Florida. 262 pages.

Bruenderman, S.A., and R.J. Neves. 1993. Life history of the endangered fine-rayed pigtoe, *Fusconaia cuneolus* (Bivalvia: Unionidae), in the Clinch River, Virginia. American Malacological Bulletin 10(1):83–91.

Bryan, J.R., T.M. Scott, and G.H. Means. 2008. Roadside Geology of Florida. Mountain Press Publishing Company, Missoula, Montana. 376 pages.

Buddensiek, V., H. Engel, S. Fleischauer-Rössing, and K. Wächtler. 1993. Studies on the chemistry of interstitial water taken from defined horizons in the fine sediments of bivalve habitats in several northern German lowland waters. II: Microhabitats of *Margaritifera margaritifera* L., *Unio crassus* (Phillipsson) and *Unio tumidus* Phillipsson. Archiv für Hydrobiologie 127:151–166.

Buell, G.R., and C.A. Couch. 1995. National Water Quality Assessment Program: environmental distribution of organochlorine compounds in the Apalachicola-Chattahoochee-Flint River basin. Pages 45–53. In: K.J. Hatcher (editor). Proceedings of the 1995 Georgia Water Resources Conference, Athens, Georgia. Institute of Ecology, University of Georgia.

Buhay, J.E., W.R. Haag, C. Lydeard, and M.L. Warren, Jr. 2003. Something's fishy with *Villosa vanuxemensis*, *V. lienosa*, and *V. ortmanni* (Bivalvia: Unionidae): fish host usage and phylogeographic analysis of morphologically similar species. Third Biennial Symposium, Freshwater Mollusk Conservation Society, 16–19 March 2003, Meeting Program and Abstracts. Page 18.

Bullen, R.P. 1958. Six sites near the Chattahoochee River in the Jim Woodruff Reservoir area, Florida. Bureau of American Ethnology, Bulletin 169, River Basin Survey Papers Number 14. Pages 321–357, plates 56–73.

Burch, J.B. 1972. Freshwater sphaeriacean clams (Mollusca: Pelecypoda) of North America. U.S. Environmental Protection Agency Biota of Freshwater Ecosystems Identification Manual Number 3. 31 pages.

Burch, J.B. 1975a. Freshwater Unionacean Clams (Mollusca: Pelecypoda) of North America. Revised edition. Malacological Publications, Hamburg, Michigan. 204 pages.

Burch, J.B. 1975b. Freshwater sphaeriacean clams (Molluca: Pelecypoda) of North America. Malacological Publications, Hamburg, Michigan. 96 pages.

Burgess, G.H., and R. Franz. 1978. Zoogeography of the aquatic fauna of the St. Johns River system with comments on adjacent peninsular faunas. The American Midland Naturalist 100(1):160–170.

Burlakova, L.E., and A.Y. Karatayev. 2007. The effect of invasive macrophytes and water level fluctuations on unionids in Texas impoundments. Hydrobiologia 586:291–302.

Butler, R.S. 1990. Distributional records for freshwater mussels (Bivalvia: Unionidae) in Florida and south Alabama, with zoogeographic and taxonomic notes. Walkerana 3(10):239–261.

Butler, R.S. 2005. Status assessment report for the rabbitsfoot, *Quadrula cylindrica cylindrica*, a freshwater mussel occurring in the Mississippi River and Great Lakes basins. U.S. Fish and Wildlife Service, Asheville, North Carolina. 205 pages.

Butler, R.S., E.J. Moyer, M.W. Hulon, and V.P. Williams. 1992. Littoral zone invertebrate communities as affected by a habitat restoration project on Lake Tohopekaliga, Florida. Journal of Freshwater Ecology 7(3):317–328.

Byers, C.F. 1930. A contribution to the knowledge of Florida Odonata. University of Florida Publication, Biological Science Series 1(1):1–327.

Cahoon, L.B., and D.A. Owen. 1996. Can suspension feeding by bivalves regulate phytoplankton biomass in Lake Waccamaw, North Carolina? Hydrobiologia 325:193–200.

Campbell, D.C., P.D. Johnson, J.D. Williams, A.K. Rindsberg, J.M. Serb, K.K. Small, and C. Lydeard. 2008. Identification of 'extinct' freshwater mussel species using DNA barcoding. Molecular Ecology Resources 8(4):711–724. doi:10.1111/j.1755–0998.2008.02108.x.

Campbell, D.C., and C. Lydeard. 2012a. The genera of Pleurobemini (Bivalvia: Unionidae: Ambleminae). American Malacological Bulletin 30(1):19–38.

Campbell, D.C, and C. Lydeard. 2012b. Molecular systematics of *Fusconaia* (Bivalvia: Unionidae: Ambleminae). American Malacological Bulletin 30(1):1–17.

Campbell, D.C., J.M. Serb, J.E. Buhay, K.J. Roe, R.L. Minton, and C. Lydeard. 2005. Phylogeny of North American amblemines (Bivalvia, Unionoida): prodigious polyphyly proves pervasive across genera. Invertebrate Biology 124(2):131–164.

Carr, A.F., Jr. 1940. A contribution to the herpetology of Florida. University of Florida Publication, Biological Science Series 3(1):1–118.

Carter, J.G., and 50 others. 2011. A synoptical classification of the Bivalvia (Mollusca). Paleontological Contributions 4:1–47. http://hdl.handle.net/1808/8287.

Cartwright, J.H.E., and A.G. Checa. 2007. The dynamics of nacre self-assembly. Journal of the Royal Society Interface 4(14):491–504.

Ceryak, R., M.S. Knapp, and T. Burnson. 1983. The geology and water resources of the upper Suwannee River basin, Florida. Report of Investigation Number 87. Florida Department of Natural Resources, Bureau of Geology, Tallahassee. 165 pages.

Chen, L.Y. 1998. The respiratory physiology and energy metabolism of freshwater mussels and their responses to lack of oxygen. Doctoral dissertation, Virginia Polytechnic Institute and State University, Blacksburg. 89 pages.

Chen, L.Y., A.G. Heath, and R.J. Neves. 2001. Comparison of oxygen consumption in freshwater mussels (Unionidae) from different habitats during declining dissolved oxygen concentration. Hydrobiologia 450:209–214.

Cherry, D.S., J.L. Scheller, N.L. Cooper, and J.R. Bidwell. 2005. Potential effects of Asian clam (*Corbicula fluminea*) die-offs on native freshwater mussels (Unionidae) I: water-column ammonia levels and ammonia toxicity. Journal of the North American Benthological Society 24(2):369–380.

Chetty, A.N., and K. Indira. 1994. Alterations in the tissue lipid profiles of *Lamellidens marginalis* under ambient ammonia stress. Bulletin of Environmental Contamination and Toxicology 53(5):693–698.

Chetty, A.N., and K. Indira. 1995. Adaptive changes in the glucose metabolism of a bivalve to ambient ammonia stress. Bulletin of Environmental Contamination and Toxicology 54(1):83–89.

Christian, A.D., E.M. Monroe, A.M. Asher, J.M. Loutsch, and D.J. Berg. 2007. Methods of DNA extraction and PCR amplification for individual freshwater mussel (Bivalvia: Unionidae) glochidia, with the first report of multiple paternity in these organisms. Molecular Ecology Notes 7:570–573.

Cicerello, R.R., and G.A. Schuster. 2003. A guide to the freshwater mussels of Kentucky. Kentucky State Nature Preserves Commission, Scientific and Technical Series, Number 7. 62 pages.

Claassen, C. 1994. Washboards, pigtoes, and muckets: historic musseling in the Mississippi watershed. Historical Archaeology 28(2):1–145.

Clarke, A.H. 1981. The tribe Alasmidontini (Unionidae: Anodontinae). Part I: *Pegias, Alasmidonta,* and *Arcidens.* Smithsonian Contributions to Zoology, Number 326. 101 pages.

Clarke, A.H. 1985. The tribe Alasmidontini (Unionidae: Anodontinae). Part II: *Lasmigona* and *Simpsonaias.* Smithsonian Contributions to Zoology, Number 399. 75 pages.

Clausen, C.J., D.D. Cohen, C. Emiliani, J.A. Holman, and J.J. Stipages. 1979. Little Salt Spring, Florida: a unique underwater site. Science 203(4381):609–614.

Clench, W.J. 1955. A freshwater mollusk survey of north Florida rivers. The Nautilus 68(3):95–98.

Clench, W.J. 1970. *Corbicula manilensis* (Phillippi) in lower Florida. Nautilus 84(1):36.

Clench, W.J., and R.D. Turner. 1956. Freshwater mollusks of Alabama, Georgia, and Florida from the Escambia to the Suwannee River. Bulletin of the Florida State Museum, Biological Sciences 1(3):97–239, plates 1–9.

Coker, R.E. 1914. The protection of fresh-water mussels. Appendix VIII to the Report of the U.S. Commissioner of Fisheries for 1912:1–23, 2 plates. [Issued separately U.S. Bureau of Fisheries Document 793.]

Coker, R.E. 1919. Fresh-water mussels and mussel industries of the United States. Bulletin of the U.S. Bureau of Fisheries 36(1917–1918):13–89, 46 plates. [Issued separately as U.S. Bureau of Fisheries Document 865.]

Coker, R.E., A.F. Shira, H.W. Clark, and A.D. Howard. 1921. Natural history and propagation of freshwater mussels. Bulletin of the U.S. Bureau of Fisheries 37(1919–1920):77–181, 17 plates. [Issued separately as U.S. Bureau of Fisheries Document 893.]

Conrad, T.A. 1831. Description of fifteen new species of recent and three of fossil shells, chiefly from the coast of the United States. Journal of the Academy of Natural Sciences of Philadelphia 6:256–268, plate 11.

Conrad, T.A. 1834. New Freshwater Shells of the United States, with lithographic illustrations; and a monograph of the genus *Anculotus* of Say; also a synopsis of the American naiades. J. Dobson, Philadelphia. 3 May 1834. 76 pages, 8 plates.

Conrad, T.A. 1835a–1840. Monography of the family Unionidae, or naiades of Lamarck, (fresh water bivalve shells) of North America, illustrated by figures drawn on stone from nature. J. Dobson, Philadelphia. Part 1(1835):1–12 [pages 13–16 not published], plates 1–5; Part 2(1836):17–24, plates 6–10; Part 3(1836):25–32, plates 11–15; Part 4(1836):33–40, plates 16–20; Part 5(1836):41–48, plates 21–25; Part 6(1836):49–56, plates 26–30; Part 7(1836):57–64, plates 32–36; Part 8(1837):65–72, plates 36–40; Part 9(1837):73–80, plates 41–45; Part 10(1838):81–94, plates 46–51; Part 11(1838):95–102, plates 52–57; Part 12(1840):103–110, plates 58–60; Part 13[1840, part 13 not dated]:111–118, plates 61–65.

Conrad, T.A. 1835b. Additions to, and corrections of, the catalogue of species of American naiades, with descriptions of new species and varieties of fresh water shells. Pages 1–8, plate 9. Appendix to Conrad 1834. New Freshwater Shells of the United States, with lithographic illustrations; and a monograph of the genus *Anculotus* of Say; also a synopsis of the American naiades. J. Dobson, Philadelphia. 3 May 1834. 76 pages, 8 plates.

Conrad, T.A. 1841. [Descriptions of three new species of *Unio* from the rivers of the United States.] Proceedings of the Academy of Natural Sciences of Philadelphia 1(2):19–20.

Conrad, T.A. 1846. Catalogue of shells inhabiting Tampa Bay and other parts of the Florida coast. [Silliman's] American Journal of Science and Arts, Series 2, 2(6):393–398.

Conrad, T.A. 1849. Descriptions of new fresh water and marine shells. Proceedings of the Academy of Natural Sciences of Philadelphia 4(7):152–155. [Subsequently published in 1849 in The Annals and Magazine of Natural History 4(22):300–303.]

Conrad, T.A. 1850. Descriptions of new fresh water and marine shells. Journal of the Academy of Natural Sciences of Philadelphia 1 (New Series):275–280, plates 37–39.

Conrad, T.A. 1853. A synopsis of the family of Naïades of North America, with notes, and a table of some of the genera and sub-genera of the family, according to their geographical distribution, and descriptions of genera and sub-genera. Proceedings of the Academy of Natural Sciences of Philadelphia 6(7):243–269.

Conrad, T.A. 1866. Descriptions of American fresh-water shells. American Journal of Conchology 2(3):278–279, plate 15.

Conrad, T.A. 1868. Description of a new *Unio* and fossil *Goniobasis*. American Journal of Conchology 4:280, plate 18.

Cooke, C.W. 1939. Scenery of Florida interpreted by a geologist. State of Florida Department of Natural Resources Geological Bulletin Number 17. 118 pages.

Cooke, C.W. 1945. Geology of Florida. Florida Geological Survey Bulletin Number 29. 339 pages.

Cooke, C.W., and S. Mossom. 1929. Geology of Florida. Twentieth Annual Report of the Florida Geological Survey. Pages 29–228, 1 plate.

Cooke, G.D., E.B. Welch, S. Peterson, and S.A. Nichols. 2005. Restoration and management of lakes and reservoirs. Third edition. CRC Press, Boca Raton, Florida. 616 pages.

Cope, W.G., R.B. Bringolf, D.B. Buchwalter, T.J. Newton, C.G. Ingersoll, N. Wang, T. Augspurger, F.J. Dwyer, M.C. Barnhart, R.J. Neves, and E. Hammer. 2008. Differential exposure, duration, and sensitivity of unionoidean bivalve life stages to environmental contaminants. Journal of the North American Benthological Society 27(2):451–462.

Cordeiro, J.R. 2007. Freshwater mussel (Bivalvia: Unionidae) causes incidental fish mortality. Veliger 49(3):129–131.

Cory, C.B. 1895. Hunting and fishing in Florida, including a key to the water birds known to occur in the state. Estes and Lauriat Company, Boston, Massachusetts. 297 pages.

Counts, C.L., III. 1986. The zoogeography and history of the invasion of the United States by *Corbicula fluminea* (Bivalvia: Corbiculidae). American Malacological Bulletin, Special Edition Number 2. Pages 7–39.

Crabtree, D.L., and T.A. Smith. 2009. Population attributes of an endangered mussel, *Epioblasma torulosa rangiana* (northern riffleshell), in French Creek and implications for its recovery. Northeastern Naturalist 16(3):339–354.

Crail, T.D., R.A. Krebs, and D.T. Zanatta. 2011. Unionid mussels from nearshore zones of Lake Erie. Journal of Great Lakes Research 37:199–202.

Crosse, H., and P.H. Fischer. 1894. Études sur les mollusques terrestres et fluviatiles du Mexique et du Guatemala. 2(15):489–576, plates 59–62. In: P.H. Fischer and H. Crosse (editors). Mission Scientifique au Mexique et dans l'Amérique Centrale. Reserches Zoologiques pour server à l'histoire de la faune de l'Amerique Centrale et du Mexique. Imprimerie Nationale, Paris.

Culp, J.J., W.R. Haag, D.A. Arrington, and T.B. Kennedy. 2011. Seasonal and species-specific patterns in abundance of freshwater mussel glochidia in stream drift. Journal of the North American Benthological Society 30(2):436–445.

Cummings, K.E., and D.L. Graf. 2010. Mollusca: Bivalvia. Pages 309–384. In: J.H. Thorp and A.P. Covich (editors). Ecology and Classification of North American Freshwater Invertebrates. Third edition. Elsevier, Inc., Amsterdam, The Netherlands.

Cvancara, A.M. 1972. Lake mussel distribution as determined with SCUBA. Ecology 53(1):154–157.

Cvancara, A.M., and P.G. Freeman. 1978. Diversity and distribution of mussels (Bivalvia: Unionacea) in a eutrophic reservoir, Lake Ashtabula, North Dakota. The Nautilus 92(1):1–9.

Cyr, F., A. Paquet, A.L. Martel, and B. Angers. 2007. Cryptic lineages and hybridization in freshwater mussels of the genus *Pyganodon* in northeastern North America. Canadian Journal of Zoology 85:1216–1227.

Cyr, H. 2008. Physical forces constrain the depth distribution of the abundant native mussel *Elliptio complanata* in lakes. Freshwater Biology 53:2414–2525.

Dall, W.H. 1898. Contributions to the Tertiary fauna of Florida with especial reference to the Miocene silex-beds of Tampa and the Pliocene beds of the Caloosahatchie River. Transactions of the Wagner Free Institute Science of Philadelphia 3:484–947, plates 23–35.

Dalrymple, G.B. 1991. The Age of the Earth. Stanford University Press, California. 492 pages.

Daraio, J.A., L.J. Weber, S.J. Zigler, T.J. Newton, and J.M. Nestler. 2010. Simulated effects of host fish distribution on juvenile unionid mussel dispersal in a large river. River Research and Applications 28(5):594–608. doi:10.1002/rra.1469.

Dartnall, H.J.G., and M. Walkey. 1979. Distribution of glochidia of the swan mussel, *Anodonta cygnea* (Mollusca), on the three-spined stickleback, *Gasterosteus aculeatus* (Pisces). Journal of Zoology (London) 189:31–37.

Daughton, C.G., and T.A. Ternes. 1999. Pharmaceuticals and personal care products in the environment: agents of subtle change? Environmental Health Perspectives 107:907–938.

Davis, G.J., and R.W. Howe. 1992. Juvenile dispersal, limited breeding sites, and the dynamics of metapopulations. Theoretical Population Biology 41:184–207.

Davis, G.M. 1983. Relative roles of molecular genetics, anatomy, morphometrics and ecology in assessing relationships among North American Unionidae (Bivalvia). Pages 193–222. In: G.S. Oxford and D. Rollinson (editors). Protein Polymorphism: Adaptive and Taxonomic Significance. Systematics Association, Special Volume 24, Academic Press, London.

Davis, G.M., and S.L.H. Fuller. 1981. Genetic relationships among Recent Unionacea (Bivalvia) of North America. Malacologia 20(2):217–253.

de Lozoya, A.V., and R. Araujo 2011. How the naiad was drawn: a pre-Linnean iconography of freshwater mussels. Malacologia 53(2):381–402.

DeMort, C.L. 1991. The St. Johns River system. Chapter 7, pages 97–129. In: R.L Livingston (editor). The Rivers of Florida. Springer-Verlag, New York. 289 pages.

DeWalle, D.R., B.R. Swistock, T.E. Johnson, and K.J. McGuire. 2000. Potential effects of climate change and urbanization on mean annual streamflow in the United States. Water Resources Research 36(9):2655–2664.

Dickinson, B.D., and B.E. Sietman. 2008. Recent observation of metamorphosis without parasitism in *Utterbackia imbecillis*. Ellipsaria 10(1):7–8.

Dimock, R.V., and A.H. Wright. 1993. Sensitivity of juvenile freshwater mussels to hypoxic, thermal, and acid stress. Journal of the Elisha Mitchell Scientific Society 109:183–192.

Dodd, B.J., M.C. Barnhart, C.L. Rogers-Lowery, T.B. Fobian, and R.V. Dimock. 2005. Cross-resistance of largemouth bass to glochidia of unionid mussels. Journal of Parasitology 91:1064–1072.

Dodd, C.K., and R.A. Seigel. 1991. Relocation, repatriation and translocation of amphibians and reptiles: are they conservation strategies that work? Herpetologica 47:336–350.

Downing, J.A., J.-P. Amyot, M. Pérusse, and Y. Rochon. 1989. Visceral sex, hermaphroditism and protandry in a population of the freshwater bivalve *Elliptio complanata*. Journal of the North American Benthological Society 8(1):92–99.

Downing, J.A., H. Van Leeuwen, and L.A. Di Paolo. 2000. Substratum patch selection in the lacustrine mussels *Elliptio complanata* and *Pyganodon grandis grandis*. Freshwater Biology 44:641–648.

Downing, J.A., P. Van Meter, and D.A. Woolnough. 2010. Suspects and evidence: a review of the causes of extirpation and decline in freshwater mussels. Animal Biodiversity and Conservation 33(2):151–185.

Downing, W.L., and J.A. Downing. 1993. Molluscan shell growth and loss. Nature 362:506.

Doyle, M.W., C.F. Rich, J.M. Harbor, and A. Spacie. 2000. Examining the effects of urbanization on streams using indicators of geomorphic stability. Physical Geography 21(2):155–181.

Dunbar, J.S. 2006. Paleoindian archaeology. Pages 403–435. In: S.D. Webb (editor). First Floridians and Last Mastodons: the Page-Ladson Site in the Aucilla River.

Eager, R.M.C. 1950. Variation in shape of shell with respect to ecological station. A review dealing with Recent Unionidae and certain species of the Anthrocosiidae in Amyot Upper Carboniferous times. Proceedings of the Royal Society of Edinburgh Section B 63:130–148.

Ellis, M.M. 1931a. A survey of conditions affecting fisheries in the upper Mississippi River. U.S. Bureau of Fisheries, Fishery Circular, Number 5. 18 pages.

Ellis, M.M. 1931b. Some factors affecting the replacement of the commercial fresh-water mussels. U.S. Bureau of Fisheries, Fishery Circular, Number 7. 10 pages.

Ellis, M.M. 1936. Erosion silt as a factor in aquatic environments. Ecology 17(1):29–42.

Englund, V.P.M., and M.P. Heino. 1994a. Valve movement of *Anodonta anatina* and *Unio tumidus* (Bivalvia, Unionidae) in a eutrophic lake. Annales Zoologici Fennici 31:257–262.

Englund, V.P.M., and M.P. Heino. 1994b. Valve movement of freshwater mussel *Anodonta anatina*: a reciprocal transplant experiment between two lakes. Annales Zoologici Fennici 31:417–423.

Engstrom, D.R., S.P. Schottler, P.R. Leavitt, and K.E. Havens. 2006. A reevaluation of the cultural eutrophication of Lake Okeechobee using multiproxy sediment records. Ecological Applications 16:1194–1206. http://dx.doi.org/10.1890/1051-0761(2006)016[1194:AROTCE]2.0.CO;2.

Evermann, B.W., and H.W. Clark. 1917. The Unionidae of Maxinkuckee. Proceedings of the Indiana Academy of Science 1917:251–285.

Fassler, C.R. 1997. The American mussel crisis: effects on the world pearl industry. Pages 265–277. In: K.S. Cummings, A.C. Buchanan, C.A. Mayer, and T.J. Naimo (editors). Conservation and Management of Freshwater Mussels II: Initiatives for the Future. Proceedings of a UMRCC Symposium, 16–18 October 1995, St. Louis Missouri. Upper Mississippi River Conservation Committee, Rock Island, Illinois.

Faught, M.K. 2004. The underwater archaeology of paleolandscapes, Apalachee Bay, Florida. American Antiquity 69(2):275–289.

Fee, E.J., R.E. Hecky, S.E.M. Kasian, and D.R. Cruikshank. 1996. Effects of lake size, water clarity, and climatic variability on mixing depths in Canadian Shield lakes. Limnology and Oceanography 41:912–920.

Fenneman, N.M. 1938. Physiography of the eastern United States. McGraw-Hill, New York. 691 pages.

Ferguson, C.D. 2009. Conservation genetics of a near threatened freshwater mussel species (*Lampsilis cardium*) and improved prospects for recovery: how nuclear and mitochondrial DNA analyses inform natural history and conservation. Doctoral dissertation, Wrights State University, Dayton, Ohio. 244 pages.

Fisher, G.R., and R.V. Dimock, Jr. 2002a. Morphological and molecular changes during metamorphosis in *Utterbackia imbecillis* (Bivalvia: Unionidae). Journal of Molluscan Studies 68(2):159–164.

Fisher, G.R., and R.V. Dimock, Jr. 2002b. Ultrastructure of the mushroom body: digestion during metamorphosis of *Utterbackia imbecillis* (Bivalvia: Unionidae). Invertebrate Biology 121(2):126–135.

Fleming, W.J., T.P. Augspurger, and J.A. Alderman. 1995. Freshwater mussel die-off attributed to anticholinesterase poisoning. Environmental Toxicology and Chemistry 14(5):877–879.

Florida Department of Environmental Protection. 2001a. Basin status report: Ochlockonee and St. Marks. Division of Water Resource Management, Florida Department of Environmental Protection, Tallahassee. 174 pages.

Florida Department of Environmental Protection. 2001b. Basin status report: Suwannee (including Aucilla, Coastal, Suwannee, and Waccasassa basins in Florida). Division of Water Resource Management, Florida Department of Environmental Protection, Tallahassee. 193 pages.

Florida Department of Environmental Protection. 2001c. Basin status report: Ocklawaha. Division of Water Resource Management, Florida Department of Environmental Protection, Tallahassee. 315 pages.

Florida Department of Environmental Protection. 2002a. Basin status report: Apalachicola-Chipola. Division of Water Resource Management, Florida Department of Environmental Protection, Tallahassee. 208 pages.

Florida Department of Environmental Protection. 2002b. Basin status report: Tampa Bay tributaries. Division of Water Resource Management, Florida Department of Environmental Protection, Tallahassee. 248 pages.

Florida Department of Environmental Protection. 2002c. Basin status report: lower St. Johns. Division of Water Resource Management, Florida Department of Environmental Protection, Tallahassee. 404 pages.

Florida Department of Environmental Protection. 2003a. Basin status report: Choctawhatchee-St. Andrew. Division of Water Resource Management, Florida Department of Environmental Protection, Tallahassee. 211 pages.

Florida Department of Environmental Protection. 2003b. Basin status report: Sarasota Bay and Peace and Myakka Rivers. Division of Water Resource Management, Florida Department of Environmental Protection, Tallahassee. 296 pages.

Florida Department of Environmental Protection. 2004a. Basin status report: Kissimmee River and Fisheating Creek. Division of Water Resource Management, Florida Department of Environmental Protection, Tallahassee. 337 pages.

Florida Department of Environmental Protection. 2004b. Basin status report: Pensacola Bay. Division of Water Resource Management, Florida Department of Environmental Protection, Tallahassee. 213 pages.

Florida Department of Environmental Protection. 2004c. Basin status report: Nassau-St. Marys. Division of Water Resource Management, Florida Department of Environmental Protection, Tallahassee. 155 pages.

Florida Department of Environmental Protection. 2005. Basin status report: Withlacoochee. Division of Water Resource Management, Florida Department of Environmental Protection, Tallahassee. 230 pages.

Florida Department of Environmental Protection. 2006. Basin status report: Perdido River and Bay. Division of Water Resource Management, Florida Department of Environmental Protection, Tallahassee. 172 pages.

Florida Department of Natural Resources. 1989. Florida rivers assessment. Florida Department of Natural Resources, Tallahassee, Florida. 452 pages.

Fobian, T. 2007. Reproductive biology of the rabbitsfoot mussel (*Quadrula cylindrica*) (Say, 1817) in the upper Arkansas River system, White River system, and Red River system. Master's thesis, Missouri State University, Springfield. 105 pages.

Folmar, L.C., H.O. Sanders, and A.M. Julin. 1979. Toxicity of the herbicide glyphosate and several of its formulations to fish and aquatic invertebrates. Archives of Environmental Contamination and Toxicology 8:269–278.

Forel, F.A. 1866. Einige Beobachungen über die Entwicklung des zelligen Muskelgewebes. Beiträge zur Entwicklungsgeschichte der Najaden. A. Studer, Würzburg. 39 pages, 3 plates.

Forsyth, D.J., and I.D. McCallum. 1978. *Xenochironomus canterburyensis* (Diptera: Chironomidae), a commensal of *Hyridella menziesi* (Lamellibranchia) in Lake Taupo; features of pre-adult life history. New Zealand Journal of Zoology 5:795–800.

Fox, R.O. 1970. The *Corbicula* story: chapter two. Third annual meeting of the Western Society of Malacologists, Stanford University, 24 June 1970. California Academy of Sciences, San Francisco. 11 pages.

Frank, H., and S. Gerstmann. 2007. Declining populations of freshwater pearl mussels (*Margaritifera margaritifera*) are burdened with heavy metals and DDT/DDE. Ambio 36(7):571–574.

Frankham, R., J.D. Ballou, and D.A. Briscoe. 2002. Introduction to conservation genetics. Cambridge University Press, Cambridge, United Kingdom. 640 pages.

Franz, R. (editor). 1982. Rare and Endangered Biota of Florida. Volume 6. Invertebrates. University of Florida Press, Gainesville. 131 pages.

Franz, R., J. Bauer, and T. Morris. 1994. Review of biologically significant caves and their faunas in Florida and south Georgia. Brimleyana 20:1–109.

Frick, E.A., D.J. Hippe, G.R. Buell, C.A. Couch, E.H. Hopkins, D.J. Wangsness, and J.W. Garrett. 1998. Water quality in the Apalachicola-Chattahoochee-Flint River basin, Georgia, Alabama, and Florida, 1992–95. U.S. Geological Survey Circular Number 1164. 38 pages.

Frierson, L.S. 1903a. The specific value of *Unio declivus*, Say. The Nautilus 17(5):49–51, plate 3.

Frierson, L.S. 1903b. Observations on the byssus of Unionidae. The Nautilus 17(7):76–77.

Frierson, L.S. 1904. Observations on the genus *Quadrula*. The Nautilus 17(10):111–112.

Frierson, L.S. 1905. New Unionidae from Alabama. The Nautilus 19(2):13–14, plate 1.

Frierson, L.S. 1911. Remarks on *Unio varicosus*, *cicatricosus* and *Unio compertus*, new species. The Nautilus 25(5):51–54, plates 2–3.

Frierson, L.S. 1912. *Unio* (*Obovaria*) *jacksonianus*, new species. The Nautilus 26(2):23–24, plate 3.

Frierson, L.S. 1927. A Classification and Annotated Check List of the North American Naiades. Baylor University Press, Waco, Texas. 111 pages.

Fritts, A.K., M.W. Fritts II, D.L. Peterson, D.A. Fox, and R.B. Bringolf. 2012. Critical linkage of imperiled species: Gulf Sturgeon as host for Purple Bankclimber mussels. Freshwater Science 31(4):1223–1232.

Fuller, S.L.H. 1971. A brief guide to the fresh-water mussels (Mollusca: Bivalvia: Unionacea) of the Savannah River system. ASB [Association of Southeastern Biologists] Bulletin 18(4):137–146, 1 plate.

Fuller, S.L.H. 1974. Clams and mussels (Mollusca: Bivalvia). Pages 215–273. In: C.W. Hart, Jr. and S.L.H. Fuller (editors). Pollution Ecology of Freshwater Invertebrates. Academic Press, New York.

Fuller, S.L.H., and D.J. Bereza. 1973. Recent additions to the naiad fauna of the eastern Gulf drainage (Bivalvia: Unionoida: Unionidae). ASB [Association of Southeastern Biologists] Bulletin 20(2):53.

Fuller, S.L.H., and D.J. Bereza. 1974. The value of anatomical characters in naiad taxonomy (Bivalvia: Unionacea). Bulletin of the American Malacological Union, Inc. 1974:21–22.

Gagné, F., C. Blaise, M. Fournier, and P.D. Hansen. 2006. Effects of selected pharmaceutical products on phagocytic activity in *Elliptio complanata* mussels. Comparative Biochemistry and Physiology Part C 143(2):179–186.

Gagné, F., B. Bouchard, C. André, E. Farcy, and M. Fournier. 2011. Evidence of feminization in wild *Elliptio complanata* mussels in the receiving waters downstream of a municipal effluent outfall. Comparative Biochemistry and Physiology Part C 153:99–106.

Gagnon, P.M., S.W. Golladay, W.K. Michener, and M.C. Freeman. 2004. Drought responses of freshwater mussels (Unionidae) in Coastal Plain tributaries of the Flint River Basin, Georgia. Journal of Freshwater Ecology 19(4):667–679.

Gagnon, P., W. Michener, M. Freeman, and J. Brim Box. 2006. Unionid habitat and assemblage composition in Coastal Plain tributaries of Flint River (Georgia). Southeastern Naturalist 5(1):31–52.

Galbraith, H.S. 2009. Reproduction in a changing environment: mussels, impoundments and conservation. Doctoral dissertation, University of Oklahoma, Norman. 149 pages.

Galbraith, H.S., D.E. Spooner, and C.C. Vaughn. 2010. Synergistic effects of regional climate patterns and local water management on freshwater mussel communities. Biological Conservation 143:1175–1183.

Galbraith, H.S., and C.C. Vaughn. 2009. Temperature and food interact to influence gamete development in freshwater mussels. Hydrobiologia 636:35–47.

Galbraith, H.S., and C.C. Vaughn. 2011. Effects of reservoir management on abundance, condition, parasitism and reproductive traits of downstream mussels. River Research and Applications 27:193–201.

Gangloff, M.M. 2011. Fat threeridge (*Amblema neislerii*) and Chipola slabshell (*Elliptio chipolaensis*) population size and depth distribution study in the Apalachicola and lower Chipola rivers. Southeastern Aquatic Research, Boone, North Carolina. 38 pages.

Gangloff, M.M., and J.W. Feminella. 2007. Stream channel geomorphology influences mussel abundance in southern Appalachian streams, U.S.A. Freshwater Biology 52(1):64–74.

Gangloff, M.M., and P.W. Hartfield. 2009. Seven populations of the southern kidneyshell (*Ptychobranchus jonesi*) discovered in the Choctawhatchee River basin, Alabama. Southeastern Naturalist 8(2):245–254.

Gangloff, M.M., K.K. Lenertz, and J.W. Feminella. 2008. Parasitic mite and trematode abundance are associated with reduced reproductive output and physiological condition of freshwater mussels. Hydrobiologia 610:25–31.

Gangloff, M.M., J.D. Williams, and J.W. Feminella. 2006. A new species of freshwater mussel (Bivalvia: Unionidae), *Pleurobema athearni*, from the Coosa River drainage of Alabama, USA. Zootaxa 1118:43–56.

Garner, J.T., T.M. Haggerty, and R.F. Modlin. 1999. Reproductive cycle of *Quadrula metanevra* (Bivalvia: Unionidae) in the Pickwick Dam tailwater of the Tennessee River. The American Midland Naturalist 141(2):277–283.

Garner, J.T., H. Blalock-Herod, A.E. Bogan, R.S. Butler, W.R. Haag, P.D. Hartfield, J.J. Herod, P.D. Johnson, S.W. McGregor and J.D. Williams. 2004. Freshwater mussels and snails. Pages 13–58. In: R.A. Mirarchi (editor). Alabama Wildlife. Volume 1. A Checklist of Vertebrates and Selected Invertebrates: Aquatic Mollusks, Fishes, Amphibians, Reptiles, Birds, and Mammals. The University of Alabama Press, Tuscaloosa.

Gascho Landis, A.M., T.L. Mosley, W.R. Haag, and J.A. Stoeckel. 2011. Influence of seston density on freshwater mussel growth and fertilization success. (Abstract). Freshwater Mollusk Conservation Society Seventh Biennial Symposium, April 2011, Louisville, Kentucky.

Gascho Landis, A.M., T.L. Mosley, W.R. Haag, and J.A. Stoeckel. 2012. Effects of temperature and photoperiod in lure display and glochidial release in a freshwater mussel. Freshwater Science 31(3):775–786. http://dx.doi.org/10.1899/11-082.1.

Geist, J., and K. Auerswald. 2007. Physicochemical stream bed characteristics and recruitment of the freshwater pearl mussel (*Margaritifera margaritifera*). Freshwater Biology 52:2299–2316.

Ghent, A.W., R. Singer, and L. Johnson-Singer. 1978. Depth distribution determined with SCUBA, and associated studies of the freshwater unionid clams *Elliptio complanata* and *Anodonta grandis* in Lake Bernard, Ontario. Canadian Journal of Zoology 56:1654–1663.

Ghiselin, M.T. 1969. The evolution of hermaphroditism among animals. Quarterly Review of Biology 44:189–208.

Glenn, L.C. 1911. Denudation and erosion in the southern Appalachian region and the Monongahela basin. U.S. Geological Survey, Professional Paper 72. 137 pages, 21 plates.

Golladay, S.W., P. Gagnon, M. Kearns, J.M. Battle, and D.W. Hicks. 2004. Response of freshwater mussel assemblages (Bivalve: Unionidae) to record drought in the Gulf Coastal Plain of southwest Georgia. Journal of the North American Benthological Society 23:494–506.

Gordon, M.E., and W.R. Hoeh. 1995. *Anodonta heardi*, a new species of freshwater mussel (Bivalvia: Unionidae) from the Apalachicola River system of the southeastern United States. Walkerana (for 1993–1994) 7(17–18):265–273.

Gordon, M.J., B.K. Swan, and C.G. Paterson. 1978. *Baeoctenus bicolor* (Diptera: Chironomidae) parasitic in unionid bivalve mollusks, and notes on other chironomid bivalve associations. Journal of the Fisheries Research Board of Canada 35:154–157.

Gough, H.M., A.M. Gascho Landis, J.A. Stoeckel. 2012. Behaviour and physiology are linked in the responses of freshwater mussels to drought. Freshwater Biology 57(11):2356–2366. doi:10.1111/fwb.12015.

Gould, A.A. 1845. [Descriptions of recent shells collected by Mr. John Bartlett in the everglades of Florida.] Proceedings of the Boston Society of Natural History 2:53.

Gould, A.A. 1850. [The following shells from the United States exploring expedition were described.] Proceedings of the Boston Society of Natural History 3(19):292–296.

Gould, A.A. 1856. [Descriptions of shells.] Proceedings of the Boston Society of Natural History 6(1):11–16.

Graf, D.L., and K.S. Cummings. 2007. Review of the systematics and global diversity of freshwater mussel species (Bivalvia: Unionoida). Journal of Molluscan Studies 73(4):291–314.

Green, R.H., R.C. Bailey, S.G. Hinch, J.L. Metcalfe, and V.H. Young. 1989. Use of freshwater mussels (Bivalvia: Unionidae) to monitor the near-shore environment of lakes. Journal of Great Lakes Research 15:635–644.

Grier, N.M. 1918. New varieties of naiades from Lake Erie. The Nautilus 32(1):9–12.

Grier, N.M. 1920a. Variation in epidermal color of certain species of najades inhabiting the upper Ohio drainage and their corresponding ones in Lake Erie. The American Midland Naturalist 6(11):247–285.

Grier, N.M. 1920b. Morphological features of certain mussel-shells found in Lake Erie, compared with those of the corresponding species found in the drainage of the upper Ohio. Annals of the Carnegie Museum 13(1–2):145–182.

Grier, N.M. 1922. Final report on the study and appraisal of mussel resources in selected areas of the upper Mississippi River. The American Midland Naturalist 8:1–33.

Grier, M.M. 1926. Notes on the naiades of the upper Mississippi drainage: III. On the relation of temperature to the rhythmical contractions of the "mantle flaps" in *Lampsilis ventricosa* (Barnes). The Nautilus 39(4):111–114.

Griffiths, N.A., and H. Cyr. 2006. Are there hot spots for *Elliptio complanata* in the shallow littoral zone of a large Canadian Shield lake? Canadian Journal of Fisheries and Aquatic Sciences 63:2137–2147.

Grizzle, J.M., and C.J. Brunner. 2007. Assessment of current information available for detection, sampling, necropsy, and diagnosis of diseased mussels. Report to Alabama Department of Conservation and Natural Resources, Montgomery. 84 pages.

Gunter, G., and G.E. Hall. 1965. A biological investigation of the Caloosahatchee Estuary of Florida. Gulf Research Reports 2(1):1–71.

Guppy, R.J.L. 1867. Description of a new freshwater bivalve found in Trinidad. Anals and Magazines of Natural History, Series 3, 19:16–161.

Gutiérrez, J.L., C.G. Jones, D.L. Strayer, and O.O. Iribarne. 2003. Mollusks as ecosystem engineers: the role of shell production in aquatic habitats. Oikos 101:79–90.

Haag, W.R. 2002. Spatial, temporal, and taxonomic variation in population dynamics and community structure of freshwater mussels. Doctoral dissertation, University of Mississippi, Oxford. 128 pages.

Haag, W.R. 2009. Extreme longevity in freshwater mussels revisited: sources of bias in age estimates derived from mark-recapture experiments. Freshwater Biology 54:1474–1486.

Haag, W.R. 2010. A hierarchical classification of freshwater mussel diversity in North America. Journal of Biogeography 37:12–26.

Haag, W.R. 2012. North American Freshwater Mussels: Natural History, Ecology, and Conservation. Cambridge University Press. 505 pages.

Haag, W.R., R.S. Butler, and P.D. Hartfield. 1995. An extraordinary reproductive strategy in freshwater bivalves: prey mimicry to facilitate larval dispersal. Freshwater Biology 34(3):471–476.

Haag, W.R., and A.M. Commens-Carson. 2008. Testing the assumption of annual shell ring deposition in freshwater mussels. Canadian Journal of Fisheries and Aquatic Sciences 65:493–508.

Haag, W.R., and A.L. Rypel. 2011. Growth and longevity in freshwater mussels: evolutionary and conservation implications. Biological Review 86:225–247.

Haag, W.R., and J.L. Staton. 2003. Variation in fecundity and other reproductive traits in freshwater mussels. Freshwater Biology 48(12):2118–2130.

Haag, W.R., and J.A. Stoeckel. 2011. Preliminary results of manipulative experiments of mussel recruitment in ponds. (Abstract). Freshwater Mollusk Conservation Society Seventh Biennial Symposium, April 2011, Louisville, Kentucky.

Haag, W.R., and M.L. Warren, Jr. 1997. Host fishes and reproductive biology of 6 freshwater mussel species from the Mobile Basin, U.S.A. Journal of the North American Benthological Society 16(3):576–585.

Haag, W.R., and M.L. Warren, Jr. 1998. Role of ecological factors and reproductive strategies in structuring freshwater mussel communities. Canadian Journal of Fisheries and Aquatic Sciences 55:297–306.

Haag, W.R., and M.L. Warren, Jr. 1999. Mantle displays of freshwater mussels elicit attacks from fish. Freshwater Biology 42(1):35–40.

Haag, W.R., and M.L. Warren, Jr. 2000. Effects of light and presence of fish on lure display and larval release behaviours in two species of freshwater mussels. Animal Behaviour 60(6):879–886.

Haag, W.R., and M.L. Warren, Jr. 2001. Host fishes and reproductive biology of freshwater mussels in the Buttahatchee River, Mississippi. Final report submitted to Mississippi Wildlife Heritage Fund. 41 pages.

Haag, W.R., and M.L. Warren, Jr. 2003. Host fishes and infection strategies of freshwater mussels in large Mobile Basin streams, U.S.A. Journal of the North American Benthological Society 22(1):78–91.

Haag, W.R., and M.L. Warren, Jr. 2007. Freshwater mussel assemblage structure in a regulated river in the lower Mississippi River alluvial basin, USA. Aquatic Conservation: Marine and Freshwater Ecosystems 17:25–36.

Haag, W.R., and M.L. Warren, Jr. 2008. Effects of severe drought on freshwater mussel assemblages. Transactions of the American Fisheries Society 137:1165–1178.

Haag, W.R., and M.L. Warren, Jr. 2010. Diversity, abundance, and size structure of bivalve assemblages in the Sipsey River, Alabama. Aquatic Conservation: Marine and Freshwater Ecosystems 20(6):655–667.

Haag, W.R., M.L. Warren, Jr., and M. Shillingsford. 1999. Host fishes and host-attracting behavior of *Lampsilis altilis* and *Villosa vibex* (Bivalvia: Unionidae). The American Midland Naturalist 141(1):149–157.

Haas, F. 1930. Über nord-und mittelamerikanische Najaden. Senckenbergiana Biologica 12(6):317–330.

Haggerty, T.M., J.T. Garner, and R. Rogers. 2005. Reproductive phenology in *Megalonaias nervosa* (Bivalvia: Unionidae) in Wheeler Reservoir, Tennessee River, Alabama, U.S.A. Hydrobiologia 539(2005):131–136.

Hall, B.M., and M.R. Hall. 1916. Second report on the water powers of Alabama. Geological Survey of Alabama Bulletin 17. 448 pages.

Hanlon, S.D., and D.G. Smith. 1999. An attempt to detect *Pyganodon fragilis* (Bivalvia: Unionidae) in Maine. Northeastern Naturalist 6(2):119–132.

Hanson, J.M., W.C. Mackay, and E.E. Prepas. 1988. The effects of water depth and density on the growth of a unionid clam. Freshwater Biology 19:345–355.

Harper, R.M. 1910. Preliminary report on the peat deposits of Florida. Florida Geological Survey Third Annual Report. Pages 197–366.

Harper, R.M. 1914. Geography and vegetation of northern Florida. Florida Geological Survey Sixth Annual Report. Pages 163–437.

Harper, R.M. 1921. Geography of central Florida. Florida Geological Survey Thirteenth Annual Report. Pages 71–307.

Harper, R.M. 1927. Natural resources of southern Florida. Florida Geological Survey Eighteenth Annual Report 18:27–206.

Hart, D.D., and N.L. Poff (editors). 2002. A special section on dam removal and river restoration. BioScience 52:653–747.

Hartfield, P., and R.S. Butler. 1997. Observations on the release of superconglutinates by *Lampsilis perovalis*. Pages 11–14. In: K.S. Cummings, A.C. Buchanan, C.A. Mayer, and T.J. Naimo (editors). Conservation and Management of Freshwater Mussels II: Initiatives for the Future. Proceedings of a UMRCC Symposium, 16–18 October 1995, St. Louis, Missouri. Upper Mississippi River Conservation Committee, Rock Island, Illinois.

Hartfield, P.D., and E. Hartfield. 1996. Observations on the conglutinates of *Ptychobranchus greeni* (Conrad, 1834) (Mollusca: Bivalvia: Unionoidea). The American Midland Naturalist 135(2):370–375.

Hastie, L.C., P.J. Boon, M.R. Young, and S. Way. 2001. The effects of a major flood on an endangered freshwater mussel population. Biological Conservation 98(1):107–115.

Hastie, L.C., P.J. Cosgrove, N. Ellis, and M.J. Gaywood. 2003. The threat of climate change to freshwater pearl mussel populations. Ambio 32(1):40–46.

Hauswald, C.L. 1997. Life history and conservation of *Elliptio crassidens* from the blue River, Indiana. Master's thesis, University of Louisville, Kentucky. 93 pages.

Hayward, D., and E.D. Estevez. 1997. Suitability of Florida waters to invasion by the zebra mussel, *Dreissena polymorpha*. Florida Sea Grant College Program Grant Number NZ36RG-0070. Mote Marine Laboratory Technical Report Number 495. 55 pages.

Headlee, T.J. 1906. Ecological notes on the mussels of Winona, Pike, and Center Lakes of Kosciusko County, Indiana. Biological Bulletin 11(6):305–316, 318–319.

Healy, H.G. 1975. Terraces and shorelines of Florida. Florida Bureau of Geology Map Series Number 71.

Heard, W.H. 1964. *Corbicula fluminea* in Florida. Nautilus 77(3):105–107.

Heard, W.H. 1966. Further records of *Corbicula fluminea* (Müller) in the southern United States. Nautilus 79(4):142–143.

Heard, W.H. 1969. Seasonal variation in gonad activity in freshwater mussels, and its systematic significance. (Abstract). American Malacological Union, Inc. Annual Reports 1969:53.

Heard, W.H. 1970. Eastern freshwater mollusks (II), the South Atlantic and Gulf drainages. Malacologia 10(1):23–31.

Heard, W.H. 1975a. Determination of the endangered status of freshwater clams of the Gulf and southeastern states. Department of Biological Sciences, Florida State University, Tallahassee. Final report prepared for the Office of Endangered Species, Bureau of Sport Fisheries and Wildlife, U.S. Department of Interior, Contract Number 14-16-000-8905. 31 pages.

Heard, W.H. 1975b. Sexuality and other aspects of reproduction in *Anodonta* (Pelecypoda: Unionidae). Malacologia 15(1):83–103.

Heard, W.H. 1977. Freshwater mollusca of the Apalachicola drainage. Pages 2021. In: R.J. Livingston and E.A. Joyce, Jr. (editors). Proceedings of the Conference on the Apalachicola Drainage System, 2324 April 1976, Gainesville, Florida. Florida Department of Natural Resources Marine Research Laboratory, St. Petersburg.

Heard, W.H. 1979a. Identification manual of the freshwater clams of Florida. Florida Department of Environmental Regulation, Technical Series 4(2):1–83.

Heard, W.H. 1979b. Hermaphroditism in *Elliptio* (Pelecypoda: Unionidae). Malacological Review 12:21–28.

Heard, W.H. 1998. Brooding patterns in freshwater mussels. Malacological Review, Supplement 7, Bivalvia 1:105–121.

Heard, W.H. 2000. Glochidial larvae of freshwater mussels and their fish hosts: early discoveries and interpretations of the association. Malacological Review, Supplement 8, Freshwater Mollusca 1:83–88.

Heard, W.H., and G. Dinesen. 1999. A history of the controversy about *Glochidium parasiticum* Rathke, 1797 (Palaeoheterodonta: Unionoida: Unionoidea). Malacological Review, Supplement 8, Freshwater Mollusca 1:89–106.

Heard, W.H., and R.H. Guckert. 1970. A re-evaluation of the recent Unionacea (Pelecypoda) of North America. Malacologia 10(2):333–355.

Heath, D., M. Hove, R. Benjamin, M. Endris, R. Kenyon, and J. Kurth. 1998. *Quadrula fragosa* exhibit unusual reproductive behaviors. Triannual Unionid Report 16:33.

Heatherington, A.L., and P.A. Mueller. 1997. Geochemistry and origin of Florida crustal basement terranes. Pages 27–37. In: A.F. Randazzo and D.S. Jones (editors). The Geology of Florida. University Press of Florida, Gainesville. 327 pages.

Heilprin, A. 1887. Explorations on the West Coast of Florida and on the Okeechobee wilderness with special reference to the geology and zoology of the Floridian Peninsula. Transactions of the Wagner Free Institute of Science of Philadelphia 1:1–134, plates 1–19.

Heinricher, J.R., and J.B. Layzer. 1999. Reproduction by individuals of a nonreproducing population of *Megalonaias nervosa* (Mollusca: Unionidae) following translocation. The American Midland Naturalist 141(1):140–148.

Helfrich, L.A., M. Zimmerman, and D.L. Weigmann. 1995. Control of suspended solids and phytoplankton with fishes and a mussel. Water Resources Bulletin 31(2):307–316.

Hemming, J.M., P.V. Winger, H. Rauschenberger, K. Herrington, P. Durkee, and D. Scollan. 2008. Water and sediment quality survey of threatened and endangered freshwater mussel habitat in the Chipola River Basin, Florida. Endangered Species Research 6:95–107.

Henderson, J. 1929. Non-marine Mollusca of Oregon and Washington. The University of Colorado Studies 17(2):47–190.

Henderson, J. 1935. Fossil non-marine Mollusca of North America. Geological Society of America, Special Papers Number 3. 313 pages.

Hendrix, S.S., M.F. Vidrine, and R.H. Hartenstine. 1985. A list of records of freshwater aspidogastrids (Tremotoda) and their hosts in North America. Proceedings of the Helminthological Society, Washington 52(2):289–296.

Henley, W.F. 2002. Evaluation of diet, gametogenesis, and hermaphroditism in freshwater mussels (Bivalvia: Unionidae). Doctoral dissertation, Virginia Polytechnic Institute and State University, Blacksburg. 109 pages.

Henley, W.F., M.A. Patterson, R.J. Neves, and A.D. Lemly. 2000. Effects of sedimentation and turbidity on lotic food webs: a concise review for natural resource managers. Reviews in Fishery Science 8(2):125–139.

Henry, J.A. 1998. Weather and climate. Pages 16–37. In: E.A. Fernald and E.D. Purdum (editors). Water Resources Atlas of Florida. Institute of Science and Public Affairs, Florida State University, Tallahassee. 312 pages.

Herring, J.L. 1951. The aquatic and semi-aquatic Hemiptera of northern Florida. Part 4: Classification of habitats and keys to species. Florida Entomologist 34:141–146.

Herrington, H.B. 1962. A revision of the Sphaeriidae of North America (Mollusca: Pelecypoda). Museum of Zoology, University of Michigan Miscellaneous Publications Number 118. 74 pages.

Higgins, F. 1858. A catalogue of the shell-bearing species of Mollusca, inhabiting the vicinity of Columbus, Ohio, with some remarks thereon. Pages 548–555. In: Twelfth Annual Report of the Ohio State Board of Agriculture with an Abstract of the Proceedings of the County Agricultural Societies, to the General Assembly of Ohio: for the Year 1857.

Hildreth, S.P. 1828. Observations on, and descriptions of the shells found in the waters of the Muskingum River, Little Muskingum, and Duck Creek, in the vicinity of Marietta, Ohio. American Journal of Science and Arts 14(2):276–291, 2 plates. [Reprinted in Sterkiana 8(1962).]

Hillis, D.M., and J.C. Patton. 1982. Morphological and electrophoretic evidence for two species of *Corbicula* (Bivalvia: Corbiculidae) in North American. The American Midland Naturalist 108:74–80.

Hinch, S.G., and R.C. Bailey. 1988. Within- and among-lake variation in shell morphology of the freshwater clam *Elliptio complanata* (Bivalvia: Unionidae) from the south-central Ontario lakes. Hydrobiologia 157:27–32.

Hinch, S.G., and R.H. Green. 1988. Shell etching on clams from low-alkalinity Ontario lakes: a physical or chemical process? Canadian Journal of Fisheries and Aquatic Sciences 45:2110–2113.

Hinch, S.G., L.J. Kelly, and R.H. Green. 1989. Morphological variation of *Elliptio complanata* (Bivalvia: Unionidae) in differing sediments of softwater lakes exposed to acidic deposition. Canadian Journal of Zoology 67:1895–1899.

Hobbs, Jr., H.H. 1942. The crayfishes of Florida. University of Florida Publication, Biological Science Series 3(2):1–179, 24 plates.

Hoeh, W.R. 1990. Phylogenetic relationships among eastern North American *Anodonta* (Bivalvia: Unionidae). Malacological Review 23(1–2):63–82.

Hoeh, W.R. 2010. Mitochondrial phylogenomics of the Bivalvia (Mollusca): searching for the origin and mitogenomic correlates of doubly uniparental inheritance of mtDNA. BMC [BioMed Central] Evolutionary Biology 10(50). 19 pages.

Hoeh, W.R., K.S. Frazer, E. Naranjo-Garcia, and R.J. Trdan. 1995. A phylogenetic perspective on the evolution of simultaneous hermaphroditism in a freshwater mussel clade (Bivalvia: Unionidae: *Utterbackia*). Malacological Review 28(1–2):25–42.

Hoggarth, M.A. 1992. An examination of the glochidia-host relationships reported in the literature for North American species of Unionacea (Mollusca: Bivalvia). Malacology Data Net 3:1–30.

Hoggarth, M.A. 1999. Descriptions of some of the glochidia of the Unionidae (Mollusca: Bivalvia). Malacologia 41(1):1–118.

Hoggarth, M.A., and A.S. Gaunt. 1988. Mechanics of glochidial attachment (Mollusca: Bivalvia: Unionidae). Journal of Morphology 198(1):71–81.

Holland-Bartels, L.E., and T.W. Kammer. 1989. Seasonal reproductive development of *Lampsilis cardium*, *Amblema plicata plicata* and *Potamilus alatus* (Pelecypoda: Unionidae) in the upper Mississippi River. Journal of Freshwater Ecology 5(1):87–92.

Holwerda, D.A., and P.R. Veenhof. 1984. Aspects of anaerobic metabolism in *Anodonta cygnea* L. Comparative Biochemistry and Physiology 78B:707–711.

Hopkins, R.L., Jr. 2009. Use of landscape pattern metrics and multiscale data in aquatic species distribution models: a case study of a freshwater mussel. Landscape Ecology 24:943–955.

Hornbach, D.J., V.J. Kurth, and M.C. Hove. 2010. Variation in freshwater mussel shell sculpture and shape along a river gradient. The American Midland Naturalist 164(1):22–36.

Houghton, W. 1862. On the parasitic nature of the fry of *Anodonta cygnea*. Quarterly Journal of Microscopical Science (New Series) 2:162–168.

Hove, M.C., and R.J. Neves. 1994. Life history of the endangered James spinymussel *Pleurobema collina* (Conrad, 1837) (Mollusca: Unionidae). American Malacological Bulletin 11(1):29–40.

Howard, A.D. 1914a. A second case of metamorphosis without parasitism in the Unionidae. Science 40(1027):353–355.

Howard, A.D. 1914b. Experiments in propagation of fresh-water mussels of the *Quadrula* group. Appendix IV to the Report of the U.S. Commissioner of Fisheries for 1913:1–52, 6 plates. [Issued separately as U.S. Bureau of Fisheries Document 801.]

Howard, A.D. 1922. Experiments in the culture of freshwater mussels. Bulletin of the U.S. Bureau of Fisheries 38(1924):63–89. [Issued separately as U.S. Bureau of Fisheries Document 916.]

Howard, J.K., and K.M. Cuffey. 2006. The functional role of native freshwater mussels in the fluvial benthic environment. Freshwater Biology 51:460–474.

Howells, R.G. 2000. Reproductive seasonality of freshwater mussels (Unionidae) in Texas. Pages 35–48. In: R.A. Tankersley, D.I. Warmolts, G.T. Watters, and B.J. Armitage (editors). Freshwater Mollusk Symposia Proceedings, Part I. Proceedings of the Conservation, Captive Care, and Propagation of Freshwater Mussels Symposium, 1998. Ohio Biological Survey, Columbus.

Howells, R.G., R.W. Neck, and H.D. Murray. 1996. Freshwater Mussels of Texas. Texas Parks and Wildlife Department, Inland Fisheries Division, Austin. 218 pages.

Hubbs, D. 2009. 2008 Statewide commercial mussel report. Tennessee Wildlife Resources Agency Fisheries Report 09-07. 44 pages.

Huddlestun, P.F. 1993. Revision of the lithostratigraphic units of the coastal plain of Georgia – the Oligocene. Georgia Geologic Survey Bulletin Number 105. 152 pages.

Huebner, J.D., and K.S. Pynnönen. 1992. Viability of glochidia of two species of *Anodonta* exposed to low pH and selected metals. Canadian Journal of Zoology 70(12):2348–2355.

Ihering, H. von. 1895. Os Unionidos da Florida. Revista do Museu Paulista 1:207–222.

Imlay, M.J. 1968. Environmental factors in activity rhythms of the freshwater clam *Elliptio complanatus catawbensis* (Lea). The American Midland Naturalist 80(2):508–528.

Imlay, M.J. 1982. Use of shells of freshwater mussels in monitoring heavy metals and environmental stresses: a review. Malacological Review 15:1–14.

International Commission on Zoological Nomenclature. 1999. International Code of Zoological Nomenclature. Fourth edition. International Trust for Zoological Nomenclature, London. 306 pages.

Isely, F.B. 1914. Experimental study of the growth and migration of freshwater mussels. Appendix III to the Report of the U.S. Commissioner of Fisheries for 1913:1–24, 3 plates. [Issued separately as U.S. Bureau of Fisheries Document 792.]

Ishibashi, R., A. Komaru, and T. Kondo. 2000. Sperm sphere in unionid mussels (Bivalvia: Unionidae). Zoological Science 17(7):947–950.

Isom, B.G. 1969. The mussel resource of the Tennessee River. Malacologia 7(2–3):397–425.

Jackson, A.P., J.F.V. Vincent, and R.M. Turner. 1988. The mechanical design of nacre. Proceedings of the Royal Society B 234:415–440.

Jansen, W., G. Bauer, and E. Zahner-Meike. 2001. Glochidial mortality in freshwater mussels. Pages 185–211. In: G. Bauer and K. Wachtler (editors). Ecology and Evolution of the Freshwater Mussels Unionoida. Ecological Studies, Volume 145. Springer-Verlag, Berlin.

Jansen, W.A., and J.M. Hanson. 1991. Estimates of the number of glochidia produced by clams (*Anodonta grandis simpsoniana* Lea), attaching to yellow perch (*Perca flavescens*), and surviving to various ages in Narrow Lake, Alberta. Canadian Journal of Zoology 69:973–977.

Jenkinson, J.J. 1976. Chromosome numbers of some North American naiads. Bulletin of the American Malacological Union, Inc. 1976:16–17.

Jenyns, R.L. 1832. A monograph of the British species of *Cyclas* and *Pisidium*. Transactions of the Cambridge Philosophical Society 4:289–311, 3 plates.

Johnson, N.A., P.J. Schofield, J.D. Williams, and J.D. Austin. In review. Comparative salinity tolerance of three freshwater mussel species (Bivalvia: Unionidae) inhabiting Gulf Coastal Plain drainages. Freshwater Science [forthcoming].

Johnson, P.D., and K.M. Brown. 1998. Intraspecific life history variation in the threatened Louisiana pearl shell, *Margaritifera hembeli*. Freshwater Biology 40:317–329.

Johnson, P.M. 2001. Habitat associations and drought responses of mussels in the lower Flint River basin, southwest GA. Master's thesis, University of Georgia, Athens. 114 pages.

Johnson, P.M., A.E. Liner, S.W. Golladay, and W.K. Michener. 2001. Effects of drought on freshwater mussels and instream habitat in Coastal Plain tributaries of the Flint River, southwest Georgia (July–October 2000). The Nature Conservancy, Tallahassee, Florida. 45 pages.

Johnson, R.I. 1956. Types of naiades (Mollusca: Unionidae) in the Museum of Comparative Zoology. Bulletin of the Museum of Comparative Zoology 115(4):102–142, plates 1–2.

Johnson, R.I. 1965. A hitherto overlooked *Anodonta* (Mollusca: Unionidae) from the Gulf drainage of Florida. Breviora 213:1–7.

Johnson, R.I. 1967a. Illustrations of all the mollusks described by Berlin Hart and Samuel Hart Wright. Museum of Comparative Zoology, Harvard University, Occasional Papers on Mollusks 3(35):1–35, plates 1–13.

Johnson, R.I. 1967b. Additions to the unionid fauna of the Gulf drainage of Alabama, Georgia and Florida (Mollusca: Bivalvia). Brevoria 270:1–21.

Johnson, R.I. 1968. *Elliptio nigella*, overlooked unionid from Apalachicola River system. The Nautilus 82(1):22–24.

Johnson, R.I. 1969a. Further additions to the unionid fauna of the Gulf drainage of Alabama, Georgia and Florida. The Nautilus 83(1):34–35.

Johnson, R.I. 1969b. Illustrations of Lamarck's types of North American Unionidae mostly in the Paris Museum. The Nautilus 83(2):52–61.

Johnson, R.I. 1970. The systematics and zoogeography of the Unionidae (Mollusca: Bivalvia) of the southern Atlantic Slope Region. Bulletin of the Museum of Comparative Zoology 140(6):263–449.

Johnson, R.I. 1971. The types and figured specimens of Unionacea (Mollusca: Bivalvia) in the British Museum (Natural History). Bulletin of the British Museum (Natural History) 20(3):75–108, plates 1–10.

Johnson, R.I. 1972. The Unionidae (Mollusca: Bivalvia) of peninsular Florida. Bulletin of the Florida State Museum, Biological Sciences 16(4):181–249.

Johnson, R.I. 1974. Lea's unionid types or Recent and fossil taxa of Unionacea and Mutelacea introduced by Isaac Lea, including the location of all the extant types. Museum of Comparative Zoology, Harvard University, Special Occasional Publication, Number 2. 159 pages.

Johnson, R.I. 1977. Monograph of the genus *Medionidus* (Bivalvia: Unionidae) mostly from the Apalachicolan region, southeastern United States. Museum of Comparative Zoology, Harvard University, Occasional Papers on Mollusks 4(56):161–187.

Johnson, R.I. 1980. The types of Unionacea (Mollusca: Bivalvia) in the Academy of Natural Sciences of Philadelphia: additions and corrections. Proceedings of the Academy of Natural Sciences of Philadelphia 132:277–278.

Johnson, R.I. 1981. Recent and fossil Unionacea and Mutelacea (freshwater bivalves) of the Caribbean Islands. Museum of Comparative Zoology, Harvard University, Occasional Papers on Mollusks 4(60):269–288.

Johnson, R.I. 1983. *Margaritifera marrianae*, a new species of Unionacea (Bivalvia: Margaritiferidae) from the Mobile-Alabama-Coosa and Escambia River systems, Alabama. Museum of Comparative Zoology, Harvard University, Occasional Papers on Mollusks 4(62):299–304, plate 41.

Johnson, R.I. 1998. Addenda to be included with Lea's figured unionid types (1974). Museum of Comparative Zoology, Harvard University. 2 pages.

Johnson, R.I. 2002. Samuel Liberty Harvey Fuller (1942–2001): A biographical sketch and his works on malacology. Nautilus 116:132–137.

Johnson, R.I., and H.B. Baker. 1973. The types of Unionacea (Mollusca: Bivalvia) in the Academy of Natural Sciences of Philadelphia. Proceedings of the Academy of Natural Sciences of Philadelphia 125(9):145–186, plates 1–10.

Johnston, M.O., B. Das, and W.R. Hoeh. 1998. Negative correlation between male allocation and rate of self-fertilization in a hermaphroditic animal. Proceedings of the National Academy of Sciences 95:617–620.

Jokela, J., and P. Palokangas. 1993. Reproductive tactics in *Anodonta* clams: parental host recognition. Animal Behaviour 46:618–620.

Jokela, J., L. Uotila, and J. Taskinen. 1993. Effect of the castrating trematodes parasite *Rhipidocotyle fennica* on energy allocation of fresh-water clam *Anodonta piscinalis*. Functional Ecology 7:332–338.

Jokela, J., E.T. Valtonen, and M. Lappalainen. 1991. Development of glochidia of *Anodonta piscinalis* and their infection of fish in a small lake in northern Finland. Archiv für Hydrobiologie 120:345–355.

Jones, D.S. 1997. The marine invertebrate fossil record of Florida. Pages 89–117. In: A.F. Randazzo and D.S. Jones (editors). The Geology of Florida. University of Florida Press, Gainesville. 327 pages.

Jones, J.W., R.J. Neves, S.A. Ahlstedt, and E.M. Hallerman. 2006a. A holistic approach to taxonomic evaluation of two closely related endangered freshwater mussel species, the oyster mussel *Epioblasma capsaeformis* and tan riffleshell *Epioblasma florentina walkeri* (Bivalvia: Unionidae). Journal of Molluscan Studies 72:267–28.

Jones, J.W., E.M. Hallerman, and R.J. Neves. 2006b. Genetic management guidelines for captive propagation of freshwater mussels (Unionoidea). Journal of Shellfish Research 25(2):527–535.

Jones, J.W., and R.J. Neves. 2011. Influence of life-history variation on demographic responses of three freshwater mussel species (Bivalvia: Unionidae) in the Clinch River, USA. Aquatic Conservation: Marine and Freshwater Ecosystems 21:57–73.

Jones, J.W., R.J. Neves, S.A. Ahlstedt, and E.M. Hallerman. 2006. A holistic approach to taxonomic evaluation of two closely related endangered freshwater mussel species, the oyster mussel *Epioblasma capsaeformis* and tan riffleshell *Epioblasma florentina walkeri* (Bivalvia: Unionidae). Journal of Molluscan Studies 72(3):267–283.

Jones, J.W., R.J. Neves, S.A. Ahlstedt, D.W. Hubbs, M. Johnson, D. Hua, and B.J.K. Ostby. 2009. Life History and demographics of the endangered birdwing pearly mussel (*Lemiox rimosus*) (Bivalvia: Unionidae). The American Midland Naturalist 163:335–350.

Kandl, K.L., H.-P. Liu, M. Mulvey, R. Butler, and W.R. Hoeh. 1997. Clarification of *Pleurobema pyriforme* as a species or species-complex and implications for the conservation of rare freshwater mussels. (Abstract). Eighty-ninth Annual Meeting of the National Shellfisheries Association, 20–24 April 1997, Fort Walton Beach, Florida.

Kandl, K.L., H.-P. Liu, R.S. Butler, W.R. Hoeh, and M. Mulvey. 2001. A genetic approach to resolving taxonomic ambiguity among *Pleurobema* (Bivalvia: Unionidae) of the eastern Gulf Coast. Malacologia 43(1–2):87–101.

Kanehl, P., and J. Lyons. 1992. Impacts of in-stream sand and gravel mining on stream habitat and fish communities, including a survey on the Big Rib River, Marathon County, Wisconsin. Wisconsin Department of Natural Resources Research Report 155, Madison. 32 pages.

Kat, P.W. 1983a. Conchiolin layers among the Unionidae and Margaritiferidae (Bivalvia) – microstructural characteristics and taxonomic implications. Malacologia 24:298–311.

Kat, P.W. 1983b. Sexual selection and simultaneous hermaphroditism among the Unionidae (Bivalvia: Mollusca). Journal of Zoology (London) 201(3):395–416.

Kat, P.W. 1983c. Genetic and morphological divergence among nominal species of North American *Anodonta* (Bivalvia: Unionidae). Malacologia 23(2):362–374.

Kaushall, S.S., G.E. Likens, N.A. Jaworski, M.L. Pace, A.M. Sides, D. Seekell, K.T. Belt, D.H. Secor, and R.L. Wingate. 2010. Rising stream and river temperatures in the United States. Frontiers in Ecology and the Environment 8(9):461–466.

Keller, A.E., and D.S. Ruessler. 1997a. Determination or verification of host fish for nine species of unionid mussels. The American Midland Naturalist 138(2):402–407.

Keller, A.E., and D.S. Ruessler. 1997b. The toxicity of malathion to unionid mussels: relationship to expected environmental concentrations. Environmental Toxicology and Chemistry 16:1028–1033.

Keller, A.E., and S.G. Zam. 1991. The acute toxicity of selected metals to the freshwater mussel, *Anodonta imbecillis*. Environmental Toxicology and Chemistry 10(4):539–546.

Kelner, D.E., and B.E. Sietman. 2000. Relic populations of the ebony shell, *Fusconaia ebena* (Bivalvia: Unionidae), in the upper Mississippi River drainage. Journal of Freshwater Ecology 15:371–377.

Kennedy, T.B., and W.R. Haag. 2005. Using morphometrics to identify glochidia from a diverse freshwater mussel community. Journal of the North American Benthological Society 24(4):880–889.

Kernaghan, N.J., D.S. Ruessler, S.E. Holm, and T.S. Gross. 2004. An evaluation of the potential effects of paper mill effluents on freshwater mussels in Rice Creek, Florida. Pages 455–463. In: D.L. Borton, T.J. Hall, R.P. Fisher, and J.E. Thomas (editors). Pulp and paper mill effluent environmental fate and effects. Fifth International Conference on Environmental Fate and Effects of Pulp and Paper Mill Effluents, 1–4 June 2003, Seattle, Washington.

Kesler, D.H., T.J. Newton, and L. Green. 2007. Long-term monitoring of growth in the Eastern Elliptio, *Elliptio complanata* (Bivalvia: Unionidae), in Rhode Island: a transplant experiment. Journal of the North American Benthological Society 26(1):123–133.

Kirk, S.G., and J.B. Layzer. 1997. Induced metamorphosis of freshwater mussel glochidia on nonhost fish. The Nautilus 110(3):102–106.

Klocker, C.A., and D.L. Strayer. 2004. Interactions among an invasive crayfish (*Orconectes rusticus*), a native crayfish (*Orconectes limosus*), and native bivalves (Sphaeriidae and Unionidae). Northeastern Naturalist 11(2):167–178.

Kołodziejczyk, A., K. Lewandowski, and A. Stańczykowska. 2009. Long-term changes of mollusc assemblages in bottom sediments of small semi-isolated lakes of different trophic state. Ekologia Polska 57(2):331–339.

Kolpin, D.W., J.E. Barbash, and R.J. Gilliom. 1998. Occurrence of pesticides in shallow groundwater of the United States: Initial results from the National Water-Quality Assessment Program. Environmental Science and Technology 32:558–566.

Kondo, T. 1993. Reproductive strategies of Japanese unionid mussels. Pages 142–146. In: K.S. Cummings, A.C. Buchanan, and L.M. Koch (editors). Conservation and Management of Freshwater Mussels. Proceedings of a UMRCC Symposium, 12–14 October 1992, St. Louis, Missouri. Upper Mississippi River Conservation Committee, Rock Island, Illinois.

Kotrla, B. 1988. Gametogenesis and gamete morphology of *Anodonta imbecillis* [sic], *Elliptio icterina*, and *Villosa villosa* (Bivalvia: Unionidae). Doctoral dissertation, Florida State University, Tallahassee. 165 pages.

Kraemer, L.R. 1970. The mantle flap in three species of *Lampsilis* (Pelecypoda: Unionidae). Malacologia 10(1):225–282.

Kryger, J., and H.U. Riisgård. 1988. Filtration rate capacities in 6 species of European freshwater bivalves. Oecologia 77:34–38.

Kuehnl, K.F. 2009. Exploring levels of genetic variation in the freshwater mussel genus *Villosa* (Bivalvia Unionidae) at different spatial and systematic scales: implications for biogeography, taxonomy, and conservation. Doctoral dissertation, Ohio State University, Columbus. 283 pages.

Kunz, G.F. 1898. A brief history of the gathering of fresh-water pearls in the United States. Bulletin of the U.S. Fish Commission 17(1897):321–330. [Issued separately as U.S. Bureau of Fisheries Document 389.]

Kurth, J.E., and M.C. Hove. 1997. Host fish suitability studies and host attracting behaviors of *Tritogonia verrucosa*, the pistolgrip. Triannual Unionid Report 12. Accessed 4 January 2013. http://ellipse.inhs.uiuc.edu/FMCS/TUR/TUR12.html#p13.

Laessle, A.M. 1968. Relationship of sand pine scrub to former shorelines. Quarterly Journal of the Florida Academy of Sciences 30:270–286.

Lamarck, J.B.P.A. 1799. Prodrome d'une nouvelle classification des coquilles, comprenent une rédaction appropriée des caractères génériques, et l'établissement d'un grand nombre de genres nouveaux. Memoirs de la Societie d'Histoire Naturelle, Paris 1:63–91.

Lamarck, J.B.P.A. 1815–1822. Histoire naturelle des Animaux sans Vertébres. 8 volumes. [Les nayades, 1819. 5:67–100.]

Lamarck, J.B.P.A. de M. de [C. de]. 1819. Les nayades. Pages 67–100. In: Histoire naturelle des animaux sans vertèbres. First edition. Paris, Volume 5.

Lane, E. 1986. Karst in Florida. Florida Geological Survey, Special Publication Number 29. 100 pages.

Lang, B.K., and P. Mehlhop. 1996. Distribution of freshwater mussels (Unionidae) of the Canadian River drainage: New Mexico and Texas. Segment I: Survey of freshwater bivalve mollusks of the Canadian River, New Mexico. National Biological Service, State Partnership Program, Final Report. 39 pages.

Layzer, J.B., M.E. Gordon, and R.M. Anderson. 1993. Mussels: the forgotten fauna of regulated rivers. A case study of the Caney Fork River. Regulated Rivers: Research and Management 8:63–71.

Layzer, J.B., and L.M. Madison. 1995. Microhabitat use by freshwater mussels and recommendations for determining their instream flow needs. Regulated Rivers: Research and Management 10(2–4):329–345.

Lea, I. 1831. Observations on the naïades, and descriptions of new species of that and other families. Transactions of the American Philosophical Society 4 (New Series) (1):63–121, plates 3–18.

Lea, I. 1834a. Observations on the naïades; and descriptions of new species of that, and other families. Transactions of the American Philosophical Society 5 (New Series) (1):23–119, plates 1–19.

Lea, I. 1834b. Observations on the genus *Unio*, together with descriptions of new genera and species in the families Naïades, Conchae, Colimacea, Lymnaeana, Melaniana and Peristomiana: consisting of four memoirs read before the American Philosophical Society from 1827 to 1834, and originally published in their Transactions. James Kay, Jun. and Company, Philadelphia. Volume 1. 4 leaves, 233 pages, plates 3–14, 3–18, 1–19.

Lea, I. 1836. A Synopsis of the Family of Naïades. Cary, Lea, and Blanchard, Philadelphia; John Miller, London. 59 pages, 1 plate.

Lea, I. 1838a. Description of new freshwater and land shells. Transactions of the American Philosophical Society 6 (New Series) (1):1–154, plates 1–24.

Lea, I. 1838b. A Synopsis of the Family of Naïades. Second edition, enlarged and improved. Philadelphia. 44 pages. [This publication represents pages 113–152 of Lea (1838a) repaginated and bound separately.]

Lea, I. 1840. Descriptions of new fresh water and land shells. Proceedings of the American Philosophical Society 1(13):284–289.

Lea, I. 1841. Continuation of paper [On fresh water and land shells]. Proceedings of the American Philosophical Society 2(17):30–35.

Lea, I. 1842. Description of new fresh water and land shells. Transactions of the American Philosophical Society 8 (New Series) (Part 2):163–250, plates 5–27.

Lea, I. 1843. Description of twelve new species of Uniones. 1 page. [Read by Isaac Lea before the American Philosophical Society on 18 August 1843, and privately published on 19 August 1843, Philadelphia.]

Lea, I. 1845. Descriptions of new fresh water and land shells. Proceedings of the American Philosophical Society 4(33):162–168.

Lea, I. 1846. Description of new freshwater and land shells. Transactions of the American Philosophical Society 9 (New Series) (Part 2):275–282.

Lea, I. 1852a. A Synopsis of the Family of Naïades. Third edition, greatly enlarged and improved. Blanchard and Lea, Philadelphia. 88 pages.

Lea, I. 1852b. Descriptions of new species of the family Unionidae. Transactions of the American Philosophical Society 10 (New Series) (Part 2):253–294, plates 12–29.

Lea, I. 1856. Description of eleven new species of uniones, from Georgia. Proceedings of the Academy of Natural Sciences of Philadelphia 8(6):262–263.

Lea, I. 1857a. Descriptions of twenty-seven new species of uniones from Georgia. Proceedings of the Academy of Natural Sciences of Philadelphia 9(1857):169–172.

Lea, I. 1857b. Description of thirteen new species of uniones, from Georgia. Proceedings of the Academy of Natural Sciences of Philadelphia 9(2):31–32.

Lea, I. 1857c. Description of eight new species of naïades from various parts of the United States. Proceedings of the Academy of Natural Sciences of Philadelphia 9(1857):84.

Lea, I. 1858a. Descriptions of seven new species of Margaritanae, and four new species of Anodontae. Proceedings of the Academy of Natural Sciences of Philadelphia 10:138–139.

Lea, I. 1858b. Descriptions of twelve new species of uniones and other fresh-water shells of the United States. Proceedings of the Academy of Natural Sciences of Philadelphia 10(1858):165–166.

Lea, I. 1858c. New Unionidae of the United States. Journal of the Academy of Natural Sciences of Philadelphia 4 (New Series) (1):51–95, plates 6–20.

Lea, I. 1859a. Descriptions of twelve new species of uniones, from Georgia. Proceedings of the Academy of Natural Sciences of Philadelphia 11(1859):170–172.

Lea, I. 1859b. Descriptions of two new species of uniones, from Georgia. Proceedings of the Academy of Natural Sciences of Philadelphia 11(1859):154.

Lea, I. 1860. Descriptions of two new species of uniones from Georgia. Proceedings of the Academy of Natural Sciences of Philadelphia 12(1860):305.

Lea, I. 1861. Descriptions of twenty-five new species of Unionidae from Georgia, Alabama, Mississippi, Tennessee and Florida. Proceedings of the Academy of Natural Sciences of Philadelphia 13(1861):38–41.

Lea, I. 1862. Descriptions of ten new species of Unionidae of the United States. Proceedings of the Academy of Natural Sciences of Philadelphia 14(1862):168–169.

Lea, I. 1865. Descriptions of eight new species of *Unio* of the United States. Proceedings of the Academy of Natural Sciences of Philadelphia 17(1865):88–89.

Lea, I. 1867a. Index to Volume I to XI of Observations on the genus *Unio*, together with description of new species of the family Unionidae. And descriptions of new species of Melanidae, Paludinae, Helicidae, etc. Volume I. Printed for the author by T.K. Collins, Philadelphia. 63 pages.

Lea, I. 1867b. Descriptions of five new species of Unionidae and one *Paludina* of the United States. Proceedings of the Academy of Natural Sciences of Philadelphia 19(1867):81.

Lea, I. 1868a. Description of sixteen new species of the genus *Unio* of the United States. Proceedings of the Academy of Natural Sciences of Philadelphia 20(1868):143–145.

Lea, I. 1868b. New Unionidae, Melanidae, etc., chiefly of the United States. Journal of the Academy of Natural Sciences of Philadelphia 6 (New Series) (3):249–302, plates 29–45.

Lea, I. 1869. Index to Volume XII and supplementary index to Volumes I to XI of Observations on the genus *Unio*, together with description of new species of the family Unionidae, and description of new species of the Melanidae, Paludinae, Helicidae, etc. Volume II. Printed for the author by T.K. Collins, Philadelphia. 23 pages.

Lea, I. 1870. A synopsis of the family Unionidae. Fourth edition, very greatly enlarged and improved. Henry C. Lea, Philadelphia. 184 pages.

Lea, I. 1872. Descriptions of twenty-nine species of Unionidae from the United States. Proceedings of the Academy of Natural Sciences of Philadelphia 24(2):155–161.

Lea, I. 1874a. Index to Volume I to XIII Observations on the genus *Unio*, together with descriptions of new species of the family Unionidae, and descriptions of new species of the Melanidae, Paludinae, Helicidae, etc. Volume III. Printed for the author by Collins Printer, Philadelphia. 29 pages.

Lea, I. 1874b. Description of seven new species of Unionidae of the United States. Proceedings of the Academy of Natural Sciences of Philadelphia 25(3):422–423.

Lea, I. 1874c. Description of three new species of Uniones of the United States. Proceedings of the Academy of Natural Sciences of Philadelphia 25(3):424.

Lee, H.G. 1979. Checklist of aquatic mollusks – Jacksonville area. Shell-O-Gram 20(11):3.

Lee, H.G. 2006. Musings on a local specimen of *Toxolasma paulum* (I. Lea, 1840), the Iridescent Lilliput. Shell-O-Gram 47(5):3–6.

Lee, H.G. 2011. The Mollusca of the Crosby Sanctuary, Clay Co., Florida. Shell-O-Gram 51(6):3–7.

Lee, H.G. 2012. Corrigenda/addenda to Crosby Sanctuary report. Shell-O-Gram 52(3):3–7.

Lee, T. 2004. Morphology and phylogenetic relationships of genera of North American Sphaeriidae (Bivalvia, Veneroida). American Malacological Bulletin 19(1 and 2):1–13.

Lee, T., and D. O'Foighil. 2003. Phylogenetic structure of the Sphaeriinae, a global clade of freshwater bivalve molluscs, inferred from nuclear (ITS-1) and mitochondrial (16S) ribosomal gene sequences. Zoological Journal of the Linnean Society 2003(137):245–260.

Lee, T., S. Siripattrawan, C.F. Ituarte, and D. O'Foighil 2005. Invasion of the clonal clams: *Corbicula* lineages in the New World. American Malacological Bulletin 20(1 and 2):113–122.

Lefevre, G., and W.C. Curtis. 1910. Reproduction and parasitism in the Unionidae. Journal of Experimental Zoology 99(1):79–115, 4 plates.

Lefevre, G., and W.C. Curtis. 1912. Studies on the reproduction and artificial propagation of fresh-water mussels. Bulletin of the U.S. Bureau of Fisheries 30(1910):105–201, 12 plates. [Issued separately as U.S. Bureau of Fisheries Document 756. Reprinted in Sterkiana 47, 48 (1972); 49, 51 (1973); 57 (1975); and 61, 63, 64 (1976).]

Leff, L.G., J.L. Burch, and J.V. McArthur. 1990. Spatial distribution, seston removal, and potential competitive interactions of the bivalves *Corbicula fluminea* and *Elliptio complanata*, in a coastal plain stream. Freshwater Biology 24(2):409–416.

Leitman, S.F., L. Ager, and C. Mesing. 1991. The Apalachicola experience: environmental effects of physical modifications for navigation purposes. Chapter 11, pages 223-246. In: R.J. Livingston (editor). The Rivers of Florida. Springer-Verlag, New York. 289 pages.

Levine, T.D., B.K. Lang, and D.J. Berg. 2009. Parasitism of mussel gills by dragonfly nymphs. The American Midland Naturalist 162:1–6.

Lewandowski, K. 1990. Unionidae of Szeszupa River and of the lakes along its course in Suwalski Landscape Park. Ekologia Polska 38:271–286.

Lewis, J.B. 1985. Breeding cycle of the freshwater mussel *Anodonta grandis* Say. Canadian Journal of Zoology 63(10):2477–2478.

Lewis, J.B., and P.N. Riebel. 1984. The effect of substrate on burrowing in freshwater mussels (Unionidae). Canadian Journal of Zoology 62(10):2023–2025.

Light, H.M., M.R. Darst, and J.M. Grubbs. 1998. Aquatic habitat in relation to river flow in the Apalachicola River floodplain, Florida. U.S. Geological Survey Professional Paper 1656A. 124 pages.

Light, H.M., M.R. Darst, F.D. Price, and S.J. Walsh. 2005. Declining water levels in the Apalachicola River and impacts on floodplain habitats. (Abstract). USGS Science in Florida, 3–5 May 2005, Orlando.

Lightfoot, J. 1786. A catalogue of the Portland Museum, lately the property of the Dutchess Dowager of Portland, deceased: which will be sold by auction by Mr. Skinner and Co. on Monday the 24th of April, 1786, and the thirty-seven following days at her late dwelling-house, in Privy-garden, Whitehall: by order of the acting executrix. Skinner, London. Pages [3]–194.

Linnaeus, C. 1758. Systema Naturae. Edition X. (Systema naturae per regna tria naturae, secundum classes, ordines, genera, species cum characteribus, differentiis, synonymis, locis. Tomus I. Edtio decima, reformata.) Holmiae. Volume 1:1–824.

Livingston, R.J. (editor). 1990. The Rivers of Florida. Springer Verlag, Ecological Studies 83. 289 pages.

Livingston, R.J., J.H. Epler, F. Jordan, Jr., W.R. Karsteter, C.C. Koenig, A.K.S.K. Prasad, and G.L. Ray. 1991. Ecology of the Choctawhatchee River system. Chapter 12, pages 247-274. In: R.L Livingston (editor). The Rivers of Florida. Springer-Verlag, New York. 289 pages.

Locke, A., J.M. Hanson, G.J. Klassen, S.M. Richardson, and C.I. Aubé. 2003. The damming of the Petitcodiac River: species, populations, and habitats lost. Northeastern Naturalist 10(1):39–54.

Lydeard, C., and R.L. Mayden. 1995. A diverse and endangered aquatic ecosystem of the southeast United States. Conservation Biology 9(4):800–805.

Lydeard, C., R.L. Minton, and J.D. Williams. 2000. Prodigious polyphyly in imperiled freshwater pearlymussels (Bivalvia: Unionidae): a phylogenetic test of species and generic designations. Pages 145–158. In: E.M. Harper, J.D. Taylor, and J.A. Crane (editors). The Evolutionary Biology of the Bivalvia. Geological Society Special Publication, Number 177.

Lynn, J.W. 1994. The ultrastructure of sperm and motile spermatozeugmata released from the freshwater mussel *Anodonta grandis* (Mollusca, Bivalvia, Unionidae). Canadian Journal of Zoology 72(8):1452–1461.

Mackie, G.L. 2007. Biology of freshwater corbiculiid and sphaeriid clams of North America. Ohio Biological Survey Bulletin New Series 15(3):1–436 pages.

Mackie, G.L., and L.A. Flippance. 1983. Relationships between buffering capacity of water and the size and calcium content of freshwater mollusks. Freshwater Invertebrate Biology 2:48–55.

Mackie, G.L., and D.W. Schloesser. 1996. Comparative biology of zebra mussels in Europe and North America: an overview. American Zoologist 36:244–258.

Marelli, D.C., and S. Gray. 1983. Conchological redescriptions of *Mytilopsis sallei* and *Mytilopsis leucophaeta* of the brackish western Atlantic (Bivalvia: Dreissenidae). The Veliger 25(3):185–193.

Marsh, W.A. 1891. Description of two new species of *Unio* from Florida. The Nautilus 5(3):29–30.

Marshall, W.B. 1890. Beaks of Unionidae inhabiting the vicinity of New Albany, N.Y. Bulletin of the New York State Museum of Natural History 2(9):167–189.

Martel, A.L., D.A. Pathy, J.B. Madill, C.B. Renaud, S.L. Dean, and S.J. Kerr. 2001. Decline and regional extirpation of freshwater mussels (Unionidae) in a small river system invaded by *Dreissena polymorpha*: the Rideau River, 1993–2000. Canadian Journal of Zoology 79(12):2181–2191.

Martin, S.M. 1997. Freshwater mussels (Bivalvia: Unionoida) of Maine. Northeastern Naturalist 4(1):1–34.

Mason, W.T., Jr. 1998. Macrobenthic monitoring in the lower St. Johns River, Florida. Environmental Monitoring and Assessment 50:101–130.

Massabuau, J.-C., B. Burtin, and M. Wheathly. 1991. How is O$_2$ consumption maintained independent of ambient oxygen in mussel *Anodonta cygnea*? Respiration Physiology 83:103–114.

Master, L.L., S.R. Flack, and B.A. Stein (editors). 1998. Rivers of Life: Critical Watersheds for Protecting Freshwater Biodiversity. The Nature Conservancy, Arlington, Virginia. 71 pages.

Matteson, M.R. 1948. Life history of *Elliptio complanata* (Dillwyn, 1817). The American Midland Naturalist 40(3):690–723.

Matthews, W.J. 1998. Patterns in Freshwater Fish Ecology. Chapman and Hall, New York. 756 pages.

Maymon, J.H., T.F. Majarov, F. Vento, M. Williams, A. Fradkin, C.T. Brown, K.M. Child, and J.C. Clarke. 1996. Phase III data recovery at Site 8LI76 for the proposed Florida gas transmission phase III expansion project, Liberty County, Florida. Final report prepared for Florida Gas Transmission Company, Houston, Texas. 414 pages.

McCorkle, S., T.C. Shirley, and T.H. Dietz. 1979. Rhythms of activity and oxygen consumption in the common pond clam, *Ligumia subrostrata* (Say). Canadian Journal of Zoology 57:1960–1964.

McDiarmid, R.W. (editor). 1978. Amphibians and Reptiles. Rare and Endangered Biota of Florida. Volume III. University Press of Florida, Gainesville. 74 pages.

McLain, D.C., and M.R. Ross. 2005. Reproduction based on local patch size of *Alasmidonta heterodon* and dispersal by its darter host in the Mill River, Massachusetts, USA. Journal of the North American Benthological Society 24(1):139–147.

McMahon, R.F. 1999. Invasive characteristics of the freshwater bivalve *Corbicula fluminea*. Pages 351–343. In: R. Claudi and J.H. Leach (editors). Nonindigenous Freshwater Organisms: Vectors, Biology, and Impacts. Lewis Publishers, Boca Raton, Florida.

McMahon, R.F., and A.E. Bogan. 2001. Mollusca: Bivalvia. Pages 331–429. In: J.H. Thorpe and A.P. Kovich (editors). Ecology and Classification of North American Freshwater Invertebrates. Second Edition. Academic Press, New York. 1056 pages.

McMaster, M.E. 2001. A review of the evidence for endocrine disruption in Canadian aquatic ecosystems. Water Quality Research Journal of Canada 36(2):215–231.

Meador, J.R., J.T. Peterson, and J.M. Wisniewski. 2011. An evaluation of the factors influencing freshwater mussel capture probability, survival, and temporary emigration in a large lowland river. Journal of the North American Benthological Society 30(2):507–521.

Meysman, F.J.R., J.J. Middleburg, and C.H.R. Heip. 2006. Bioturbation: a fresh look at Darwin's last idea. Trends in Ecology and Evolution 21:688–695.

Mikkelsen, P.M., and R. Bieler. 2008. Seashells of Southern Florida: Living Marine Mollusks of the Florida Keys and Adjacent Regions: Bivalves. Princeton University Press, New Jersey. 503 pages.

Milanich, J.T. 1974. Life in a 9th century Indian household. A Weeden Island Fall-Winter site on the upper Apalachicola River, Florida. Division of Archives, History, and Records Management, Department of State, Tallahassee, Florida. Bureau of Historical Sites and Properties Bulletin 4:1–44.

Mills, E.L., G. Rosenberg, A.P. Spidle, M. Ludyanskiy, Y. Pligin, and B. May. 1996. A review of the biology and ecology of the quagga mussel (*Dreissena bugensis*), a second species of freshwater dreissenid introduced to North America. American Zoologist 36:271–286.

Mock, K.E., J.C. Brim Box, M.P. Miller, M.E. Downing, and W.R. Hoeh. 2004. Genetic diversity and divergence among freshwater mussel (*Anodonta*) populations in the Bonneville Basin of Utah. Molecular Ecology 13:1085–1098.

Moles, K.R., and J.B. Layzer. 2008. Reproductive ecology of *Actinonaias ligamentina* (Bivalvia: Unionidae) in a regulated river. Journal of the North American Benthological Society 17(1):212–222.

Montagna, P.A., E.D. Estevez, T.A. Palmer, and M.S. Flannery. 2008. Meta-analysis of the relationship between salinity and molluscs in tidal river estuaries of southwest Florida, U.S.A. American Malacological Bulletin 24:101–115.

Morales, Y., L.J. Weber, A.E. Mynett, and T.J. Newton. 2006. Effects of substrate and hydrodynamic conditions on the formation of mussel beds in a large river. Journal of the North American Benthological Society 26:664–676.

Morell, V. 2008. Into the wild: reintroduced animals face daunting odds. Science 320:742–743.

Morris, A. 1995. Florida Place Names. Pineapple Express Inc., Sarasota, Florida. 291 pages.

Morris, T.J., and L.D. Corkum. 1996. Assemblage structure of freshwater mussels (Bivalvia: Unionidae) in rivers with grassy and forested riparian zones. Journal of the North American Benthological Society 15(4):576–586.

Morris, T.J., and L.D. Corkum. 1999. Unionid growth patterns in rivers of differing riparian vegetation. Freshwater Biology 42:59–68.

Morrison, J.P.E. 1973. The families of the pearly freshwater mussels. Bulletin of the American Malacological Union 1973:45–46.

Mosley, T.L., W.R. Haag, and J.A. Stoeckel. 2011. Fertilization efficiency of *Lampsilis straminea* in relation to distance and water flow. (Abstract). Freshwater Mollusk Conservation Society Seventh Biennial Symposium, April 2011, Louisville, Kentucky.

Mossa, J. 1998. Surface water. Pages 64–81. In: E.A. Fernald and E.D. Purdum (editors). Water Resources Atlas of Florida. Institute of Science and Public Affairs, Florida State University. 312 pages.

Moyer, E.J., M.W. Hulon, R.S. Butler, and V.P. Williams. 1992. Lake Tohopekaliga muck removal project. Pages 18–25. In: C.E. Watkins, H. McGinnis, and K.J. Hatcher (editors). Proceedings of the First Annual Southeastern Lakes Management Conference. North American Lake Management Society, Alachua, Florida.

Müller, O.F. 1773–1774. Vermium terrestrium et fluviatilium, seu animalium Infusoriorum, Helminthicorum et Testaceorum, non marinorum, succincta historia, &c. Havniæ; Lipsiæ, 2 volumes.

Mulvey, M., H.-P. Liu, and K.L. Kandl. 1998. Application of molecular genetic markers to conservation of freshwater bivalves. Journal of Shellfish Research 17(5):1395–1405.

Mulvey, M., C. Lydeard, D.L. Pyer, K. Hicks, J. Brim-Box, J.D. Williams, and R.S. Butler. 1997. Conservation genetics of North American freshwater mussels *Amblema* and *Megalonaias*. Conservation Biology 11(4):868–878.

Murphy, D.D., K.E. Freas, and S.B. Weiss. 1990. An environmental-metapopulation approach to population viability analysis for a threatened invertebrate. Conservation Biology 4:41–51.

Naiman, R.J., and H. Decamps. 1997. The ecology of interfaces: riparian zones. Annual Review of Ecology and Systematics 28:621–658.

Naimo, T.J. 1995. A review of the effects of heavy metals on freshwater mussels. Ecotoxicology 4(6):341–362.

Nalepa, T.F., B.A. Manny, J.C. Roth, S.C. Mozley, and D.W. Schloesser. 1991. Long-term decline in freshwater mussels (Bivalvia: Unionidae) of the western basin of Lake Erie. Journal of Great Lakes Research 17:214–219.

Negishi, J.N., H. Doi, I. Katano, and Y. Kayaba. 2011. Seasonally tracking vertical and horizontal distribution of unionid mussels (*Pronodularia japanensis*): implications for agricultural drainage management. Aquatic Conservation: Marine and Freshwater Ecosystems 21:49–56.

Neves, R.J. 1986. Recent die-offs of freshwater mussels in the United States: an overview. Pages 7–18. In: R.J. Neves (editor). Proceedings of a Workshop, June 1986, on Die-offs of Freshwater Mussels in the United States. Davenport, Iowa.

Neves, R.J. 1993. A state-of-the-unionids address. Pages 1–10. In: K.S. Cummings, A.C. Buchanan, and L.M. Koch (editors). Conservation and Management of Freshwater Mussels. Proceedings of the UMRCC Symposium, 12–14 October 1992, St. Louis, Missouri. Upper Mississippi River Conservation Committee, Rock Island, Illinois.

Neves, R.J., and S.N. Moyer. 1988. Evaluation of techniques for age determination of freshwater mussels (Unionidae). American Malacological Bulletin 6(2):179–188.

Neves, R.J., and J.C. Widlak. 1988. Occurrence of glochidia in stream drift and on fishes of the upper North Fork Holston River, Virginia. The American Midland Naturalist 119(1):111–120.

Newton, T.J. 2003. The effects of ammonia on freshwater unionid mussels. Environmental Toxicology and Chemistry 22(11):2543–2544.

Newton, T.J., and M.R. Bartsch. 2007. Lethal and sublethal effects of ammonia to juvenile *Lampsilis* mussels (Unionidae) in sediment and water-only exposures. Environmental Toxicology and Chemistry 26(10):2057–2065.

Newton, T., S. Zigler, R. Kennedy, A. Hunt, M. Davis, and P. Ries. 2011. Response of native mussels to water level manipulation in the upper Mississippi River. (Abstract). Freshwater Mollusk Conservation Society Seventh Biennial Symposium, April 2011, Louisville, Kentucky.

Nezlin, L.P., R.A. Cunjak, A.A. Zotin, and V.V. Ziuganov. 1994. Glochidium morphology of the freshwater pearl mussel (*Margaritifera margaritifera*) and glochidiosis of Atlantic salmon (*Salmo salar*): a study by scanning electron microscopy. Canadian Journal of Zoology 72(1):15–21.

Nichols, S.J., H. Silverman, T.H. Dietz, J.W. Lynn, and D.L. Garling. 2005. Pathways of food uptake in native (Unionidae) and introduced (Corbiculidae and Dreissenidae) freshwater bivalves. Journal of Great Lakes Research 31:87–96.

Nichols, S.J., and D.A. Wilcox. 1997. Burrowing saves Lake Erie clams. Nature 389:921.

Nordlie, F.G. 1990. Rivers and Springs. Chapter 12, pages 392–425. In: R.L. Myers and J.J. Ewel (editors). Ecosystems of Florida. University of Central Florida Press, Orlando. 765 pages.

Nott, M., E. Rogers, and S. Pimm. 1995. Modern extinctions in the kilo-death range. Current Biology 1995(5):14–17.

O'Brien, C.A., and J. Brim Box. 1999. Reproductive biology and juvenile recruitment of the Shinyrayed Pocketbook, *Lampsilis subangulata* (Bivalvia: Unionidae), in the Gulf Coastal Plain. The American Midland Naturalist 142(1):129–140.

O'Brien, C.A., and J.D. Williams. 2002. Reproductive biology of four freshwater mussels (Bivalvia: Unionidae) endemic to eastern Gulf Coastal Plain drainages of Alabama, Florida, and Georgia. American Malacological Bulletin 17(1 and 2):147–158.

O'Brien, C.A., J.D. Williams, and M.A. Hoggarth. 2003. Morphological variation in glochidia shells of six species of *Elliptio* from Gulf of Mexico and Atlantic Coast drainages in the southeastern United States. Proceedings of the Biological Society of Washington 116(3):719–731.

O'Connell, M.T., and R.J. Neves. 1999. Evidence of immunological responses by a host fish (*Ambloplites rupestris*) and two non-host fishes (*Cyprinus carpio* and *Carassius auratus*) to glochidia of a freshwater mussel (*Villosa iris*). Journal of Freshwater Ecology 14(1):71–78.

Oesch, R.D. 1995. Missouri Naiades. A Guide to the Mussels of Missouri. Missouri Department of Conservation, Jefferson City. 271 pages.

Ohio State University Museum of Biological Diversity, Division of Molluscs. Host-Mussel database. Accessed 13 December 2013. http://www.biosci.ohio-state.edu/~molluscs/OSUM2/.

Ormerod, S.J., M. Dobson, A.G. Hildrew, and C.R. Townsend. 2010. Multiple stressors in freshwater ecosystems. Freshwater Biology 55 (Supplement 1):1–4.

Ortmann, A.E. 1909a. The destruction of the fresh-water fauna in western Pennsylvania. Proceedings of the American Philosophical Society 48(191):90–110.

Ortmann, A.E. 1909b. The breeding season of Unionidae in Pennsylvania. The Nautilus 22(9):91–95; 22(10):99–103.

Ortmann, A.E. 1910a. A new system of the Unionidae. The Nautilus 23(9):114–120.

Ortmann, A.E. 1910b. The discharge of the glochidia in the Unionidae. The Nautilus 24(8):94–95.

Ortmann, A.E. 1911. A monograph of the najades of Pennsylvania. Parts I and II. Memoirs of the Carnegie Museum 4(6):279–347, 4 plates.

Ortmann, A.E. 1912. Notes upon the families and genera of the najades. Annals of the Carnegie Museum 8(2):222–365, plates 18–20.

Ortmann, A.E. 1913. The Alleghenian Divide and its influence upon the freshwater fauna. Proceedings of the American Philosophical Society 52(210):287–390, plates 12–14.

Ortmann, A.E. 1914. Studies in najades (continued). The Nautilus 28(2):20–22; 28(3):28–34; 28(4):41–47; 28(5[sic]):65–69.

Ortmann, A.E. 1916. Studies in najades (concluded). The Nautilus 30(5):54–57.

Ortmann, A.E. 1919. A monograph of the naiades of Pennsylvania. Part III: systematic account of the genera and species. Memoirs of the Carnegie Museum 8(1):1–384, 21 plates.

Ortmann, A.E. 1920. Correlation of shape and station in freshwater mussels (Naiades). Proceedings of the American Philosophical Society 59(4):269–312.

Ortmann, A.E. 1923a. The anatomy and taxonomy of certain Unioninae and Anodontinae from the Gulf drainage. The Nautilus 36(3):73–84; 36(4):129–132.

Ortmann, A.E. 1923b. Notes on the anatomy and taxonomy of certain Lampsilinae from the Gulf Drainage. The Nautilus 37(2):56–60.

Ortmann, A.E. 1924a. Notes on the anatomy and taxonomy of certain Lampsilinae from the Gulf drainage. The Nautilus 37(3):99–105; 37(4):137–144.

Ortmann, A.E. 1924b. Mussel Shoals. Science 60(1564):565–566.

Ortmann, A.E. 1925. The naiad fauna of the Tennessee River system below Walden Gorge. The American Midland Naturalist 9(8):321–372.

Ortmann, A.E., and B. Walker. 1922. A new genus and species of American naiades. The Nautilus 36(1):1–6, plate 1.

Österling, M.E., B.L. Arvidsson, and L.A. Greenberg. 2010. Habitat degradation and the decline of the threatened mussel *Margaritifera margaritifera*: influence of turbidity and sedimentation on the mussel and its host. Journal of Applied Ecology 47:759–768.

Ostrovsky, I., M. Gophen, and I. Kalikhman. 1993. Distribution, growth, and ecological significance of the clam *Unio terminalis* in Lake Kinneret, Israel. Hydrobiologia 271:49–63.

Page, L.M., and B.M. Burr. 2011. Peterson Field Guide to Freshwater Fishes of North America North of Mexico. Houghton Mifflin Harcourt, New York. 633 pages.

Pallas, P.S. 1771, 1773, 1776. Reise durch verschiedene Provinzen des russischen Reichs. St. Petersburg, Gedruckt bey der kayserlichen Academie der Wissenschaften.

Pandolfo, T.J., W.G. Cope, C. Arellano, R.B. Bringolf, M.C. Barnhart, and E. Hammer. 2010. Upper thermal tolerances of early life stages of freshwater mussels. Journal of the North American Benthological Society 29(3):959–969.

Parker, G.G., G.E. Ferguson, S.K. Love, and others. 1955. Water resources of southeastern Florida, with special reference to geology and groundwater of the Miami area. U.S. Geological Survey Water-Supply Paper 1255. 965 pages.

Parker, R.S., C.T. Hackney, and M.F. Vidrine. 1984. Ecology and reproductive strategy of a south Louisiana freshwater mussel, *Glebula rotundata* (Lamarck) (Unionidae: Lampsilini). Freshwater Invertebrate Biology 3(2):53–58.

Parmalee, P.W., and A.E. Bogan. 1998. The Freshwater Mussels of Tennessee. The University of Tennessee Press, Knoxville. 328 pages.

Parmesan, C. 2006. Ecological and evolutionary responses to recent climate change. Annual Review of Ecology, Evolution, and Systematics 37:637–669.

Patzner, R.A., and D. Müller. 2001. Effects of eutrophication on unionids. Pages 327–335. In: G. Bauer and K. Wächtler (editors). Ecology and Evolution of the Freshwater Mussels Unionoida. Ecological Studies, Volume 145. Springer-Verlag, Berlin.

Paul, M.J., and J.L. Meyer. 2001. Streams in the urban landscape. Annual Review of Ecology and Systematics 32:333–365.

Pauley, G.B. 1968. A disease of the freshwater mussel, *Margaritifera margaritifera*. Journal of Invertebrate Pathology 12(3):321–328.

Payne, B.S., and A.C. Miller. 2000. Recruitment of *Fusconaia ebena* (Bivalvia: Unionidae) in relation to discharge of the lower Ohio River. The American Midland Naturalist 144(2):328–341.

Pepi, V.E., and M.C. Hove. 1997. Suitable fish hosts and mantle display behavior of *Tritogonia verrucosa*. Triannual Unionid Report 11:5.

Percy, G.W. 1976. Salvage investigations at the Scholz Steam Plant Site (8JA104), a middle Weeden Island habitation site in Jackson County, Florida. Division of Archives, History, and Records Management, Department of State, Tallahassee, Florida. Bureau of Historic Sites and Properties, Miscellaneous Projects Report Series Number 35. 150 pages.

Perles, S.J., A.D. Christian, and D.J. Berg. 2003. Vertical migration, orientation, aggregation, and fecundity of the freshwater mussel *Lampsilis siliquoidea*. Ohio Journal of Science 103:73–78.

Peterjohn, W.T., and D.L. Correll. 1984. Nutrient dynamics in an agricultural watershed: observations on the role of a riparian forest. Ecology 65(5):1466–1475.

Peterson, J.T., J.M. Wisniewski, C.P. Shea, and C.R. Jackson. 2011. Estimation of mussel population response to hydrologic alteration in a southeastern U.S. stream. Environmental Management 48(1):109–122.

Pfitzenmeyer, H.T., and K.G. Drobeck. 1964. The occurrence of the Brackish Water Clam, *Rangia cuneata*, in the Potomac River, Maryland. Chesapeake Science 5(4):209–212.

Pharris, G.L., C.C. Chandler, and J.B. Sickel. 1982. Range extension for *Plectomerus dombeyanus* (Bivalvia: Unionidae) into Kentucky. (Abstract). Transactions of the Kentucky Academy of Science 43(1–2):95–96.

Pilarczyk, M.M., P.M. Stewart, D.N. Shelton, H.N. Blalock-Herod, and J.D. Williams. 2006. Current and recent historical freshwater mussel assemblages in the Gulf Coastal Plains. Southeastern Naturalist 5(2):205–226.

Pilarczyk, M.M., P.M Stewart, D.N. Shelton, W.H. Heath, and J.M. Miller. 2005. Contemporary survey and historical freshwater mussel assemblages in southeast Alabama and northwest Florida and life history and host fish identification of two candidate unionids (*Quincuncina burkei* and *Pleurobema strodeanum*). Report to the U.S. Fish and Wildlife Service, Panama City, Florida, Contract Number 401214G049. 93 pages, appendices.

Pilsbry, H.A. 1953. Fresh-water mollusks. Part III-B. Pages 439–447, plate 65. In: A.A. Olsson and A. Harbison. Pliocene Mollusks of Southern Florida. Academy Natural Sciences Philadelphia Monograph 8.

Pinder, M.J., E.S. Wilhelm, and J.W. Jones. 2003. Status survey of the freshwater mussels (Bivalvia: Unionidae) in the New River drainage, Virginia. Walkerana 12(29–30):189–223.

Plumb, J.A., M.L. Wise, and W.A. Rogers. 1986. Modular effects of temperature on antibody response and specific resistance to challenge of channel catfish, *Ictalurus punctatus*, immunized against *Edwardsiella ictaluri*. Veterinary Immunology and Immunopathology 12:297–304.

Poff, N.L., P.L. Angermeier, S.D. Cooper, P.S. Lake, K.D. Fausch, K.O. Winemiller, L.A.K. Mertes, M.W. Oswood, J. Reynolds, and F.J. Rahel. 2001. Fish diversity in streams and rivers. Pages 315–349. In: F.S. Chapin, E. Sala, and E. Huner-Sannwald (editors). Global Biodiversity in a Changing Environment. Springer, New York.

Poff, N.L., J.D. Olden, D.M. Merritt, and D.M. Pepin. 2007. Homogenization of regional river dynamics by dams and global biodiversity implications. Proceedings of the National Academy of Sciences 104:5732–5737.

Poli, I.X. 1791. Testacea utriusque Siciliae eorumque Historia et Anatome, tabulis aeneis illustrata. Regio Typographeio, Parmae. Volume 1. 74 pages.

Poole, K.E., and J.A. Downing. 2004. Relationship of declining mussel biodiversity to stream-reach and watershed characteristics in an agricultural landscape. Journal of the North American Benthological Society 23:114–125.

Portell, R.W., and B.A. Kittle. 2010. Mollusca: Bermont Formation (middle Pleistocene). Florida Fossil Invertebrates Part 13. 40 pages.

Poulin, R.R., R.A. Paterson, C.R. Townsend, D.M. Tompkins, and D.W. Kelly. 2011. Biological invasions and the dynamics of endemic diseases in freshwater ecosystems. Freshwater Biology 56(4):676–688.

Preister, L. 2008. Life history observations and determination of potential host fish species for Chipola Slabshell, *Elliptio chipolaensis*. Master's thesis, Columbus State University, Georgia. 48 pages.

Price, J.E., and C.B. Eads. 2011. Brooding patterns in three freshwater mussels of the genus *Elliptio* in the Broad River in South Carolina. American Malacological Bulletin 29:121–126.

Price, R.E., and F.R. Schiebe. 1978. Measurements of velocity from excurrent siphons of freshwater clams. The Nautilus 92(2):67–69.

Prime, T. 1852. New species of Cycladidae, with descriptions. Proceedings of the Boston Society of Natural History 4:155–165.

Prime, T. 1853. Notes on the species of *Cyclas* found in the United States. Proceedings of the Boston Society of Natural History 4:271–286.

Prime, T. 1865. Monograph of the American Corbiculadae (Recent and fossil). Smithsonian Miscellaneous Collections, Number 145. 80 pages.

Pritchard, J. 2001. An historical analysis of mussel propagation and culture: research performed at the Fairport Biological Station. Final report to U.S. Army Corp of Engineers, Rock Island, Illinois. 76 pp. Accessed 14 January 2013. http://www.fws.gov/midwest/mussel/documents/an_historical_analysis_of_mussel_propagation_and_culture.pdf.

Puri, H.S., and R.O. Vernon. 1964. Summary of the geology of Florida and guidebook to the classic exposures. Florida Geological Survey Special Publication Number 5 (revised). 312 pages.

Purdum, E.D., L.C. Burney, and T.M. Swihart. 1998. History of water management. Surface water. Pages 156–169. In: E.A. Fernald and E.D. Purdum (editors). Water Resources Atlas of Florida. Institute of Science and Public Affairs, Florida State University. 312 pages.

Pynnönen, K. 1995. Effect of pH, hardness and maternal pre-exposure on the toxicity of Cd, Cu and Zn to the glochidial larvae of a freshwater clam *Anodonta cygnea*. Water Resources 29:247–254.

Rafinesque, C.S. 1819. Prodrome de 70 nouveaux Genres d'Animaux découverts dans l'intérieur des États-Unis d'Amérique, durant l'année 1818. Journal de Physique, de Chimie, d'Histoire Naturelle et des Arts 88:417–429.

Rafinesque, C.S. 1820. Monographie des coquilles bivalves fluviatiles de la Rivière Ohio, contenant douze genres et soixante-huit espèces. Annales générales des sciences Physiques, a Bruxelles 5(5):287–322, plates 80–82.

Rafinesque, C.S. 1831. Continuation of a monograph of the bivalve shells of the River Ohio, and other rivers of the western states. By Prof. C.S. Rafinesque. (Published at Brussels, September 1820.) Containing 46 species, from Number 76 to Number 121. Including an appendix on some bivalve shells of the rivers of Hindustan, with a supplement on the fossil bivalve shells of the Western states, and the Tulosites, a new genus of fossils. Philadelphia. 8 pages.

Raikow, D.F., and S.K. Hamilton. 2001. Bivalve diets in a midwestern U.S. stream: a stable isotope enrichment study. Limnology and Oceanography 46(3):514–522.

Randazzo, A.F. 1997. The sedimentary platform of Florida: Mesozoic to Cenozoic. Pages 39–56. In: A.F. Randazzo and D.S. Jones (editors). The Geology of Florida. University Press of Florida, Gainesville. 327 pages.

Rashleigh, B., and D.L. DeAngelis. 2007. Conditions for coexistence of freshwater mussel species via partitioning of fish host resources. Ecological Modelling 201:171–178.

Récluz, C.A. 1849. Description de quelques nouvelles especes de coquilles. Revue et Magazin de Zoologie Pure et Applique (2), 1:64–71.

Régnier, C., B. Fontaine, and P. Bouchet. 2009. Not knowing, not recording, not listing: numerous unnoticed mollusk extinctions. Conservation Biology 23(5):1214–1221.

Reuling, F.H. 1919. Acquired immunity to an animal parasite. Journal of Infectious Diseases 24:337–347.

Ricciardi, A., and J.B. Rasmussen. 1999. Extinction rates of North American freshwater fauna. Conservation Biology 13:220–222.

Ricciardi, A., F.G. Whoriskey, and J.B. Rasmussen. 1996. Impact of the *Dreissena* invasion on native unionid bivalves in the upper St. Lawrence River. Canadian Journal of Fisheries and Aquatic Science 53(6):1434–1444.

Rice, A.N. 2004. Diet and condition of American alligators (*Alligator mississippiensis*) in three central Florida lakes. Master's thesis, University of Florida, Gainesville. 89 pages.

Richard, P.E., T.H. Dietz, and H. Silverman. 1991. Structure of the gill during reproduction of the unionids *Anodonta grandis*, *Ligumia subrostrata* and *Carunculina parva texasensis*. Canadian Journal of Zoology 69(7):1744–1754.

Richards, H.G. 1968. Catalogue of invertebrate fossil types at the Academy of Natural Sciences of Philadelphia. Academy of Natural Sciences of Philadelphia Special Publication 8. 222 pages.

Richmond, G.M., and D.S. Fullerton. 1986. Summation of Quaternary glaciations in the United States of America. Quaternary Science Reviews 5:183–196.

Richter, B.D., D.P. Braun, M.A. Mendelson, and L.L. Master. 1997. Threats to imperiled freshwater fauna. Conservation Biology 11(5):1081–1093.

Richter, B.D., R. Mathews, D.L. Harrison, and R. Wigington. 2003. Ecologically sustainable water management: managing river flows for ecological integrity. Ecological Applications 13(1):206–224.

Ridgeway, R. 1912. The Color Standards and Color Nomenclature. Privately published, Washington, DC. 43 pages.

Ridgway, I.D., C.A. Richardson, and S.N. Austad. 2011. Maximum shell size, growth rate, and maturation age correlate with longevity in bivalve mollusks. Journal of Gerontology: Biological Sciences 11(1):83–90.

Roback, S.S., D.J. Bereza, and M.F. Vidrine. 1979. Description of an *Ablabesmyia* [Diptera: Chironomidae: Tanypodinae] symbiont of unionid fresh-water mussels [Mollusca: Bivalvia: Unionacea], with notes on its biology and zoogeography. Transactions of the American Entomological Society 105(4):577–620.

Roberts, A.D., and M.C. Barnhart. 1999. Effects of temperature, pH, and CO_2 on transformation of the glochidia of *Anodonta suborbiculata* on fish hosts in vitro. Journal of the North American Benthological Society 18:477–487.

Roberts, N.C., and S.W. Moorhead. 1914. History of Lee County, Iowa. Volume II. S.J. Clarke Publishing Company, Chicago. 467 pages.

Rodgers, L., M. Bodle, D. Black, and F. Laroche. 2011. Status of nonindigenous species in the South Florida environment. Chapter 9, pages 9-1–9-76. In: 2008 South Florida Environmental Report. Volume I. South Florida Water Management District, West Palm Beach.

Rodland, D.L., B.R. Schöne, S. Baier, Z. Zhang, W. Dreyer, and N.A. Page. 2009. Changes in gape frequency, siphon activity and thermal response in the freshwater bivalves *Anodonta cygnea* and *Margaritifera falcata*. Journal of Molluscan Studies 75(1):51–57.

Roe, K.J., and P.D. Hartfield. 2005. *Hamiota*, a new genus of freshwater mussel (Bivalvia: Unionidae) from the Gulf of Mexico drainages of the southeastern United States. The Nautilus 119(1):1–10.

Roe, K.J., P.D. Hartfield, and C. Lydeard. 2001. Phylogenetic analysis of the threatened and endangered superconglutinate-producing mussels of the genus *Lampsilis* (Bivalvia: Unionidae). Molecular Ecology 10(9):2225–2234.

Rogers, J.S. 1933. The ecological distribution of crane-flies of northern Florida. Ecological Monographs 3(1):1–74.

Roper, C.S., and D.W. Hickey. 1994. Population structure, shell morphology, age and condition of the freshwater mussel *Hyridella menziesi* (Unionacea: Hyriidae) from seven lake and river sites in the Waikato River system. Hydrobiologia 284:205–217.

Rosenau, J.C., and G.L. Faulkner. 1975. An index to springs of Florida. Florida Department of Natural Resources, Bureau of Geology, Map Series Number 63.

Rosenau, J.C., G.L. Faulkner, C.W. Hendty, Jr., and R.W. Hull. 1977. Springs of Florida (revised). Geological Bulletin Number 31. Florida Department of Natural Resources, Bureau of Geology.

Rosenzweig, M.L. 2001. The four questions: what does the introduction of exotic species do to diversity? Evolutionary Ecology Research 3:361–367.

Rothra, E.O. 1995. Florida's pioneer naturalist. The life of Charles Torrey Simpson. The University Press of Florida, Gainesville. 232 pages.

Rugel, K., C.R. Jackson, J.J. Romeis, S.W. Golladay, D.W. Hicks, and J.F. Dowd. 2011. Effects of irrigation withdrawals on streamflows in a karst environment: lower Flint River Basin, Georgia, USA. Hydrological Processes 26(4):523–534. doi:10.1002 /hyp.8149.

Rypel, A.L. 2008. Field observations of the nocturnal mantle-flap lure of *Lampsilis teres*. American Malacological Bulletin 24:97–100.

Rypel, A.L., W.R. Haag, and R.H. Findlay. 2008. Validation of annual growth rings in freshwater mussel shells using cross dating. Canadian Journal of Fisheries and Aquatic Sciences 65:2224–2232.

Rypel, A.L., W.R. Haag, and R.H. Findlay. 2009. Pervasive hydrologic effects of freshwater mussels and riparian trees in southeastern floodplain ecosystems. Wetlands 29:497–504.

Saha, S., and J.B. Layzer. 2008. Evaluation of a nonlethal technique for determining sex of freshwater mussels. Journal of the North American Society of Benthologists 27(1):84–89.

Salomons, W., N.M. de Rooij, H. Kerdijk, and J. Bril. 1987. Sediments as a source for contaminants? Hydrobiologia 149:13–30.

Samad, F., and J.G. Stanley. 1986. Loss of freshwater shellfish after water drawdown in Lake Sebasticook, Maine. Journal of Freshwater Ecology 3(4):519–523.

Sanchez, C., H. Arribart, and M.M. Giraud-Guille. 2005. Biomimetism and bioinspiration as tools for the design of innovative materials and systems. Nature Materials 4:1–12.

Say, T. 1817. Article conchology. [No pagination, 14 pages, plates 1–4]. In: W. Nicholson (editor). American Edition of the British Encyclopedia or Dictionary of Arts and Sciences, Comprising an Accurate and Popular View of the Present Improved State of Human Knowledge. Volume 2. First edition. Samuel A. Mitchel and Horace Ames, Philadelphia.

Say, T. 1818. Description of a new genus of fresh water bivalve shells. Journal of the Academy of Natural Sciences of Philadelphia 1(11):459–460.

Say, T. 1822. Description of univalve terrestrial and fluviatile shells of the United States. Journal of the Academy of Natural Sciences of Philadelphia 2(2):370–381.

Say. T. 1824. Appendix. Part 1. Natural history. [Section] 1. Zoology. Pages 254–267, plates 14–15. In: W.H. Keating. Narrative of an Expedition to the Source of St. Peter's River, Lake Winnepeek, Lake of the Woods, etc.: Performed in the Year 1823, by Order of the Hon. J.C. Calhoun, Secretary of War, Under the Command of Stephen H. Long, Major U.S.T.E. Compiled from the notes of Major Long, Messrs. Say, Keating and Colhoun [sic]. In 2 volumes. H.C. Carey and I. Lea, Philadelphia.

Say, T. 1829. Descriptions of some new terrestrial and fluviatile shells of North America. The Disseminator of Useful Knowledge; containing hints to the youth of the United States, from the School of Industry, New Harmony, Indiana 2(19):291–293, 23 September 1829; 2(20):308–310, 7 October 1829; 2(21):323–325, 21 October 1829; 2(22):339–341, 4 November 1829; 2(23):355–356, 18 November 1829.

Say, T. 1830–1834. American Conchology, or descriptions of the shells of North America. Illustrated by colored figures from original drawings executed from nature. School Press, New Harmony, Indiana. Part 1 (1830); Part 2 (April 1831); Part 3 (September 1831); Part 4 (March 1832); Part 5 (August 1832); Part 6 (April 1834); Part 7 (1834?, published after Say's death, edited by T.A. Conrad).

Schenk, E.R., C.R. Hupp, and A. Gellis. 2011. Sediment dynamics in the restored reach of the Kissimmee River basin, Florida: a vast subtropical riparian wetland. River Research and Applications 28(10):1753–1767. doi:10.1002/rra.1577.

Schmidt, W. 1997. Geomorphology and physiography of Florida. Pages 1–12. In: A.F. Randazzo and D.S. Jones (editors). The Geology of Florida. University Press of Florida, Gainesville. 327 pages.

Schneider, R.F. 1967. Range of the Asiatic clam in Florida. Nautilus 81(2):68–69.

Schöne, B.R., N.A. Page, D.L. Rodland, J. Fiebig, S. Baier, S.O. Helama, and W. Oschmann. 2007. ENSO-coupled precipitation records (1959–2004) based on shells of freshwater bivalve mollusks (*Margaritifera falcata*) from British Columbia. International Journal of Earth Sciences 96(3):525–540.

Schuchert, C. 1905. Catalogue of the type specimens of fossil invertebrates in the Department of Geology, United States National Museum. Bulletin of the U.S. National Museum Number 53, Part I. 704 pages.

Schwalb, A.N. 2009. Host infection strategies determine dispersal abilities in freshwater mussels (Bivalvia: Unionidae). Doctoral dissertation, University of Guelph, Ontario, Canada. 198 pages.

Schwalb, A.N., and M.T. Pusch. 2007. Horizontal and vertical movements of unionid mussels in a lowland river. Journal of the North American Benthological Society 26:261–272.

Schwartz, M.L., and R.V. Dimock, Jr. 2001. Ultrastructural evidence for nutritional exchange between brooding unionid mussels and their glochidia larvae. Invertebrate Biology 120(3):227–236.

Schwegman, J.E. 1998. Lure behavior in *Toxolasma texasensis*. Triannual Unionid Report 14:35.

Scott, T.M. 1992. A geological overview of Florida. Florida Geological Survey Open File Report Number 50. 78 pages.

Scott, T.M. 1997. Miocene to Holocene history of Florida. Pages 57 68. In: A.F. Randazzo and D.S. Jones (editors). The Geology of Florida. University Press of Florida, Gainesville. 327 pages.

Serb, J., J. Buhay, and C. Lydeard. 2003. Molecular systematics of the North American freshwater bivalve genus *Quadrula* (Unionidae: Ambleminae) based on mitochondrial ND1 sequences. Molecular Phylogenetics and Evolution 28:1–11.

Şereflişan, H., Ş. Çek, and M. Şereflişan. 2009. Histological studies on gametogenesis, hermaphroditism and the gametogenic cycle of *Anodonta gabillotia pseudodopsis* (Locard, 1883) in the Lake Golbaşi, Turkey (Bivalvia: Unionidae). Journal of Shellfish Research 28(2):337–344.

Sethi, S.A., A.R. Selle, M.W. Doyle, E.H. Stanley, and H.E. Kitchel. 2004. Response of unionid mussels to dam removal in Koshkonong Creek, Wisconsin (USA). Hydrobiologia 525:157–165.

Shadoan, M.K., and R.V. Dimock, Jr. 2000. Differential sensitivity of hooked (*Utterbackia imbecillis*) and hookless (*Megalonaias nervosa*) glochidia to chemical and mechanical stimuli (Bivalvia: Unionidae). Pages 93–102. In: R.A. Tankersley, D.I. Warmolts, G.T. Watters, and B.J. Armitage (editors). Part I. Proceedings of the Conservation, Captive Care, and Propagation of Freshwater Mussels Symposium. Ohio Biological Survey Special Publication, Columbus.

Shea, C.P., J.T. Peterson, J.M. Wisniewski, and N.A. Johnson. 2011. Misidentification of freshwater mussel species (Bivalvia: Unionidae): contributing factors, management implications, and potential solutions. Journal of the North American Benthological Society 30(2):446–458.

Sheehan, R.J., R.J. Neves, and H.E. Kitchel. 1989. Fate of freshwater mussels transplanted to formerly polluted reaches of the Clinch and North Fork Holston Rivers, Virginia. Journal of Freshwater Ecology 5(2):139–149.

Sheldon, R., and K.F. Walker. 1989. Effects of hypoxia on oxygen consumption by two species of freshwater mussel (Unionacea: Hyriidae) from the River Murray. Australian Journal of Marine and Freshwater Research 40:491–499.

Shelton, D.N., P.M. Stewart, and J.M. Miller. 2006. Candidate mussel surveys in the Escambia, Yellow, and Choctawhatchee Rivers in southern Alabama and northwest Florida. Report to U.S. Fish and Wildlife Service, Project Number 2586-023. 71 pages.

Silverman, H., W.T. Kays, and T.H. Dietz. 1987. Maternal calcium contribution to glochidial shells in freshwater mussels (Eulamellibranchia: Unionidae). Journal of Experimental Zoology 242(2):137–146.

Simpson, C.T. 1889. What is a species? The Nautilus 3(7):78–80.

Simpson, C.T. 1892a. Notes on the Unionidae of Florida and the southeastern states. Proceedings of the U.S. National Museum 15:405–436, plates XLIX–LXXIV.

Simpson, C.T. 1892b. On a revision of the American Unionidae. The Nautilus 6(7):78–80.

Simpson, C.T. 1893a. Contributions to the Mollusca of Florida. Proceedings of the Davenport Academy of Natural Sciences 5:45–72.

Simpson, C.T. 1893b. On the relationships and distribution of the North American Unionidae, with notes on the west coast species. American Naturalist 27:353–358.

Simpson, C.T. 1895. The classification and geographical distribution on the pearly fresh-water mussels. Proceedings of the U.S. National Museum 18:295–343.

Simpson, C.T. 1896. Notes on the *parvus* group of Unionidae and its allies. The Nautilus 10(5):57–59.

Simpson, C.T. 1899. The pearly fresh-water mussels of the United States; their habits, enemies, and diseases, with suggestions for their protection. Bulletin of the U.S. Fish Commission 18(1898):279–288.

Simpson, C.T. 1900a. Synopsis of the naiades, or pearly fresh-water mussels. Proceedings of the U.S. National Museum 22(1205):501–1044.

Simpson, C.T. 1900b. New and unfigured Unionidae. Proceedings of the Academy of Natural Sciences of Philadelphia 52(1):74–86, plates 1–5.

Simpson, C.T. 1914. A Descriptive Catalogue of the Naiades, or Pearly Fresh-water Mussels. Parts I–III. Bryant Walker, Detroit, Michigan. 1540 pages.

Sinclair, R.M. 1971. Annotated bibliography on the exotic bivalve *Corbicula* in North America, 1900–1971. Sterkiana 43:11–18.

Sinclair, R.M., and B.G. Isom. 1963. Further studies on the introduced Asiatic clam (*Corbicula*) in Tennessee. Tennessee Stream Pollution Board, Tennessee Department of Public Health, Nashville. 75pages.

Singer, E.E., and M.M. Gangloff. 2011. Effects of a small dam on freshwater mussel growth in an Alabama (U.S.A.) stream. Freshwater Biology 56(9):1904–1915.56. doi:10.1111/j.1365–2427.2011.02608.x.

Siripattrawan, S., J.-K. Park, and D. O´Foighil. 2000. Two lineages of the introduced Asian freshwater clam *Corbicula* occur in North America. Journal of Molluscan Studies 66:423–429.

Smith, D.G. 1985. Recent range expansion of the freshwater mussel *Anodonta implicata* and its relationship to clupeid fish restoration in the Connecticut River system. Freshwater Invertebrate Biology 4:105–108.

Smith, D.G. 2000. Investigations on the byssal gland in juvenile unionids. Pages 103–107. In: R.A. Tankersley, D.I. Warmolts, G.T. Watters, and B.J. Armitage (editors). Part I. Proceedings of the Conservation, Captive Care, and Propagation of Freshwater Mussels Symposium. Ohio Biological Survey, Columbus, Ohio.

Smith, D.G., B.K. Lang, and M.E. Gordon. 2003. Gametogenetic cycle, reproductive anatomy, and larval morphology of *Popenaias popeii* (Unionoida) from the Black River, New Mexico. Southwestern Naturalist 48(3):333–340.

Smith, D.L., and K.M. Lord. 1997. Tectonic evolution and geophysics of the Florida basement. Pages 13–26. In: A.F. Randazzo and D.S. Jones (editors). The Geology of Florida. University Press of Florida, Gainesville. 327 pages.

Smith, D.L., and A.F. Randazzo. 1989. History of seismological activity in Florida: evidence of an extremely stable basement. Pages 2-37–2-58. In: J.C. Stepp (editor). Second Symposium on Current Issues Related to Nuclear Power Plant Structures, Equipment, and Piping, with Emphasis on Reduction of Seismic Issues in Low-seismicity Regions. Seismicity Owners Group and Electric Power Research Institute.

Smith, J.M., C. Bogan, and A.E. Bogan. 2007. Museum of Fluviatile Mollusks collection donated to North Carolina State Museum of Natural Sciences, Raleigh. Ellipsaria 9(3):12–14.

Smock, L.A., A.B. Wright, and A.C. Benke. 2005. Atlantic coast rivers of the southeastern United States. Chapter 3, pages 73–122. In: A.C. Benke and C.E. Cushing (editors). Rivers of North America. Elsevier Academic Press, Burlington, Massachusetts. 1144 pages.

Sneen, M.E., K.S. Cummings, T. Minarik, Jr., and J. Wasik. 2009. The discovery of the nonindigenous, mottled fingernail clam, *Eupera cubensis* (Prime 1865) (Bivalvia: Sphaeriidae) in the Chicago Sanitary and Ship Canal (Illinois River drainage), Cook County, Illinois. Journal of Great Lakes Research 35(4):627–629.

Sonenshein, R.S. 1996. Delineation of saltwater intrusion in the Biscayne Aquifer, eastern Dade County, Florida, 1995. U.S. Geological Survey Water Resources Investigations Report 96–4285.

South Florida Water Management District. 2009. Climate change and water management in south Florida. South Florida Water Management District, West Palm Beach. 20 pages.

Sowerby, G.B. (I). 1820–1834. The Genera of Recent and Fossil Shells, for the Use of Students in Conchology and Geology Commenced by J. Sowerby, and Continued by G.B. Sowerby. Sowerby, London. [Without title page or index, plates not numbered, genera and plates arranged alphabetically and bound in two volumes.]

Sparks, B., and D. Strayer. 1998. Effects of low dissolved oxygen on juvenile *Elliptio complanata* (Bivalvia: Unionidae). Journal of the North American Benthological Society 17(1):129–134.

Spooner, D.E., and C.C. Vaughn. 2006. Context-dependent effects of freshwater mussels on stream benthic communities. Freshwater Biology 51:1016–1024.

Spooner, D.E., and C.C. Vaughn. 2008. A trait-based approach to species' roles in stream ecosystems: climate change, community structure, and material cycling. Oecologia 158:307–317.

Spooner, D.E., and C.C. Vaughn. 2009. Species richness and temperature influence mussel biomass: a partitioning approach applied to natural communities. Ecology 93(3):781–790.

Spooner, D.E., M.A. Xenopoulos, C. Schneider, and D.A. Woolnough. 2011. Coextirpation of host-affiliate relationships in rivers: the role of climate change, water withdrawal, and host-specificity. Global Change Biology 17(4):1720–1732.

Stamey, T.C. 1996. Streamflow characteristics at selected sites in southwestern Georgia, southeastern Alabama, and northwestern Florida, near Lake Seminole. U.S. Geologic Survey Open File Report 95-455, Atlanta, Georgia. 11 pages.

Stanley, S.M. 1970. Relation of shell form to life habits in the Bivalvia. Memoirs of the Geological Society of America 125:1–296.

Stanley, S.M. 1975. Why clams have the shape they have: an experimental analysis of burrowing. Paleobiology 1:48–58.

Stansbery, D.H. 1971. Rare and endangered freshwater mollusks in eastern United States. Pages 5–18f. In: S.E. Jorgensen and R.E. Sharp (editors). Proceedings of a Symposium on Rare and Endangered Mollusks (Naiads) of the U.S. Region 3, Bureau Sport Fisheries and Wildlife, U.S. Fish Wildlife Service, Twin Cities, Minnesota.

Stansbery, D.H. 1976. Naiad mollusks. Pages 42–52. In: H.T. Boschung (editor). Endangered and Threatened Plants and Animals of Alabama. Alabama Museum of Natural History Bulletin 2.

Stein, C.B. 1969. Gonad development in the three-ridge naiad, *Amblema plicata* (Say, 1817). (Abstract). American Malacological Union, Inc. Annual Reports 1969:30.

Steingraber, M.T., M.R. Bartsch, J.E. Kalas, and T.J. Newton. 2007. Thermal criteria for early life stage development of the winged mapleleaf mussel (*Quadrula fragosa*). The American Midland Naturalist 157(2):297–311.

Sterki, V. 1891a. A byssus in *Unio*. The Nautilus 5(7):73–74.

Sterki, V. 1891b. On the byssus of Unionidae. II. The Nautilus 5(8):90–91.

Sterki, V. 1898a. Some observations on the genital organs of Unionidae with reference to classification. The Nautilus 12(2):18–21; 12(3):28–32.

Sterki, V. 1898b. *Anodonta imbecillis*, hermaphroditic. The Nautilus 12(8):87–88.

Stern, E.M., and D.L. Felder. 1978. Identification of host fishes for four species of freshwater mussels (Bivalvia: Unionidae). The American Midland Naturalist 100(1):233–236.

Steuer, J.J., T.J. Newton, and S.J. Zigler. 2008. Use of complex hydraulic variables to predict the distribution and density of unionids in a side channel of the Upper Mississippi River. Hydrobiologia 610(1):67–82.

Stimpson, W. 1851. Shells of New England; a revision of the synonymy of the testaceous mollusks of New England, with notes on their structure, and their geographical and bathymetrical distribution, with figures of new species. Phillips, Sampson and Company, Boston, Massachusetts. 58 pages, 2 plates.

Stokstad, E. 2007. Feared Quagga Mussel turns up in western United States. Science 315(5811):453.

Strayer, D.L. 1999a. Effects of alien species on freshwater mollusks in North America. Journal of the North American Benthological Society 18(1):74–98.

Strayer, D.L. 1999b. Use of flow refuges by unionid mussels in rivers. Journal of the North American Benthological Society 18(4):468–476.

Strayer, D.L. 2006. Challenges for freshwater invertebrate conservation. Journal of the North American Benthological Society 25(2):271–287.

Strayer, D.L. 2008. Freshwater mussel ecology: a multifactor approach to distribution and abundance. Freshwater Ecology Series, University of California Press, Berkeley. 204 pages.

Strayer, D.L. 2010. Alien species in fresh waters: ecological effects, interactions with other stressors, and prospects for the future. Freshwater Biology 55 Supplement 1:152–174.

Strayer, D.L., J.A. Downing, W.R. Haag, T.L. King, J.B. Layzer, T.J. Newton, and S.J. Nichols. 2004. Changing perspectives on pearly mussels, North America's most imperiled animals. BioScience 54(5):429–439.

Strayer, D.L., and D. Dudgeon. 2010. Freshwater biodiversity conservation: recent progress and future challenges. Journal of the North American Benthological Society 29(1):344–358.

Strayer, D.L., D.C. Hunter, L.C. Smith, and C.K. Borg. 1994. Distribution, abundance, and roles of freshwater clams (Bivalvia: Unionidea) in the freshwater tidal Hudson River. Freshwater Biology 31:239–248.

Strayer, D.L., and H.M. Malcom. 2007. Shell decay rates of native and alien freshwater bivalves and implications for habitat engineering. Freshwater Biology 52:1611–1617.

Surber, T. 1912. Identification of the glochidia of freshwater mussels. Report and Special Papers of the U.S. Bureau of Fisheries 1912:1–10, plates 1–3. [Issued separately as U.S. Bureau of Fisheries Document 771.]

Surber, T. 1913. Notes on the natural hosts of fresh-water mussels. Bulletin of the U.S. Bureau of Fisheries 32(1912):101–116, plates 29–31. [Issued separately as U.S. Bureau of Fisheries Document 778.]

Surber, T. 1915. Identification of the glochidia of fresh-water mussels. Appendix V to the Report of the U.S. Commissioner of Fisheries for 1914:3–9, 1 plate. [Issued separately as U.S. Bureau of Fisheries Document 813.]

Swift, C.C., C.R. Gilbert, S.A. Bortone, G.H. Burgess, and R.W. Yerger. 1986. Zoogeography of the freshwater fishes of the southeastern United States: Savannah River to Lake Pontchartrain. Pages 213–285, Chapter 7. In: C.H. Hocutt and E.O. Wiley (editors). The Zoogeography of North American Freshwater Fishes. John Wiley & Sons, New York.

Swingle, H.A., and D.G. Bland. 1974. Distribution of the estuarine clam *Rangia cuneata* Gray in coastal waters of Alabama. Alabama Marine Research Bulletin 10:9–16.

Tankersley, R.A., and R.V. Dimock, Jr. 1992. Quantitative analysis of the structure and function of the marsupial gills of the freshwater mussel *Anodonta cataracta*. Biological Bulletin 182(1):145–154.

Tankersley, R.A., and R.V. Dimock. 1993. The effect of larval brooding on the respiratory physiology of the unionid mussel *Anodonta cataracta*. The American Midland Naturalist 130:146–163.

Taskinen, J., and M. Saarinen. 1999. Increased parasite abundance associated with reproductive maturity of the clam *Anodonta piscinalis*. Journal of Parasitology 85(3):588–591.

Tevesz, M.J.S., and J.G. Carter. 1980. Environmental relationships of shell form and structure in Unioncean bivalves. Pages 295–322. In: D.C. Rhoads and R.A. Lutz (editors). Skeletal Growth of Aquatic Organisms. Plenum Press, New York.

The Catena Group. 2010. Chattahoochee River baseline mussel surveys, Russell and Lee counties, Alabama, Muscogee County, Georgia. Report to Uptown Columbus, Inc. The Catena Group Job Number 3282. 40 pages.

Theologidis, I., S. Fodelianakis, M.B. Gaspar, and E. Zouros. 2008. Doubly uniparental inheritance (dui) of mitochondrial DNA in *Donax trunculus* (Bivalvia: Donacidae) and the problem of its sporadic detection in Bivalvia. Evolution 62(4):959–970.

Thompson, F.G. 1968. The Aquatic Snails of the Family Hydrobiidae of Peninsular Florida. University of Florida Press, Gainesville, Florida, USA. 268 pages.

Thompson, F.G. 1982. Phylum Mollusca, Class Bivalvia. Pages 23–27. In: R. Franz (editor). Rare and Endangered Biota of Florida. Volume 6. Invertebrates. University of Florida Press, Gainesville.

Thompson, F.G., and R. Hershler. 1991. Two new hydrobiid snails (Amnicolinae) from Florida and Georgia, with a discussion of the biogeography of freshwater gastropods of south Georgia streams. Malacological Review 24:55–72.

Tilman, D., R.M. May, C.L. Lehman, and M.A. Nowak. 1994. Habitat destruction and the extinction debt. Nature 371:65–66.

Torak, L.J., J.A. Painter, and M.F. Peck. 2010. Geohydrology of the Aucilla-Suwannee-Ochlockonee River basin, south-central Georgia and adjacent parts of Florida: U.S. Geological Survey Scientific Investigations Report 2010-5072. 78 pages.

Traill, L.W., B.W. Brook, R.R. Frankham, and C.J.A. Bradshaw. 2010. Pragmatic population viability targets in a rapidly changing world. Biological Conservation 143:28–34.

Trautman, M.B. 1981. The Fishes of Ohio with Illustrated Keys. Second edition. Ohio State University Press, Columbus. 782 pages.

Trimble, S.W., and A.C. Mendel. 1995. The cow as a geomorphic agent: a critical review. Geomorphology 13(1–4):233–253.

Tucker, M.E. 1928. Studies on the life cycle of two species of fresh-water mussels belonging to the genus *Anodonta*. Biological Bulletin 54(2):117–127.

Tudorancea, C. 1972. Studies on Unionidae populations from the Crapina-Jijila complex of pools (Danube zone liable to inundation). Hydrobiologia 39:527–561.

Turgeon, D.D., A.E. Bogan, E.V. Coan, W.K. Emerson, W.G. Lyons, W.L. Pratt, C.F.E. Roper, A. Scheltema, F.G. Thompson, and J.D. Williams. 1988. Common and Scientific Names of Aquatic Invertebrates from the United States and Canada: Mollusks. American Fisheries Society, Special Publication 16. 277 pages, 12 plates.

Turgeon, D.D., J.F. Quinn, A.E. Bogan, E.V. Coan, F.G. Hochberg, W.G. Lyons, P. Mikkelsen, R.J. Neves, C.F.E. Roper, G. Rosenberg, B. Roth, A. Scheltema, F.G. Thompson, M. Vecchione, and J.D. Williams. 1998. Common and Scientific Names of Aquatic Invertebrates from the United States and Canada: Mollusks. Second edition. American Fisheries Society, Special Publication 26. 526 pages.

U.S. Fish and Wildlife Service. 1984. Endangered and threatened wildlife and plants; review of invertebrate wildlife for listing as endangered or threatened species. Federal Register 49(100):21664–21675.

U.S. Fish and Wildlife Service. 1989. Endangered and threatened wildlife and plants; animal notice of review. Federal Register 54(4):554–579.

U.S. Fish and Wildlife Service. 1991. Endangered and threatened wildlife and plants; animal candidate review for listing as endangered or threatened species. Federal Register 56(225):58804–58836.

U.S. Fish and Wildlife Service. 1993. Endangered and threatened wildlife and plants; determination of endangered status for eight freshwater mussels and threatened status for three freshwater mussels in the Mobile River drainage. Federal Register 58:14330–14339.

U.S. Fish and Wildlife Service. 1994. Endangered and threatened wildlife and plants; animal candidate review for listing as endangered or threatened species. Federal Register 59(219):58982–59028.

U.S. Fish and Wildlife Service. 1998. Endangered and threatened wildlife and plants; determination of endangered status for five freshwater mussels and threatened status for two freshwater mussels from eastern Gulf slope drainages of Alabama, Florida, and Georgia. Federal Register 63(50):12664–12687.

U.S. Fish and Wildlife Service. 2000. Mobile River Basin aquatic ecosystem recovery plan. U.S. Fish and Wildlife Service, Atlanta, Georgia. 128 pages.

U.S. Fish and Wildlife Service. 2003. Recovery plan for endangered Fat Threeridge (*Amblema neislerii*), Shinyrayed Pocketbook (*Lampsilis subangulata*), Gulf Moccasinshell (*Medionidus penicillatus*), Ochlockonee Moccasinshell (*Medionidus simpsonianus*), Oval Pigtoe (*Pleurobema pyriforme*), and threatened Chipola Slabshell (*Elliptio chipolaensis*), and Purple Bankclimber (*Elliptoideus sloatianus*). U.S. Fish and Wildlife Service, Atlanta, Georgia. 142 pages.

U.S. Fish and Wildlife Service. 2004. Endangered and threatened wildlife and plants; designation of critical habitat for three threatened mussels and eight endangered mussels in the Mobile River basin. Federal Register 69(126):40084–40171.

U.S. Fish and Wildlife Service. 2006. Biological opinion and conference report on the U.S. Army Corps of Engineers, Mobile District, interim operating plan for Jim Woodruff Dam and the associated releases to the Apalachicola River. U.S. Fish and Wildlife Service, Panama City Field Office, Florida. 164 pages.

U.S. Fish and Wildlife Service. 2007. Endangered and threatened wildlife and plants; designation of critical habitat for five endangered and two threatened mussels in four northeast Gulf of Mexico drainages. Federal Register 72(220):64286–64340.

U.S. Fish and Wildlife Service. 2012. Endangered and threatened wildlife and plants; determination of endangered species status for the Alabama Pearlshell, Round Ebonyshell, Southern Kidneyshell, and Choctaw Bean, and threatened species status for the Tapered Pigtoe, Narrow Pigtoe, Southern Sandshell, and Fuzzy Pigtoe, and designation of critical habitat. Federal Register 77(196):61664–61719.

Utterback, W.I. 1915. The naiades of Missouri. The American Midland Naturalist 4(3):41–53, 4(4):97–152, 4(5):181–204, 4(6):244–273.

Utterback, W.I. 1916a. Breeding record of Missouri mussels. The Nautilus 30(2):13–21.

Utterback, W.I. 1916b. The naiades of Missouri (continued). The American Midland Naturalist 4(7):311–327, 4(8):339–354, 4(9):387–400, 4(10):432–464, plates 1–28.

Utterback, W.I. 1931. Sex behavior among naiades. West Virginia Academy of Science 5:43–45.

Valdovinos, C., and P. Pedreros. 2007. Geographic variations in shell growth rates of the mussel *Diplodon chilensis* from temperate lakes of Chile: implications for biodiversity conservation. Limnologica 37:63–75.

Valenciennes, A. 1827. Coquilles fluviatiles bivalves du Nouveau-Continent, recueillies pendant le voyage de MM. De Humboldt et Bonpland. In: A. von Humboldt and A.J.A. Bonpland. Recueil d'observations de zoologie et d' anatomie compare, faites dans l'ocean Atlantique, dans l'intérieur du nouveau continent et dans la mer du sud pendant les années 1799, 1800, 1801, 1802 et 1803; par Al. de Humbodt et A. Bonpland. J. Smith and Gide, Paris. 2(13):225–237, plates 48, 50, 53, 54.

Valentine, B.D., and D.H. Stansbery. 1971. An introduction to the naiades of the Lake Texoma region, Oklahoma, with notes on the Red River fauna (Mollusca: Unionidae). Sterkiana 42:1–40.

van der Schalie, H. 1933. Notes on the brackish water bivalve, *Polymesoda caroliniana* (Bosc). University of Michigan Occasional Papers of the Museum of Zoology 11(258):1–8.

van der Schalie, H. 1934. *Lampsilis jonesi*, a new naiad from southeastern Alabama. The Nautilus 47(4):125–127, plate 15.

van der Schalie, H. 1939. *Medionidus mcglameriae*, a new naiad from the Tombigbee River, with notes on other naiades of that drainage. Occasional Papers of the Museum of Zoology, University of Michigan, Number 407. 6 pages.

van der Schalie, H. 1940. The naiad fauna of the Chipola River, in northwestern Florida. Lloydia 3(3):191–208.

van der Schalie, H. 1966. Hermaphroditism among North American freshwater mussels. Malacologia 5(1):77–78.

van der Schalie, H. 1970. Hermaphroditism among North American freshwater mussels. Malacologia 10(1):93–112.

van der Schalie, H., and F. Locke. 1941. Hermaphroditism in *Anodonta grandis*, a fresh-water mussel. Occasional Papers of the Museum of Zoology, University of Michigan, Number 432. 7 pages, 3 plates.

van der Schalie, H., and A. van der Schalie. 1950. The mussels of the Mississippi River. The American Midland Naturalist 44(2):448–466.

Van Hyning, T.H. 1917. The distinctive characters of *Lampsilis minor* and *L. villosa*. The Nautilus 31(1):14–15.

Van Hyning, T. 1925. *Amblema neisleri* (sic) nest located. The Nautilus 38(3):105.

Van Hyning, T.H. 1940. A check-list of the Mollusca of Florida. Unpublished manuscript. On file in the Mollusk Division Library at the Smithsonian Institution, Washington, DC. 30 pages.

Vanatta, E.G. 1915. Rafinesque's types of *Unio*. Proceedings of the Academy of Natural Sciences of Philadelphia 67(1915):549–559.

Vannote, R.L., and G.W. Minshall. 1982. Fluvial processes and local lithology controlling abundance, structure, and composition of mussel beds. Proceedings of the National Academy of Sciences 79(13):4103–4107.

Vaughn, C.C. 1993. Can biogeographic models be used to predict the persistence of mussel populations in rivers? Pages 117–122. In: K.S. Cummings, A.C. Buchanan, and L.M. Koch (editors). Conservation and Management of Freshwater Mussels. Proceedings of the UMRCC Symposium, 12–14 October 1992, St. Louis, Missouri. Upper Mississippi River Conservation Committee, Rock Island, Illinois.

Vaughn, C.C. 1997. Regional patterns of mussel species distributions in North American rivers. Ecography 20:107–115.

Vaughn, C.C. 2010. Biodiversity losses and ecosystem function in freshwaters: emerging conclusions and research directions. BioScience 60(1):25–35.

Vaughn, C.C., and K.B. Gido, and D.E. Spooner. 2004. Ecosystem processes performed by unionid mussels in stream mesocosms: species roles and effects of abundance. Hydrobiologia 527:35–47.

Vaughn, C.C., and C.C. Hakenkamp. 2001. The functional role of burrowing bivalves in freshwater ecosystems. Freshwater Biology 46:1431–1446.

Vaughn, C.C., S.J. Nichols, and D.E. Spooner. 2008. Community and foodweb ecology of freshwater mussels. Journal of the North American Benthological Society 27(2):409–423.

Vaughn, C.C., and D.E. Spooner. 2006a. Unionid mussels influence macroinvertebrate assemblage structure in streams. Journal of the North American Benthological Society 25:691–700.

Vaughn, C.C., and D.E. Spooner. 2006b. Scale-dependent associations between native freshwater mussels and invasive *Corbicula*. Hydrobiologia 568(1):331–339.

Vaughn, C.C., D.E. Spooner, and H.S. Galbraith. 2007. Context-dependent species identity effects within a functional group of filter-feeding bivalves. Ecology 88:1654–1662.

Vaughn, C.C., D.E. Spooner, and B.W. Hoagland. 2002. River weed growing epizootically on freshwater mussels. Southwestern Naturalist 47:604–605.

Vaughn, C.C., and C.M. Taylor. 1999. Impoundment and the decline of freshwater mussels: a case study of an extinction gradient. Conservation Biology 13(4):912–920.

Vaughn, C.C., and C.M. Taylor. 2000. Macroecology of a host-parasite relationship. Ecography 23(1):11–20.

Vidrine, M.F. 1989. A summer of the mollusk-mite associations of Louisiana and adjacent waters. The Louisiana Environmental Profesional 6(1):30–63.

Vidrine, M.F. 1993. The Historical Distributions of Freshwater Mussels in Louisiana. Gail Q. Vidrine Collectibles, Eunice, Louisiana. 225 pages, 20 plates.

Vidrine, M.F. 1996. North American *Najadicola* and *Unionicola*: Diagnoses and Distributions. Gail Q. Vidrine Collectibles, Eunice, Louisiana. 365 pages.

Vorosmarty, C.J., P.B. McIntyre, M.O. Gessner, D. Dudgeon, A. Prusevich, P. Green, S. Glidden, S.E. Bunn, C.A. Sullivan, C. Reidy Liermann, and P.M. Davies. 2010. Global threats to human water security and river biodiversity. Nature 467:555–561.

Wächtler, K., M.C. Dreher-Mansur, and T. Richter. 2001. Larval types and early postlarval biology in naiads (Unionoida). Pages 93–125. In: G. Bauer and K. Wächtler (editors). Ecology and Evolution of the Freshwater Mussels Unionoida. Ecological Studies, Volume 145. Springer-Verlag, Berlin.

Walker, A.D., and J.W. Geissman (compilers). 2009. Geologic time scale. Geological Society of America. doi:10.1130/2009.cts004r2c.

Walker, B. 1901. A new species of *Strophitus*. The Nautilus 15(6):65–66.

Walker, B. 1905a. List of shells from northwestern Florida. The Nautilus 18(12):133–136, plate 9.

Walker, B. 1905b. A new species of *Medionidus*. The Nautilus 18(12):136–137, plate 9.

Walker, B. 1913. The unione fauna of the Great Lakes. The Nautilus 27:18–23.

Waller, D.L., S. Gutreuter, and J.J. Rach. 1999. Behavioral responses to disturbance in freshwater mussels with implications for conservation and management. Journal of the North American Benthological Society 18(3):381–390.

Waller, D.L., and B.A. Lassee. 1997. External morphology of spermatozoa and spermatozeugmata of the freshwater mussel *Truncilla truncata* (Mollusca: Bivalvia: Unionidae). The American Midland Naturalist 138(1):220–223.

Waller, D.L., and L.G. Mitchell. 1989. Gill tissue reactions in walleye *Stizostedion vitreum vitreum* and common carp *Cyprinus carpio* to glochidia of the freshwater mussel *Lampsilis radiata siliquoidea*. Diseases of Aquatic Organisms 6(2):81–87.

Walsh, S.J., and J.D. Williams. 2003. Inventory of fishes and mussels in springs and spring effluents of north-central Florida state parks. Final report to Florida Park Service, Tallahassee. 94 pages.

Wang, N., C.G. Ingersoll, C.D. Ivey, D.K. Hardesty, T.W. May, T. Augspurger, A.D. Roberts, E. van Genderen, and M.C. Barnhart. 2010. Sensitivity of early life stages of freshwater mussels (Unionidea) to acute and chronic toxicity of lead, cadmium and zinc in water. Environmental Toxicity and Chemistry 29(9):2053–2063.

Ward, G.M., P.M. Harris, and A.K. Ward. 2005. Gulf coast rivers of the southeastern United States. Chapter 4, pages 125–178. In: A.C. Benke and C.E. Cushing (editors). Rivers of North America. Elsevier Academic Press, Burlington, Massachusetts. 1144 pages.

Warren, G.L. 1997. Nonindigenous freshwater invertebrates. Chapter 6, pages 101–108. In: D. Simberloff, D.C. Schmitz, and T.C. Brown (editors). Strangers in Paradise: Impact and Management of Nonindigenous Species in Florida. Island Press, Washington, DC. 407 pages.

Warren, G.L., D.A. Hohlt, C.E. Cichra, and D. VanGenechten. 2000. Fish and aquatic invertebrate communities of the Wekiva and Little Wekiva Rivers: a baseline evaluation in the context of Florida's minimum flows and levels statutes. St. Johns River Water Management District Special Publication SJ2000-SP4.

Warren, G.L., and M. Vogel. 1991. Invertebrate communities of Lake Okeechobee. Pages 107–177. In: L.A. Bull and G.L. Warren (editors). Lake Okeechobee-Kissimmee River-Everglades resource evaluation. Completion Report to U.S. Department of the Interior, Wallop-Breaux Project Number F-52. Florida Game and Fresh Water Fish Commission, Tallahassee.

Waters, T.F. 1995. Sediment in streams: sources, biological effects, and control. American Fisheries Society Monograph 7, Bethesda, Maryland.

Watters, G.T. 1993. Some aspects of the functional morphology of the shell of infaunal bivalves (Mollusca). Malacologia 35(2):315–342.

Watters, G.T. 1994a. An annotated bibliography of the reproduction and propagation of the Unionoidea (primarily of North America). Ohio Biological Survey, Miscellaneous Contributions, Number 1. 158 pages.

Watters, G.T. 1994b. Form and function of unionoidean shell sculpture and shape (Bivalvia). American Malacological Bulletin 11(1):1–20.

Watters, G.T. 1996. Small dams as barriers to freshwater mussels (Bivalvia, Unionoida) and their hosts. Biological Conservation 75:79–85.

Watters, G.T. 1999. Morphology of the conglutinate of the kidneyshell freshwater mussel, *Ptychobranchus fasciolaris*. Invertebrate Biology 118(3):289–295.

Watters, G.T. 2008. The morphology of conglutinates and conglutinate-like structures in North American freshwater mussels: a scanning-electron microscopy survey. Novapex 9:1–20.

Watters, G.T. 2000. Freshwater mussels and water quality: a review of the effects of hydrologic and instream habitat alterations. Pages 261–274. In: R.A. Tankersley, D.I. Warmolts, G.T. Watters, B.J. Armitage, P.D. Johnson, and R.S. Butler (editors). Special Contribution. Freshwater Mollusks as Indicators of Water Quality. Freshwater Mollusk Symposia Proceedings. Ohio Biological Survey, Columbus.

Watters, G.T. 2001. The evolution of the Unionacea in North America, and its implications for the Worldwide fauna. Pages 281–307. In: G. Bauer and K. Wächtler (editors). Ecology and Evolution of the Freshwater Mussels Unionoida. Ecological Studies, Volume 145. Springer-Verlag, Berlin.

Watters, G.T., M.A. Hoggarth, and D.H. Stansbery. 2009. The freshwater mussels of Ohio. Ohio State University Press, Columbus. 421 pages.

Watters, G.T., and S.H. O'Dee. 1996. Shedding of untransformed glochidia by fishes parasitized by *Lampsilis fasciola* Rafinesque, 1820 (Mollusca: Bivalvia: Unionidae): evidence of acquired immunity in the field? Journal of Freshwater Ecology 11(4):383–389.

Watters, G.T., and S.H. O'Dee. 1998. Metamorphosis of freshwater mussel glochidia (Bivalvia: Unionidae) on amphibians and exotic fishes. The American Midland Naturalist 139:49–57.

Watters, G.T., and S.H. O'Dee. 2000. Glochidial release as a function of water temperature: beyond bradyticty and tachyticty. Pages 135–140. In: R.A. Tankersley, D. Warmolts, G.T. Watters, and B. Armitage (editors). Part I. Proceedings of the Conservation, Captive Care, and Propagation of Freshwater Mussels Symposium. Ohio Biological Survey Special Publication, Columbus.

Watters, G.T., S.H. O'Dee, and S. Chordas, III. 2001. Patterns of vertical migration in freshwater mussels (Bivalvia: Unionoida). Journal of Freshwater Ecology 16(4):541–549.

Watters, G.T., and B.A. Wolfe. 2011. Sperm balls and how to explode them. (Abstract). Freshwater Mollusk Conservation Society Seventh Biennial Symposium, April 2011, Louisville, Kentucky.

Webb, S.D. (editor). 2006. First Floridians and Last Mastodons: The Page-Ladson Site in the Aucilla River. Springer, The Netherlands. 588 pages.

Weiss, J.L., and J.B. Layzer. 1995. Infestations of glochidia on fishes in the Barren River, Kentucky. American Malacological Bulletin 11(2):153–159.

Wheeler, C.L. 1893. The *Unio* muddle. The Nautilus 7(1):9–10.

Wheeler, H.E. 1935. Timothy A. Conrad. Bulletins of American Paleontology 23(77):1–157.

Whelan, N.V., A.J. Geneva, and D.L. Graf. 2011. Molecular phylogenetic analysis of tropical freshwater mussels (Mollusca: Bivalvia: Unionoida) resolves the position of *Coelatura* and supports a monophyletic Unionidae. Molecular Phylogenetics and Evolution 61:504–514. doi:10.1016/j.ympev.2011.07.016.

White, M.P., H.N. Blalock-Herod, and P.A. Stewart. 2008. Life history and host fish identification for *Fusconaia burkei* and *Pleurobema strodeanum* (Bivalvia: Unionidae). American Malacological Bulletin 24:121–125.

White, N.M. 1994. Archaeological investigations at six sites in the Apalachicola River Valley, northwest Florida. National Oceanic and Atmospheric Administration Technical Memorandum NOS SRD 26. Marine and Estuarine Management Division, Washington, DC. 270 pages, 5 appendices.

White, W.A. 1970. The geomorphology of the Florida peninsula. State of Florida Department of Natural Resources Geological Bulletin Number 51. 164 pages.

Williams, J.D., A.E. Bogan, J. Brim Box, R.S. Butler, N.M. Burkhead, A. Contreras-Arquieta, K.S. Cummings, J.T. Garner, J.L. Harris, R.G. Howells, S.J. Jepsen, N.A. Johnson, T.J. Morris, T.L. Myers, E. Naranjo García, and J.M. Wisniewski. In review. Conservation Status of North American Freshwater Mussels. Fisheries [forthcoming].

Williams, J.D., A.E. Bogan, and J.T. Garner. 2008. The Freshwater Mussels of Alabama and the Mobile Basin of Georgia, Mississippi, and Tennessee. University of Alabama Press, Tuscaloosa, Alabama. 908 pages.

Williams, J.D., A.E. Bogan, and J.T. Garner. 2009. A new species of freshwater mussel, *Anodonta hartfieldorum* (Bivalvia: Unionidae), from the Gulf Coastal Plain drainages of Alabama, Florida, Louisiana and Mississippi, USA. The Nautilus 123(2):25–33.

Williams, J.D., and R.S. Butler. 1994. Freshwater bivalves. Pages 53–128. In: M. Deyrup and R. Franz (editors). Rare and Endangered Biota of Florida. Volume IV. Invertebrates. University Press of Florida, Gainesville.

Williams, J.D., R.S. Butler, and J.M. Wisniewski. 2011. Annotated synonymy of the recent freshwater mussel taxa of the families Margaritiferidae and Unionidae described from Florida and drainages contiguous with Alabama and Georgia. Bulletin of the Florida Museum of Natural History 51(1):1–84.

Williams, J.D., and A. Fradkin. 1999. *Fusconaia apalachicola*, a new species of freshwater mussel (Bivalvia: Unionidae) from pre-Columbian archeological sites in the Apalachicola Basin of Alabama, Florida, and Georgia. Tulane Studies in Zoology 31(1):51–62.

Williams, J.D., S.L.H. Fuller, and R. Grace. 1992. Effects of impoundments on freshwater mussels (Mollusca: Bivalvia: Unionidae) in the main channel of the Black Warrior and Tombigbee rivers in western Alabama. Alabama Museum of Natural History Bulletin 13:1–10.

Williams, J.D., M.L. Warren, Jr., K.S. Cummings, J.L. Harris, and R.J. Neves. 1993. Conservation status of the freshwater mussels of the United States and Canada. Fisheries 18(9):6–22.

Willoughby, H.L. 1898. Across the Everglades. F.B. Lippincott Company, Philadelphia, Pennsylvania. 192 pages.

Wilson, C.D., G. Arnott, N. Reid, and D. Roberts. 2010. The pitfall with PIT tags: marking freshwater bivalves for translocation induces short-term behavioural costs. Animal Behaviour 81(1):341–346.

Wing, E.S., and L. McKean. 1987. Preliminary study of the animal remains excavated from the Hontoon Island site. The Florida Anthropologist 40(1):30–46.

Winger, P.V., D.P. Schultz, and W.W. Johnson. 1985. Contamination from battery salvage operations on the Chipola River, Florida. Pages 139–145. In: Proceedings of the Thirty-Ninth Annual Conference of the Southeastern Association of Fish and Wildlife Agencies, 27-30 October, 1985, Lexington, Kentucky.

Wood, E.M. 1974. Development and morphology of the glochidium larva of *Anodonta cygnea* (Mollusca: Bivalvia). Journal of Zoology (London) 173(1):1–13.

Woodburn, K.D. 1962. Clams and oysters in Charlotte County and vicinity. Florida State Board of Conservation Marine Laboratory, FSBCML Number: 62-12, CS Number: 62-1. 29 pages.

Woody, C.A., and L. Holland-Bartels. 1993. Reproductive characteristics of a population of the wash-board mussel *Megalonaias nervosa* (Rafinesque 1820) in the upper Mississippi River. Journal of Freshwater Biology 8:57–66.

Woolnough, D.A. 2006. The importance of host fish in long range transport of unionids in large rivers. Doctoral dissertation, Iowa State University, Ames. 144 pages.

Wright, B.H. 1883. A new *Unio* from Florida. Proceedings of the Academy of Natural Sciences of Phila-delphia 35(1883):58.

Wright, B.H. 1888. Descriptions of new species of Uniones from Florida. Proceedings of the Academy of Natural Sciences of Philadelphia 40(1888):113–120.

Wright, B.H. 1892. A new Florida *Unio*. The Nautilus 5(11):124–125.

Wright, B.H. 1893. The *Unio* muddle. The Nautilus 6(10):113–116.

Wright, B.H. 1896. New Florida Unios. The Nautilus 9(11):121–122.

Wright, B.H. 1897a. New Unios. The Nautilus 11(5):55–57.

Wright, B.H. 1897b. New Unios. The Nautilus 11(4):40–41.

Wright, B.H. 1897c. A new plicate *Unio*. The Nautilus 11(8):91–92.

Wright, B.H. 1898a. New Unionidae. The Nautilus 12(1):5–6.

Wright, B.H. 1898b. Description of a new *Unio*. The Nautilus 11(10):111–112.

Wright, B.H. 1898c. A new *Unio*. The Nautilus 12(3):32–33.

Wright, B.H. 1899. New southern Unios. The Nautilus 13(1):6–8; 13(2):22–23; 13(3):31; 13(4):42–43; 13(5):50–51; 13(6):69; 13(7):75–76; 13(8):89–90.

Wright, B.H. 1900. New southern Unios. The Nautilus 13(12):138–139.

Wright, B.H. 1933. A new species of Florida *Unio*. The Nautilus 47(1):17–18.

Wright, B.H. 1934a. A new Florida pearly freshwater mussel. The Nautilus 48(1):28–29.

Wright, B.H. 1934b. New Florida pearly mussels. The Nautilus 47(3):94–95.

Wright, B.H., and B. Walker. 1902. Check List of North American Naiades. Privately published, Detroit, Michigan. 19 pages.

Wright, S.H. 1891. Unionidae of Ga., Ala., S.C., and La., in south Florida. The Nautilus 4(11):125.

Wright, S.H. 1888. A new *Unio*. The Western American Scientist 4:60.

Wright, S.H. 1897. Contributions to a knowledge of United States Unionidae. The Nautilus 10(12):136–139; 11(1):4–5.

Wright, S.H., and B.H. Wright. 1887. Notes on the Unionidae of southern Florida. The Nautilus 2(5):67; 2(7):95.

Wright, S.H., and B.H. Wright. 1888. Notes on the Unionidae of Florida. The Nautilus 2(7):104–105; 2(9):111–112.

Wyman, J. 1868. An Account of the Fresh-water Shell-heaps of the St. Johns River, East Florida. Essex Institute Press, Salem, Massachusetts. 27 pages.

Yeager, B. 1993. Dams. Pages 57–92. In: C.F. Bryan and D.A. Rutherford (editors). Impacts on Warmwater Streams: Guidelines for Evaluation. American Fisheries Society, Little Rock, Arkansas.

Yeager, B.L., and R.J. Neves. 1986. Reproductive cycle and fish hosts of the rabbit's foot mussel, *Quadrula cylindrica strigillata* (Mollusca: Bivalvia: Unionidae) in the upper Tennessee River drainage. The American Midland Naturalist 116(2):329–340.

Yeager, M.M., D.S. Cherry, and R.J. Neves. 1994. Feeding and burrowing behaviors of juvenile rainbow mussels, *Villosa iris* (Bivalvia: Unionidae). Journal of the North American Benthological Society 13(2):217–222.

Yeager, M.M., R.J. Neves, and D.S. Cherry. 2000. Competitive interactions between early life stages of *Villosa iris* (Bivalvia: Unionidae) and adult Asian clams (*Corbicula fluminea*). Pages 253–259. In: P.D. Johnson and R.S. Butler (editors). Part II. Proceedings of the First Symposium of the Freshwater Mollusk Conservation Society. Ohio Biological Survey, Columbus.

Yokley, P., Jr. 1972. Life history of *Pleurobema cordatum* (Rafinesque, 1820) (Bivalvia: Unionacea). Malacologia 11(2):351–364.

Young, F. 1954. The water beetles of Florida. University of Florida Studies, Biological Science Series 5(1):1–238.

Zale, A.V., and R.J. Neves. 1982a. Reproductive biology of four freshwater mussel species (Mollusca: Unionidae) in Virginia. Freshwater Invertebrate Biology 1(1):17–28.

Zale, A.V., and R.J. Neves. 1982b. Fish hosts of four species of lampsiline mussels (Mollusca: Unionidae) in Big Moccasin Creek, Virginia. Canadian Journal of Zoology 60(11):2535–2542.

Zanatta, D.T., and R.W. Murphy. 2006. The evolution of active host-attraction strategies in the freshwater mussel tribe Lampsilini (Bivalvia: Unionidae). Molecular Phylogenetics and Evolution 41:195–208.

Zanatta, D.T., A. Ngo, and J. Lindell. 2007. Reassessment of the phylogenetic relationships among *Anodonta*, *Pyganodon*, and *Utterbackia* (Bivalvia: Unionoida) using mutation coding of allozyme data. Proceedings of the Academy of Natural Sciences of Philadelphia 156:211–216.

Zimmerman, G.F., and F.A. de Szalay. 2007. Influence of unionid mussels (Mollusca: Unionidae) on sediment stability: an artificial stream study. Fundamental and Applied Limnology 168:299–306.

Zimmerman, L.L., and R.J. Neves. 2002. Effects of temperature on duration of viability for glochidia of freshwater mussels (Bivalvia: Unionidae). American Malacological Bulletin 17(1 and 2):31–35.

Ziuganov, V., E. San Miguel, R.J. Neves, A. Longa, C. Fernández, R. Amaro, V. Beletsky, E. Popkovitch, S. Kaliuzhin, and T. Johnson. 2000. Life span variation of the freshwater pearl shell: a model species for testing longevity mechanisms in animals. Ambio 29(2):102–105.

General Index

Systematic Index